AF517911

AGROBIOLOGY

Trofim Denisovich Lysenko

AGROBIOLOGY

ESSAYS ON PROBLEMS OF GENETICS, PLANT BREEDING AND SEED GROWING

by
T.D. Lysenko

2005
Biotech Books
Delhi - 110 035

ISBN 81-7622-146-5

Published by : **BIOTECH BOOKS**
1123/74, Tri Nagar,
Delhi - 110 035
Phone: 27382765
e-mail: biotechbooks@yahoo.co.in

Showroom : 4762-63/23, Ansari Road, Darya Ganj,
New Delhi - 110 002
Phone: 23245578, 23244987

Printed at : **Chawla Offset Printers**
New Delhi - 110 052

PRINTED IN INDIA

CONTENTS

BIBLIOGRAPHY AND INDEX

THE THEORETICAL PRINCIPLES OF VERNALIZATION

INTRODUCTION

THE THEORY of phasic development is a general biological theory. For that very reason it holds for all branches of agrobiological science and is widely applied in socialist agriculture. A far from complete list of deductions from the theory of phasic development which are already being applied in socialist agriculture includes the following: a) reduction of the vegetative period of cereal plants under field conditions as a method of combating hot dry wind; b) vernalization of potatoes and the planting of eyes of vernalized tubers as a means of economizing planting material, and at the same time obtaining larger yields; c) the discovery of different degrees of winter hardiness at different stages of development and, following from this, measures for preventing winter crops from perishing in the winter; d) the method of producing varieties of winter crops by selection from populations by means of planting undervernalized seed; e) discovery of the causes of the degeneration of potatoes in the South and the planting of this crop in the summer as a means of preventing the degeneration of seed potatoes in the dry steppe regions; f) the theoretical principles of the deliberate choice of parental pairs for crossing in breeding varieties of different crops; g) discovery and formulation of the laws of segregation in conformity with the length of the vegetative periods as the theoretical basis of new methods of culling in the breeding process and an entirely new presentation of the problems of seed growing.

The breeding in two and a half years, on the basis of the theory of development by phases, of a previously planned variety of spring wheat for districts of the Odessa Region is, in our opinion, one of the most striking victories of the theory of phasic development. Not so long ago, the soundness of the principles on which this work was conducted was challenged by many scientists.[1] The theory of phasic development, as a general biological theory, and the deductions that followed from it were put to the severest test, the test of practice, and everywhere emerged victorious.

---

[1] At a meeting at the Seed-Growing Farms Union, January 16, 1934.

All this was achieved at an ever-increasing tempo of work which, in the shape of new methods of growing plants, propagating seeds, etc., is an inseparable aspect of work based on the theory of phasic development.

The progress which the general theory of biology has made in the mainstream of our practical life once again shatters the bourgeois falsehood that there are two truths—theoretical and practical.

"...the solution of theoretical contradictions is possible *only by practical* means, only as a result of the practical energy of man, and is, therefore, under no circumstances the task of cognition alone, but an *actual* task of life.... The assumption that there is one basis for life and another for science is a priori false," wrote Marx.[1] The theory of phasic development conquers in practice and by practice. This theory holds that everything in a plant, every one of its properties, characters, etc., is the result of the development of the hereditary foundation under concrete environmental conditions. The hereditary foundation, in its turn, is the result of the whole of preceding phylogenetic history. This biological history, which proceeded by the selection of adaptations to definite conditions of existence, resulted in the demands which the plant organism, throughout the whole period of its individual history, beginning from the zygote, makes upon the definite conditions of its development. These demands are the reverse side of the adaptations worked out in the historical process.

The phylogenetic history of the organic world did not, however, proceed in a straight line. For that reason, the biology of the individual plant organism is by no means uniform in its adaptations, and hence, in its demands. It has turning points, phases of definite duration. These phases constitute the most general biological stages in the individual development of the plant. Concentrated at the growing points of the plant's stem, the phasic processes, being specifically the development of the hereditary foundation, are the innermost processes in the life of plants. To discover the biophysics and biochemistry of the phasic processes means discovering the biophysics and biochemistry of the most intimate processes of the life of the plant cell. This most difficult problem will, in time, be solved, but the course of investigation that leads to this solution is an intricate one. What a vulgarly simplified conception of this problem is held by those investigators who, finding, or, more often, making attempts to find, changes in the chemical reactions of plants at one or other stage of development, think that they have discovered the innermost "nature" of vernalization, of the photo phase, etc.! The chemical indicators of phases are only a few of the many indicators, and although they are undoubtedly extremely important, they are still a long way from the "ultimate substance of life."

The offhand way in which some "discoverers" of the biochemical nature of the phasic processes approach this problem is due to the fact that they

[1] Marx-Engels, *Gesamtausgabe*, Erste Abteilung. Band 3. Berlin 1932, S. 121, 122.

regard "vernalization" as "stimulation," as a method which "thrusts" into the normal process of development something that is alien, supplementary, that can easily be chemically isolated and extracted.

Actually, however, vernalization, and the other phases of development, are necessary, normal processes of development of the same type and nature both under artificially created conditions (the presowing vernalization of seeds which have begun to germinate) and under field conditions. After all, it is possible to induce in any variety of winter crop a process of development of the whole cycle (from seed to seed), a process of plant development, similar to that of a spring plant, both by vernalizing the sowing material and by choosing the suitable field conditions (choice of district and time of sowing).

We are in favour of studying the chemical, physical, morphological and all other indicators of phasic development. In this respect, the Institute of Selection and Genetics already has some achievements to its credit (different chemical reactions to staining in sectional preparations of vernalized and unvernalized cuttings at the growing points;[1] differentiation in the growing points as a morphological indicator of vernalization, etc.). But we are opposed to the view that these indicators constitute the sum and substance of phasic development.

Above all, we are in favour of studying the *biology of development,* of studying what in development constitutes the specific character of biological relationships. Just as it would be absurd to say that since the physico-chemistry of the hermit crab and actinia is still obscure, therefore the essential nature of their interrelationships is still obscure, so it would be absurd not to regard the study of the biology of the phasic processes as being the study of their essential nature. We do not agree with the theoreticians of the "mechanics of development" who have deserted Darwinism, who operate on the principle: "I shall exert an influence with some agent and see what comes of it," who disregard the role of adaptedness in development (the mechanists), or those who play on this adaptedness and convert it into some sort of immaterial principle (the vitalists). We stand by the principles of Darwinism and study the biological stages of development, which are characterized primarily by turning points in the adaptive requirements of definite conditions of existence.

We well remember K. A. Timiryazev's statement that "we must have a historical conception of present-day organisms,"[2] because "an organ, i.e., the *adaptive form,* is the result of a historical factor—selection."[3]

---

[1] The work of M. A. Bassarskaya, a specialist at the Odessa Institute of Selection and Genetics.

[2] К. А. Тимирязев, *Исторический метод в биологии*, 1921 г., стр. 36-37.

[3] K. A. Timiryazev, editorial comment on p. 57 of G. Klebs's *Willkurliche Entwicklungsanderungen bei Pflanzen.* Translation, preface and comments by K. A. Timiryazev, 1905.

Following I. V. Michurin, we are developing this Darwinist idea further, not confining ourselves to the morphological formulation of the connection between phylogenesis and ontogenesis, but establishing their *biological* connection. The course of the phylogenetic process, the creation of organic forms by the *selection of adaptations,* cannot but affect primarily the very biology of the individual development of the hereditary foundation by determining the latter's demands upon the environmental conditions of its development. The concrete phasic development of the hereditary foundation and the development of organs and characters on the basis of the phases involved proceed only through a given pattern arranging the satisfaction of these demands throughout the entire course of individual development. Various factors not necessarily required for development are also operative here.

The biology of individual development must be studied as the individualization and concretization of the development of the historically formed hereditary foundation, which is the varietal, specific, etc., basis of the development of a plant individual. That is why we are opposed both to the preformism that exists in contemporary genetics and looks for the *direct* predetermination of *characters* in the hereditary foundation, skipping the biological stages of development, as well as to the mechanistic epigenesis characteristic of the "mechanics of development," which fails to see that the hereditary foundation is *the generic basis of the individual.* The hereditary foundation determines the general background, the *general* character of the plant's *cycle* of individual development.

The organism has no concretely given characters, nor does it undergo arbitrary changes of form. Winter habit, spring habit, winter hardiness, considerable or inconsiderable tillering, awnedness, colour, etc., are not given in the hereditary foundation, but are the result of the *development* of the hereditary foundation under the different environmental conditions that take part in the very formation of the organism's concrete characters. At the same time, however, external conditions are not free to turn development in any direction, are not free to turn it back, are not free to cancel the *demands* made by the given hereditary foundation upon the given conditions of development at any stage. The individual development of a plant organism proceeds on the basis of the biological requirements of a given stage of development of the hereditary foundation itself.

It is precisely because phases of development constitute general biological stages in the individual development of the hereditary foundation itself that these phases constitute the *basis* of development of each of the plant's characters. Spring habit, winter habit, frost resistance, drought hardiness, pest resistance, length of vegetative period, tillering, etc., cannot be studied apart from the general phases of development, because the formation of all these characters will vary if the course of this or that phase varies (owing to differences in external conditions) and they will differ in the different phases.

Of course, phases constitute only the *general* basis for the development of characters; for the latter develop under their *own* external environmental conditions and are subject to their own influencing factors.

The biology of development is the theoretical basis of all branches of agronomic science. The scourge of these departments, to this day, is their peculiar abiologization and divorcement from each other.

The biology of development must connect the disconnected and provide the general background for the study of all the various laws that govern plant organisms. This imposes upon all branches of agrobiological science, upon plant breeding, genetics, physiology, agrotechnique, etc., the task of critically examining their scientific stock-in-trade from the point of view of the theory of development.

Our socialist agriculture needs concrete knowledge. Planned economy demands, and gives rise to, the planning of the development of science. But the planned development of agrobiology can be ensured solely by employing the one and only scientific methodology, that of dialectical materialism, which is the theory that treats of the general laws of development.

The *theoretical* value of work on vernalization lies in the fact that this method marks the beginning of the conscious control of the *development of field crops*. However, until recently agricultural science lacked methods of controlling the rapidity of development of field crops. The various species and varieties of field crops whose development did not fit in with the climatic and geographical conditions of a given district were simply cast aside.

The starting point, the theoretical premise of our work on the vernalization of agricultural plants, is the law of the phasic development of plants that we have discovered. The principles of this law of phasic development of plants and some examples of its practical application are expounded in this book.

THE HISTORY OF THE PROBLEM OF VERNALIZATION

In practical farming *varieties* of annual crops which, although sown in the spring at the same time as other varieties of the same crop, proceed to fruit and ripen earlier than the latter, are called early ripeners. Varieties with a comparatively lengthy period from sowing to ripening are called late ripeners. Lastly, there are crops which, planted in the spring, do not proceed to fruit (to form reproductive organs) even in the autumn. These plants are called winter crops.

Many investigators, here and abroad, sought for that which distinguishes the winter group from the spring group in order to ascertain why winter crops fail to fruit when sown in the spring. Different investigators explained this problem in different ways. Some arrived at the conclusion that winter crops fail to fruit when sown in the spring because these plants need a definite dormancy period, i.e., a halt in their development for a definite time. When sown in the autumn, these winter plants obtain this dormancy period in the winter; if planted in the spring, however, winter plants continue to grow and therefore, in the opinion of these investigators, cannot bear fruit.

The fallacy of this assertion can be proved fairly easily. For this purpose it is sufficient to sow the seeds of several varieties of winter wheat under conditions where for a period of one and a half to two months the temperature ranges from 5°-10° C.; the plants will then grow all the time and, skipping the dormancy period, will proceed to ear, and later, under a higher temperature, begin to ripen (Fig. 1).

Other investigators, judging by the method of cultivating winter crops in districts with a frosty winter, assumed that in order to be able to fruit; winter plants must be chilled. This assumption also proved to be wrong. It can be disproved by experiments in growing winter plants under conditions where they will not be chilled and yet, in many cases, will fruit. This is confirmed by practical experience in districts where there is no, or scarcely any, frost in winter, and yet winter crops, sown on hundreds of thousands of hectares, ear and fruit after wintering.

A number of other investigators have advanced other explanations for the failure of winter crops to ear if sown in the spring. Lastly, a German professor named Gassner arrived at the conclusion from experiments he had conducted (the results of which he published in 1918) that winter varieties need a cold spell during their first stage of development. This gave rise to the method called the "cold germination" of winter crops. In the cold-germination method, the seeds of winter plants (wheat, rye) must be made to germinate at a temperature slightly above 0° C. The germination is allowed to go on until the rootlets reach a length of 2.5-3 cm. The plants brought to germination in this way can, in some cases, after planting, begin to ear. In the U.S.S.R., this method was tested in the physiological laboratory of the All-Union Institute of Plant Industry in Leningrad by Professor Maximov and Poyarkova. Professor Maximov and Poyarkova conducted their experiments during the winter, in a greenhouse. These experiments showed that with late greenhouse planting (in May, i.e., when in the Leningrad district, where these researches were conducted, the sowing of spring grain crops begins in the fields) winter crops do not ear uniformly and fully, irrespective of whether they germinated in cold or warmth. Only greenhouse planting in the early spring, when the snow still lies in the fields, gave advantage in earing to plants raised from seed that had germinated in the cold.

The facts obtained led these investigators to the conclusion that the cold germination of winter crops does not always produce a corresponding effect, and results in earing only when the sowing is done at definite times.

Thus, these experiments would appear to refute Professor Gassner's argument that winter crops need a cold spell.

Fig. 1. Left to right—winter wheats: Lutescens 0329 (pot No. 191); Stepnyachka (pot No. 183); rye: Petkusskaya (pot No. 167); Tulunskaya. (pot No. 175); Yeliseyevskaya (pot No. 125); and Vyatka (pot No. 157)

Sown with ordinary seeds on December 14, 1929, in a greenhouse at a temperature of 5°-10° C. At the end of February the rye plants eared. In the middle of April, after the temperature in the greenhouse had been raised, the wheat too eared

After our investigations, it may be said definitely that the facts obtained by the experiments conducted by Maximov and Poyarkova speak against the "cold-germination" method only. *They do not contradict Gassner's argument that in the first stage of their development winter plants need low temperatures.* In principle, Gassner's assertion that at a definite stage of development winter plants require low temperatures need not be challenged. The cold-germination method, as such, however, is wrong. By this method it is not always possible under greenhouse conditions to induce winter varieties to ear normally when sown in the spring; and it is almost altogether impossible when the spring sowing is done in a warm period under open field conditions.

In addition to the above explanations of the difference between winter and spring crops, and of why winter crops fail to ear if sown in the spring, a number of other investigators could be quoted. But none of them, whether foreign or Russian, provides a definite solution of the problem as to why winter crops fail to ear when sown in the spring. Not one of these investigators has been able to propose a method that will induce any winter variety to ear when sown in the spring. *They have failed to work out any such method not only for the spring sowing of winter varieties under ordinary farming conditions* but even for sowing in the open on a square metre plot at research institutions.

The chief defect in the work of the majority of the investigators of this problem was that they failed to set themselves the proper target. The main task they set themselves was not to induce winter varieties to ear when sown in the spring, but to "explain" why they failed to do so; and in their explanations, these investigators proceeded from the erroneous premise that in every district crops like wheat and rye are divided into separate and distinct groups—winter and spring varieties. Actually, however, the varieties of these crops, formed in the process of development of the genus and species, are often in this regard not separate and distinct groups, but an uninterrupted series of transitional forms from greater winter habit to lesser winter habit, i.e., to spring habit. Moreover, these investigators regarded winter habit and spring habit as properties belonging to, or what is the same thing, inherent in, the hereditary foundation of the seed germ, and not as properties acquired in the process of the plant's ontogenetic development.

These investigators lost sight of the fact that in the process of development, the germ or the hereditary foundation (the genotype) may give rise to the property of spring habit under some conditions, and to the property of winter habit under other conditions.

The term *vernalization*[1] appeared in the middle of 1929, when for the first time in the history of agricultural science the winter wheat Ukrainka, following suitable treatment of the sowing material, eared fully and uniformly after being sown in the spring under practical farming conditions. (D. N. Lysenko in the Poltava Region.) The yield of this spring-sown Ukrainka was 24 c. per ha. The fact that a spring-sown winter variety not only eared but produced a good yield seemed extraordinary at first. It seemed to contradict the nature of winter plants. The spring-sown plants, which by their nature have always been winter varieties in our districts, but which behaved like spring varieties (eared), came to be called *vernalized* plants by the Soviet public. The method of treating the seeds of winter varieties for spring sowing was called *vernalization*. Similarly, the work we did in studying the causes of the length of the vegetative period of agricultural plants, on the basis of which the vernalization method was elaborated, was referred to as the *vernalization of agricultural plants*.

[1] The Russian term is *yarovizatsia*.—*Tr.*

Many investigators believe that our work of studying the causes of the length of the vegetative period of agricultural plants consists solely in preparing the seeds of winter crops for spring sowing. This conception of our theoretical and practical work is incomplete and inexact not only today, when research work on the basis of the theory of phasic development is being conducted on a fairly extensive scale; it did not reflect the state of our scientific researches even in 1929.

Although the term "vernalization" appeared only in 1929, I had commenced to study the causes of the length of the vegetative period of agricultural plants at the Kirovabad (Ganja) Plant-Breeding Station (Azerbaijan) as early as 1926. This work marked the beginning of our researches into the vernalization of winter and spring varieties of different crops.

At the end of 1925, at the newly organized plant-breeding experimental station in Kirovabad, I was entrusted with work on breeding legumes for sideration and fodder. The cultivation of agricultural crops in the lowlands of the Azérbaijan S.S.R. calls for irrigation. In the summer dependence on irrigation water was one of the factors that restricted the introduction of southern leguminous plants (mung bean, vigna, and others), as crops for green manure. These plants need high temperature for their vegetation, and, therefore, they can be cultivated only in the summer. But in the summer, cotton—the staple crop in these parts—needs watering and absorbs the available water supply.

From September to the beginning of April the demand for irrigation water diminishes considerably. We decided to make an attempt to select from the family of leguminous plants such genera and species as could develop and provide the green mass necessary for fodder in the autumn-winter and early spring period, i.e., when irrigation water is available. This seemed all the more possible because many districts of Azerbaijan have a long autumn and a comparatively warm winter. Morning frosts do, indeed, go down to 6°-12° C. below zero, but there are only ten days in the year when the average temperature is below zero.

In the autumn of 1925, a collection of legume varieties was sown. For sowing we chose legumes that require relatively low temperatures for vegetation and can withstand morning frost. In the main, we took such crops as peas, vetch, horse beans and lentils. We placed most hope in the early, and not the late and medium, ripeners. The result was not bad. The peas and vetch, in general, came up well. Our assumptions were confirmed. All that was necessary now was to select and improve the varieties needed for this purpose; but in these sowings our attention was drawn to a phenomenon that seemed to us unusual at the time. Some of the varieties of peas, which in ordinary spring sowing—in Belaya Tserkov (Ukrainian S.S.R.), for example—were earliest ripening, behaved like latest ripeners when cultivated in the autumn and winter in Kirovabad. The Victoria variety (a medium ripener under normal conditions of cultivation) turned out here to be the earliest ripener. This variety flowered early and produced a green mass suitable for mowing, or for ploughing in. The above-described facts induced us to investigate the length of the vegetative period of agricultural plants. A solution of this problem was needed for our work of selecting and creating a variety of pea suitable for cultivation in the autumn and winter season.

As the result of a study of the length of the vegetative period of agricultural plants, it was proved experimentally that this period depends on the variety as well as on the external environmental conditions under which that variety is grown. It was confirmed over and over again that some varieties were early ripeners under certain conditions of cultivation and late ripeners under other conditions; conversely, some late varieties became early ripeners when their conditions of cultivation were changed.

It became clear to us that *different varieties of the same crop may demand different external conditions for their growth and development.* The less the environmental conditions suit the nature of the development of the plants of a given variety, the longer will it take those plants to develop, the longer will be the period from the sowing of the seeds to the ripening of the new seeds. If the external conditions are totally unsuitable for the nature of the development of the plants of the given variety, the plants will be unable to complete their development. They will neither flower nor bear fruit. In practical farming plants of crops which, when sown in the spring, germinate and develop leaves, but do not form reproductive organs even in the autumn (such as wheat, rye, barley, vetch, rape, etc.), are called winter plants.

Thus, we arrived at the conclusion *that the question of the spring or winter habit of plants is part of the general question of the length of their period of vegetation.*

After this, the question of winter or spring habit inevitably had to be included in our investigation into the causes of the length of the vegetative period of agricultural plants.

In the course of our experimental work we were able to prove that the plants of any variety of wheat, depending upon the conditions of cultivation, may behave like early spring, late spring, and even winter plants, i.e., such as always form only leaves, do not develop stems and do not ear.

In our experiments we observed that *plants of the same variety, when grown under different conditions*, may, depending upon these conditions, be winter, early-spring or late-spring plants, and that the behaviour of plants of different varieties, when grown under the same *definite* conditions, may be different. Some varieties of wheat may behave like winter wheat, others like late-spring varieties, and others again like early-spring varieties. From all the material we obtained in our experiments in 1927, we arrived at the conclusion that *the length of the vegetative period of plants from the sowing of the seeds to the ripening of the new seeds depends upon the interaction between the plant organism and environmental conditions*. By changing external conditions it is possible to change the behaviour of different plants of the same variety. Late-ripening forms may become early ripeners, winter forms may become spring forms, and spring forms winter forms.

Even before 1929, we observed in our experiments that definite groups of varieties can be of winter or spring habit only if sown under definite environmental conditions. For example, field experiments were conducted to ascertain the best sowing dates for different varieties of cereals (rye, wheat, barley). They were sown at intervals of ten days, from August 24, 1926, to August 27, 1927, and from October 1, 1927, to June 1, 1928. The results showed that there is no definite sowing time limit after which all the varieties that behave like winter varieties in other districts begin to behave in Kirovabad, Azerbaijan, like winter varieties, i.e., cease to develop stalks, and spring varieties, on the contrary, continue to develop stalks. Different varieties begin to display winter properties at different periods, and in 2 years (1927 and 1928) they did so on different calendar dates.

March 1928 was colder than the same month in 1927 and for that reason many varieties eared in 1928 even when sown on later dates than in 1927. Table 1 shows the last calendar dates for sowing different varieties; when sown after these dates the varieties failed to develop stalks.

Table 1

Variety	Date of sowing		Variety	Date of sowing	
	1927	1928		1927	1928
Rye 3	12/2	3/3	Tr. speciosissimum 1348/5 .	—	27/3
Kooperatorka 963	22/2	3-10/3	„ coerulescens 60/2 . . .	—	27/3
Tr. erythrosp. 1325/6 . . .	—	3/3	„ apulicum 35/1	—	27/3
H. pallidum 133/2	—	10/3	„ apulicum 44/1	—	4/4
H. nigrum 174/2	—	10/3	„ leucurum 1273	1/4	4/4
H. pallidum 419	12/3	10/3	„ leucurum 160/5	—	11/4
Tr. barbarossa 70/1	—	10/3	„ apulicum 2634	1/4	11/4
„ nigrobarbatum 1345/1 .	—	19/3	Av. grisea	—	11/4
„ niloticum 1229/1 . . .	—	9/3	„ brunnea 569	—	23/4
„ ferrugineum 1338/1 . .	—	19/3	„ byzantina 952	—	23/4
„ erythrospermum 2627 .	3/3	19/3			

As is evident from the table, the given collection of varieties, sown on different calendar dates, divides into the winter group and spring group in different ways. Each variety behaves in its own particular way. In some varieties the property of winter habit reveals itself when they are sown on earlier dates; in others, when sown on later dates.

From the large assortment of cereals which, in our experiments, were sown on different dates from winter to summer a striking series of consecutive transitions from spring forms to winter forms could be drawn. In this series it is impossible to distinguish where the winter forms end and where the spring ones begin without reference to the definite dates of sowing, because, depending on the date of sowing, the same forms may be of either spring or winter habit.

From this it follows that the varieties of wheat, and also of rye and barley, that exist in nature are not divided into separate and distinct groups—one winter and the other spring. They are connected by transitional series, from greater winter to lesser winter habit, that is, to spring habit. The winter forms, represented by a correspondingly chosen series of varieties, gradually pass into the spring forms and, vice versa, spring into winter forms. Definite groups of varieties may be winter or spring, if they are sown under definite conditions. It cannot be said that this or that variety is a winter or a spring form unless its inherent properties are linked with the concrete climatic conditions of the district (it would be more correct to say—the conditions of the postsowing period) in which the plants of this variety are to be grown. Now that we have brought into our researches in vernalization as many as 7,000 varieties, collected by the All-Union Institute of Plant Industry from nearly all countries in the world, it is easy to point to thousands of varieties which, when sown in the spring in some districts of the Soviet Union, will behave like spring varieties (i.e., will ear). The same varieties, sown in the same year in other districts, behave like winter varieties (will not ear). Thus, out of the 1,427 specimens of Azerbaijan wheat we sowed in Kazakhstan in the spring of 1932, 79.9% eared (without presowing vernalization). That is to say, under the conditions prevailing in that district, 79.9% of all the Azerbaijan wheats sown behaved like spring forms, and only 20.1% like winter forms. In the same year, 1932, at the Gigant State Farm in the North Caucasus, the same collection of wheats presented a different picture: only 4.8% (as against 79.9%) proved to be of spring habit; the rest, 95.2% (as against 20.1%), turned out to be of winter habit.

The same applies to the question of early- or late-ripening varieties. The behaviour of a definite group of varieties may be such in some districts that for practical farming purposes they will be early varieties, i.e., will ripen earlier than others. In other districts, the same varieties may be late-spring-ripeness. For example, in Table 2 we give a number of wheat varieties that originated in Finland and India. Their behaviour in respect to maturity varies rather sharply with the district in which they are sown. As a rule, in Kirovabad the Indian wheats ear 11 to 19 days earlier than the Finnish; in Odessa, the difference is only 2 to 11 days. When sown in Khibiny, most of the Finnish wheats ear simultaneously with, or even 5 days earlier than the Indian wheats.

Proceeding from this premise, it is easy to arrive at the conclusion that *it is wrong to divide all varieties of wheat (or any other plant) into a winter group and a spring group, into an early- and a late-ripening group, without taking into consideration the concrete conditions prevailing in the district where these varieties are to be grown.*

As has been already stated, all properties, qualities and characters, including, of course, winter habit, spring habit, early ripening, late ripening, and others, are the concrete result of the interaction between the plant organism and environmental conditions. The fact that it is wrong to divide varieties into winter and spring, early- and late-ripening varieties without linking this division with the concrete conditions of the district (i.e., the conditions of cultivation), does not in the least mean that all varieties are, by their nature, equally early or late ripeners, or equally winter and spring varieties. Different varieties (of wheat, for example) *differ in their natures. The conditions of cultivation in different districts also differ.* However, winter habit and spring habit, early and late ripening, are the result of the interaction between the natures of plants and external environmental conditions. *Therefore, for definite conditions of cultivation* (for definite districts) *all varieties not only can but must be divided into winter and spring, early- and late-ripening, and other forms.*

To ascertain whether a variety in this or that district will be a winter or spring form, the nature of the variety must be experimentally studied. This we have already done in respect to the majority of varieties of cereals.

The object of studying the causes that determine the length of the vegetative period of agricultural plants, which naturally included the problem of winter and spring habit, as well as the early and late ripening of varieties, was to find a method of treating the sowing material in order to change the behaviour of the plants—to convert late-ripening into early-ripening forms, varieties with winter behaviour into spring forms.

In 1928, at the Kirovabad Plant-Breeding Station, we conducted a number of experiments in this direction, under laboratory as well as under field conditions, with different varieties of wheat, rye and barley. We established that the causes of the late earing of many varieties of these crops and the failure of a number of other varieties to ear under field conditions when sown in the spring are in many cases phenomena of the same order. The cause of this phenomenon with spring sowing under field conditions was found to be the excessive temperature of the postsowing period, which prevented the plant from passing through a definite stage of development. Different varieties may pass through this stage of development (this phase of development) in different lengths of time and under different temperatures. Moreover, it was found that *plants may pass through this phase of development even when still in the seed state, i.e., when the embryo has just begun to grow and has not yet broken through the seed integument.* All that such sowing material needs is that definite external conditions (suitable temperature, moisture, access to air) be created for it for a definite period of time, depending on the variety. After being sown in the spring under field conditions, the plants of late-ripening varieties obtained from such sowing material may become early ripening, and plants of winter forms may become spring forms. One of the most important facts revealed by these experiments was that varieties of different natures (heredities) require different lengths of time and different conditions (moisture and temperature) of presowing treatment, i.e., *the plants of different varieties were found to possess different degrees of winter habit.* Under definite conditions it is sufficient to subject the seeds of some varieties to presowing treatment for

Table 2

Place of origin	Variety	No. in catalogue of All-Union Institute of Plant Industry	Date of earing; number of days Finnish wheats were later (+) or earlier (—) in earing than Indian wheats at points of sowing		
			Ganja	Odessa	Khibiny
Finland	ferrugineum	5512	21/5	24/6	18/7
India	turcicum	24406	7/5	19/6	21/7
			+14	+5	−3
Finland	ferrugineum	13313	23/5	25/6	20/7
India	erythroleucon	26586	4/5	20/6	19/7
			+19	+5	+1
Finland	erythrospermum	5694	18/5	21/6	16/7
India	erythroleucon	26598	4/5	18/6	21/7
			+14	+3	−5
Finland	erythrospermum	5382	16/5	27/6	19/7
India	graecum	25715	4/5	16/6	16/7
			+12	+11	+3
Finland	lutescens	5696	21/5	21/6	16/7
India	graecum	25715	4/5	16/6	16/7
			+17	+5	0
Finland	milturum	25702	23/5	25/6	21/7
India	alborubrum	23731	4/5	17/6	18/7
			+19	+8	+3
Finland	erythrospermum	5694	18/5	21/6	16/7
India	turcicum	24406	7/5	19/6	21/7
			+11	+2	−5
Finland	lutescens	5693	21/5	23/6	18/7
India	erythroleucon	26598	4/5	18/6	21/7
			+17	+5	−3
Finland	lutescens	5696	21/5	21/6	16/7
India	anglicum	23842	4/5	17/6	19/7
			+17	+4	−3
Finland	erythrospermum	5702	21/5	24/6	19/7
India	erythroleucon	26586	4/5	20/6	19/7
			+17	+4	0

a period of only 5 days. The seeds of other varieties require 10, 15, 20, 25, up to 60 days of suitable treatment (depending upon the variety) to make these plants behave like spring instead of winter forms in contrast to the simultaneously sown untreated seeds of the same forms. Thus, in our experiments, the problem of winter habit and spring habit, like that of early and late ripening, emerged from the problem of the length of the vegetative period.

Fig. 2. Leucurum 160/5

Sown in Ganja on April 15, 1929; number of days of presowing vernalization for normal earing 20; with fewer days of vernalization either does not ear at all, or ears late

Fig. 3. Nigrobarbatum 1348/10

Sown in Ganja on April 15; presowing vernalization lasts 36 days, after which the wheat ears normally

The result of these researches was reported to the All-Union Genetics Congress in Leningrad (January 1929).

The report on our investigation into the causes of the failure of winter varieties to ear when sown in the spring and on the connection between this problem and the length of the vegetative period introduced nothing definite or new in the conceptions of the participants in that congress. As was stated above, quite a number of reasons had been advanced hitherto for the failure of winter varieties to ear when sown in the spring, and at best, the upshot of our report was that still another explanation was put forward. It was difficult for the audience to determine which of these explanations was correct. One of the principal objections raised against our propositions was that if applied in other districts the results might turn out to be different from those we obtained at the plant-breeding station in Azerbaijan, just as was the case with the "*cold-germination*" method, which in Professor Maximov's experiments proved effective (relative to earing) only when the seeds were sown at definite dates.

In the spring and summer of 1929 we continued our research work on this problem at the plant-breeding station in Azerbaijan on a fairly extensive scale, without divorcing it from the general problem of the length of the vegetative period of agricultural plants. In the same summer (1929) the Soviet public learned from the press of the full and uniform earing of winter wheat sown in the spring under practical farming conditions in the Ukraine.[1]

This practical sowing confirmed the chief conclusions of our researches, after which they received general recognition. The Soviet public came out in support of our explanation of the length of the vegetative period of plants. By order of the People's Commissariat of Agriculture, a special laboratory, and later a department, was established at the Ukrainian Institute of Selection and Genetics (Odessa) to study this problem. To test and further elaborate our idea of controlling the length of the vegetative period of agricultural plants, hundreds of collective-farm experimenters and state-farm workers were drawn into the work in 1930, besides the opening of the laboratory. Had this not been done, not only would our laboratory researches have gotten no further than the four walls of the laboratory, not only would they not have been applied to the fields, but the working out of the theory that treats of the problem would not have been as successful as it is today.

In 1935, the experimental and practical sowing of vernalized spring cereals alone was carried out by over 40,000 collective and state farms on a total area of 2,100,000 ha. Practice has shown that collective- and state-farm experiments on the problem of vernalization, when properly linked up with the work of research institutions, yield results, both theoretical and practical, that cannot be expected when the work is conducted by the research institutions alone.

---

[1] This sowing was not accidental. On my suggestion it was done by my father, D. N. Lysenko, on his farm.

Table 3

SPRING SOWING OF VERNALIZED 1930 WINTER WHEAT (ODESSA)

No.	Variety	Number of days necessary for vernalization, temperature ranging from 0° to+1° C.[1]
1	Winter wheat 808 (1/26), Verkhnyachka station	16
2	Novokrymka 0204	36
3	Kooperatorka	36
4	Erythrospermum 917, Kharkov station	36
5	Ukrainka	41
6	Stepnyachka 0464	41
7	Hostianum 237	46
8	Lutescens 329, Saratov station	46
9	„ 1060/10, „ „	46
10	Dürabl, Ivanovskaya station	46
11	Winter wheat 037, Belaya Tserkov station	46
12	Minchardi	46
13	No. 15, Ukr. Inst. of Selection and Genetics	52
14	No. 14, „ „ „ „ „ „	52
15	White awned 040	52
16	Winter wheat 564/115 (1/26) Verkhnyachka station	53
17	No. 2, Ukr. Inst. of Selection and Genetics	57
18	Winter wheat Erythrospermum 132/5, Ganja station	57

The collective- and state-farm experiments of 1930 clearly demonstrated that all the winter plants of wheat, rye, rape, vetch, and others, can be induced not only to bear under practical farming conditions when sown in the spring, but also, in many cases, to produce a fairly good yield. In the experimental sowing of winter varieties in the spring of 1930 at the Ilyich Commune of the former Mariupol Okrug, spring-sown Ukrainka yielded 29.5 c. on 1.1 ha. (27.3 c. per ha.); at the Batrak Ukrainy Artel, vernalized Ukrainka, sown on 1.5 ha., yielded 32.6 c. (21.4 c. per ha.). At the Pervoye Maya Commune 6.9 c. were obtained on two-fifths of a ha. (17.2 c. per ha.). On the Oktyabrskaya Revolutsia State Farm in the former Stalino Okrug, spring-sown Ukrainka yielded 13.3 c. per ha. A number of other examples could be cited of experiments in the spring sowing of winter wheat made in 1930 on collective and state farms resulting in good yields. It must be emphasized, however, that these examples do not by any means show that any vernalized winter variety can be sown in any district and produce good yields. Not every winter variety will produce a good crop in every district.

[1] In Odessa the spring of 1930 was cool and prolonged. In years of hotter and shorter springs, the seed of each variety must be vernalized before sowing for 5 days longer than shown in the table.

Fig. 4. Ferrugineum 1388/1

Sown in Ganja on April 15 presowing vernalization lasts 51 days, after which the wheat ears

Fig. 5. Winter vetch. Spring sown in Ganja

The vernalized plants flowered; the unvernalized plants did not flower

The yield will depend upon the variety taken for vernalization, as well as upon the conditions under which the given variety is grown.

By quoting the above examples of yields obtained from the spring sowing of the vernalized winter wheat Ukrainka, we merely emphasize that the rather important fundamental scientific problem of the causes responsible for the failure of winter varieties to ear when sown in the spring was definitely solved by Soviet science in the collective- and state-farm fields, i.e., under practical farming conditions. In the process of vernalizing seeds under practical farming conditions, the technique of preparing the sowing material of winter varieties for spring sowing was worked out. This technique can already be employed for practical purposes. The method of vernalizing winter varieties employed for the first time (by D. N. Lysenko) in 1929 (seeds beginning to germinate were collected in a sack and buried in snow) and proposed in 1930 has now been considerably altered. The vernalization of sowing material is now done not in sacks, and not in snow, but in ordinary barns, granaries and sheds.[1]

The technique of vernalizing spring varieties of cereals was worked out simultaneously with the technique for winter varieties.

Thanks to the mass state- and collective-farm experiments, not only was the technique of vernalizing winter and spring cereal crops worked

[1] For a brief description of the technique of vernalization see pp. 32-33.

out, but considerable progress was made in elaborating the theory of controlling the vegetative period of different agricultural plants.

Today many people already know that it is possible to vernalize not only winter but also spring varieties of rye, wheat, barley, vetch, rape and other crops. Moreover, it is possible to vernalize plants like millet, cotton and a number of other crops, which in practical farming are never called winter crops. By suitably treating (vernalizing) the seeds it is possible to grow many so-called cold-loving plants when sown under hot spring conditions; or to grow certain thermophilic plants in districts where the temperature is not high enough for them; or to grow "short-day" plants under "long-day" conditions. All this has become possible only because of the creative initiative of collective-farm experimenters, combined with the work of research institutions in studying the development of the plant organism (from the sowing to the ripening of seeds).

Thus, our research work on the vernalization of agricultural plants did not begin in 1929 (the year when the term "vernalization" appeared). It is organically connected with our earlier work (in 1926-27) on changing, under field conditions, the behaviour of late-ripening varieties of different crops into that of early-ripening ones.

The sowing of vernalized Ukrainka under field conditions by D. N. Lysenko in 1929 served as valuable confirmation of our three years' study at the plant-breeding station in Azerbaijan of the length of the vegetative period.

Let us now deal in brief outline with the chief general features in the development of annual seed plants as we see them today.

THE DEVELOPMENT OF A SEED PLANT AND ITS GROWTH ARE NOT IDENTICAL PHENOMENA

The growth of a seed plant and its development are often taken as synonymous terms, signifying one and the same phenomenon in the life of the plant. Observation of plant life, however, has shown that growth and development are not one and the same thing, that they are not identical aspects of a plant organism's life.

Rye or wheat plants, grown by the wayside from accidentally dropped seeds, may ripen, i.e., may fully complete their development, just like plants of the same varieties grown in a cultivated, well-tilled field. But the height and general vigour of these plants, as well as the size and quality of their yield, may differ considerably. The height of the first-mentioned plants may not exceed 10-15 cm. Their miniature spikes may contain only 1 or 2 tiny seeds. The fact that these seeds have ripened shows that development—the ordinary path of life of a plant—is fully completed. The height of the plants mentioned second, those grown in a cultivated, well-tilled field, may be 200 cm. and over instead of 10-12 cm. The grains in the spikes may number not 1 or 2, but 60 to 80. And yet the life span of the plants may be the same in both cases.

Experimentally, it is fairly easy under a given set of conditions to obtain ripe plants one-hundredth the size and weight (i.e., of the vigour) of plants of the same variety at the same stage of development but grown under different conditions. Hence, when the development of plants (of the same variety) is completed, their growth and size, their vigour, and also the size and quality of their yield, may differ very considerably.

By the development of a seed plant we mean the chain of necessary qualitative changes of the cell content and the organ-forming processes

Fig. 6. Esparcet (*Onobrychis satwa*). Spring sown in Ganja in 1929

Left—sown with ordinary seeds, did not flower. Right—sown with vernalized seeds, flowered in the normal way

the plant undergoes from seed to seed. It may be observed that a given plant, under certain conditions, may not be able to yield ripe seeds, or even begin to develop reproductive organs. The cause here may be that one or other organ has failed to develop, or that in the course of the plant's development the cell content of a given organ has not acquired that indispensable quality without which the given plant cannot proceed to the formation of the different organs and, in the end, of seeds.

Fig. 7. Maize (*Morelos tepoxtean*)

Sown in Odessa in 1931. The plants were unable to go through the photo phase (the Odessa day is too long for these plants). The plants reached the height of 3 m. with no initiation of reproductive organs

By the growth of a plant we, in our work, mean what this usually implies in practical farming, namely, increase in the plant's weight and size, apart from the form-building processes. *By growth we mean increase in the plant's mass*, irrespective of the organs or characters whose development served to increase the plant's mass. Growth is one of the properties of a plant's development. The property of growth may manifest itself in different degrees, depending on the nature of the plant, on the external environmental conditions, and on the plant's phase of development. Not only under experimental conditions, but also in ordinary practical farming, it is fairly easy to observe:

a) rapid growth (increase in mass) and slow development, the plant's slow progress towards the ultimate point, towards the formation of seeds;
b) slow growth and accelerated development;
c) rapid growth and rapid development;
d) slow growth and slow development.

In other words, *the rapidity of a plant's development, i.e., the rate at which the whole plant develops from seed to seed, and also the rate at which its different organs develop, depend not only on the speed with which the plant accumulates mass.* The rapidity of a plant's development, as well as the rapidity of its growth during a given phase of development, is inseparably connected with the environmental conditions. The sets of external conditions required for the development of the plant as a whole as well as of its separate organs, and those required for the plant's growth, i.e., for increasing its mass at the expense of the development of separate organs and parts of the plant, often do not coincide. This applies not only to the relative dosage of factors needed for growth on the one hand and development on the other. In the case of many plants there is a difference in the very factors that combine to make up the respective sets of external conditions required for development and for growth. In a suitable external environment plants may grow, increase in weight and volume, for an indefinitely long time and yet remain in one phase of development without passing into the next. Thus, winter cereals and other winter plants, if sown in the spring, after starting to develop, keep on growing, accumulate (develop) green mass, but do not fruit even in the autumn. In this case, the plants remain in the first phase of development and do not pass into the next phase owing to the absence of vernalization changes, an absence caused by the high temperatures during the spring and summer periods. Many varieties of millet, soya bean and other "short-day" plants, when grown in continuous light under thermal conditions favourable for these plants (20°-25° C.), cannot fruit owing to the absence of the darkness (long nights) required for passing through the processes of one of their phases of development (the photo phase). These plants, however, continue to grow all the time.

In practical farming, it may fairly often be observed that the field conditions of a given district are not suitable for a plant's whole cycle of development (from seed to seed). In such cases, the plants from the sown seeds reach only that phase of development beyond which suitable environmental conditions are lacking. The development of such plants comes to a halt, i.e., they do not pass to the next stage, although the growth of these plants may continue. Plants whose development has stopped do not produce seeds until they receive the conditions necessary for their further development. Hence, many plants may grow but cannot fruit under the climatic conditions of a given district. Such plants are usually not sown there for consumption.

A given district may lack various factors essential to the development of various plants: low or high temperatures, suitably long spring and summer days or nights, etc.

We have already stated above that when winter plants are sown in the spring they do not fruit because during a definite period of development (the vernalization phase) they need a fairly long spell (depending on the variety) of low temperature (below 10° C.). The optimal temperature for them will be from 3°-0° C. Winter plants sown in the field in the autumn utilize the low thermal conditions they require for a definite period, depending on the variety. Once the winter plants have developed under low thermal conditions for a definite time, i.e., after they have passed through the phase of development for which low temperatures are essential, their further development no longer requires low temperatures.

Fig. 8. Cotton

These plants were grown under different conditions. The plants in the pot on the right grew slowly and developed quickly, forming flowers and bolls. The plants in the middle pot grew quickly and developed quickly, forming flowers and bolls. The plants in the pot on the left grew quickly but developed slowly; they are not yet ready to form buds and flowers

Many varieties of cotton sown in the south of the Ukrainian S.S.R. will grow, but are late in fruiting, because the temperature in the spring and beginning of the summer is not sufficiently high for the development of these varieties. If the relatively high temperatures (20°-30° C.) required for the development of cotton plants for a definite length of time are not forthcoming they cannot fruit. If, on the other hand, the high temperatures are available for a definite length of time, cotton can subsequently form buds and fruit even under lower temperature conditions (15°-20° C.)

Knowing the conditions a plant requires for rapid or slow development, and knowing the conditions required for rapid or slow growth, it is possible to induce plants to develop at different rates while simultaneously retarding or accelerating their growth in different degrees.

Thus, a) the terms plant development and plant growth are not synonymous. Growth is one of the properties of development; the degree to which

Fig. 9. Turning over a layer of wheat during vernalization

growth manifests itself depends on the plant's phase of development and on external environmental conditions; b) for many plants the sets of environmental conditions required for development on the one hand and for growth on the other do not coincide. The rapidity of a plant's development does not always depend on the rapidity of its growth.

All this served as one of the theoretical premises for working out the method of vernalizing the seeds of a number of plants. In practical farming, vernalization is achieved under artificially created conditions, under which the plants (embryos which have just slightly begun to grow) pass through one of the phases of their development (the vernalization phase) at an extremely slow rate of growth, almost invisible to the naked eye (germination of seeds).

The method and technique of vernalizing winter cereals and other winter plants, and also of vernalizing spring wheat, have been worked out quite fully, and they can already be employed on collective and state farms after a suitable choice of variety for each district. In 1933, the vernalization of spring wheat was already practised by collective and state farms on a total area of nearly 200,000 ha. In 1934, vernalized seed was used by collective and state farms on a total area of over 500,000 ha. In 1935, vernalized seed was used on a total area of 2,100,000 ha. In 1936, according to plan, 4,900,000 ha. are to be sown with vernalized seed.

The method and technique of vernalizing thermophilic plants (cotton) have been worked out less fully, and lastly, the vernalization of "short-day" plants, except millet, has been worked out even less.

Briefly, the technique of vernalizing winter and spring wheat employed at collective and state farms is as follows.[1] On every 100 weight units of seeds spread out on the floor, the following weight units of water are poured in three doses: winter varieties, 37; spring late ripening, 33; spring early ripening, 31. After soaking, the seeds of winter varieties are kept at a temperature ranging from 0°-3° C. for a period of 35 to 50 days, depending on the variety. The seeds of spring wheat, after soaking, are kept from 5 to 15 days at a temperature ranging from 5°-12° C. The number of days of vernalization and the temperature at which the vernalization of spring varieties is conducted depend on the wheat variety. The vernalization of the seeds of winter as well as of spring wheat is performed in sheds, barns, granaries or any other grain repository. The temperature necessary for the vernalization of seed is regulated by the thickness of the layer of seeds, and also by turning the seeds over.

If weather conditions at the end of the vernalization period are unfavourable for sowing in the field (rain or belated spring), or if it is necessary to transport the vernalized seeds a long distance before sowing, they must be dried until they are as dry as the air. As a rule, the work of vernalization must be so timed as to permit of sowing immediately on the expiration of the vernalization period, without drying and further storage. The drying and storage of vernalized seed affect to some extent both the percentage and vigour of germination. The vernalization of spring wheat must be commenced not earlier than 2 or 3 days before the beginning of spring field work. The vernalization of winter varieties must be commenced from 30 to 50 days (depending on the variety) before the beginning of spring field work. Vernalized seed is quite suitable for sowing with ordinary seed drills, especially through the upper feed (upper sowing).

PHASES IN THE DEVELOPMENT OF PLANTS

The development of plants requires a definite set of factors, which, in addition to mineral nutrition, includes temperature, light, moisture, suitable length of daylight or night, etc. If any of the conditions enumerated are unsuited to the nature of the development of the given plants, the latter cannot produce a good crop. That is why it is often observed that some plants grow fairly well, but begin to flower and fruit late, or fail to do so altogether.

For normal growth and development, different plants need different climatic conditions. The climatic conditions required for winter varieties of cereals are unsuitable for thermophilic plants, like cotton. That is why winter cereals are sown in the autumn, and cotton and many other plants

[1] For vernalizing large quantities of seeds, see T. D. Lysenko's pamphlet *Яровизация сельскохозяйственных растений*. It contains instructions on how to vernalize wheat, oats and barley. Selkhozgiz, 1936.

Fig. 10. A layer of wheat during vernalization

are sown in the spring, when it is warm in the fields. Moreover, winter plants, like rye and wheat, are also cultivated in the North, whereas cotton is cultivated only in the South.

Most plants require different external conditions during the different periods in the course of their lives, from the sowing of the seed to the ripening of the new seeds. For example, plants like our spring cereals, and to a larger extent our winter cereals, require lower temperatures at the beginning of their development than at the end (during the ripening of the seeds). Many varieties of winter wheat, if grown continuously under conditions of relatively high temperatures (say, higher than 10°-12° C.), fail to fruit. Nor will these plants be able to fruit if, from the time of sowing, the temperature is constantly lower than 10°-12° C. For our winter varieties, the temperature must range from 0°-10° C. for a definite period at the beginning of their development (from 20 to 50 days, depending upon the variety); for their further development these plants require a higher temperature (Figs. 11 and 12). It is immaterial whether in the initial period of development of these plants the days are long or short, provided the temperature remains approximately within the range of 0°-10° C. and there is, of course, moisture and access to air. These plants can usually pass through the next stage of their development at a higher temperature (above 5°-10° C.) but only under long-day conditions; under short-day conditions the development of most varieties of wheat, barley, rye and others is halted, i.e., they fail to undergo those qualitative changes which bring the plants to the formation (development) of reproductive organs.

Cotton requires much more warmth at the beginning of its development than at the end, during the ripening of the bolls. The changes in the en-

Fig. 11. Three pots on left—wheat Kooperatorka; three pots on right—Hostianum 0237

The plants in the pots on the left of both varieties (Nos. 121 and 152) were grown all the time at a temperature of 2°-8° C. The plants were vernalized, but owing to low temperature did not ear. The plants in the middle pots (Nos. 118 and 150) of both varieties were grown all the time at a temperature of 8°-15° C. The plants were vernalized and eared. The plants in the pots on the right (Nos. 116 and 148) were grown all the time at a temperature over 15° C.; they could not vernalize and did not ear. Sown on December 14, Hostianum 0237 eared on March 25; Kooperatorka eared on March 25. All the plants were grown under continuous light.

Fig. 12. Plants of winter wheat Lutescens 0329—the three pots on left and Ukrainka—the three pots on right

The plants grew and behaved in the same way as those described in Fig. 11

vironmental conditions which developing plants need indicate that the development of annual seed plants from the germination of the seed to the ripening of the new seeds is itself not uniform in type, not uniform in quality. *The development of plants consists of separate qualitatively different stages, or phases.* To pass through the different phases of development, the plants require different external conditions (different nutriment, light, temperature, etc.). Phases are definite, *necessary* stages in the plant's development, and serve as the basis of the development of all the plant's *particular forms*—its organs and characters. The different organs and characters can develop only at definite phases. On the basis of a given phase various organs and characters may form, but by no means all those the plant is known to possess.

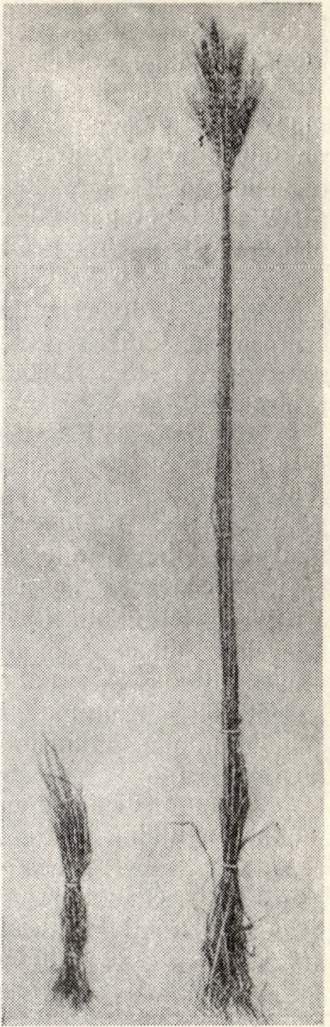

Fig. 13. Spring sowing of winter wheat Ukrainka: Left—from unvernalized seeds; Right—from vernalized seeds

The plants of sheaf on left could not go through the vernalization phase and did not ear

The appearance of new morphological changes observed in plants is not always the result of the transition of these plants to a new phase of development. To a certain extent, morphological changes may take place also on the basis of the old phase of development. In passing to a new phase, plants may not at once develop external morphological changes visible to the eye. Only subsequently, on the basis of this new phase, can the organs and characters corresponding to it develop.

Thus, by phases of development we do not mean the formation (development) of the different organs and parts of plants—leaves, stalks, etc.—but those stages and qualitative turning points in the development of plants that take place at the growing points of the stem and which are essential for their normal progress through the formation of different organs and characters to fruiting. On the basis of these qualitative changes, i.e., phases of development (it would be more correct to say—out of the latter), the plants' parts and organs, and their different properties and qualities, develop. Some of the latter are the result of the plant's passing through only one phase of development; others are characteristic of several and even of all the phases of development.

In the general chain of development of annual seed plants, we have, so far, disclosed only the first two phases of development, two qualitatively different stages in their development.

One of the first phases of development, called the vernalization phase, may begin as soon as the embryo within the seed starts growing, if the environmental conditions for it (a relatively definite range and correlation of temperature, air and moisture) exist. If these conditions are absent, the plant will not pass through the vernalization phase, notwithstanding the fact that the plant's growth, its increase in weight and volume, may proceed normally. Plants that have not passed through the vernalization phase cannot proceed further in their development (namely, the development that brings them to the point of seed formation). In consequence, the corresponding organs and their characters will not develop and hence there will be no fruiting.

The changes that take place in the embryo when the seed is vernalized are not something specific in the sense that it can take place in the plant only when, externally, it still resembles the seed. Moreover, as will be seen later on, these changes, while actually specific, are so not in the sense that they are inherent only in the embryo that has just begun to grow. Their specific character lies in the inability of winter varieties—and, we think, of other seed plants too—to fruit without these changes. The changes that take place in the embryos as a result of vernalization can take place only under suitable external environmental conditions. If these conditions are absent, the sowing material will not go through the process of vernalization. Winter plants obtained from such sowing material will not fruit if sown in the spring.

Autumn-sown winter varieties, after passing the winter, ear and fruit at the beginning of the summer without undergoing presowing vernalization. It is perfectly clear that in the autumn, winter varieties, especially if sown early, cannot vernalize before the appearance of shoots and for a fairly long time after, because at that time the temperature in the fields is above 10° C. In spite of that, the plants of autumn-sown seeds fruit after passing the winter. Thus, the presowing vernalization of winter varieties is essential only for spring sowing. The plants of autumn-sown seeds, after passing the winter, fruit without previous vernalization.

Can autumn-sown winter plants fruit without undergoing those qualitative changes that take place in the germinating seeds vernalized before sowing? Experiments in this direction show that unless they pass through the processes of vernalization, winter plants, whether sown in the spring or in the autumn, cannot fruit.

All our winter varieties of plants such as rye, wheat, barley and rape germinate uniformly when sown in the spring in our districts and in the course of the spring and summer produce much foliage, but to the end of the vegetative period these plants usually do not form reproductive organs—stalks and ears (except in years when the spring is cold and prolonged).

A number of experiments we have conducted show that plants like cabbage, beetroot and other root crops, transplanted from the open to a heated greenhouse in the early autumn, and replanted in the open in the spring, continue to grow (develop new leaves and roots) but cannot fruit. The same kind of plants, when planted in the spring, after passing the winter not in a greenhouse but in cellars or rooms where the temperature is lower than in greenhouses, reach normal fruiting.

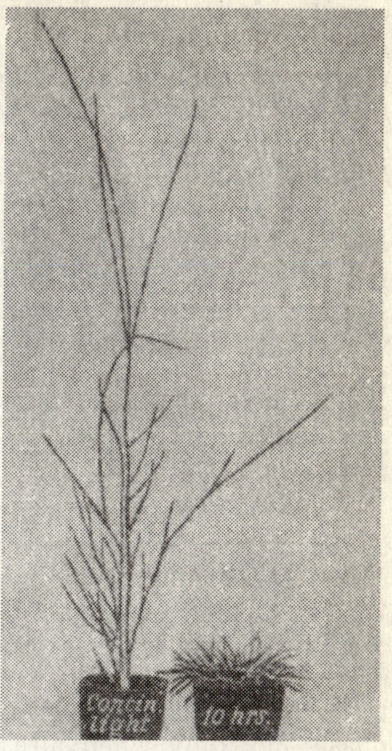

Fig. 14. Wheat-agropyron hybrid

The plants in pot on left were grown in continuous light. They went through the photo phase in 30 days and eared. The plants in the pot on the right were grown for two years under 10-hour-day conditions, were unable to go through the photo phase and therefore did not ear

The above facts show that without the changes that take place in the germinating seed when the sowing material is being vernalized, the plants of winter varieties cannot fruit. Absence of fruiting is observed not only in spring-sown plants, but also in the plants of wheat and of many other autumn-sown winter crops if such plants, after sowing, are grown under warm conditions (15°-20° C.).

That the changes which take place as a result of the vernalization of seeds (sowing material) are identical with the *changes* taking place in green plants is proved by a number of our experiments. It was found that in order to induce fruiting in winter wheat, rye, barley, or any other winter crop, under conditions prevailing in spring and summer, it is absolutely necessary to keep these plants for not less than a specified period (its length depending on the variety) under low temperature conditions (from 0°-10° C., better still from 0°-2° C.). After that, such plants may reach fruiting under higher temperature conditions, i.e., under spring and summer conditions. Moreover, the rapidity with which the vernalization phase is traversed does not depend on the size or age of the experimental plants. The rapidity with which the vernalization processes are passed through depends entirely on the variety (genotype) and on the external environmental conditions. In our experiments, the embryo of the winter wheat Kooperatorka which had just begun to grow and had not yet broken through the seed integument, and three- to four-month-old plants of the same variety which had tillered considerably but had not gone through the processes of vernalization (the plants were grown in a heated greenhouse), demanded the same external conditions and the same amount of time to pass through the vernalization phase. The time required for passing through the vernalization phase may change with a change in external conditions, and it will do so equally for plants in the embryo as well as for tillering plants. *From the practical aspect this is one of the fundamental rules for the vernalization of agricultural crops.* Its importance lies in the fact that *plants can pass through the vernalization phase—without which cereals cannot fruit—not only in the field when in the green state, but also in the seed when the embryo has just begun to grow.*

Experiments have established that if the seeds have only just swelled and the germs have not yet begun to grow, the vernalization processes cannot take place. The embryos must be induced to start growing if only so slightly as not even to break through the seed integument. After that, the vernalization processes can take place under the same external conditions and with the same rapidity as in green plants. This gives us grounds for assuming that seeds just beginning to germinate already cease to be seeds and become plants with the same properties for passing through the vernalization phase as green plants.

The fact that plants can pass through the vernalization phase, firstly, in the embryo and, secondly, independently of the plant's *rate of growth,* makes it possible to employ the vernalization method in practical farming. We have tested this experimentally on wheat and other winter plants under laboratory conditions as well as in practical farming on collective and state farms. Experiments with cotton have shown that this plant too can pass through one of the periods preliminary to fruiting—the vernalization phase—not only when already green, but also when outwardly it is still sowing material. We have cited the example of cotton merely to show that the vernalization phase is one of the necessary stages of development not only of winter plants (wheat, clover, esparto grass, and others), and not only of spring cereals, the vernalization of which is already being practised in experimental farming. Thermophilic plants like cotton also pass through this phase of development.

Unless plants pass through the vernalization phase (specific qualitative changes) they cannot bear fruit. Plants can pass through this phase while still in the form of sowing material, i.e., the changes that characterize the vernalization phase can take place in embryos that have just begun to grow, but have not yet, or have barely, broken through the seed integument. If plants do not pass through the vernalization phase when the embryo is just beginning to grow because of the absence of the conditions necessary for this phase of development of the given plant variety, they can do so in the green state when the suitable conditions arise. At the same time the plants may continue to grow slowly or quickly, i.e., may accumulate mass (develop leaves, roots, tillering nodes).

Thus, the vernalization of a given plant before sowing consists in the interaction between the plant organism (the awakening embryo in the seed) and the environmental conditions suitable for the given plant. In the course of this interaction, those qualitative changes take place in the embryo in the seed without which the plant cannot proceed further in its development leading to the formation and ripening of seeds. Some plants either cannot under the field conditions of a particular district undergo these changes at all, or take too long to do so, depending upon the extent to which environmental conditions favour the passage of the given plant variety through the vernalization phase. When plants have not passed through the vernalization phase at all, they will behave like winter plants (the results of which will be absence of fruiting). If the passage through the vernalization phase is slow, the plants will ripen late.

An example of the technique of vernalizing winter and spring wheat before sowing was cited above. It is necessary to emphasize, however, that in vernalizing particular species and particular varieties, care must be taken to create the environmental conditions suitable for each species or variety. The conditions favouring a given plant's passage through the processes of vernalization will be the same as those which the particular plant requires for passing through the same processes when in the green state under field or greenhouse conditions. Therefore, in making up the set of factors suitable for the vernalization of a given plant, it is first of all necessary to know what environmental conditions the particular plant requires for passing through the vernalization phase during its period of growth.

When elaborating a method for the presowing vernalization of any plant, one ought to take as one's starting point the particular environmental conditions under which that plant is cultivated in practical farming. It must be borne in mind that if, under definite environmental conditions, natural or artificially created, the given plant reaches fruiting (and, consequently, passes through the vernalization phase), it does not mean that this environment is the best and the only one in which the plant can pass through the vernalization phase. It is not essential to create this environment in full for the vernalization of the sowing material. The *environment* in which the given plant develops and *the conditions the plant requires* for passing through its whole cycle of development as well as through its individual phases *are far from identical.* Thus, the environment in which plants pass through the vernalization phase under field conditions, contains, besides the set of conditions they need to traverse this phase of development, many components they do not require in the least for passing through this phase. Moreover, the correlation of the individual factors necessary to enable the given plants to pass through the vernalization phase under field conditions may be far from optimal. In many cases, this correlation can be greatly improved artificially. Therefore, the environment in which the given plants pass through the vernalization phase under natural conditions should be only the first step, the starting point, for the investigator. It is necessary to ascertain by analysis, experimentally, which of the conditions of this environment the given plant really requires for passing through the vernalization phase.

The external conditions plants require for passing through separate phases of development, as well as through the whole cycle of development, from seed to seed, are determined by the nature of the given plant (the genotype). Different varieties of wheat require different temperatures for passing through the vernalization phase. If we divide all varieties of wheat into winter, semiwinter and spring, we will find that to pass through the vernalization phase the winter varieties require, in addition to other conditions, a temperature not lower than —2° and not higher than 10° C.; semiwinter varieties—not lower than 3° and not higher than 15° C.; and spring varieties—from 5°-20° C. and higher. In each of the groups named the varietal differences as regards thermal requirements for passing through the vernalization phase, are rather considerable. The ranges of temperature given above are expressed in average figures (i.e., for the majority of the varieties in each group, not for all), and they are given only in order to show that to pass through the same phase of development, different varieties of wheat require different dosages of separate factors.

We stated above that it is wrong to divide the multitude of varieties of wheat and of other plants (as is usually done in agricultural science) into two absolutely separate groups—winter and spring. This division had its origin in the practical cultivation of the different varieties. The varieties which in a given district fail to fruit when sown in the spring will be winter varieties for that given district, but not for all districts. The varieties which in a given district fruit when sown in the spring are spring varieties for that given district, but not for all districts. Our experiments have proved that the absence of fruiting in winter varieties when sown in the spring and the presence of fruiting in spring varieties are due to the fact that the first-mentioned varieties cannot pass through the vernalization phase under spring thermal conditions, whereas the latter can. The different varieties, however, in the winter group as well as in the spring group, require different temperatures and have to be kept under these temperatures for different lengths of time to pass through this phase of development. The thermal conditions, and also the duration of the spring, differ in different districts. Therefore, cases are not rare when spring varieties from some districts behave like winter varieties when sown in other districts.

The lower the temperature the plants of a given variety require for passing through the vernalization phase and the longer they take to pass through that phase, the greater is the degree of "winter habit" of that variety compared with the plants of another variety which require a higher temperature and take less time to pass through the vernalization phase. Our experiments show that different varieties in the winter group and in the spring group require different temperatures and different lengths of time for vernalization (i.e., time necessary for passing through the vernalization phase). Hence, not all winter varieties *are such to an equal degree*, nor *are* all spring varieties *such to an equal degree*. We express the degree of winter habit and spring habit of a variety by the temperature the given variety requires for traversing the vernalization phase and by the time it takes to do so. Knowing the degree of winter habit and spring habit of this or that variety, and the spring temperature (during the sowing and postsowing period of cereals) in a given district for a considerable number of years, it is possible to tell even without sowing whether the plants of a given variety will behave like a winter or spring variety in the given district, and what degree of a spring habit they will possess, i.e., how long the plants will be kept in the vernalization phase, and consequently, retarded to some extent in their further development, for example, earing.

SIGNIFICANCE OF THE INDIVIDUAL FACTORS IN THE SET OF EXTERNAL CONDITIONS PLANTS NEED FOR TRAVERSING THE VERNALIZATION PHASE

For passing through the vernalization phase, as well as for passing through other phases of development, plants require not individual external factors, such as temperature, air, moisture, light, darkness, etc., but a set of factors. The composition of the set and the correlation of the factors within it are determined by the natural properties of the plants. Depending upon whether the plants are wheat or millet, they will require different environmental conditions for passing through the same phases of development—phases that are the same, yet specifically their own, peculiar to their particular natures. Moreover, for traversing different phases of development, the same plant may require different sets of external conditions.

Many investigators (physiologists and others) often confuse the significance of individual external factors with the significance of the entire set of external conditions that plants require for traversing the vernalization phase. This confusion often brings it about that in practice, when employing presowing vernalization, these investigators fail to achieve the anticipated results. It often happens that so-called "vernalized" winter wheat does not ear, or ears slowly and not uniformly. Hence, the incorrect, although in the opinion of these investigators "legitimate," conclusion that not all varieties of winter wheat can be induced by vernalization to ear when sown in the spring.

In the majority of cases, the plants of many of our winter varieties of wheat, rye and other crops, when sown in the fields in the spring, cannot completely pass through the vernalization phase (hence the failure to ear) solely because of the relatively high temperatures of the sowing and postsowing periods.[1] From this some investigators drew the incorrect conclusion that for vernalization "in general," including vernalization of winter and spring cereal seeds, the only thing needed is a temperature suitable for the vernalization of the given variety. Hence, we often read in the literature on the subject that vernalization is

[1] The spring low temperatures, which nearly every year are to be observed only for a short period of time, do not last long enough to enable the plants to complete their passage through the vernalization phase and therefore the latter do not ear.

the thermal stimulation of the sowing material, or that the vernalization of grains is the result of thermal treatment, whereas the vernalization of millet is the result of the operation of the darkness factor, etc. These investigators link the plant's passage through its phases of development with separate factors. It must be borne in mind, however, that although the high temperature of spring under field conditions is the sole reason for the absence of vernalization in winter varieties, nevertheless, it is impossible to vernalize wheat or other crops by temperature alone (any temperature—high, low or variable).

To pass through the vernalization phase, the plants of winter and spring crops require not only the thermal factor, but the thermal factor within a set of other factors. The components of this set known to us at the present time are: temperature, moisture, air. A certain quantitative combination of doses (depending on the variety) of these factors makes it possible (and in all the cases we know of, this possibility becomes actuality) for any variety of cereals, and for many other plants, to go through the vernalization phase.

Some investigators agree that plants need a set of factors and not only the thermal factor in order to pass through the vernalization phase; but they assign the principal role to the thermal factor. Their grounds for doing so are that the plants of spring-sown winter varieties cannot traverse the vernalization phase exclusively because of the high temperature in the spring and beginning of the summer.

If it were possible, in practice, artificially to create the conditions for the vernalization of plants on wide field spaces, the thermal factor would indeed become the principal factor. In the main, it would be the only thing needing regulation to suit the requirements of the plants. In plant-breeding practice, when growing several generations of winter varieties of plants in one year, we recommend the vernalization of plants in pots, and not of seeds. The potted plants should be kept for 6 weeks or 2 months in a temperature of 2°-6° C. under ordinary daylight. In that case, the vernalization of the plant will be more advantageous than the vernalization of the seed, for simultaneously with its passage through the vernalization phase shoots will be obtained which will have had time to gain vigour. In ordinary practice, however, when plants are grown under field conditions, it is impossible to regulate the thermal factor for the purpose of vernalizing the plants. Therefore, for field conditions, it is necessary to vernalize the seed. For plants in the field, or in pots, the chief factor that determines passage through the vernalization stage is temperature. But when these plants are still in the seed form, not yet sown in the ground, their passage through this phase of development, i.e., vernalization, depends, in practice, mainly on the moisture factor. Plants in the field or in pots nearly always possess sufficient moisture to enable them, given the suitable temperature, to go through the processes of vernalization. With presowing vernalization, however, the embryos in the seeds often lack sufficient moisture to enable them to go through the vernalization processes even under the optimal temperature for the given variety.

The main object pursued by presowing vernalization is to compel the plants (still in the seed form) to go through the vernalization phase and at the same time to be economically suitable sowing material (i.e., to prevent the *complete germination* of the seeds). Therefore, in determining the doses of the separate factors comprising the set of conditions for the vernalization of the seed, the aim must be to enable the embryos just beginning to grow in the seed to pass through the vernalization phase and at the same time to make the given conditions least suitable for the complete germination of the seed. Hence, in state- and collective-farm practice, *the moisture factor is at the relative minimum in presowing vernalization more often than the thermal factor.* The amount of water with which the seeds are soaked when preparing them for vernalization is only enough to make them just begin to germinate. The further growth of the seeds should almost cease automatically owing to the shortage of moisture for the growth processes when the temperature is such that the seeds of the given variety will vernalize. For the vernalization of our winter varieties on our collective and state farms we recommend from 0°-2° C., for our spring, late-ripening varieties from 3°-5° C. and for our spring, early-ripening varieties—from 10°-12° C.; in other words, for the vernalization of different groups of varieties, different temperatures are recommended. When preparing the seeds for vernalization, they must be moistened with a suitable quantity of water, depending on the temperature at which they are to be vernalized. The degree of moisture in the seeds of winter wheat must be brought up to 55% of their absolute dry weight. To bring the moisture from 12% (the usual normal degree of moisture in seeds) up to 55%, it is necessary to pour 37 kg. of water on every 100 kg. of seed. The moisture in the seeds of spring, late-ripening wheat must, during vernalization, be brought up to 50% of their absolute dry weight. With 12% of moisture in the seeds, 33 kg. of water for every 100 kg. of seed is required. The moisture in the seeds of spring, early-ripening wheat must be brought up to 48%. Every 100 kg. of seed requires 31 kg. of water. Different percentages of moisture (55 for winter, 50 for spring late-ripening, and 48 for spring early-ripening varieties) are given when preparing seeds for vernalization, not because the seeds of winter varieties require more moisture for swelling than the seeds of spring varieties, but because the seeds of different varieties must be vernalized under different temperatures. If the moisture in the seeds of spring varieties, which must be vernalized in a temperature of 10°-12° C., is brought up to 55% (instead of the 48%) the vernalization processes in the embryos of these seeds will be traversed better than with a 48% moisture content; but at this temperature and with 55% of moisture, the seeds will sprout very considerably. *The lower the temperature at which the vernalization of the sowing material of the particular variety is to proceed, the higher must be the percentage of seed moisture.*

A moisture content of winter wheat seeds not exceeding 50% (of absolute dry weight) is a guarantee that, when being vernalized, the seeds will not germinate at a temperature of 0°-2° C., nor even at a temperature of 3°-5° C. But with a moisture content of 50% and the temperature recommended for the vernalization of winter wheat (from 0°-2° C.), the vernalization processes will not take place, or will take place at an extremely slow pace. A higher temperature (from 3°-5° C.) and a seed moisture content of 50% will be sufficient to enable the vernalization processes to take place. But at a temperature of 3°-5° C., the sowing material of our winter varieties may, during the 40-50 days of vernalization, suffer to some extent from fungus and other diseases. Therefore, notwithstanding the fact that a temperature of from 0°-2° C. is in itself less favourable for the vernalization of winter varieties than a temperature of 3°-5° C., it is, in practice, more profitable to vernalize these varieties at a temperature of 0°-2° C., and to bring the moisture content of the seeds up to 55%. With such a moisture content, a temperature of 0°-2° C. will be no less effective for winter varieties than a temperature of 3°-5° C. with a smaller moisture content (50%). The development of fungous microflora, however, will be ever so much less.

Thus individual factors of the set of conditions necessary for passage through the vernalization processes, when in the minimal state, i.e., when they are, as it were, limiting factors, can be brought out of the minimal state by altering the doses of other factors of the set.

For some plants the combination of doses of the individual factors composing the set of conditions necessary to enable the seeds to go through the vernalization processes does not fully coincide with the combination of doses of the same factors for the germination of the seeds of these same plants. For different plants, and even for different varieties, the doses of the factors for growth and for passing through the vernalization phase differ. The bigger the difference in the doses, the easier is it, in practice, to carry out the presowing vernalization of the given plant. A wider range of variation in the doses of individual factors may be permitted when vernalizing seeds.

In cereals, the discrepancy in the doses of moisture factors for the vernalization and the growth processes, respectively, is not very great. In the vernalization of spring wheat, a moisture content of over 50% (of the absolute dry weight) leads to the germination of the seeds; with a moisture content of less than 45%, the vernalization processes take place at an extremely slow pace (almost come to a stop). Therefore, in practical farming, when vernalizing spring wheat, and more so when vernalizing winter wheat, the collective and state farms *must pay most attention to the moisture in the seeds.* In the presowing vernalization of cereals under collective- and state-farm conditions, *the passage through the processes of vernalization depends mainly on the moisture factor.* On the other hand, the vernalization of cereals and of many other plants in the fields (in the open) *depends in the main on the thermal factor of spring and beginning of summer.* In general, however, to go through any phase of development, including the vernalization phase, a plant requires not individual factors—temperature, moisture, light, darkness, mineral nutrition—but *a set of factors, in a certain combination.* By altering the doses of some factors it may become possible to alter the doses of other factors, without reducing the effectiveness of the set as a whole. By altering the dose of one factor it is possible to make an ineffective set effective.

SEQUENCE OF PHASES IN THE DEVELOPMENT OF PLANTS

It was stated above that the changes that take place during the vernalization of winter cereals and other plants constitute necessary phases of their development. Until plants have undergone these changes (under artificial conditions when the seed is vernalized, or under natural conditions after sowing), they cannot proceed with their further development and, as a result, cannot fruit.

To be able to fruit, plants must go through other qualitative (phasic) changes besides the vernalization phase. The qualitative changes that take place in passing through the vernalization phase are not enough to enable plants to fruit. For example, after completing the vernalization phase, the plants of any variety of wheat, rye and other winter crops can normally continue their development, ear and fruit, only if sown under the conditions of spring or the beginning of summer. If, however, vernalized winter or spring plants are sown in the latter half of the summer, or in greenhouses in the winter (up to February), such vernalized plants do not fruit and in external appearance differ in no way from unvernalized plants. This example shows that under the conditions of spring or of the first half of summer, these plants, after passing through the vernalization phase, undergo certain other changes which they cannot experience under the conditions of the second half of summer, autumn or winter, and without which they cannot fruit. Thus, under the conditions of the latter half of the summer, autumn or winter, notwithstanding the favourable artificial or natural temperature created for the plants, certain conditions are lacking for the normal development even of vernalized plants. The deficient factor in this case for vernalized plants is inadequate duration of daylight.

By artificially lengthening (by means of electric light) the short autumn and winter days, and by suitably raising the temperature (to 15°-25° C.), it is possible to induce cereals to develop normally when sown in the autumn and winter. In such cases it is only necessary to vernalize the seed of winter varieties before sowing, or to vernalize the plants after sowing. For traversing the vernalization phase, the photo factor and the duration of daylight play no role at all. Plants can go through the vernalization phase with equal success under long-day conditions and short-day conditions, no matter how short the day may be (even in continuous darkness), as long as the

Fig. 15. Winter wheat Novokrymka 0204

Sown in greenhouse on August 6. Photographed on September 26. The plants in the first pot on the left (sown with ordinary seeds) were grown in continuous light. The plants in the second pot from the left (sown with vernalized seeds) were also grown in continuous light. The plants in the third pot from the left (also sown with vernalized seeds) were grown in continuous light for the first 17 days and for the next 32 days under 10-hour-day conditions. The plants in the fourth pot from the left (sown with ordinary seeds) were grown under 10-hour-day conditions. The plants in the fifth pot from the left (sown with vernalized seeds) were also grown under 10-hour-day conditions

seeds are sufficiently moist and the temperature does not exceed the range permissible for the given moisture of the given plant. At a temperature below 0° C., none of the varieties of cereals known to us can pass through the vernalization phase; at a temperature exceeding 10° C., it is practically impossible to vernalize the majority of winter varieties of cereals known to us: they vernalize too slowly.

In greenhouses, under the conditions of artificially shortened spring or summer days, or under those of natural short autumn or winter days, the growth of the vernalized plants of cereals does not cease. Given suitable mineral nutrition, these plants produce a fairly large amount of green mass, but they do not fruit. In our experiments we have been able to keep barley plants under the conditions of a 10-hour day for as long as two years. These plants produced new leaves all the time but did not fruit. The same variety of plants can be forced to ear and flower in 25 to 30 days if given the same thermal conditions but placed in continuous light instead of getting light only 10 hours a day.

These observations indicate that cereals can continue to grow under short-day conditions; but then their further development (after the vernalization phase), their advance towards fruiting, is inhibited, or else proceeds very slowly.

To solve the problem of whether short-day conditions are unsuitable for the whole of a plant's further development after the vernalization phase or only for a part of it, only for some of its phases of development, E. P. Melnik conducted the following experiment in our laboratory. In 1932, she sowed the winter wheat Novokrymka 0204 in a vernalized and in an unvernalized state. The experiment was conducted under high summer thermal conditions, which made it impossible for the plants from the unvernalized seeds to vernalize. Therefore, in spite of the continuous light that was given to the experimental plants, i.e., in spite of conditions most favourable for the fruiting of vernalized wheat, the plants of the unvernalized sowing material grew for a long time, developed numerous leaves, but did not fruit. The plants of the vernalized sowing material, however, under these very same conditions, proceeded to ear and fruit fairly quickly (on the 35th day after sowing). Another series of experimental plants from vernalized and unvernalized sowing material was grown under the same thermal conditions, but with a 10-hour day instead of continuous light. Under these conditions, the plants of both the vernalized and unvernalized seeds failed to fruit. In external appearance, the vernalized plants in no way differed from the unvernalized plants.

Some of the potted plants from vernalized and unvernalized seeds were put in continuous light after having been kept under 10-hour-day conditions for different numbers of days. When this change had been effected all the vernalized plants, irrespective of the length of time they had been kept under the short-day conditions, fruited rapidly. The plants of the unvernalized sowing material, however, regardless of the number of days they had been subjected to short-day conditions, failed to ear when put under continuous light, as was the case with the unvernalized plants which had been kept under continuous light all the time. This is further proof that passage through the vernalization phase does not at all depend upon variation in the duration of exposure to daylight.

The following question naturally arises: Are plants which fail to fruit under short-day conditions qualitatively the same whether they are derived from vernalized or unvernalized seeds? In external appearance, plants from vernalized sowing material grown under short-day conditions differ in no way from those derived from unvernalized seeds grown under the same conditions. Both the former and the latter tiller fairly vigorously under these conditions but they do not develop stalks. Placed under long-day conditions, or still better, in continuous light, the plants from vernalized seeds rapidly began to differ in external appearance from those grown from unvernalized seeds. Consequently, the plants from vernalized sowing material which had grown for a long time under short-day conditions and had been unable to fruit, although not differing in external appearance from the plants of unvernalized seeds, nevertheless differed from them qualitatively. This leads us to the conclusion that the changes *taking place in the cells of the just slightly grown embryos during presowing vernalization are not lost, but are transmitted to the plant's new cells formed in the process of growth, no matter how long its further development is retarded.*

The chief object of the experiment we are discussing was to ascertain whether a long day or continuous light is a necessary condition for a plant's entire further development after the vernalization phase, or only for separate phases of development. Therefore, in the experiment referred to, variations were made in the treatment of the vernalized sowing material. In the first postsowing period, which was of different duration for different plants (from 2-6 to 40 days), some plants were grown under continuous light and then transferred to a 10-hour day. After being kept under continuous light for 20 days and then transferred to a 10-hour day, the plants from the vernalized sowing material formed stalks, eared and completed their development just as quickly as those plants which had been kept under continuous light all the time. This shows that wheat plants *do not need a long day or continuous light for the processes of development and the growth of stalks*, although the latter do not appear when the plants are grown (from the moment the seeds sprout) under short-day conditions. Moreover, it also shows that long-day or continuous light conditions are not essential for the plants' *entire* cycle of development after the vernalization phase. Wheat requires a long day, or more correctly, continuous light, only for a part of its cycle of development. This stage of development has been named the photo phase. It has been ascertained that cereals pass through the photo phase, which follows immediately after the vernalization phase, best of all in continuous light or at least under a long day, in combination with the other factors (temperature, moisture, air).

On the basis of this experiment, we believe, by analogy with the vernalization phase, that when passing through this (photo) phase of development, which requires prolonged light, the plants undergo qualitative changes which are later transmitted to all the new cells formed in the plants when subsequently grown under short-day (10-hour) conditions. *Under the influence of a suitable set of external conditions, which includes continuous light, or a long day, changes take place in vernalized plants that are transmitted to the new cells formed in the process of the plants' growth, in the same way as are the changes characteristic of the vernalization phase.*

The qualitative changes characteristic of the photo phase can take place only after the vernalization phase. This has been proved by numerous experiments conducted in the vernalization of winter cereals and other plants. Late spring sowing conducted not only with unvernalized, but even with slightly undervernalized winter crop seeds, results in failure to fruit and ear. Hence, in spite of favourable external conditions (long spring and summer days) for the plants' passage through the photo phase, such plants cannot proceed to that phase. The reason for this is that the qualitative changes that take place with presowing vernalization have not been completed (the sowing material was undervernalized), and vernalization fails to continue after sowing because of the high temperature. If such plants are

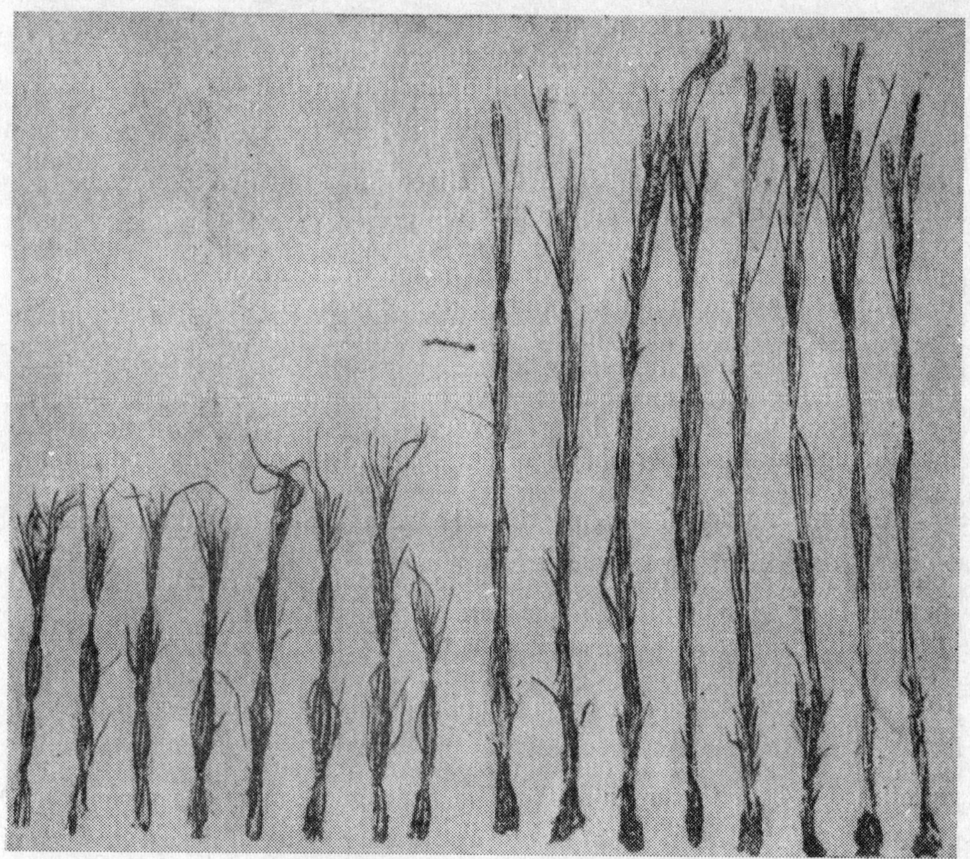

Fig. 16. Winter wheat Erythrospermum 1325/5

The plants in all the sheaves were sown simultaneously on March 30, 1930, at the Ukrainian Institute of Selection and Genetics (Odessa). The plants in the first sheaf on the left were grown from unvernalized seeds. The plants in all the other sheaves were grown from vernalized seeds. The respective periods of vernalization were as follows, from left to right: 7, 11, 17, 21, 26, 31, 36, 41, 46, 52, 57, 62, 67, 72 and 77 days. 41 days (9th sheaf from left) of presowing vernalization were enough for the fruiting of the given variety. If vernalized for fewer days, this wheat does not ear

vernalized, or if their vernalization is completed, in the autumn, and they are then placed under short-day conditions, they will neither ear nor fruit. Hence, during the long days of spring and summer the undervernalized plants do not traverse the photo phase. After vernalization these plants will need long-day conditions to enable them to undergo the qualitative changes that are characteristic of the photo phase.

The photo phase cannot be traversed before or during the vernalization phase. It can be traversed only after the vernalization phase has been passed through.

Figure 16 shows sheaves of Erythrospermum 1325/5 winter wheat that was sown in the spring of 1930. The plants in the first sheaf on the left are from ordinary seeds. The plants of all the other sheaves are from sowing material that had been vernalized for different numbers of days (7, 11, 17, up to 77 days). It can be seen from the illustration that ears were produced only by the plants grown from seeds that had been subjected to presowing vernalization for 41 days and longer. The plants of all the other variants grown from seeds that had been vernalized for a smaller number of days failed to ear. The external appearance and behaviour of these plants (from undervernalized seeds) differ in no way from those of winter plants grown from ordinary spring-sown seeds. (In the illustration, first sheaf on the left.) These same plants, obtained from undervernalized sowing material, can be brought to fruiting under spring and summer conditions. For this purpose it is necessary to complete their vernalization, which is done by giving them an additional vernalization period corresponding to the deficiency they suffered before sowing. After this they will be able to pass through all the other phases of development under spring and summer conditions.

The time needed for completing the vernalization of plants does not depend upon the length of the interval between the first period of vernalization and the beginning of the second. Plants from undervernalized seeds can continue their vernalization in the field immediately after sowing, bearing in mind the low temperature conditions of early spring. If, however, the external conditions are unsuitable for passing through the vernalization phase it will not be completed. It will be completed only when those conditions are forthcoming. The method of completing the vernalization of undervernalized seeds is often practised in our laboratory. We vernalize the seeds of many varieties of wheat that require 56-57 days to go through the vernalization phase in 5, 10, 15 and up to 40 days. Such sowing material is kept in a semi-dried state (up to 15 or 20%) and is used for various experiments as required. To obtain fruiting plants from such sowing material it is necessary to vernalize it for the additional length of time it would have required to complete the vernalization during the first period.

Consequently, during the vernalization of seeds or plants, an accumulation of changes takes place. These changes remain in the cells in which they have taken place and are also transmitted to all the new cells formed from them. If these changes in the cells are not completed, i.e., if the given phase of development does not reach its end, the accumulation of changes may continue in the newly-formed cells up to a certain limit, which is a sign that the passage through the given phase of the development has come to an end. After this, changes in the cells in this direction no longer take place, no matter how long the plants are kept under the influence of the external factors which caused these changes to take place earlier. In addition to different dosages of external factors for passing through the vernalization phase, the winter plants of different varieties of wheat, rye, barley, etc., also require different lengths of time during which these factors operate. Thus, for vernalization at 55% of moisture and 0°-2° C. temperature, the various varieties of winter wheat require the following: Erythrospermum 808 1/26 needs 18 days; Kooperatorka, 40 days; Stepnyachka,' 45 days; Ukrainka, 50 days. If vernalized for shorter periods under these conditions, and if sown under high-temperature conditions, these varieties fail to ear. Longer periods of presowing vernalization do not, however, produce earlier earing in plants from seeds of these varieties than occurs in plants from sowing material vernalized for only the number of days required for the given variety.

Thus, when passing through the vernalization phase, the accumulation of qualitative changes in the just slightly grown embryo, or in the green plants, takes place only up to a certain limit. Beyond this limit, the accumulation of changes in the given direction ceases entirely. If, however, this limit has not been reached, the plants cannot pass to the next phase of development, i.e., cannot undergo the other changes that characterize the next phase, even though the external conditions are favourable for these changes. *In the development of plants, a sequence of phases* (stages of development) *is observed.* Normally developing (i.e., hereditarily unaltered) plants cannot skip any phase of development.

PHASIC CHANGES IN PLANTS TAKE PLACE AT THE GROWING POINTS OF THE STEMS

The *phasic changes* that take place in a plant, or in its separate organs, are irreversible, i.e., *cannot be turned back.* Our numerous experiments show that undervernalized plants can always be brought to complete vernalization. This applies not only to sowing material, but also to plants grown from undervernalized seeds. The vernalization of such plants can be completed at the time of sowing or whenever thereafter suitable external environmental conditions are created for these plants. The possibility of completing the vernalization of the sowing material and plants of wheat has already been mentioned above. Other plants behave in the same way as wheat in this respect. Let us cite the case observed in the work of D. A. Dolgushin.[1] In 1930, at the Ganja station, he vernalized cabbage seeds for sowing with the object of obtaining flowering and fruiting plants. Not one of the experimental plants developed flowering shoots in the first year of life, neither did the control plants obtained from ordinary seeds of the same variety. In the autumn, a few score of experimental and control plants were transplanted to sand in the laboratory, where they passed the winter, and in the spring were planted in the field. All the plants from the vernalized seeds which had not developed flowering shoots in their first year produced a flowering shoot in their second year. The plants from the ordinary seeds failed to develop flowering shoots even in their second year. In our opinion,

[1] A collaborator at our laboratory.

two explanations are possible of the failure of the cabbage plants grown from vernalized seeds to develop flowering shoots in the first year of life.

Firstly, that the cabbage plants grown from the vernalized seeds may not have been completely vernalized. Therefore, they were unable to develop flowering shoots, but formed heads. In the winter these plants completed their vernalization in the laboratory, in spite of the relatively high temperature at which the processes characteristic of the vernalization phase could only take place slowly. The plants from the ordinary seeds did begin to go through the vernalization phase in the laboratory in the winter, but owing to the relatively high temperature, the process was slow, and therefore they failed to complete the passage in the course of the winter. For this reason these plants failed to develop flowering shoots also in the second year of life.

Secondly, the cabbage plants from the vernalized seeds may have been fully vernalized before sowing, but owing to the high temperature and the late emergence of the shoots above ground in the spring, these plants were unable to pass through the photo phase, without which such plants cannot develop flowering shoots and reproductive organs.

A number of other special experiments that have been conducted show that the processes characteristic of the vernalization phase cannot be reversed. Whereas incompletely vernalized plants can be completely vernalized if placed under suitable conditions, we know of no case of the unvernalization of vernalized plants. *Plant cells possessing the qualities of the vernalization phase cannot be turned back to the initial (prevernalization) state.*

At the same time, we know of a number of perennial plants that require vernalizing every year. If the conditions for passing through the vernalization stage are absent, some perennial plants that have already fruited cannot fruit again. For example, many varieties of perennial rye or barley, transplanted in the winter or spring from the ground into pots to grow in the greenhouse, proceed to ear and flower at the end of the spring or the beginning of the summer. Later, these plants produce ripe grains, i.e., they complete their cycle of development. Simultaneously (or later), as the old straw below withers, these plants form new shoots, which do not develop into stalks that year (the straw does not develop). The further behaviour of these plants differs in no way from the behaviour of the ordinary annual unvernalized winter plants of rye, barley or wheat. Until these plants are placed under low temperature conditions (from 0°-10°C.) for the purpose of passing through the vernalization phase, they cannot proceed further in the direction of developing new stalks and reproductive organs. Behaviour similar to that of the plants of perennial rye and barley is observed among the summer shoots of some annual plants of ordinary winter varieties of wheat sown in the autumn, which in the spring normally proceeded to ear. In the summer, the shoots that grow from the roots of these plants behave like typical unvernalized winter plants.

All these examples of the behaviour of the plants of perennial rye and barley, or of the behaviour of shoots that appear in the spring from the lower parts of winter wheat that has passed through the winter, seem to contradict the above-formulated proposition that a plant's passage through its phases of development is not reversible. On the one hand, cells that possess the properties of vernalization cannot be turned back to the initial (prevernalization) state—in their individual development plants can only move forward. On the other hand, the plants of perennial rye and of annual winter wheat which have produced ripe grains and, therefore, have passed not only through the vernalization and the photo phases, but also all other succeeding phases of development, can form shoots in the lower parts, which (in the phasic sense) begin their development anew. First they must pass through the vernalization phase, then the photo phase, and so forth. All this seems contradictory; but the contradiction is only apparent. To be able to understand this seeming contradiction it is first of all necessary to determine: a) in what parts of the plant the qualitative changes characteristic of the separate phases of the given plant's development take place; b) how these changes are transmitted from cells to cells.

To answer the question as to the parts of a plant in which the qualitative changes related to the phasic changes take place—whether, under given external conditions, in the entire plant or only in definite parts of it—we conducted a number of experiments, chiefly with soya bean and cotton. Plants obtained from cuttings taken at successive points along the stem of a soya bean behave differently as regards initiation of fruiting (flowering). All the plants from cuttings taken above the point of insertion of the first flowering shoot of the mother plant come into flower with extreme rapidity (as soon as the cuttings take root), whereas the plants from cuttings taken below the point of insertion of the first flowering shoot (of the mother plant) are late in flowering. The lower the point of the main stem from which the cutting was taken, the later was the flowering. The same was observed in our experiment with cotton. After passing the winter with fallen leaves in a cold greenhouse (from 0°-5° C.), the old tufts of cotton plant that had already fruited began to develop new young leaves in the spring, when warm days arrived. Simultaneously, sympodial (fruiting) branches began to appear in the axils of the young leaves. The sympodial branches did not appear in all the axils, but only in the axils of the leaves situated above the point of insertion (along the main stem) of the previous year's first sympodial branch. No new sympodial branches appeared in the axils of the leaves below the first, old sympodial branches; in them monopodial (vegetative growth) branches appeared.

Usually, there is a greater flow of nutritive substances to the upper buds of the stem than there is to the lower ones. Therefore, to determine whether this alone, in the case in question, explains why fruit buds appear in the upper parts of the cotton stem and growth buds in the lower, the tops of a number of plants were cut off at the point of insertion of the previous year's first sympodial branch. This caused an increased flow of nutritive substances to the remaining parts of the plants. Nevertheless, all the buds of these truncated plants gave rise only to vegetative growth shoots (monopodial branches) but not to flowering shoots. Consequently, in this case, the appearance of vegetative growth or fruiting branches depended not upon nutrition, but upon the tissue cells from which buds were formed.

Figure 17 shows two soya bean plants grown from cuttings in continuous light. The plant on the left is from a cutting taken from the top of a sterile plant; before the cutting was taken it had grown in continuous light, which usually prevents soya beans from fruiting. The plant on the right is also from a cutting taken from the top of a plant, but one that had fruited. Before the cutting was taken the plant had grown under the ordinary conditions of alternate day and night. Before the cuttings were planted, their leaves and all their buds were removed. After taking root, the plant from the cutting taken from the sterile plant (Fig. 17, left) did not flower. Under the same continuous-light conditions the cutting taken from the plant that had fruited formed buds as soon as it took root and later flowered and produced beans. Thus, on the basis of our experiments we arrived at the conclusion that the failure to form reproductive organs (fruit buds) may not depend in many cases upon the flow of nutritive substances to the section of the given tissue (experiment in cutting the tops of cotton plants). The failure to form reproductive organs may also not depend upon the location of the given buds. It may be immaterial whether they are situated on lower, middle or upper parts of the stem (experiment in taking cuttings from fruiting and sterile soya bean plants). The formation of reproductive organs depends first of all upon whether the cells of the given tissue have gone through those qualitative, phasic changes without which they cannot assimilate the external conditions necessary for the formation (development) of reproductive organs. At the same time the above example of the growing of soya bean plants from cuttings taken at successive points along the stem of the mother plant appears to show that the time of fruiting depends upon the location of the tissue along the stem (the part of the stem from which the cutting is taken). In this experiment the lower, and hence the older, the part of the stem of the mother plant from which the cutting was taken, the later the plants proceeded to flower. In this case it was found that the tissue of the lower part of the stem was less ready to form fruit buds than that of the younger upper part.

Thus *at different points along the stem the tissue cells may possess different phasic qualities. Different parts of the stem tissue may be in different phases of development.* The tissues of the lower part of the stem are in a younger phase of development than those of the upper parts. The lower part of the stem may possess the properties of the vernalization phase; the upper parts may possess the properties of the next phase, the photo phase, etc.

The proposition that the tissue at different points along the stem may possess different properties as regards readiness for fruiting is supported not only by our experiments but also by a number of facts met with in practical farming and in the literature on the subject. The lower down the main

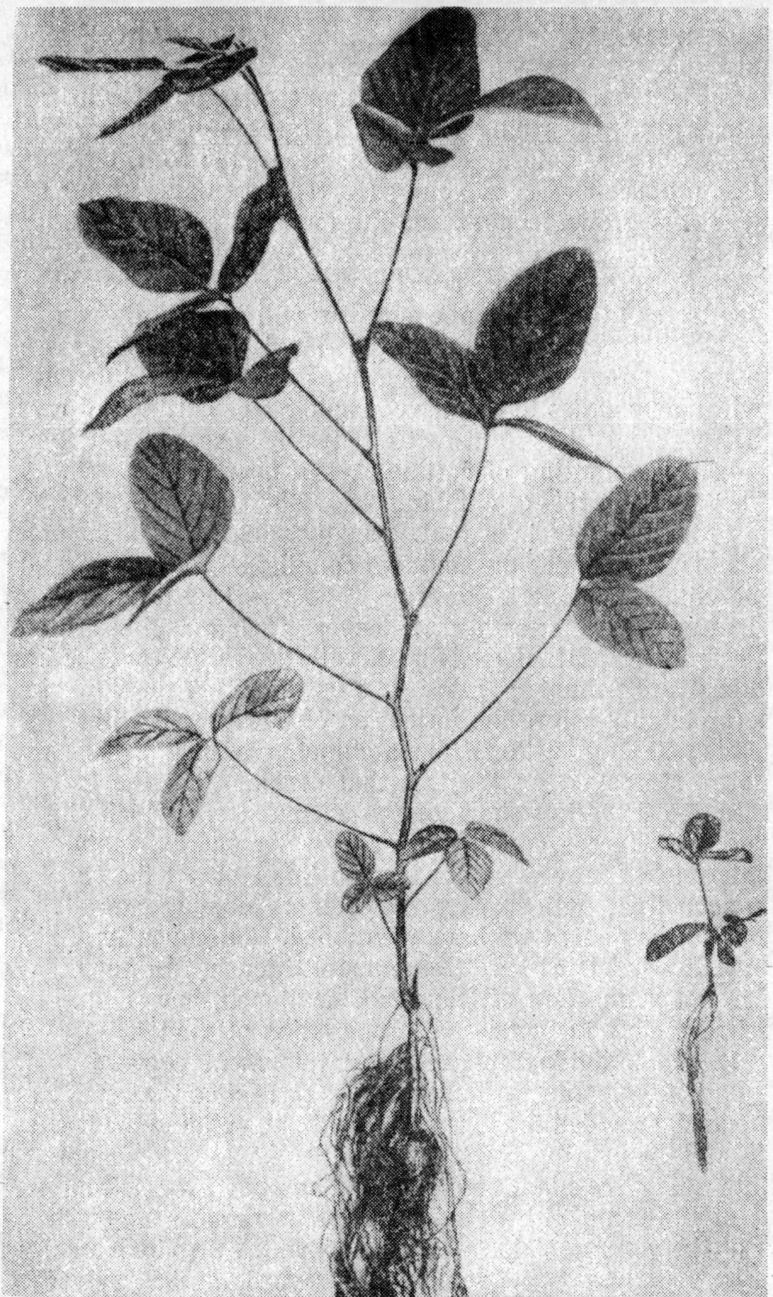

Fig. 17. Soya Bean

Both plants were grown in continuous light; the plant on the left was grown from a cutting taken from a sterile plant; the plant on the right was grown from a cutting taken from a fruiting plant. The plant on the left, grown in continuous light, did not flower. The one on the right, after throwing out a new stem 2 mm. long, formed a flower bud and fruited. Hence, the tissues of the two cuttings were not of like quality

stem a cutting is taken from a fruit tree (apple or pear) grown from a seed and not from a graft, the younger will be the phase of the new shoots, and the more years will pass before these shoots proceed to fruit. Low-cut stumps of forest trees produce shoots as young (as regards readiness for flowering) as one-year shoots grown from seeds. The fact that such shoots, possessing a strong old root system, will grow more quickly and vigorously than a one-year seedling is another matter; and therefore, of course, the quality of the wood of such trees cut for economic purposes will also be different.

N. P. Krenke (*Plant Surgery,* pp. 264-78, 1928) deals in considerable detail with the different qualities of cuttings taken from different parts of plants. It will not be amiss to cite the example of the cuttings of ivy (Hedera helix) that Krenke mentions in his book. "Here it is appropriate to mention," writes Krenke, "the planting of cuttings taken from the flowering shoots of ivy and of several varieties of climbing *ficus.* The strands of ivy, when trailing on the ground, easily take root. Usually such strands do not develop flowering shoots. But, under the southern conditions that are normal for ivy, if the strand climbs a support, flowering shoots will form on such branches. A special feature of the latter is their leaves. Their margins are entire and they are of acuminate-ovate shape, whereas all the rest (except the first leaves of the seedling) are palmately, lobed. If a cutting from the flowering shoots is planted (preferably before the formation of flowers), it will grow into an *erect* tree, whereas any cutting from a climbing branch will reproduce the climbing form. Moreover, on the tree that develops, all the leaves will be like those on the flowering shoot of the cutting from which this tree had grown. True, with increased nutrition and water supply, individual shoots with palmately lobed leaves will appear on this tree. At the same time, the seeds from such trees will produce the ordinary climbing form of ivy. The same applies to the *ficuses* we have mentioned. Consequently, this phenomenon belongs to what is called "Dauermodificationen." In our opinion, this, of course, is not a question of Dauermodificationen, but of phasic changes which are irreversible in vegetative propagation of plants.

Thus, 1) the individual phases of development proceed in strict sequence; 2) each developmental phase can be passed only after the preceding phase has ended, provided the external conditions suitable for it exist; 3) the tissue cells may be passing through different phases of development at different points along the plant stems. If a plant has grown from a seed, the lower part of the stem, although the oldest as regards age, possesses the properties of the youngest phase of development. And vice versa, the upper part of the stem, although the youngest as regards age, may be in an older phase of development.

From this we draw the conclusion that when a plant, or individual parts of it, go through the vernalization or other phase of development, changes take place only in the cells at the growing points of the stem. The changes that take place in the cells at the growing points as a result of cell division are transmitted to the newly-forming cells. Given the suitable external conditions, the young cells go on changing until these changes reach their limit, i.e., until the given phase of development is completed. Thereafter, under other necessary external conditions, they begin to pass through the next phase of development. This explains the different degrees of readiness for fruiting manifested by the tissues at different points along the stem.

When external conditions are suitable for a plant's rapid progress through its phases of development and unsuitable for its rapid growth, cases may often be observed of a plant's rapid progress through its phases of development although its growth is extremely restricted. When the plant has completely passed through its first phase of development it passes on to the next phase if the external conditions are suitable for this, and so on until the seeds ripen. The quicker a plant, under the suitable external conditions, passes through its phases of development and the slower it grows under these conditions, the lower will be the point on the main stem at which the tissue will be ready to form (develop) fruit buds. In our experiments, the height at which the first fruit buds appeared in the axils of the leaves on the main stem (counting from bottom up) of plants like cotton, brown hemp and soya bean, varied greatly, depending upon the conditions under which the plants were grown. In the practical farming of cotton (upland) the first sympodial branches appear in the axil of the fourth or fifth leaf. In our experiments with plants of this variety of cotton, some variants had the first sympodia in the axil of the second leaf. The plants of other variants developed 25-30 regular leaves and were unable to form sympodial branches. The tissue cells of the stems of these plants did not undergo the corresponding changes. The same may be easily observed in brown hemp, soya bean, and other plants.

When vernalizing seeds artificially, special conditions are created which inhibit the growth of the seeds and favour the rapid progress of the embryos that are just beginning to develop through the vernalization phase. We can already induce several plants (millet, soya beans) before they are sown, but after passing through the vernalization phase, to go through the next phase of development, the photo phase.

In soya bean plants grown from vernalized seeds, the slightly-developed embryos of which have gone through the vernalization and photo phase processes, it is possible to observe not only early flowering, but the appearance of the first fruit bud at a low level of the stem.

Cases are not rare in soya bean plants vernalized before sowing of the first buds appearing in the axil of the first leaf. In such plants, the stem tissue, even before sowing, may be quite ready (as regards phasic qualities) for fruit buds to form (develop) from its cells if the conditions for this are suitable.

The readiness of plants, as regards phase of development, for fruiting does not necessarily mean that these plants will bear fruit. It merely means that out of the qualitatively, phasically ready cells reproductive organs may develop. The development of these organs, like everything else in a plant, calls for specific external conditions. By mineral feeding or light alone it is easily possible to create such an environment for cotton (and for many other crops) that this plant, although phasically ready for fruiting, will not only be unable to develop fruit buds, but will lose the buds, flowers and ovaries (bolls) that it may already have developed.

THE LOCALIZATION OF PHASIC CHANGES

From the foregoing we arrive at the conclusion that the phasic changes in plants taking place in the cells at the growing points of the stem are transmitted by division to all the new cells that are formed from them. Can the phasic changes be transmitted in some way other than cell division, i.e., can the phasic changes that take place at the growing point of a stem be transmitted to the cells of a lower part of the same stem, and also to the cells of other stems or branches above them?

A number of observations, and also special experiments, show that phasic changes are localized in the cells in which they take place. They can be transmitted only to the new cells that are formed from them, i.e., these changes are transmitted only from the mother cells to the daughter cells.

When vernalized seeds of winter varieties are used in practical farming, cases occur when the plants from such material produce one or two (usually central) flowering shoots and a rosette of winter, nonflowering shoots. The explanation of this phenomenon is that the cells of the growing point of the central bud in the embryo were vernalized, whereas the cells of the growing points of the other (not central) buds were not vernalized (or were undervernalized). They cannot acquire the quality of vernalization from the neighbouring vernalized cells (Figs. 18 and 19). This, too, explains why perennial rye requires annual vernalization of the stems that grow below out of dormant buds.

In proof of the proposition that phasic changes are localized and that they can be transmitted only to the cells that are formed out of the changed cells (i.e., by their division), a number of facts can be cited.

Some of the stems of the same winter wheat plant (Fig. 20) that had wintered in the field (and consequently had gone through the vernalization phase) did not fruit in the spring, after being put under short, 8-hour-day conditions. Under these conditions, the changes characteristic of the next, the photo, phase cannot take place in the growing points of these stems. Under 24-hour-day conditions, however, the other stems in the same cluster fruited without exercising any phase-changing influence upon the neighbouring stems.

The *perennial cotton plant* No. 01632/2 (of Abyssinian origin) cannot fruit under the conditions prevailing in Central Asia, and still less under those prevailing in the Ukrainian S.S.R. Owing to the excessively long day

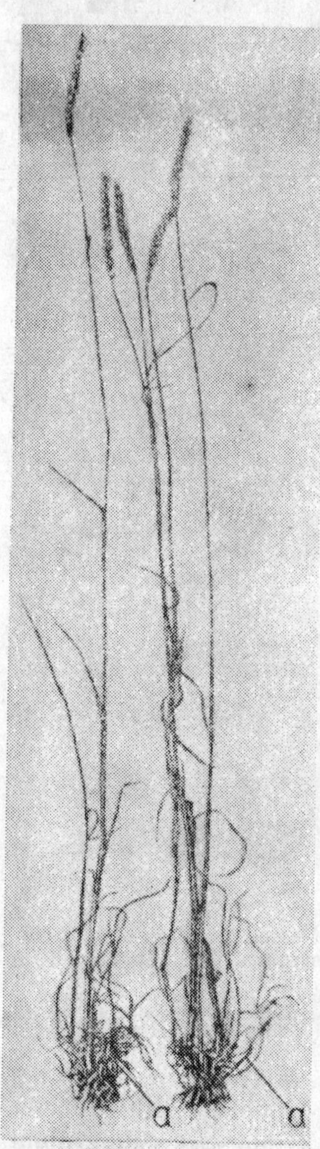

Fig. 18. Winter rye Tarashchanskaya, autumn sowing:

a—sterile shoots grown in late spring from dormant unvernalized buds at the tillering nodes of normally fruiting rye

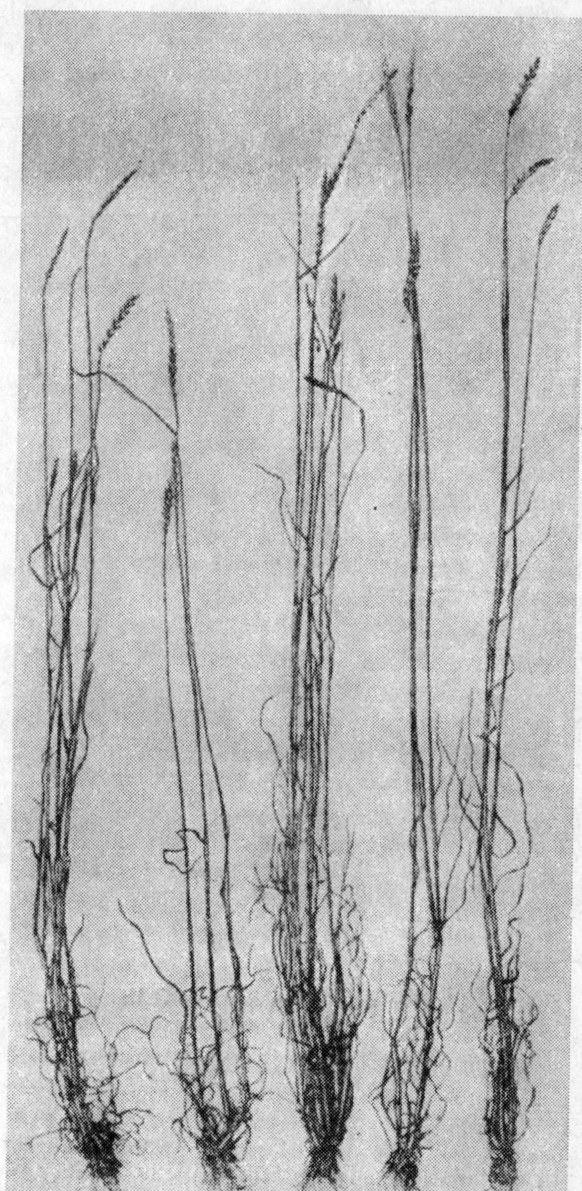

Fig. 19. Winter wheat sown in autumn

In the spring has at the lower parts of its stem winter-grown unvernalized shoots from the dormant buds at the tillering nodes

Fig. 20. Winter wheat Byelokoloska Ostistaya 0719, autumn sown

Taken from the field on April 23 and grown in greenhouse. The left half of the tillers was grown in continuous light; the right—under 8-hour-day conditions

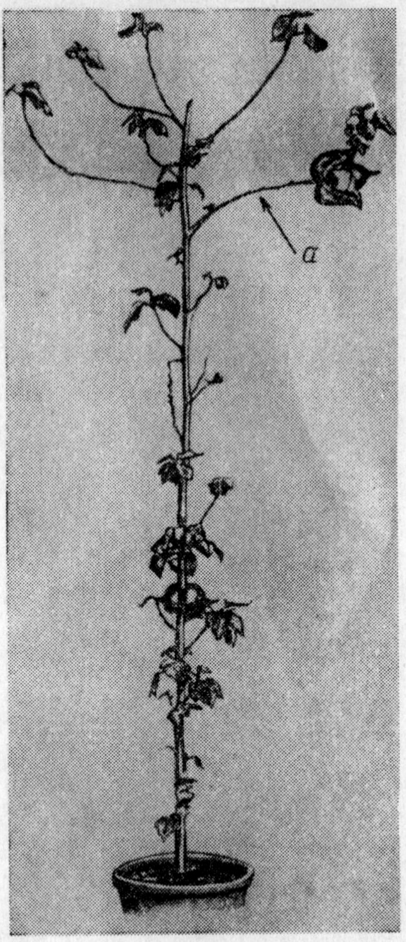

Fig. 21. Perennial cotton plant from Abyssinia, No. 01632/2

Sown on May 15, 1932. This variety cannot go through the photo phase in view of the long day in Odessa

in our districts, the plants of this variety of cotton cannot pass through the phase that follows the vernalization phase.[1] Figure 21 shows a cotton plant of this variety that has been grown for two years in continuous light.

In the first year of this plant's life, one of its branches (branch *a*) was kept in the dark for 14 hours during 30 consecutive days. After that, this

[1] For this phase of development, many of the so-called short-day plants require prolonged darkness (long nights).

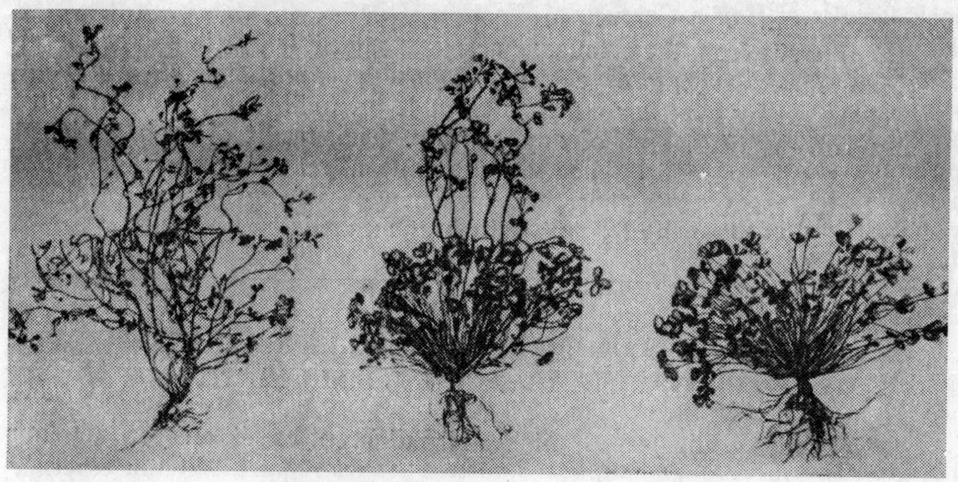

Fig. 22. Shabdar (annual clover)

The bush on the left was grown from spring-sown vernalized seeds; in the summer it produced flower buds, flowered and seeded. The bush on the right was grown from ordinary seeds; sown in the field at the same time as the first, it could not vernalize and therefore did not form reproductive organs. The middle bush was grown from undervernalized seeds. Before sowing, a group of embryo cells of this plant was able to complete the vernalization phase. Out of some of the cells flowering shoots were formed, and out of the adjacent unvernalized embryo cells nonflowering vegetative buds were formed

branch was kept under the same conditions as all the others. Later, buds formed on this branch. It was the only one on the whole plant that fruited in the second as well as in the first year of its life. The entire mass of nonflowering branches of this cotton plant could exercise no influence on the fruiting branch, nor could the latter transmit its properties to the neighbouring branches.

All these facts show that phasic changes are localized in the cells and cannot be transmitted to the neighbouring, adjacent cells.

It is not infrequently stated in the literature on this subject that if cuttings from one- or two-year apple or pear seedlings are grafted on to the crowns of fruit-bearing trees, their own fruiting is greatly accelerated, i.e., a property of the fruit-bearing trees is, as it were, transmitted to the grafts, which, according to our interpretation, are phasically not ready to form fruit buds. If these statements were true, our interpretation of the phasic changes that take place at the growing points, and of the localization of these changes, would be wrong. There is every ground for doubting the truth of these statements on this question, the more so since they are merely repetitions (to be found in many textbooks) and contain no references to sources, no indication as to when and by whom it was observed that the fruiting of one- or two-year seedlings is accelerated by grafting on to the crowns of fruit-bearing trees. The authoritative view expressed by I. V. Michurin on this question contradicts these unsupported statements. In his book *Results of Half a Century of Work*, I. V. Michurin writes: "But we get the very opposite from the erroneous assertion that it is possible to accelerate the inception of fruiting of a young hybrid seedling in its early stage of development by grafting cuttings from it on to the crown of an adult, already fruit-bearing, tree of any variety. The result of such action is quite the contrary—not an acceleration, but a retardation of the inception of fruiting, except in those cases when we perform this operation not with young hybrid seedlings, but with branches adult as regards age and time of fruiting."

BRIEF CONCLUSIONS CONCERNING THE PHASIC DEVELOPMENT OF ANNUAL SEED PLANTS

1. Not only do different plants require different conditions for their normal growth and development, but the same plants require different external conditions in the course of their lives from seed to seed. The fact that a plant requires different environmental conditions for its development shows that the plant's development from seed to seed is itself different in different periods. *The development of annual plants consists of separate stages—phases of development.*

The phases of development of a seed plant must be understood to mean not the formation (development) of the different organs and parts, but the qualitative turning points and stages characteristic of and conditioned primarily by the changes in the environmental conditions the developing plant requires.

The need (including assimilation) of relatively definite conditions and also the changes in this need during the course of the individual life of a plant are conditioned by the entire generic, specific and varietal history of each individual case of a seed embryo examined by us.

The development of all preceding generations gives fairly definite direction to the development of a plant from a given, concrete seed. As special experiments and observations of plant life have shown, the relatively directed development characteristic of an embryo in the seed proceeds in stages, or phases.

By the growth of a plant we mean what this is usually taken to mean in practical farming, namely, the increase in the plant's weight and volume, irrespective of the organs and parts of the plant at whose expense this increase is obtained.

We apply the same conception to the growth of the separate organs and parts of a plant. For example, by the growth of the root of a sugar beet we mean the increase in the root's mass and volume.

The term growth does not characterize the qualitative state, the state of maturity of a plant, or of its organs. *Growth is the increase* in the mass of a plant in a given phase of development. The property of growth can be expressed in different degrees, depending on the nature of a plant, on its phase of development, and also on its environmental conditions.

2. The sets of external conditions a plant requires for passing through its phases of development and for growth in a given phase of development often do not coincide. This incongruity applies not only to the difference in the dosages of factors required for growth and in those required for development. In the case of many plants there is a difference in the very factors that constitute the sets required for development and growth respectively. Hence, in the life of individual plants the following may often be observed: a) rapid growth and slow development, slow progress towards fruiting; b) slow growth and accelerated development; c) rapid growth and rapid development.

3. When seeds are vernalized under artificial conditions in the laboratory or in the collective-farm barn, conditions are created under which the plants (the just slightly-developed embryos) pass through one of their phases of development (the vernalization phase) with growth extremely slow, barely perceptible.

4. The changes that take place in the just slightly-grown embryos during presowing vernalization constitute one of the phases of a seed plant's development. Without these changes the plants of winter varieties (and of all spring varieties, it must be assumed) cannot fruit.

Given the suitable external conditions, a plant can begin to go through the vernalization phase immediately after the embryo barely begins to grow. If the external conditions suitable for passing through the vernalization phase do not exist, the plants do not traverse this phase of development until the necessary conditions appear; but the growth of such plants (in this case the development of leaves and roots) can continue.

Experimental data show that to pass through the vernalization phase plants in the shape of just slightly-grown embryos and 5- to 8-month-old plants of the same variety of winter wheat require the same environmental conditions and the same length of time. Hence, the speed with which a plant goes through the vernalization phase does not depend upon its size and age. The speed of this passage depends upon the nature of the plant and upon its environmental conditions.

The plants of winter varieties of wheat, rye and other crops sown in the autumn usually go through the vernalization phase not in the shape of just slightly-grown embryos but as green tillering plants.

5. To pass through the other phases of development as well as through the vernalization phase plants require not separate external factors such as temperature, air, moisture, light, darkness, etc., but sets of factors, the components of which are determined by the natural properties of the plants in question. Wheat and millet, for example, require different conditions for passing through the same phases of development, phases that are specifically their own and inherent in the nature of the respective plants. Moreover, to pass through the different phases of its development, the same plant requires different sets of external conditions.

In the majority of cases, when the plants of our winter varieties of wheat, rye and other crops, are sown in the field in the spring, they cannot complete their vernalization (hence their failure to ear), *solely because of the relatively high temperature in the sowing and postsowing periods.* This does not mean that for vernalization "in general," including presowing vernalization, winter and spring cereals require only suitable temperature. To pass through the vernalization phase the plants of winter and spring crops require not simply the thermal factor, but that factor in a set of other factors. The known components of this set are: moisture, temperature, air (and also the plastic nutritive substances to be found either in the seed or in the green plant). A certain quantitative combination of doses of these factors (according to variety) enables any variety of cereals, and of many other plants, to traverse the vernalization phase.

If it were possible in practical farming artificially to regulate the temperature over wide field spaces, the thermal factor would be the chief factor in the spring sowing of winter plants. It would only be necessary to regulate it to suit the plants' requirements for going through the vernalization phase. All the other factors of the set required for vernalizing cereals under the spring field conditions prevailing in our districts are always present in the necessary doses. In practical farming, plants have to be vernalized in the seed state under artificial conditions before sowing. In artificial vernalization it is necessary to create not only the temperature required by the given plants, but also a number of other necessary conditions. In the presowing vernalization of cereals, as well as of other plants, the chief and decisive factor is usually moisture.

6. In the development of annual seed plants, a definite sequence is observed in their passage through their different phases of development.

Passage from one phase of development to the next can begin only after the one phase has been completed, and only if the environmental conditions suitable for the next phase are present. For example, wheat plants can begin to go through the photo phase only after the preceding phase of development, namely, the vernalization phase, has been fully completed, and only under long-day conditions (still better, in continuous light).

PLANT BREEDING AND THE THEORY OF PHASIC DEVELOPMENT OF PLANTS[1]

THE INDIVIDUAL DEVELOPMENT OF THE HEREDITARY FOUNDATION OF PLANTS

THE PARTY and the Government have set our plant-breeding science the task of creating new varieties of plants at the shortest date. At a number of institutes and plant-breeding stations, the most up-to-date laboratories, greenhouses, refrigerator installations, thermic chambers, etc., equipped with the most precise technical apparatus, have been built or are in the course of erection. Nevertheless, the science of plant breeding continues to lag behind and there is no guarantee that this socialist task will be carried out within the appointed time.

We are firmly convinced that the root of this evil lies in the critical state of the science of plant biology that we inherited from methodologically impotent bourgeois science.

Plant breeding must be based on the science of genetics, the science of heredity. Moreover, it presents very definite demands to genetics, namely, to develop that department of this science which will make it possible consciously to direct the form-building of varieties and breeds along economically valuable lines, to work out the theoretical principles of determining the properties and characters of varieties, breeds, etc. Important as many of the achievements of genetics may be for breeding (methods of inducing mutations, the theories of pure lines, of homozygosity and heterozygosity, establishment of the fact of dominance, establishment of the fact of the segregation of the properties of hybrids in a series of generations, etc.), this science has completely divorced itself from the study of the laws governing the individual development of plants. As a result the branch of genetics that deals with the law governing the hereditary determination of characters has been largely converted into one advocating *a theory of the simple transmission, combination and segregation of the rudiments of characters and of the characters themselves, over a number of generations.* It is along these lines that genetics seeks to establish the laws of the inheritance of characters, drawing a direct, in no way mediated, mental projection

[1] Written in collaboration with I. I. Prezent.

from characters and groups of characters to the rudiments, the "genes," and establishing a similarly unmediated correlation between sections of chromosomes, their location and interrelation, and the characters of organisms. In this, genetics has, as it were, unbiologized itself, has divorced itself from the Darwinian biological study of the hereditary "factors." Genetics takes no part whatever, and its official representatives are not interested, in the study of the law of development of characters but tries to discover the law governing their simple presence or absence on the basis of the abstract mathematical probability of factors "meeting."

But "characters," including those of economic importance that interest the plant breeder, such as drought resistance, frost hardiness, surplus moisture resistance, resistance to various pests, duration of vegetative period, size of grain, vitreousness and farinaceousness, crown the edifice. They are the most concrete and complex formations in the plant organism. These "characters" are only the *final* result of the organism's development. The development of organs and characters is linked with the conditions they require, with the concomitant influence of varied factors upon the form-building of organs and their characters. Moreover, the development of organs, groups of organs and their characters and properties proceeds on the basis of separate *phases* of development (biologically necessary stages in the life of the organism). The latter, in their turn, develop in interaction with the special conditions of existence which they biologically need. Unable, however, to declare that development is concrete and full of contradictions, present-day bourgeois genetics evades the complex laws governing the formation of characters and groups of characters and tries to deduce their presence or absence directly from the genotype. Precisely here we find proof that the science of genetics reflects the general anarchy prevailing in the development of bourgeois science and, consequently, has divorced itself from the study of the laws of individual development (leaving this study to a special science, the "mechanics of development," divorced in its turn from the laws of heredity and from philogenesis); that it has mapped out for itself a historical path which takes it far away from the inherent dialectical logic of cognition, a logic objectively dictated by its subject. Instead of proceeding from the establishment of the laws of development of the hereditary foundation into the *general* law of ontogenesis, into phases, and only of the latter into organs and characters; instead of proceeding only in the final analysis to the ascertainment of the laws governing the most complex result of development, namely characters, genetics in its course of development has gone straight to find out the laws governing the correlation of characters and rudiments (genes). It is but natural that, following this path, the science of genetics should become largely formal as regards its chief theoretical constructions and should be unable, in an adequate degree, to be for plant breeding what it necessarily *must* be—the theoretical basis of a *guide to action.* Failing to receive concrete guidance to action from genetics, plant breeding was obliged to solve many of its problems as if the science of genetics did not exist.

To produce a variety, plant breeding must choose a parental pair for crossing. By what should it be guided in this? Genetics is silent on this score, and plant breeding is obliged to grope in the dark in the endeavour, empirically, in the lottery of crossings, to stumble upon the needed result. As we shall show later on, breeders ought *to know beforehand, before the crossing*, which of the parents' characters will dominate in the offspring, but geneticists can tell us nothing about this.

Incidentally, genetics is not always silent. Sometimes it reacts to these fundamental requirements of plant breeding. Which characters will be dominant in the offspring? We will tell you, but only after we have made the crossing and see what the offspring looks like. What pairs should be taken for crossing? Take as many as possible with the particular characters in which the plant breeder is interested, in the hope that, by chance, the needed combination may result.

But this needed combination will result (if it results at all) only in one case out of many thousands. To obtain an economically valuable variety, one better than any now extant in a given district, it is necessary to combine in it quite a large number of favourable characters: required vegetative period, resistance to frost, to excessive moisture, to drought and to various pests, no lodging, no shattering, etc.

On what scale must work be conducted in order to obtain the needed combination in the course of segregation? True, we are able to enlarge the scale of our work; but only up to a certain limit. One must not lose one's sense of proportion. In order to add only 10 genes to Kooperatorka, for example, it is necessary, according to Academician Sapegin, to grow hundreds of thousands of plants: only in that case will there be a chance of one desirable combination being obtained among the segregates. But is there any guarantee that this combination will be detected among the millions? It is *highly* probable that this *slightly* probable combination (one chance in hundreds of thousands) will be overlooked.

Does this give profound theoretical guidance to quick and precise practical action? Empirical plant breeding can "work" in this way even without genetics. And, unfortunately, it is obliged to work in this way, only very, very rarely producing varieties which after being tested are found to be useful in some districts, but in most cases only promising to produce a variety in the more or less distant future.

In many cases, under these circumstances, the State Cereal Variety Trial Commission has to serve as the real "plant breeder" and to locate a suitable district for a variety that was found unsuitable for the district to which it was originally assigned. For example, the winter wheat Kooperatorka, raised by the Odessa station, is now grown in an extremely limited area in the Ukrainian S.S.R., but is extensively grown in the Caucasus and Transcaucasus. The winter wheat Stepnyachka, likewise raised by the Odessa station, is not grown in the Ukrainian S.S.R., but thrives in some of the districts in the North Caucasus. Hordeiforme 010, raised by the Dniepropetrovsk station, finds no place among the crops of the Ukrainian S.S.R., but occupies a considerable area in the Trans-Urals. The winter wheat Dürabl, raised by the Ivanovskaya station and intended for the eastern forest-steppe region of the Ukrainian S.S.R., was found to be unsuitable here, but suitable for the northern regions (Kirov). The spring wheat Melanopus 069, raised by the Krasnokutsk station, is extensively grown in the steppe region of the Ukrainian S.S.R. Milturum 162, raised by the Kharkov station for the forest-steppe region of the Ukrainian S.S.R., is recommended for the Odessa Region. Numerous other cases could be cited to show that most of the varieties raised by plant-breeding institutions for the districts they serve are found to be unsuitable for the particular district, but are extensively grown in quite another, often sharply differing district, i.e., are "sent out into the world" not by the plant breeders, but by the State Cereal Variety Trial Commission. All this is because plant breeding lacks a sufficiently firm theoretical basis for its activities.

Here we must emphasize that plant breeding has been responsible for a large number of good varieties of different crops. But in many cases these varieties have been raised by breeders who have made a deep study of plant life and are guided by their own long years of experience, which quite often is at variance with the rules laid down by official science. Of these plant breeders, however, only a few, like I. V. Michurin, built up and developed a theory of plant breeding on the basis of their many years of experience.

In most cases, the experience which has enabled veteran plant breeders to raise different varieties has not been analyzed or generalized theoretically, has been kept outside the realm of science, as it were, has remained the local, personal knowledge of individual plant breeders.

Failing to analyze the process of plant breeding, or confining this analysis to a calculation of segregation, failing to obtain sufficiently effective advice from genetics, plant breeding is proceeding along empirical lines. As regards genetics "forecasting" dominant properties, this too is done after empirical crossing, and thus remains simply an empirically-inductive statement of fact.

Failure to analyze observed phenomena makes it impossible actually to predict before crossing what will dominate and to ascertain the levers that control this process.

We cannot keep marking time on the old genetic plant-breeding positions. A bold and determined change in the very methods of research is needed here.

Of course, we must assimilate the legacy of the whole of the preceding development of science. At the same time, however, we must firmly bear in mind the methodologically metaphysical line of thought of the protagonists of bourgeois science. We must fight uncompromisingly for the reconstruction of the genetic plant-breeding theory, for the building of our own genetic plant-breeding theory on the basis of the materialist principles of *development*, which actually reflect the dialectics of heredity. Only by consciously building such a theory can the breeding process be provided with real guidance suitable for the requirements of socialist production.

To overcome the formalism of direct projection from "characters" to "genes" and vice versa, it is necessary to discover and reveal the path of development of an organism's properties, to analyze them *according to degree of concretization*, to ascertain the law of *development* of the hereditary foundation in a series of generations, and not go out in search of a simple transmission of factors and characters over a series of generations. This work of bridging the gulf between the study of the hereditary foundation and the study of the individual development of this foundation was undertaken by the Laboratory of the Physiology of Plant Development of the Institute of Selection and Genetics (Odessa), which began to breed varieties in a new way. We proceeded from the following theoretical propositions.

The development of present-day plant organisms always starts from some structural foundation—the hereditary basis (genotype) which bears the "impress" of all preceding phylogenetic history. This "impress" serves as the general background and determines the progress of the necessary stages of individual development, relatively sets the general tone of the entire cycle of the organism's development (in plants, from seed to seed). Being the starting point of the plant organism's development, the hereditary basis (genotype) accordingly *determines the necessity within the frame of which the entire subsequent individual development proceeds.* The organism is not free to choose its hereditary basis. It begins its development from this basis as something already given.[1]

The organism's hereditary foundation, created by the fusion of two ancestral bases—maternal and paternal—which differ in some degree (heterozygote), is made complex by the union of the two lines of the phylogenetic history of its ancestors. Thus, this complex hereditary foundation (heterozygote) is richer in aspects and, consequently, in potentialities of development, than either of the parent homozygotes.

The zygote thus formed (by the fusion of two gametes) contains a wealth of *potentialities* for the development of the organism's properties in the shape of an impress of the phylogenetic line. The heterozygote combines the potentialities of both the maternal and paternal lines, with all the wealth of aspects of their hereditary bases, which they contributed to the heterozygote.

How will development proceed? What will determine and condition it? Will all the aspects of the complex (heterozygote) hereditary foundation make themselves equally felt in the individual history of the first generation (F_1), and what will determine the *actual* development of the different aspects

[1] This, of course, does not mean that it is impossible consciously to influence the building up of the hereditary basis. The very opposite is the case, but we are not discussing this now.

of the hereditary foundation? All this must be ascertained in order to become proficient in the knowledge of the law of genetics and place it at the service of *socialist, planned breeding of varieties.*

The hereditary foundation (zygote) contains in its multifarious aspects only the *possibility* of a plant's development from one phase to another. True, this possibility is absolutely *real*: not just any organic form, of any nature and sequence of phases, can develop from a *given* hereditary foundation. The hereditary foundation contains the general regularities of a plant's varietal nature only as a *real possibility*. For this possibility to develop into actuality, into the plant's developmental phases (and for the latter to lead to the formation of organs and characters), *conditions of existence* suitable to the nature of the plant are required.

When speaking of "conditions of existence" and acting accordingly in practice, we distinguish the "conditions of existence" of the process of development from the plant's "habitat" as well as from the external "influencing factors." Not everything in the plant's "habitat" is a factor that really influences the organism's course of development. And not every "influencing factor" is a "condition of existence" of the organism's development.

The "conditions of existence" of a plant's cycle of development are those essential conditions without which there is no development of phases, organs and characters in the plant's progress towards reproduction. The "conditions of existence" of a plant's cycle of development are the result of natural selection in the course of the many thousand years' history of organisms. The organism's interrelation with the conditions of existence of a plant's stages of development includes the organism's relative *adaptedness* to these conditions created by natural selection and, consequently, includes the requirement of these conditions by the organism as an *indispensable* premise for the stages of its individual *development*. Only complete disregard of Darwin's scientific legacy can explain why Klebs and other "mechanists of development" reject the adaptive requirements of plants in the process of form-building, and are unable to distinguish, in their theory and scientific experimental practice, "conditions of existence" from "factors that influence" form-building.

In the process of development of a plant's hereditary foundation and the formation of the phases of its cycle, organs and characters (from seed to seed), the plant can be influenced by various factors: electricity, ionization, temperature, water, etc. But by no means all these factors are necessary *conditions of existence* of a plant's development from its hereditary foundation to reproduction; not all are essential conditions called for by the very nature of the plant organism as a result of the *adaptation* of its varietal, specific, etc., nature to these conditions of its development.

Accordingly, every phase of a plant's development calls for *special* conditions of existence during the whole period of the given phase. A plant cannot pass through the first phase (vernalization) unless the just slightly-sprouted seed or the green plant is given a suitable dosage (specific for each variety) of thermic and moisture conditions (together with the other components: access of air, etc.). The second phase requires other conditions of existence, including without fail light in doses suitable for the given variety. During the first phase of development (vernalization), there may or may not be light or darkness, although, of course, they will always be present, for it is impossible to avoid keeping the just slightly-sprouted seeds either in light (of some degree) or in darkness. But neither light nor darkness are essential conditions for the whole of this phase of a plant's development. As regards the second phase (the "photo" phase), suitable light (in specified doses for each variety and species) is an essential condition of its existence. Of course a definite degree of moisture, thermic factors, light, and many other elements not needed at all, or not needed at the given moment for a plant's development, may sometimes, instead of being inactive elements of the habitat, become really active factors, influencing some of the processes in a plant when going through a given phase. But if the plant's nature does not require these factors, the plant can pass through that phase without them. However, without the *conditions of existence* to which the given phase is adapted, the phase will not develop at all. "Winter" varieties are such only because in the spring their vernalization phase does not find in the fields the conditions of development required by their varietal nature.

Absolute proof and documentary confirmation of this is found in numerous experiments conducted in our laboratory for the purpose of ascertaining the conditions of existence of various plant phases. Seeds of the winter wheat Stepnyachka, which usually does not fruit in one period of vegetation, were kept with 55% moisture content at 0°-2° C. temperature for 45 days (the conditions of vernalization for this variety) and subsequently, when sown, they were kept under a 9-hour day. As a result, the plants *failed to ear* and, consequently, did not go through their whole cycle of development from seed to seed (Fig. 23, second pot, left to right). When, however, seeds of the same variety were kept under the same moisture and thermal conditions and for the same length of time as those in the first case, and then, after being sown, were kept in continuous light for 30 days, the plants *eared* and ripened, i.e., they passed through the whole cycle of development nature had assigned for them (Fig. 23, first pot on right). When seeds of the same variety of winter wheat were kept in continuous light for the same length of time as those mentioned above, without providing them, before sowing, with moisture (55%) and thermal conditions (0°-2° C.), i.e., without creating the conditions for passing through the vernalization phase, the plants failed to ear. The records of these and similar experiments (Fig. 24) serve as indisputable documentary proof that there are phases, stages, in the development of plants which call for their own specific conditions of existence.

To offer these conditions to another phase (Fig. 24, first pot on left, and Fig. 23) means ignoring the plant's biological adaptational

Fig. 23. In each pair of pots, the plants in the left pots are from unvernalized seeds and in the right pots from vernalized seeds.

The plants in the two pots on the left were grown under short-day conditions, those in the two pots on the right were grown in continuous light. Only the plants grown from vernalized seeds in continuous light eared (pot on extreme right). This experiment shows that continuous light cannot serve as a substitute for the low temperature needed for passage through the vernalization phase. The experiment also shows that after passing through the vernalization phase, wheat plants need the long day for passing the photo phase

requirements and, at best, converting "conditions of existence" into simple "influencing factors," and even into mere elements of an indifferent environment.

Therefore, all attempts to force a plant to go through a given phase of development by *substituting any other* conditions for those the plant requires, for example, by substituting light for temperature in the first phase (vernalization), as the "repudiators" of the phasic development of plants try to do, are misguided and methodologically fallacious in their very conception.[1]

---

[1] The methodological mistakes made in experiments in "vernalization" by means of light were examined by Lysenko in the paper he read at the All-Union Conference on Winter Hardiness on June 24, 1934, and at a scientific conference of the Institute of Genetics of the Academy of Sciences of the U.S.S.R. on January 6, 1935.

Fig. 24. Winter wheat Novokrymka 0204

This experiment shows the same as was shown in the case of Stepnyachka. The plants in the first pot on the left (from ordinary seeds) were grown in continuous light. The plants in the second pot from the left (from vernalized seeds) were grown in continuous light. The plants in the third pot from the left (from vernalized seeds) were grown in continuous light for the first seventeen days and under 10-hour-day conditions for the next 32 days The plants in the fourth pot from the left (from ordinary seeds) were grown under 10-hour-day conditions. The plants in the fifth pot (from vernalized seeds) were grown under 10-hour-day conditions

Confusing "conditions of existence of development" with "influencing factors" also leads to confusing and identifying "vernalization" with "stimulation." It is possible and necessary to find numerous means of accelerating different processes in the life of plants. But to prove that it is correct to *identify* "vernalization" with "stimulation," the stimulators must force known winter wheats to fruit by methods of their own invention without introducing, and even directly eliminating, the established conditions of existence of the phases of development. Perhaps the adherents of stimulation will undertake to do this? Otherwise we have a right to demand that they and the "vernalizers" should not irresponsibly introduce confusion into the work of vernalization and into the principles underlying methods already being applied on hundreds of thousands of hectares of crop fields, where their soundness has been proved. Confusion in such a matter is far from being the private affair of this or that author.

A plant's phases of development and conditions of existence constitute a *general law*, a general type in its developmental cycle. But in each of these phases of its life the plant enters, *and cannot avoid entering*, into innumerable connections with many other factors of inorganic and organic nature. The latter always somewhat deflect and individualize the general scheme of development of a plant of a given variety, species, etc. A plant's different organs and characters that develop on the basis of definite phases, require in their turn, their own conditions of existence and are also inevitably subjected to the influence of particular factors. A plant's concrete characters are the result of this highly complex connection. Variations of these characters, however, always find their *measure* in the general phasic law. Therefore, to seek the law governing these characters *outside* of the organism's *phases* of development, to deduce them straight from the genotype (as the formal geneticists do) or from external factors (as the mechanists of the "mechanics of development" do) means completely disregarding the immediate *basis* of the laws governing the formation of characters.

Is it possible, for example, to ascertain the law governing the development of such economically important characters as the "vegetative period," "winter habit," and "spring habit," and accordingly to control their formation in a variety not on the basis of phases, but by deduction *straight* from the genotype? Notwithstanding all reservations concerning the "norm of reaction," which varies in different environments, it is *impossible to understand* the law governing the period of vegetation of any given variety without a prior phasic analysis.

The usual genetic methods may, perhaps, reveal that the vegetative period of a given variety is *different* when sown in different *localities*. But how are we to *explain* the presence of this difference in some and its absence in other cases of plants sown in different environments, and, in accordance with this, how are we to control the vegetative period? The usual genetic methods are powerless to explain this. At best, these methods can help us only to certify the presence or absence of such a difference (in the norm of reaction), but they cannot help us to *foresee* the limits of this "norm," or the pattern of the "norm of reaction" of the genotype under different conditions, to forecast this pattern before the seeds are sown at any given point on the map. It is not surprising that, notwithstanding the immense wealth of knowledge accumulated by present-day genetics, varieties are still divided into "winter" and "spring" forms irrespective of the conditions of individual development of the genotype. At the same time the ordinary phenotypical indices of "spring habit" and "winter habit" are projected into the genotype, although different genotypes are merely different foundations for the development of spring habit and winter habit under definite conditions.

If, however, we understand that the characters "vegetative period," "spring habit" and "winter habit" are directly based on phases of development, we are able to really foresee the pattern of a variety's vegetative period under various conditions (to foresee the "norm of reaction" of the genotype). We are likewise able to control this period on the basis of a preliminary analysis of the conditions of existence of the *phases* of development, by directing the development of the hereditary foundation now along the channel of "spring habit," and now along the channel of "winter habit" (Figs. 25 and 26).

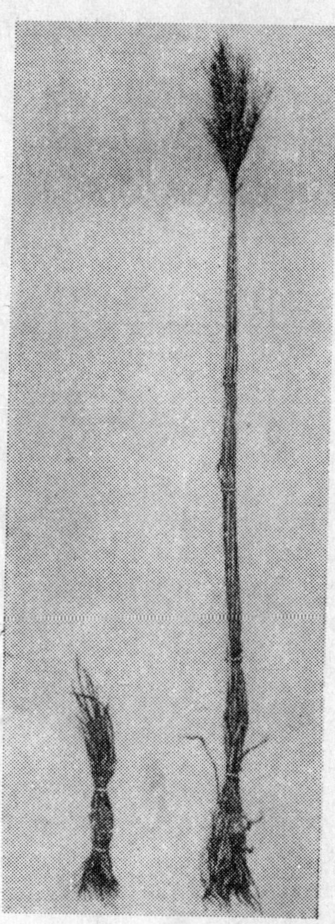

Fig. 25. Winter wheat Ukrainka

This illustration shows the direct dependence of the characters "winter habit" and "spring habit" on the course of the vernalization phase. For the Ukrainka to pass through the vernalization phase, its hereditary foundation needs conditions which are usually absent in the spring in the localities where this variety is grown. As a result it turns out to be a winter plant since it cannot, under these conditions, pass through its cycle of development in the course of one summer. Under these conditions Ukrainka develops neither straw nor ears, i.e., is a "winter" plant (sheaf on left). The Ukrainka sheaf on the right passed through the vernalization phase (in the seed state, before sowing) and, finding in the field the conditions for the development of all succeeding phases, the tufts behaved like spring plants

Today it is possible, and necessary, to speak of the plant organism's phases of development as the immediate basis for the development of numerous economically important characters. Passage through a given phase of development does not by itself guarantee the appearance of particular organs and characters, because both organs and characters, which develop on the basis of phases, are limited, like the latter, by *their own* conditions of existence. But if a plant does not go through a given phase, the very *basis* for the formation of organs and characters corresponding to that basis fails to come into existence, and the presence of the "external conditions of existence" of these organs and characters will not help to form them. Thus, if a plant is not

Fig. 26. Millet

The plant does or does not form panicles, depending on the course of the photo phase of development. On the left is a sheaf grown from ordinary seeds. On the right is a sheaf grown from seeds which had undergone five days' treatment before sowing. The plants were grown under long-day conditions

given the conditions of existence of the vernalization and photo phases, there will be no flowering or earing nor any development of the corresponding characters. The development of the size and weight of the grain may be different, depending on the time the plants pass through their phases. For example, in 1933 the absolute weight of the grains of the wheat Girka 0274, developed under Odessa conditions, was 17.8 gr. The grains from a Girka 0274 sown in the same field and on the same day as those previously mentioned, but which had passed the photo phase quicker (thanks to additional light provided after sunset) had an absolute weight of 28 gr. The colour and shape of grains also depend on the time the plant passes through its phases.

The characters "farinaceousness" and "vitreousness" are also by no means predetermined entirely in the genotype, but develop in the field in different ways if the course of the phases is different. Immunity and nonimmunity to fungi will likewise be different if the phases take a different course.

The character "frost hardiness" has been studied by geneticists and especially by many plant breeders. But no guidance to action can be obtained in this matter from the ordinary genetic analysis. No matter what analysis the chromosome set may be subjected to, it will not reveal the causes of plants' resistance or nonresistance to frost. From the standpoint of the theory of plant development, however, it was possible to foretell that the character "frost resistance" is also based on the differences in the natures of plants' developmental phases. (Lysenko's paper read at the All-Union Conference on Winter Hardiness in 1933.)

On the basis of this forecast a number of investigators (Kuperman, Saltykovsky, Timofeyeva, Melnik) conducted extensive work and the soundness of the proposition was proved.

From all this it follows that genes, and the genotype as a whole, are by no means the *direct* cause of the concrete way in which the different characters are formed. Only if one-type conditions are maintained in the laboratory, where so-called "other equal conditions" can be eliminated is it possible to establish the correlation between "genes" and "characters" (which geneticists succeed in doing in a number of cases). If, however, the genotype is, so to speak, carried out to the field (when the seeds are sown), not a trace of these "other equal conditions" is left. Here, one must know the specific role these field conditions play in the development of the genotype. When that is known, it becomes clear that the nature of a given character is directly determined by the concrete course of the phases, or stages, of development, under the definite conditions of existence of these characters. Moreover, the action of most diverse factors not required by the cycle of development inevitably intrudes here and only the whole of this chain moulds the given concrete characters. The genotype is merely the inherited general foundation of the development of the species, variety, etc., which determines the general direction of a plant's development and the character of its varietal requirements of conditions of existence; it does not predetermine the concrete nature of characters. The hereditary foundation is the ancestral basis, which is relatively conservative and within definite limits preserves its specific structure.[1] But it itself undergoes development

[1] The changes in the structure of the hereditary foundation (mutation), i.e., the changes in the ancestral basis, must be distinguished from the individual development of a given hereditary foundation, as being the reproduction in offspring of biological form-building processes similar to, but by no means identical with, those that took place in ancestors. The distinction laid down by Weismann, de Vries and others, between "inherited" and "acquired" characters must be purged of its metaphysical interpretation. Some of the changes that take place in the organism are hereditarily significant for future generations and some are not, but there is not a single purely "inherited" or purely "acquired" character. Every character is the result of a concrete path of *individual* development of the *ancestral* basis (hereditary foundation).

in the course of the plant's individual life, developing through phases and stages into the plant's characters.

The important thing for the practical plant breeder is neither the genotype as such nor phases as such, but characters; however to the latter, and, consequently, to their creation, there is no direct, *immediate* passage from the genotype. The path from the genotype to the characters runs through the phases of development and the conditions of existence of both the phases and the characters themselves.

Genetic investigations, however, being abstracted from the laws governing the individual development of the genotype into phases, and confined to the search for the "tele-action of the gene," cannot provide a concrete theoretical basis for plant-breeding work.

It is precisely on the established facts of such an absolutely definite sequence of connections in the development of the hereditary foundation into phases, and of the latter into organs and characters, that the work of the Institute of Selection and Genetics (Odessa) is based.

THE CHOICE OF PARENTAL PAIRS, THE LAW OF DOMINANCE, AND THE NATURE OF HETEROSIS AS EXPRESSED IN DURATION OF VEGETATIVE PERIODS

After first investigating the laws of the individual development of plant forms, and studying the necessary conditions and character of the individual development of each prospective choice for crossing, we, to a certain extent, learn the hidden nature of the hereditary foundation of the picked candidates. If we guide ourselves by the theory of the development of the hereditary foundation (genotype), we shall be able *intelligently to choose the parental pairs.*

This intelligent choice of pairs for crossing includes the possibility:

a) of foretelling, *before crossing*, the pattern of dominance of the developmental phases of one or the other parent, and hence, of foreseeing the length of the vegetative period of F_1 compared with that of both parents;

b) of deliberately changing the pattern of the dominance of vegetative periods;

c) of foreseeing, from the length of the vegetative period of F_1, the duration of vegetative growth of the earliest ripeners among the potential segregates and of culling the combinations accordingly, beginning with F_1;

d) of foretelling the general pattern of the likely segregation of heterozygotes according to length of vegetative period from the data of any generation;

e) of not confining ourselves to the chance appearance of heterosis and the accidental breeding of early-ripening varieties from the crossing of late ripeners, but of deliberately creating them under given conditions by means of an intelligent choice of parental pairs.

Usually plant breeders, juggling with combinations of individual characters and ignorant of the immediate basis of character development—the law of phasic development—stumble upon the combinations needed for producing a particular variety only by chance, and hence very rarely. With the vegetative period, too, plant breeders operate simply as if it were a "character," losing sight of the fact that this "character" is the final result of an entire course of development, the immediate basis of which is the law of phasic development. After all, it is the length of time required to complete the given phases of development under the conditions of existence required by the phases, as well as delay in the passage from one phase to the next as a consequence of the absence of the conditions of existence for the new phase, that, in the main, determines the length of the vegetative period, determines whether the variety will be an early-spring, late-spring, semiwinter, winter, or some other variety. The extremely diverse influencing factors that inevitably invade a plant's life in the field, of course, alter somewhat, shorten or lengthen, the ripening period. But these deviations are insignificant and keep within the limits of the period determined by the law of phasic development.

Juggling with lengths of vegetative period as if this were a character, and ignorant of the causes that determine it, the plant breeder cannot know for certain which pair to take for crossing in order to obtain a radical change in the duration of this period. If he tries to get the shortest vegetative period from the available varieties he will inevitably drag into the heterozygote, along with the short period, any number of bad indices. This is just what the plant breeder is usually compelled to do.

He takes into account the immunity to smut, rust, etc., of one variety and the shorter ripening period of another, with the result that in his quest for a short ripening period he introduces into the heterozygote susceptibility to various diseases, susceptibility to Hessian and frit flies, low resistance to drought, bad milling and baking properties, etc. That is why the plant breeder cannot know beforehand when, if at all, he will get from his combination the variety he picked and whether its yield will be better or worse. Only subsequent tests will reveal to him the results obtained.

Knowing no other way out and wishing to protect the zygote from deterioration, wishing to deal with a smaller number of unknown quantities in the segregation of the heterozygote, plant breeders have advanced the principle of producing varieties by inbreeding. However, in doing so they deprive themselves of the opportunity to enrich the hereditary basis, of the opportunity, for example, to eliminate a bad (long) vernalization phase by creating heterozygotes from a cross with another variety possessing a good (shorter) vernalization phase. Therefore, *when it is necessary to create a variety with a shorter vegetative period*, inbreeding is not only of no avail but even harmful.

We, however, proceed from entirely different principles in breeding a variety with a relatively guaranteed yield in a given district.

From the available varieties it is necessary to choose such specimens for crossing as have the potentiality of showing good indices in respect to all types of hardiness and productivity, provided inability to traverse the first phase (vernalization) or the second (photo) phase as well as delay in passing through either of these phases under the conditions of a given district are eliminated from the hereditary basis.

But how are these potentialities to be discovered? How can the potential yield of a variety be ascertained if it usually does not ripen, or ripens late and produces grain of bad quality?

Since they do not investigate the laws governing individual development, ordinary plant breeding and formal genetics are incapable of penetrating the latent potentialities of the hereditary foundation. They underrate these potentialities, regard them merely as characters which usually come to the surface of their own accord when plants are sown in geographically different localities or when hybridological analyses are made. These latent potentialities can be partly revealed by inducing them to operate in a plant's individual development. This can be done artificially by creating for each of the parents the conditions necessary for their phasic development which are absent in the fields of the given district. For example, one of the parents may have to be vernalized, and suitable photo conditions created for the other. The result will be that in reproduction each parent will go through its whole cycle in its own way, and this will enable people to ascertain the whole course and period of its development, and also its yield. That is precisely what is meant by investigating the potentialities of a parent's hereditary foundation, by getting at the latent potentialities of the genotype. Discovering the parents' potential ripening period and potential yield—provided there is a continuous, undelayed development of their hereditary foundations through all phases to reproduction—and thereby ascertaining the one weak spot (different in each parent) in the hereditary bases under the given conditions, make it possible deliberately to create a heterozygote in which both these weak spots are certain to be eliminated. If one of the parents belongs to a variety which possesses all economically valuable properties but requires that its sole weak spot—poor adaptation of the *vernalization phase* to the conditions of the given district—be eliminated, this weak spot can be eliminated by introducing into the zygote the gametes of the other parent, which possesses a well-adapted vernalization phase and also all economically valuable indices *but requires that its own weak spot*—poor adaptation of the photo phase to the conditions of the given district—*be eliminated.* Thus, by crossing these two parents a heterozygote will be created possessing a *real possibility* of developing, and rapidly at that, under the conditions of the given district, both the first and second phase, and at the same time having all the parental gametes' other indices that are desirable under the conditions of the given district. That is how the Laboratory of the Physiology of Plant Development of the Institute of Selection and Genetics approaches the choice of parental pairs, and it formulates this approach in the following way:

Parental pairs for crossing must be chosen not according to the maximum number of favourable characters observed in the parents, but according to the minimum number of unfavourable characters; that is, of weak spots in the potentialities of adaptation of the hereditary foundation which restrict the yield in the given district. This for the purpose of creating in the heterozygote the possibility of the mutual elimination of the parents' weak spots.

In the case of plants like cereals, soya bean, cotton and a number of others, the yields of which are restricted in a vast number of districts by the time required for maturation, parental pairs must be chosen from the entire world collection in such a way that the parents shall have not the same, but different sole "weak spots" as regards potential development of the hereditary foundation (vernalization phase in one parent, photo phase in the other).

These sole weak spots will be eliminated in the process of segregation of the heterozygote by the mutual substitution in the hereditary basis of the bad quality of one form by the analogous good quality of the other, and vice versa.

The creation of a zygote from deliberately chosen pairs, among whose offspring forms appear from which the parental "weak spots" have been eliminated as a result of segregation, makes it possible to know even before crossing what the general character of the development of the hereditary foundation's potentialities will be, makes it possible to ascertain which of these potentialities will be realized and, consequently, what the character of the *development of the first generation* (F_1) will be as regards the flowering period of cotton, ripening period of wheat, etc. It also makes it possible to forecast with a great degree of probability what all the other economic indices will be.

If two forms are crossed, one with a short vegetative period and the other with a longer period, or if we cross a winter and a spring pair, a spring and a semiwinter pair, a spring late ripener and a semiwinter plant, etc.—*which will be the dominant*? This highly important question can be answered correctly only from the standpoint of development, by presenting and solving the problem of the general law of dominance.

Already Gregor Mendel noted (and this was one of the greatest achievements of genetics) that in the first hybrid generation (F_1) of an alternative pair of characters, those of only one parent appear. Particular characters of one parent are, as it were, "swallowed up" for the time being by the "alleles" of the other. In conformity with this, Mendel established the rule of dominance, the correctness of which has been confirmed in many cases by subsequent observations and researches in genetics. But neither Mendel himself nor all the subsequent trends in present-day genetics have discovered what is at the bottom of this; they have confined themselves to the mere statement of the fact that there is such a thing as dominance.

From the standpoint of the theory of development, however, which points to the role necessarily played in development by the external conditions of existence and emphasizes the part of *adaptation* in biological development, it is possible to find out the necessity of dominance, to understand its biological essence, and thus make considerable progress towards controlling the law of dominance.

In the zygote that is formed by hybridization, two ancestral bases, two hereditary foundations, unite and introduce all their features into the zygote as a basis for the potential development of phases. These, in their turn, determine the potential concrete formation of the organism's characters. But which of the pairs of potentialities (allelomorphs) inherited from the different parents will be materialized, will become realities, will develop?

This is not predetermined in any particular direction in the zygote itself. It depends upon the extent to which external conditions correspond to the biological requirements of a given feature of the zygote to be adapted to the conditions of existence of its development. Therefore, in F_1 that allelomorphic feature of the heterozygote will develop which finds suitable conditions of existence for its development. Precisely for this reason, under the given conditions, only that feature of the zygote will develop, only that possibility of the hereditary foundation will become a reality, which finds the conditions that it needs, and is best adapted to the *available* conditions of existence. It is this that determines which of the two allelomorphic features of the zygote will develop, i.e., *will be dominant.*

We must at once parry the objection that may be raised that in a number of cases very definite characters dominate, regardless of the fact that the organisms develop in different environments. This possible objection is based on the failure to distinguish between "environment" and "conditions of existence" in the course of development. We must ascertain from the plant and not simply guess at random whether transferring it to a different habitat is the same as transferring it to different "conditions of existence." Big differences in a plant's environment may turn out to be of no importance to the plant. At the same time seemingly unimportant changes in environment which happen to be the conditions of existence at any stage of a plant's life may cause most important changes in it.

Figure 27 illustrates the difference between some barley plants and others. Some of the plants are earing; others are creeping on the ground, not having even stalks. Yet they are all of the same variety, grown in the same field from ordinary, unvernalized seeds. The only difference is that some of them were sown in the field on March 12, 1928, whereas the others were sown next to them two days later (March 14). The difference of two days radically affected the entire behaviour of the plants. The reason for the difference between them is now perfectly well understood: some of the plants (those sown on March 12) secured the conditions of existence that

Fig. 27. Barley Pallidum 419

Sown at intervals of two days from March 1 to autumn 1928 in Ganja. The plants in all the sowings, including that of March 12, proved to be of spring habit. All the plants sown from March 14 onward (the March 14 planting is not shown in the illustration) proved to be of winter habit and did not ear. The difference of two days in the date of sowing (March 12 and 14) played a decisive role in the development of the plants' organs and characters

their vernalization phase required, because, during those two days the temperature was 2°-3° lower than throughout the rest of the vegetative period. The subsequent much bigger changes in temperature were only passive environmental elements in the development of the vernalization phase.

We maintain that in all cases when a hybrid plant is given really different conditions of existence for its development, this causes corresponding changes in dominance: *the dominant character will be the one that has more favourable conditions for adapting itself to its development.* We repeat that dominance is not predetermined in one particular direction in the zygote. Combining both parental hereditary foundations, the zygote contains the *potentiality* of developing both allelomorphs. The question of dominance is decided by the adaptedness of either allelomorph to develop under the given conditions of existence. Consequently, what will be dominant under one set of conditions will be recessive under another.

The conception of dominance from the standpoint of development was expressed long ago by Ivan Vladimirovich Michurin and was applied in his work. He wrote:

"The qualities of every hybrid grown from the seed of fruit obtained from the crossing of two progenitors are a combination of only those hereditarily transmitted characters of the parent plants, namely, father, mother, and their kin, the development of which at the earliest stage of the hybrid's growth was favoured by the environmental conditions (i.e., temperature of the surrounding air and soil, degree of activity of atmospheric electricity, direction and velocity of the prevailing winds, intensity of light, composition of the soil, moisture content, etc.)"[1]

It was this conception of the zygote as of something that develops, and of dominance as the development of only those features of the zygote which are favoured by the given conditions, that guided Ivan Vladimirovich in all his immense and extremely fruitful work. It was these theoretical premises that enabled him to raise a vast number of new varieties and unprecedented plant forms. Proceeding from these theoretical premises, he propounded *distant hybridization* as one of the principles of his work. But Michurin's principle differs fundamentally from that of the present-day distant hybridizers. Michurin advanced the principle of distant hybridization not merely for the purpose of increasing the diversity of plant forms, irrespective of whether these forms will be interesting playthings or something that some day may perhaps, "by chance," turn out to be useful. He did not propose the crossing of just any plants as long as the pairs were remote species and geographically distant from each other.

I. V. Michurin *deliberately* picked parental pairs, including such as were quite remote from each other as well as from the locality for which the future variety was intended, very carefully taking into account the difference in their conditions of existence. And he did this quite concretely in order to inhibit, when necessary, the development of definite properties that was dominant owing to adaptedness to the *local* conditions of development, and to create the conditions required for the dominance in development of the properties he had planned to impart to the variety he was producing. In choosing hereditary foundations for crossing, Michurin always kept in mind their historically formed biological adaptational requirements. He calculated *beforehand* how the hereditary foundation would develop under definite conditions of existence and under definite influencing factors, choosing these conditions beforehand in order to obtain the desirable pattern of dominance. In other words, he made a blueprint in advance of his work with respect to the entire path of development with its numerous contradictions, from the hereditary foundation to the characters. To explain his principles, he cited the following example of his work, one of a vast number of similar cases:

"By crossing foreign varieties of winter pears with our Tonkovetka, Limonka and other hardy varieties, we obtained hybrids, which, while superior in flavour, were all summer ripeners and had small-sized fruit. This was the result of the *dominating development of the characters of our local varieties, due to the climatic and other conditions in our locality which were suitable for them, and to which they were habituated.* On the other hand, when I crossed foreign winter pears with a wild Ussurian pear which I had grown from seeds obtained from North Manchuria, I obtained hy-

[1] И. В. Мичурин, *Выведение новых улучшенных сортов плодовых и ягодных растений*, Сельхозгиз, 1933, стр. 27.

brids half the number of which produced large-sized fruit of excellent flavour, with the qualities of winter ripening in storage and ability of all the above-ground parts of the trees fully to resist our frosts. . . ."

Ivan Vladimirovich formulated this theoretical principle of his in the following way:

"The farther apart the crossed parental pairs are in respect to place of origin and environmental conditions, the more easily the hybrid seedlings adapt themselves to the environmental conditions of the new locality. I attribute this to the fact that, in this case, when the properties of the father or mother and their immediate kin hereditarily transmitted to the hybrids *fail to find the environmental conditions habitual in their native locality, they will be unable to dominate too strongly in the transmission of certain of their properties in the development of the hybrid organisms, which is of enormous importance for this work.*"[1]

Dominance as development determined by greater adaptability to conditions—such was the conception entertained by Ivan Vladimirovich, such the view by which he was guided and which was one of his theoretical principles that placed in his hands a powerful weapon and tool in aid of his struggle and work.

Michurin advanced this principle very long ago in opposition to the formal conception of dominance as something determined in the same way in the zygote itself independently of the conditions of adaptation. He had already come forth as an opponent of such formalism in his earlier works, where he wrote:

"It seems to me that all of our Mendelists are disinclined to take into account the powerful influence of these factors on the development of the hybrid plant, beginning with the formation of the seed from the cross of two individuals and continuing during the first few years of growth of the young seedling right up to the stage of maturity. . . ." (Earlier in this article I. V. M. enumerated the chief of the above-mentioned factors: "atmospheric pressure, the temperature conditions, the amount of moisture, the intensity of the sunlight and the activity of the atmospheric electricity; the separate effects of each of these factors as well as the combined effects rendered by their various combinations. . . .") "The following fact may serve as an example: I pollinated the flowers of *Pyrus elaegnifolia* with the pollen of a well-known pear named Bessemyanka [Seedless]. When rearing these seedlings I observed that whenever they were given better nourishment externally in all their parts the hybrid seedlings invariably deviated towards the Bessemyanka type. The leaf blades became broader and had a glossy surface, the shoots became thicker and their bark acquired a colour resembling that of the shoots of Bessemyanka. On the other hand, if the seedlings were subjected to some hardship, such as replanting or the insufficient water supply in the beginning of the vegetation period due

[1] *Ibid.*, p. 28.

to summer drought, the leaves of the hybrid plants grew narrow and elongated in shape. Similar phenomena have been recorded in hybrids from other crosses as well, whenever a cultivated variety was crossed to a wild species.... I have also carried out numerous other experiments to determine the effect of the composition of the soil on the constitution of growing hybrid plants and each time I became convinced of the considerable influence exerted by this factor. This influence was particularly pronounced in those cases when I succeeded in providing for the hybrid seedlings such a soil that was closely similar in composition to that on which one of the two parental plant varieties involved in the cross had successfully developed for a long period of time, or, so to say, had been formed, whereas the type of the other parent had been developed on a soil of an entirely different composition. In almost all such cases the hybrid seedlings were observed to resemble in type the first parent. Thus, I *used to order several poods of soil to be brought from the environs of Vladimir* to grow the hybrids obtained in crosses between one of our cherry varieties raised in the Samara steppe region (*Prunus Chamaecerasus*) and the Roditeleva cherry from Vladimir. The soil ordered was the very one on which the Roditeleva cherry—a well-known Vladimir variety of cherry—was grown in its native locality. Although by means of this substitution of soil I succeeded only in partly approximating the environmental conditions in which these hybrids were reared to those of the Roditeleva cherry's native habitat, nevertheless the few specimens of hybrid seedlings that were given a mixed soil containing a high proportion of Vladimir soil, showed a pronounced trend towards the Roditeleva cherry and markedly differed from the rest of the seedlings brought up on the ordinary soil of our locality. And just to think that this result has been obtained in experiments, in which so many necessary conditions were missing! These hybrid seedlings ought really to have been planted in Vladimir, not in Kozlov, and grown in the native locality of the Roditeleva cherry, because (besides soil composition) other factors such as the composition of the subsoil, and of the subsoil water, the depth of the subsoil water table level, the lay of the site, the difference in the climatic conditions, etc., play an important role. And if even in the absence of the influence of these important factors the supplying of the native soil alone was enough to produce so marked a deviation towards the maternal plant, then how is it possible to make correct estimates of the proportion of plants in a hybrid progeny that would deviate towards one or the other parental type and of the degree of this deviation merely on the basis of the hereditary transmission of the latter's properties?"[1]

We see that Ivan Vladimirovich *directed* the development of dominance both by choosing pairs with definite requirements of adaptation to the conditions of development and by creating the conditions of existence required for development in the desired direction. It was precisely the practical re-

[1] I. V. Michurin, *Selected Works*, Eng. ed., Moscow 1950, pp. 101-103.

alization of theoretical principles founded on the development of the hereditary basis under the conditions of existence that led Ivan Vladimirovich to such grand achievements in the breeding of varieties.

We were obliged to deal here in a cursory way, as part of our exposition, with one of the theoretical principles and premises of Michurin's work. We cannot, of course, give here an exhaustive analysis of the immense wealth of his thought.

But one thing must be said: Michurin was guided by the search for the laws governing the development of plants *during the whole course of their lives* and he found the way to control many of these laws.

His conception of dominance as development, in contrast to the whole of official formal science, and his conscious direction of dominance, are enough in themselves to refute the arguments of some "theoreticians" about the scientific "illegitimacy" of Michurin's work.

The real science of hybridization is to be found in the works of Michurin, but not everybody is able to understand him. To be able to do so one must really take the stand of materialist development.

Of the developmental potentialities of the hereditary foundation (allelomorphs) that are of uniform type as regards their nature and demands on conditions of existence, but differ in the specific character of the demands they make on the uniform-type conditions of existence, only one can actually develop (short or long "vernalization" phase, short or long "photo" phase, etc.). We are not discussing the "dominance" of developmental potentialities that are alien to each other in nature, not the "dominance" of long over black, yellow over short, etc., but paired and allelomorphic potentialities (long—short, round—wrinkled, etc.). It is clear, therefore, that in these cases the actual development of such potentialities can only be mutually exclusive. And precisely that one of the mutually exclusive potentialities will develop which finds the surrounding conditions more favourable for the requirements of its development, which is better adapted for development under the given conditions of existence. Therefore, the adaptability or inadaptability of the heterozygote for development under the given conditions will certainly make itself felt already in F_1. If the heterozygote possesses the potentiality to develop fully under the given conditions of a district, this potentiality will become a reality in the first generation. This is what determines dominance in the development of the potentialities of the hereditary foundation.

If we regard dominance as the development of one of the paired potentialities of the heterozygote because it fits the requirements better, because of the greater adaptedness of the development of this potentiality of the zygote to the given conditions of existence, it is possible beforehand, before crossing, to visualize the pattern of dominance with respect to given allelomorphic inclinations (potentialities).

For this it is necessary to study beforehand the potentialities of development of the parents' hereditary foundation; to ascertain these potentialities by finding the conditions required by each of the parents for its phasic development and for the development, on this basis, of the plant's organs and characters; to ascertain which of these requirements are best satisfied by the given conditions of the place in which the plant is grown (conditions of the district).

Phasic analysis must, therefore, *precede* both *hybridological analysis* and hybridization itself. Only then will hybridological analysis not be based on blind groping for segregates, only then will it become possible to control the dominance of "characters," just as it is already possible, in the main, to control those that are the direct final result of phasic development. The length of the *vegetative period* is one such character. Further progress in the analysis of the conditions of existence of the process of forming organs and their characters, the development of which is a particular form of existence of the general laws governing phasic development, will create ever-increasing possibilities for dominance control.

Thus, the general law of dominance may be formulated as follows:

Dominance is the development of either of the two allelomorphic features of the hereditary foundation (heterozygote), provided the conditions of existence are suitable for its requirements and the other feature of the heterozygote cannot be developed owing to the absence of the required conditions, or to less favourable developmental conditions.

As, however, either feature of the hereditary foundation is merely the tangible vehicle of *potential* development, the same law of dominance may be formulated as follows:

Dominance is the transformation into reality of one of the paired (allelomorphic) and mutually exclusive developmental potentialities of the hereditary foundation owing to the existence of suitable conditions, or to lesser adaptability to these developmental conditions on the part of the other allelomorphic potentiality.

Proceeding from this law, it is possible to know in advance, even *before crossing*, what the dominant ripening period will be on crossing a winter and spring pair, a pair with a short and long vegetative period, a semiwinter and spring pair, etc.

For this it is necessary to make a *phasic analysis*, i.e., to ascertain why, thanks to which phases, the parents develop as winter, spring, early-ripening, etc., varieties.

Having ascertained what conditions each of the parents needs for the development of each of these phases in order that development from phase to phase may proceed more quickly and continuously, it is possible, after studying the conditions of the district, to forecast how the heterozygote will develop in F_1 into a winter, spring, semiwinter, late, early, or other form, and to act accordingly in choosing pairs for crossing and the conditions of existence (choice of district) for obtaining the necessary pattern of dominance with respect to length of ripening period in the first generation (Figs. 28 and 29).

Fig. 28. Illustration shows two late-ripening spring varieties of wheat (under Odessa conditions), Hordeiforme 2508 and Melanopus 069

Presowing vernalization of the seeds of Hordeiforme 2508 under Odessa conditions accelerates earing. Consequently, this variety, under Odessa conditions, is a late ripener owing to slow passage through the vernalization phase. The plants in the sheaf on left, grown from vernalized seeds, eared; those in the second sheaf on the left were grown from ordinary seeds. Presowing vernalization of Melanopus 069 seeds does not accelerate earing. But under long-day conditions, plants of this variety do ear more quickly. Consequently, under Odessa conditions this variety is a late ripener owing to slow passage through the photo phase. The hybridization of these two forms creates a hereditary foundation from which the two weak spots of these parents are eliminated. The phasic analysis of the two parental forms made it possible, already before crossing, to foresee that their offspring would develop as early-ripening forms. Note eared plants of first hybrid generation (F_1), second sheaf from right

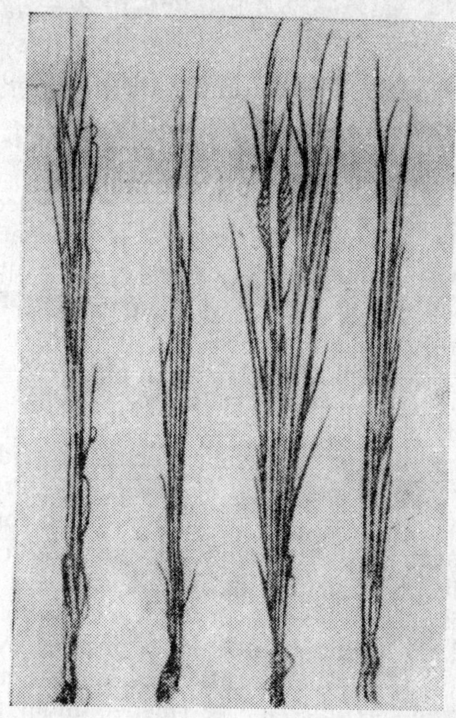

Fig. 29. The phasic analysis of Hordeiforme 2506 and of Melanopus 069 showed that these two late ripeners will produce an early-ripening hybrid (see second sheaf from right); Hordeiforme has (under Odessa conditions) a long vernalization phase, Melanopus has a long photo phase

Hybridization creates a hereditary foundation containing the potentiality of rapid passage through both phases (under Odessa conditions)

When crossing two varieties, one of which develops as a winter variety and the other as a spring variety under the given conditions of existence (conditions of the district), a heterozygous hereditary foundation is obtained which contains a real potentiality for spring development. The hybrid seed when sown in the spring, will find the necessary conditions of existence for the development of the "spring" potentiality present in the hereditary basis.

Thus, "spring habit" will dominate over "winter habit." It must always be borne in mind, however, that this springness of either of the parents will be preserved only under the *given* conditions of existence; under other conditions it will become "winter habit." Therefore, the first generation resulting from the cross of these same parents may turn out to be of winter habit if developed under other conditions.

Such a mistake can be avoided only by a prior phasic analysis of the parents to be crossed. It is also necessary to ascertain before crossing what conditions of existence are required by each parent for the vernalization phase. After that it will be possible to foretell whether the first generation will be of winter or spring habit if sown in any district and not only in the district where the parents were grown.

If, when crossing two parents, only one develops as a spring variety under the given conditions, we always know beforehand that under the same conditions the development of hybrid F_1 will also be of the spring type, i.e., we know beforehand what the pattern of dominance will be.

If a semiwinter and a spring form are taken for crossing, then, for the same reason, namely, the possession by the heterozygote of a real potentiality for "spring" development under the given conditions of existence, that is to say, if sown in the warm season of the year, F_1 will be of spring habit, i.e., "spring habit" will dominate over "semiwinter habit."

When two forms are taken for crossing, one of which is an early ripener and the other a late ripener under the given conditions of a district, then, since the heterozygote will possess a real potentiality for early development under the given conditions, and since these conditions are present, F_1 will develop as an early ripener. Thus, in this case, early ripening will dominate over late ripening.

It is possible to formulate the following law of the duration of vegetative periods, which accords with and is a concretization of the general law of dominance.

In the first generation of hybrids (F_1) from two parents, one of which ripens early and the other later, early ripening will dominate, other conditions being equal. The first generation will always be as early ripening as, or even more early ripening than, the most early ripening of the parents. (Tables 1 and 2 on pp. 94-95.)

Cases of F_1 being earlier ripening than either of the homozygous parents have long been known to modern genetics, and in our opinion they are of the same common nature as cases of the more vigorous growth of F_1 than either of the parents, being a form of so-called "heterosis." But failing to analyze the regularities of the development of characters on the basis of phasic development, formal genetics is unable to explain heterosis. From the standpoint of phasic development, however, heterosis as regards early ripening becomes understandable, and this enables us to foretell such heterosis and deliberately to create it. We proceed from the proposition that

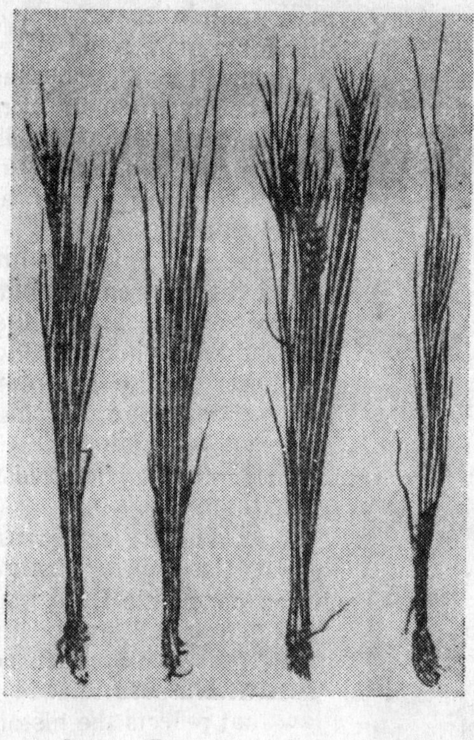

Fig. 30. "Heterosis" in respect to early ripening created by choosing pairs on the basis of a phasic analysis of wheats

Fig. 31. Early-ripening "heterosis" created by choosing pairs on the basis of a phasic analysis of barley

Pallidum 883/6 from the Azerbaijan station is not adapted for passing through the vernalization phase when sown in the spring under Odessa conditions (see sheaf on left). Pallidum 046 from the Odessa station (see sheaf on right) is badly adapted for traversing the photo phase under these conditions of spring sowing. The hybridization of these forms creates an early-ripening "heterosis"

early-ripening heterosis *is the resultant of dominance in the development of several more rapidly passing phases during hybridization.*

When hybridizing two hereditary bases, one with a more favourable (more rapid) phase of vernalization under the given conditions, but with a bad (prolonged) photo phase, and the other with a more favourable development of the photo phase under the same conditions, but with a bad (prolonged) vernalization phase, we, owing to the law of dominance, will necessarily obtain early-ripening heterosis in F_1.

Take, for example, two late-ripening forms and make a phasic analysis of each of them. If you find that they are late ripening owing to a difference in phases, you will, by crossing them, create a hereditary foundation which

Fig. 32. Sesame

Both parents have long vegetative periods. Phasic analysis showed that their late ripening was conditioned by different phases. Choice of pairs on the basis of a phasic analysis showed that it was possible to create an early-ripening form out of these late-ripening forms

will contain a real potentiality for a more rapid development in both phases. Under the given conditions of existence the hybrid obtained from two late ripeners will develop like an early ripener.

Thus, if we take into account the different conditions of existence available for the development of the phases of the parents that we choose for crossing, we can deliberately create early-ripening heterosis (Figs. 30-33).

As we have already indicated, the developmental phases constitute the general biological law of the individual development of plants from seed to seed, a law that reflects the historical action of the natural selection of adaptations in the demands which the developmental phases make upon the conditions of existence.

These phases, in which the difference in the nature of the demands made by different organic forms upon their conditions of existence are based on the initial difference in the hereditary foundations of these forms, are themselves the base on which the organs and their specific characters develop. The latter, however, as well as their base (the phases), while in general developing in their turn under their own conditions of existence, concretely develop under the inevitable influence of extremely diverse and sometimes not immediately calculable external influencing factors, which are by no means inactive in the formation of the characters of each separate organism as the individual representative of the variety, for they condition specific differences in the form-building process.

Although the rapidity and duration of passage through the phases determine the rapidity and duration of shooting, budding, flowering, fruiting, etc., they do not by any means do so entirely. Therefore, changes in the

Pallidum (Azerbaijan) F_1 Nutans (Sweden)

Fig. 33. Deliberately-created "heterosis," the result of choosing phasically analyzed pairs for crossing (from world collection)

duration of passage through the phases, and changes in the date of budding, flowering, fruiting, etc., will necessarily diverge somewhat, and this divergence will be different in different plants, being determined by the specific nature of each plant form and by the specific development of its organs, and also by the different influences exerted by the external factors. All this notwithstanding, the *general law* that budding, flowering and fruiting dates depend upon the duration of passage through the phases of development remains valid.

With all that has been said above as its theoretical premises, the Laboratory of the Physiology of Plant Development of the Institute of Selection and Genetics has been carrying out experiments under both field and laboratory conditions. The results obtained *fully confirm the correctness of these theoretical premises, fully confirm the correctness of the law of dominance in vegetative periods that we have formulated.* (Tables 1 and 2.)

By taking the above-formulated law of dominance in periods of vegetation as a basis, it became possible to cull combinations in the very first hybrid generation, which plant breeders usually ignore. From this law of dominance it follows that the best pattern of development (as regards early

Table 1[1]

SOWN IN FIELD MARCH 26, 1934

No.	Combination	Earing			
		F1	Maternal plants from seeds		Paternal plants
			Ordinary	Vernalized	Ordinary seeds
1	Leucur. (1241)×Melanop. 069	27/5	29/5	25/5	6/6
2	„ (1301)× „ 069	2/6	6/6	3/6	6/6
3	„ (138)× „ 069	1/6	1/6	26/5	6/6
4	„ (1406)× „ 069	2/6	6/6	1/6	6/6
5	„ (1413)× „ 069	2/6	2/6	1/6	10/6
6	Hordeif. (1522)× „ 069	27/5	27/5	25/5	6/6
7	Ersp. (2038)× „ 069	2/6	6/6	2/6	6/6
8	Apulic. (2236)× „ 069	2/6	2/6	1/6	7/6
9	Hordeif. (2506)× „ 069	2/6	7/6	2/6	7/6
10	„ (2508)× „ 069	2/6	7/6	2/6	6/6
11	„ (2753)× „ 069	2/6	6/6	1/6	6/6
12	„ (2577)× „ 069	2/6	11/6	2/6	6/6
13	„ (2813)× „ 069	3/6	8/6	5/6	6/6
14	Apulic. (3418)× „ 069	25/5	26/5	24/5	6/6

[1] The numbers in parentheses are taken from the catalogue of the All-Union Institute of Plant Industry.

Table 2

SOWN IN GREENHOUSE AUGUST 24, 1933

No.	Combination	Earing		
		F1 (in different pots)	Maternal plants	Paternal plants
1	Ukrainka×Lutesc. (2167)	9/10, 10/10	Winter	8/10
2	Ferr. (818)×Girka 0274	29/9, 29/9, 29/9, 29/9	10/10	4/10, 6/10, 6/10
3	Ferr. (818)×Lutesc. 062	29/9	10/10	2/10, 2/10, 4/10
4	Erinac. (991)×Ferr. (2166)	29/9	10/10	29/9
5	Erinac. (1506)×Lutesc. 062	29/9	20/10	2/10, 2/10, 4/10
6	Ersp. (2038)×Lutesc. 062	30/9, 28/9, 29/9, 29/9	Winter	2/10, 2/10, 4/10
7	Ferr. (2146)×Lutesc. 062	29/9, 27/9, 28/9 29/9, 28/9, 28/9	„	2/10, 2/10, 4/10
8	Ersp. (2150)×Lutesc. 062	29/9, 30/9, 30/9	„	2/10, 2/10, 4/10
9	Apulic. (2236)×Lutesc. 062	30/9, 4/10	6/10	2/10, 2/10, 4/10
10	Ersp. (2551)×Lutesc. 062	30/9	29/9	2/10, 2/10, 4/10
11	Ferr. (2705)×Lutesc. 062	29/9, 30/9, 30/9, 1/10, 1/10	Winter	2/10, 2/10, 4/10
12	Ferr. (2705)×Girka 0274	3/10, 3/10, 4/10, 4/10	„	4/10, 6/10, 6/10
13	Ferr. (2707)×Lutesc. 062	1/10, 1/10, 1/10	„	2/10, 2/10, 4/10
14	Ferr. (2707)×Girka 0274	30/9, 30/9, 1/10 3/10, 3/10, 3/10	„	4/10, 6/10, 6/10
15	Ersp. (2752)×Lutesc. 062	29/9, 29/9	„	2/10, 2/10, 4/10
16	Ersp. (2752)×Girka 0274	30/9, 2/10	„	4/10, 6/10, 6/10
17	Ersp. (2774)× Ferr. 2166 m	25/9, 25/9	„	28/9
18	Ersp. (2781)×Lutesc. 062	27/9, 29/9, 28/9, 29/9	„	2/10, 2/10, 4/10

Table 2 continued

SOWN IN GREENHOUSE AUGUST 24, 1933

No.	Combination	Earing		
		F_1 (in different pots)	Maternal plants	Paternal plants
19	Ersp. (2781)×Girka 0274	29/9, 20/9, 30/9	Winter	4/10, 6/10, 6/
20	Apulic (3418)×Lutesc. 062	25/9, 26/9	3/10	2/10, 2/10, 4/10
21	Ersp. 534/1×Lutesc. 062	30/9, 3/10, 3/10, 3/10, 3/10 3/10, 3/10, 3/10, 2/10, 3/10, 4/10, 3/10, 4/10 2/10, 3/10, 4/10, 4/10	Winter "	2/10, 2/10, 4/10
22	Ersp. 534/1×Girka 0274	2/10, 2/10, 3/10, 2/10, 2/10, 3/10, 3/10, 3/10, 3/10, 3/10, 3/10, 3/10, 3/10, 3/10, 3/10, 3/10, 3/10, 3/10, 3/10	Winter	4/10, 6/10, 6/10
23	1316/2×Lutesc. 062	4/10, 4/10	"	2/10, 2/10, 4/10
24	1637/1×Lutesc. 062	3/10, 3/10, 2/10, 3/10, 3/10, 2/10		2/10, 2/10, 4/10
25	1637/1×Girka 0274	3/10, 2/10, 2/10, 2/10, 2/10, 2/10, 2/10, 3/10	"	4/10, 6/10, 6/10 "
26	2522/1×Lutesc. 062	5/10, 8/10	"	2/10, 2/10, 4/10
27	2522/1×Girka 0274	30/9, 2/10, 2/10, 2/10, 3/10, 2/10	"	4/10, 6/10, 6/10
28	3060/15×Lutesc. 062	4/10, 3/10, 3/10, 6/10, 3/10, 4/10	"	2/10, 2/10, 4/10
29	3060/15×Girka 0274	30/9, 3/10, 4/10, 1/10, 3/10, 3/10, 3/10, 2/10, 3/10, 3/10	"	4/10, 6/10 6/10

ripening) will be presented by F_1, which has a more favourable basis for richer development potentialities than all succeeding generations.

The adaptability of a given potentiality of the hereditary foundation to the available conditions, if that potentiality is really present, must already make itself felt in F_1 and if F_1 is unable to develop fully under the given conditions, then it is useless to expect a better development (in this case, a shorter ripening period) in the succeeding generations, the hereditary foundations of which are poorer in potentialities owing to segregation.

Consequently, if F_1 comes out in a form that does not satisfy the range of requirements of the variety for the given district as regards ripening period, further breeding work for obtaining segregates of the given combination will be superfluous.

None of the succeeding generations can develop a form that will be earlier ripening than F_1.

Experimental hybridization work, now already carried on up to F_7, fully confirms this without a single exception.

From this same conception of the hereditary foundation as the vehicle of developmental potentialities it follows that in all succeeding generations the potentialities of earlier development may be impoverished as a result of segregation, but they can never be enriched (without mutations). Therefore, F_2 cannot be more early ripening than F_1, F_3 cannot be more early ripening than F_2, and so forth.

All the experiments conducted in the Laboratory of the Physiology of Plant Development and all the data given in the literature on the subject and examined for this purpose fully confirm the correctness of this thesis.

PRACTICAL METHODS OF SELECTION WORK BASED ON THE THEORY OF DEVELOPMENT

The Institute of Selection and Genetics (Odessa) has resolutely proceeded to put the theoretical premises here briefly enunciated into practice—*to reorganize the work of genetic breeding from the standpoint of the theory of development.*

In this reorganization we strive to take the Marxist-Leninist theory of development, which is the only true theory, as our guide. It is our aim to apply this theory specifically to the problem that we are working on.

All this has enabled us now not only to conduct extensive work in controlling the individual development of plants, but also, in theory and in practice, to trace a new path for plant-breeding work, to take the course of vanquishing formalism and empiricism in genetic-breeding research.

EARING

Combination \ Date of Earing	25/V	26	27	28	29	30	31	1/VI	2	3	4	5	Total earing earlier than Lutesc. 062	6	7	8
Ersp. 534/1×0274 Girka	—	—	—	—	—	—	4	11	3	33	5	120	176	31	172	116
Ersp. 1637/1×062 Lutesc.	—	—	—	—	1	—	1	5	6	1	4	16	34	—	9	72
Ersp. 1637/1×0274 Girka	—	—	—	—	2	—	27	23	51	35	10	197	345	17	403	104
Ersp. 2522/1×062 Lutesc.	—	—	—	—	—	—	6	25	34	26	12	128	231	—	36	128
Ersp. 2522/1×0274 Girka	—	—	—	—	1	—	4	15	44	12	—	123	199	15	121	37
Ersp. 3060/15×062 Lutesc.	1	—	—	3	2	—	25	93	81	17	7	202	431	1	27	232
Ersp. 3060/15×0274 Girka	—	—	—	—	—	—	9	15	11	57	1	97	190	186	224	43
Ferrug. 1316/2×062 Lutesc.	—	—	—	—	—	—	—	—	1	—	1	—	2	5	10	8
Total													1608			

These premises have enabled us to indicate exactly and according to plan how long it will take to raise varieties of spring wheat for the Odessa district. These premises mark a considerable step forward in overcoming the element of chance in breeding work, for they make it possible to plan the output of varieties with ever-improving indices according to schedule. The approach to the hereditary foundation from the standpoint of development creates real possibilities of making the entire work of plant breeding a much more intelligent process because, if one bases himself on the laws already ascertained one can keep constant check on the work and, in the course of it, note and rectify errors in that process.

After choosing the parental forms for crossing with the object of obtaining forms more early ripening than either parent, the correctness of this choice can be tested by the plants of the very first generation. If F_1 is later ripening than intended for the future desired variety, it is a sure sign that a mistake has been made in the choice of parental forms. On the assumption that the segregates of F_2 and of succeeding generations cannot be earlier ripening than F_1, it is possible in F_1 to cull the combinations with respect to length of vegetative period.

Thus, no variety of spring wheat that ripens later than Lutescens 062 from the Saratov station is suitable for Odessa conditions. This, of course, does not mean that any early-ripening variety will be suitable here.

Hence, when it proceeded to breed a spring variety for Odessa conditions, the Institute of Selection and Genetics resolved beforehand that the future variety of spring wheat must under no circumstances ripen later than Lutescens 062. It must ripen at least 3-4 days earlier.

F_2 1934 **Table 3**

9	10	11	12	13	14	15	16	17	18	19	20	21	22	23	24	25	26	27	28	29	30	Did not ear	Total plants earing later than Lutesc. 062	Total eared
35	9	12	3	71	3	34	3	14	4	12	4	—	2	5	—	1	1	1	—	2	—	44	579	711
8	3	5	12	9	2	3	—	—	—	—	—	—	—	—	—	—	—	—	—	—	—	29	142	147
55	35	28	11	92	18	52	—	2	—	9	—	—	—	1	—	1	—	2	—	2	—	186	1,021	1,180
34	88	18	6	53	8	3	3	—	—	2	1	—	—	—	—	1	1	1	—	—	—	26	409	614
34	5	17	7	49	3	2	5	—	—	5	—	—	—	2	—	4	—	—	—	—	—	24	330	505
38	50	22	12	87	50	5	22	9	—	2	3	—	2	1	—	5	1	—	—	—	—	57	626	1,000
32	35	8	6	57	65	11	11	18	1	7	3	1	1	5	3	4	1	4	2	—	1	28	757	919
—	2	—	—	7	—	2	—	—	—	1	—	—	—	—	—	—	—	—	—	—	—	2	37	37
																							3,901	5,113

This line of action makes it possible to perform the first culling of the breeding material in the first generation—something which plant breeders usually did not do before. If there are several combinations of F_1 plants in the field, all the combinations which in F_1 are later ripening than Lutescens 062, or have the same length of vegetative period, must be culled. Only those combinations must be left which in F_1 are earlier-ripening than 062.

We will remind the reader that such culling has become possible only because of the law that we have discovered, namely, that not a single segregate in all succeeding generations can be earlier ripening than the initial heterozygous forms of the preceding generation, and cannot be earlier ripening than F_1. By culling combinations in F_1, the plant breeder can, to some extent, save himself work and bother with combinations which obviously will not give him segregates with the necessary length of vegetative period, as he will definitely know beforehand that from such combinations he will not get a variety suitable for his district, in this case, Odessa.

After sowing F_2, which will segregate in numerous characters, including length of vegetative period, as regards the course of the phases as well as the organs and characters that develop on the basis of the phases, it is necessary, first of all, to cull while still in the field all the segregates that ripen later than the variety it is intended to breed. Segregates which under the conditions of our district have the same ripening period as Lutescens 062, or are later ripening, will in any case be unable (in the absence of an inner basis for segregation) in succeeding generations to produce forms with the requisite vegetative period. After that it is necessary to proceed to the next breeding job, namely, to choose specimens from the remaining small percentage of segregates. In the breeding work conducted at the Institute of Selection and Genetics the following procedure has been adopted: in F_2, only those plants are left which ripen at least two days earlier than 062.

EARING

Combination \ Date of Earing	29/Y	30	31	1/VI	2	3	4	5	Total earing earlier than Lutesc. 062	6	7	8	9	10
Ersp. 534/1×0274 Girka	—	—	—	1	2	—	1	2	6	9	16	9	3	4
Ersp. 1637/1×062 Lutesc.	1	—	1	—	1	—	—	2	5	4	6	8	1	3
Ersp. 1637/1×0274 Girka	—	—	—	—	—	2	—	6	8	13	12	51	9	11
Ersp. 2522/1×062 Lutesc.	—	—	—	—	1	—	1	1	3	11	5	9	5	2
Ersp. 2522/1×0274 Girka	1	—	1	—	6	—	2	—	10	14	—	18	5	2
Ersp. 3060/15×062 Lutesc.	—	—	—	—	5	1	—	—	6	5	2	4	5	—
Ersp. 3060/15×0274 Girka	—	—	—	—	5	-	2	4	11	25	7	27	14	5
Ferrug. 1316/2×062 Lutesc.	—	—	—	—	1	—	1	1	3	1	—	—	—	—
Total									52					

For example, in the combination given in Table 3 Erythrospermum 534/1×0274 Girka of local breeding, out of 755 F_2 plants, only 176 eared earlier than Lutescens 062 (plants enumerated on the left side of the table); the remaining 579 were culled. In the combination Erythrospermum 1637/1×062, out of 176 plants, only 34 were left. In the combination Erythrospermum 1637/1×0274 Girka, out of 1,366 F_2 plants only 345 were left as regards length of vegetative period; the rest were culled when still in the field, etc. The small percentage of plants left after culling according to vegetative period is then subjected to the ordinary culling practised in plant breeding, which, of course, greatly reduces the number of remaining plants (degenerate, obviously unproductive, and other such forms are rejected), depending on the approach and on what is required of the variety that is being raised. The same is done in F_3, F_4, etc., until the constant requisite form for the conditions of the given district is obtained (Table 4).

Proceeding in this way, one may at once encounter the following obstacles: parental forms, properly chosen for crossing in regard to length of vegetative period, as confirmed by the first generation, may sometimes, in F_2, the succeeding generation, fail to produce a single segregate, or give rise to exceedingly few that remain (in view of range of vegetative period) after culling in the manner described above. It becomes obvious to the plant breeder that the needed variety cannot be raised from the remaining forms, but at the same time the phasic analysis shows that the variety can be obtained from this parental pair. In such cases it is perfectly evident that such a combination must not be rejected. It merely indicates that in this case F_2 was sown on too small a scale, and that this combination, the sowing of F_2, must be repeated on a larger scale. The same will apply to all the succeeding generations.

Thus, the propositions we advance in aid of the practical work of the

Table 4

F_3 1934

11	12	13	14	15	16	17	18	19	20	21	22	23	24	25	26	27	28	29	Did not ear	Total plants earing later than Lutesc. 062	Total eared
5	—	5	2	3	—	3	—	1	1	1	—	—	—	—	—	—	—	4	13	79	72
1	—	5	2	—	1	—	—	—	—	1	—	—	—	—	—	—	—	1	—	33	38
17	1	11	3	2	—	15	—	1	—	1	—	1	—	—	—	—	—	1	11	160	157
9	—	1	1	—	1	1	—	—	—	—	—	—	—	—	—	—	1	1	1	48	50
4	1	3	—	—	—	1	—	—	—	1	—	—	—	—	—	—	—	—	—	48	58
—	—	1	—	—	—	2	—	—	—	1	—	—	—	—	—	—	—	1	—	21	27
8	—	6	1	—	—	2	—	2	—	1	2	1	—	1	—	1	—	1	5	109	115
—	—	—	—	—	—	—	—	—	—	—	—	—	—	—	—	—	—	—	—	1	4
																				499	521

Fig. 34. Early-ripening F_1 obtained from two late-ripening parents. Choice of pair made on basis of phasic analysis

Fig. 35. Pallidum 574/1 (Azerbaijan)

Late ripening owing to vernalization phase (sheaf on left). 943 late ripening owing to photo phase (sheaf on right). Phasic analysis showed that F_1 should be of earlier habit than either parent, and it turned out to be so

plant breeder enable him to work all the time with ten and even a hundred times less plants than he has worked with up till now (or at least than were required by the genetic theory of hybridological analysis). At the same time he can now enlarge the scale of work tenfold, subjecting to selection combinations of all the existing varieties of the given crop, and greatly reducing the time required for raising varieties and the area of nurseries needed for this. Owing to the small number of new forms of potential varieties, each of them will receive ever so much more attention from the breeder than was the case before, under preliminary variety testing, when he had to deal with hundreds of claimants to be the future variety. It must be pointed out that choosing parental pairs for crossing according to phases of development, i.e., taking phases as one of the criteria in choosing parental pairs, does not mean that for crossing purposes those forms are chosen, or given preference, out of which a new form may be obtained phasically best suited to the conditions of the given district. If this course is followed it may safely be predicted that any plant breeder will very easily breed at request only early ripeners. But as yet they will not be the varieties of, say, spring wheat needed for the steppes of the Ukrainian S.S.R., or for the Far North (where early ripeners are also needed). In practical work, it will become perfectly obvious to every plant breeder that in large collections analyzed according to developmental phases the choice of crossing material should not be confined to those forms that are phasically best suited for this purpose. Different forms may be phasically suitable for the conditions of a given district, but these same forms, with their excellent phases of development, may be unable to stand the dry conditions of the Odessa district, or other climatic and soil conditions, or may not meet the demands socialist farming in a given district makes upon the given crop. Therefore, when choosing parental forms by phasic analysis, it is necessary, before crossing, to ascertain the economic indices of these forms by phasic analysis under the conditions of the given district.

Fig. 36. Choice of sesame pair for producing an early-ripening form from two late-ripening forms. Choice made on basis of phasic analysis

In the course of its work, the Laboratory of the Physiology of Plant Development (Institute of Selection and Genetics) crossed thirty pairs of wheat forms chosen according to phases of development. But before the crossing it was arranged that for practical purposes, i.e., for raising a variety for the Odessa district, the laboratory should cross only two pairs of forms: Erythrospermum 534/1×0274 Girka of local breeding, and Erythrospermum 543/1×Lutescens 062 from the Saratov station. The bulk of the other crossings was made exclusively for the purpose of testing the possibility of combining new forms according to the length of the vegetative period, taking into account the phasic development of the parents; they were made for the purpose of testing the propositions put forward above. These crossings whose purpose was methodological, we, in jest, called "crossings for the camera," illustrations of which are given in this essay (Figs. 34-40). As is evident

Pallidum (Azerbaijan) F_1 Medicum (Azerbaijan)

Fig. 37. Choice of parental pair of barley for producing early-ripening form from two late-ripening forms

Choice made by phasic analysis of hereditary foundation

from the illustrations, the pictures obtained were not bad. Every combination of crosses resulted in a form that was much earlier ripening than either parent.

We did not choose the first two pairs of combinations for breeding the variety because they were the best in the entire world collection. We are convinced that the collection contains better combinations of all possible kinds, but they have not yet been discovered. It is important for us, however, to ascertain the potentialities and developmental conditions of the hereditary foundation of the two components that are taken for crossing against the background of the length of the vegetative period of our future variety under the conditions of the district for which it is being raised. As these conditions vary from year to year in each district, although they remain relatively constant (Odessa conditions in Odessa, Kharkov conditions in Kharkov, etc.), we must, before crossing, find out the economic indices of the parent forms under the district's conditions of growth over a number of years against the background of the vegetative period of the future variety.

If not vernalized, the Erythrospermum variety we took from the Azerbaijan station as the initial form for obtaining a variety of spring wheat for the Odessa district is, under the conditions of that district, an extremely late ripener and its yield is small. After vernalization, this variety, under Odessa conditions, produces a fairly good yield for three years. Girka 0274 of local breeding is less productive under Odessa conditions than Lutescens 062 from the Saratov station; and it ripens 5 days later than Lutescens 062. Both Lutescens 062 and Girka produce exceedingly low yields under Odessa conditions. A phasic analysis of these forms of wheat, however, shows that if their growth is accelerated (in the photophase) by 6-8 days, Lutescens 062 and Girka 0274 show good indices as regards fulness and quality of grains, i.e., the chief yield indices of spring wheat for our southern districts.

Fig. 38. Segregates of F_3 are earlier ripening than either parent

Fig. 39. Segregates of F_3. Earlier ripening than parents (middle sheaves)

In view of the fact that Girka is a local variety, there are more grounds for assuming that its hereditary foundation, if its "weak spot"—the photophase—is removed, will prove to be better adapted to the variations of the climatic conditions of the Odessa district. Lutescens 062 is not a local variety (it was raised at the Saratov station). But it was produced from Poltavka by individual selection; moreover, this variety has already been grown in the districts of Odessa Region for quite a number of years. This explains why we took these two combinations in hand in order to obtain quickly the variety we want. During the initial period, these two combinations are much safer insurance for us than other combinations, which, though more tempting as regards their phases of development, provide no guarantee as regards economic indices. The constant forms obtained from these combinations under Odessa field conditions ripened 10 days earlier than Girka (one of the parents) and 6 days earlier than Lutescens 062 from the Saratov station (the variety recommended for Odessa Region). We believe that these forms will be the varieties we have been seeking.

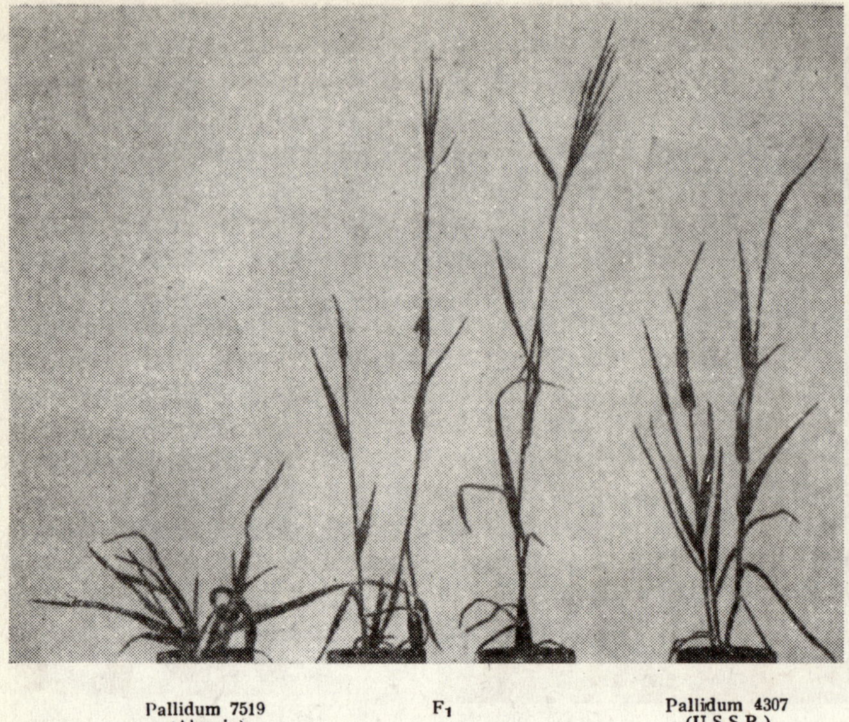

Pallidum 7519 (Algeria) — F_1 — Pallidum 4307 (U.S.S.R.)

Fig. 40. Pair chosen on basis of phasic analysis to produce an early-ripening first generation from a winter and a late-ripening form

The plant-breeding work on the new lines which the Laboratory of the Physiology of Plant Development of the Institute of Selection and Genetics has been conducting since 1933, was begun with the growing of the parents. It has already resulted in constant forms and has been carried to F_7. Beginning with the spring of 1935, the forms that turn out to be the best under the given conditions of development will be subjected to various field tests. At the same time, three or four forms, the indices of which we believe to be the best on theoretical grounds and which are being tested and verified in the course of different generations, will be reproduced, and by the autumn of 1935 each will be brought up to 10 c. of grain (thus passing in 2 years and 10 months from the parental forms to 10 c. of grain).

The proposed seemingly excessive forcing of the reproduction of a variety which has not gone through the usual preliminary varietal test is, in our opinion, not only possible but necessary. This is so because the theoretical propositions by which the Laboratory of the Physiology of Plant Development is guided in selecting parental pairs for crossing, and the theoretically assumed general pattern of segregation, have been confirmed by the whole course of segregation up to the appearance of the expected constant

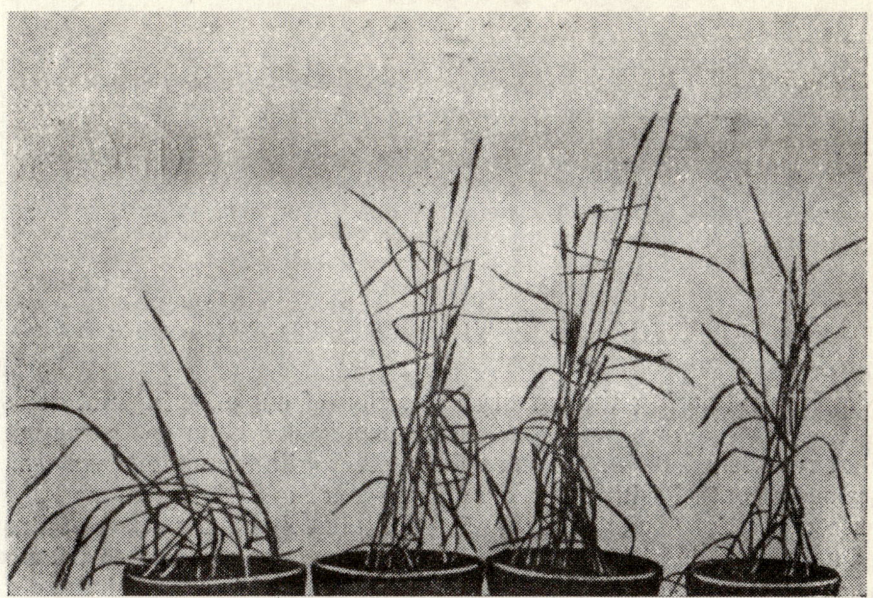

Fig. 41. Erythrosp. 2781 (Azerbaijan) ripens very late under Odessa conditions (pot on extreme left)

When vernalized shows good economic indices. Lutescens 062, late ripening under Odessa conditions (pot on extreme right). Under favourable conditions for passage through second (photo) phase becomes early ripening and shows good economic indices. When these are crossed a hereditary basis is obtained from which the sole "weak spots" have been eliminated. The cross should result in a variety with good economic indices, ascertained by means of the phasic analysis of each of these forms. Centre (two pots): F_1 hybrids

forms and their characters. This imbues us with a certain amount of confidence that the field varietal tests will further confirm our assumptions concerning the varieties' economic qualities. These, we think, should be higher under conditions analogous to those of the district in which they were raised than the economic qualities of the standard varieties of the given district, not to speak of the original parental forms (Girka and Ersp. 534/1), which in many respects are inferior to the standard variety of spring wheat.

The course and state of development of the hybrids on the experimental fields of the Institute of Selection and Genetics make us feel confident that the problem of breeding the varieties needed by our socialist agriculture will shortly be solved.

THE REORGANIZATION OF SEED GROWING[1]

COMRADES, during the cursory inspection of our experimental fields we made yesterday, we were not able to examine everything as closely as we might have wished. As we acquainted ourselves with the various experiments, we came in contact with numerous branches of agrobiological science. We discussed questions concerning the breeding of self- and of cross-pollinators and a new method of breeding, and in this connection we touched upon questions of genetics and physiology. We also discussed in detail the degeneration of the planting material of early varieties of potatoes.

Diverse as the questions were that we touched upon during our excursion, we were, nevertheless, unable to inspect a number of important departments of the work of our Institute, as, for example, the breeding of new varieties of cotton by the deliberate choice of pairs for crossing, or the breeding of a variety of cotton by crossing hundreds of combinations with the view to culling the vast majority of the combinations in F_1.

You, of course, already know that we claim that it is not absolutely necessary, when breeding new varieties of certain crops (for example, of cotton in our district), to make a phasic analysis of the parents taken for the cross. This has become possible only because we are now able to tell in any hybrid generation, beginning with F_1, whether the requisite constant variety can be obtained from the given heterozygous plant.

About two and a half years ago, at a conference on plant breeding held at the Seed-Growing Union, I raised the question of employing new methods of breeding self-pollinating spring crops. I cannot help recalling that at that time my theoretical propositions on hybridization with phasic analysis as their basis were sharply opposed by a number of specialists in genetics and plant breeding. You have now convinced yourselves that this is quite possible and feasible. The new varieties that we have bred during this very short period of time have fully confirmed the correctness of our assumptions. Moreover, in the course of breeding new varieties these assumptions became

---

[1] A paper read at the Visiting Session of the Lenin Academy of Agricultural Sciences of the U.S.S.R. held in Odessa, June 26, 1935.—*Ed.*

more definite. After all, any job becomes much clearer as the work proceeds. And so in the course of our work, the assumptions we made two and a half years ago were not only confirmed but even exceeded by the further development of the theory itself. We are now solving many problems of plant breeding in a simpler and better way. If I were now commissioned to breed an entirely new variety of spring wheat, I would certainly do so in less than two and a half years and would, most likely, produce a better variety.

Many of the propositions on the basis of which we did our breeding are now a past stage for us. They were the principles that made it possible for us to produce, according to plan and within the unprecedentedly short period of time assigned, the varieties of spring wheat you saw yesterday, principles which have been expounded in our recently published works: *The Theoretical Principles of Vernalization* by T. D. Lysenko and *Plant Breeding and the Theory of Phasic Development of Plants* by T. D. Lysenko and I. I. Prezent. Many of you, however, especially the geneticists, regard these propositions as a new discovery and, moreover, as controversial and far from being proved. These propositions are so new to many people that they still have to study them. It seems to me that many scientists challenge the new approaches to plant breeding proposed on the basis of our theory of plant development not because these propositions are fundamentally wrong, but because these comrades have not studied them theoretically in detail and, what is most important, have not subjected their relative truth to the test of practice. Indeed, take the crossbreeding—according to plan and within a set period (two-three years)—of new varieties of spring crops. Yesterday you saw that in conformity with a previously worked-out plan, we obtained varieties of spring wheat in the course of two and a half years, counting from the date of sowing the parents. Would that have been possible without the new theory that we are advocating? Until now there have been no cases in practical plant breeding of new varieties being raised by crossing in two and a half years.

The object of our work in breeding a variety of spring wheat was not the variety itself, but the working out of the methods of breeding spring crops by crossing. It is impossible, however, to work out methods of breeding varieties by conjuring up theories (even if out of a good head), or merely by assimilating the whole of world experience contained in literature. For this it is necessary to engage in the practical work of breeding varieties.

It is beyond doubt that as in every other kind of theoretical work, so in the elaboration of the methods of plant breeding, mental work, the assimilation of the wealth of human experience contained in literature, and what is most important, practical work in breeding varieties, are all required. Not only did the new variety we bred confirm the correctness of our methods, but in the process of breeding it in the theoretical premises—which are component parts of the methods—underwent a change, became more precise.

By this I want to emphasize that new methods of breeding varieties, more effective than existing ones, can be worked out only in the process of breeding new varieties. Hence, although our object was to work out *methods* of breeding spring crops, we knew very well, and know now, that if we failed to produce a new variety of one of the most difficult crops for the South—spring wheat—the methods would be useless. That spring wheat is one of the most difficult crops to breed under the conditions of the south of the Ukrainian S.S.R. is a point that does not have to be laboured. It is proved by the one fact alone that although work on breeding spring wheat has been going on for twenty years (and over), so far not a single hectare of spring wheat is grown in practical farming in the Odessa district.

We now have every right to say that the new methods of breeding spring crops have been worked out in the main, proof of which are the varieties of spring wheat you saw yesterday.

To fully appreciate the importance of these varieties you must remember that they have been produced by hybridization in a fabulously short space of time—two and a half years. They are already being tested for yield and resistance to smut and stinking rust; next month they will be tested for their baking qualities. In two weeks they will be sown in the field again for further reproduction. In addition to all this you must remember that not one of the hybrid generations of all these varieties *occupied more than two square metres of greenhouse space.* These varieties began to occupy a relatively large amount of space only from the time they were reproduced; and we conducted this reproduction in a way different from that in which wheat varieties have been reproduced up till now. Yesterday, I showed you a plot planted with a new variety of wheat from which, in a few days' time, a crop of 25-30 kg. will be harvested, and on which only 20 gr. of seed were used. The coefficient of yield we set out to achieve, and did achieve, was not 8- or 10-fold, as is usually the case, but 1,500-fold.

Up till now, the vast majority of plant breeders have been working on hundreds of new competing varieties in preliminary tests to determine the best among them. Unfortunately it often happens that in the end the plant breeder is unable to find among these numerous varieties even one that is superior to the standard district variety. We, however, did not take a hundred new competitors, but only four. We knew from our theory (which is challenged by so many geneticists), that all the plants of the different hybrid generations that we rejected, that we did not develop to the constant stage, let alone to the testing stage, were useless for the purpose of breeding a variety. Plant breeders who work according to the theories of the Morgan school of genetics cannot resort to such culling of heterozygous plants. Unable to distinguish the valuable hybrid plant from those that are useless for the purpose of breeding a variety, they burden themselves with hundreds, and even thousands, of competing varieties.

Of the four varieties we took for testing, one, No. 1160, is, in our opinion, useless as a variety. That was already clear to us when it was but a single spike on the hybrid plot. We did not reject it for the sole reason that L. P. Maximchuk had taken a liking to it. I always respect his opinion, although in my plant-breeding work I have fairly often acted contrary to it. Whether No. 1160 proves to be a variety of wheat suitable for the district remains to be seen. I will be very glad to lose out to Maximchuk if this variety turns out to be a good one. For a good variety one can, and should, sacrifice a theory, if one wants a sound theory.

In my opinion, the fundamental achievements of our breeding work are not only the new varieties of spring wheat, but also the further development of the theory of breeding varieties. Our quite efficient staff can now not only choose in advance parental pairs for crossing with the object of obtaining constant forms possessing definite lengths of vegetative periods specified in advance, and other characters and properties. The chief thing is that we can already control our plant-breeding work at any stage. We are able to judge from any hybrid generation whether a variety can be produced from the given hybrid material or not. Up till now, genetics, as a science, has not given such guidance to plant breeders.

I will try to formulate here some of the principal propositions of our theory, which guide us in our plant-breeding work, although this is not the immediate subject of my paper.

1. We regard the following proposition as definitely confirmed and borne out by practice: that F_1 obtained from the crossing of any parental pair, in the main, cannot be later in flowering than the earlier parent; but F_1 can be earlier than either parent. If a phasic analysis is made before crossing, it is possible to foresee to what degree F_1 will be earlier flowering than the early parent.

2. Now we maintain that at the time of earing, or flowering (if we are not dealing with cereals), F_1 cannot be less vigorous than the more vigorous parent; but it can be more vigorous than the vigorous parent.

These propositions of ours were questioned and met with objection two years ago, and even yesterday, when we inspected the objects (the new varieties) which irrefutably confirmed the assumptions we had then expressed. I think that the comrades who challenge the propositions I have just cited forget that what they saw yesterday is not a series of ordinary results accidentally obtained in an experimental crossing of plants, and then "explained" from the standpoint of "infallible" Mendelism and the Morganist theory. So far, the results you saw require no other explanation than the propositions I have just formulated, and also a number of others that are expounded in our works, if only for the reason that none of these results were obtained by chance, but had been planned deliberately, were brought about by putting into practice the theoretical propositions that you are challenging.

Not so long ago the phasic analysis of parental pairs chosen for crossing also raised doubts in the minds of many theoreticians in genetics and plant breeding. Nobody in this hall today expresses any such doubts. More than that, I myself am obliged to call your special attention to the fact that the phasic analysis of parents that are to be crossed is no longer essential for all the crops in a given district. We have advanced the science of plant breeding and genetics, and can now state that in some cases phasic analysis of parental pairs will be harmful, as it will cause the plant breeder to waste a year or two of his precious time.

For we now claim that in many cases it is possible to dispense with a phasic analysis of parental forms. To spend a year or two on phasic analysis is often unprofitable not only for our socialist practical farming, but also for the plant breeder.

What does the loss of two years mean to the plant breeder? How long, on the average, can a plant breeder work? He usually graduates from the institute at about 30 years of age, after which he is allowed to spend some time to gain practical experience in his work (subject). And so the plant breeder reaches the age of 35. Well, how many years can we work? Until the age of 50 or 60. A few lucky ones live to be 80 or 90. As the average (by no means exact, of course), we take the age of 50. In other words, we can take it that the plant breeder has 15 vegetative periods for real work. The plant breeder can sow crops in the field 15 times. Now, if we consider that formerly (and in 99 cases out of 100 it is the same today) it took plant breeders 10 to 15 years to breed a variety, it turns out that the majority of plant breeders, after spending all their lives on this work, never live to see the varieties they have bred actually growing in the fields. We, however, can breed a variety in two and a half years. Therefore, we have no right, legal or moral, to waste one or two years on phasic analysis in cases where we can dispense with it. But, comrades, you must not take my statement to mean that the phasic analysis of parental forms, which many of you have not yet learnt, is now altogether superfluous. In many cases of plant breeding it is still essential.

We must employ the most up-to-date methods, the methods that produce the quickest and most effective results. In every practical step we take, it is always our duty to reflect and utilize the most up-to-date theoretical achievements of our agricultural science.

Judging from the discussion we had yesterday, during our excursion to the fields, I may be asked: how can we refrain from making phasic analyses when choosing parental pairs, at least for some plants (crops) when we are in doubt as to whether the first hybrid generation will enable us to judge the fitness of the given combination for breeding a variety? Is it possible to tell from the individual plants of the second, third, etc., generation that no variety can be bred from them? And the comrades who raise these objections declare that they fully accept the theory of phasic development, but do not agree with the principles of culling we propose.

In other words, the majority of geneticists already recognize the principle of choosing parental pairs for crossing on the basis of phasic analysis, the more so that in the opinion of many geneticists there is nothing fundamentally new in this. All this easily fits in with Mendelist-Morganist symbolics. The short vernalization-phase genes of one of the parents combine with the short photo-phase genes of the other parent, with the result that two late-ripening parents bring forth an early-ripening progeny. This seemed perfectly "explicable" from the Morganist standpoint, and it is only because of this that many geneticists now accept it. But I must remind you that the Morganist standpoint "can explain" any result obtained from a cross. Everything can be explained from the Morganist standpoint, but nothing can be *foreseen* with certainty. One cannot, for example, foresee what will come of a cross between two plants that have never been crossed before. The one fact alone that it is possible, by the phasic analysis of parental plant forms, to ascertain before parents are crossed what their progeny will be like after crossing shows that this must be explained not from the Morganist standpoint, but from the standpoint of the theory of plant development that we are elaborating. Our propositions have their roots in and obtain their theoretical nutriment from Darwinism, from the works of I. V. Michurin.

The appraisal of the hereditary basis of the heterozygous plant that we propose, the appraisal of the possibility of obtaining the requisite segregates (constants) in succeeding generations, is nothing more nor less than the further logical development of the phasic analysis of parental forms that you accept. To refuse to recognize in theory that segregates cannot be earlier-ripening than the initial heterozygous forms means refusing to recognize the possibility of foreseeing the length of the vegetative period in the first hybrid generation obtained from the crossing of forms that have been subjected to a preliminary phasic analysis. If, however, phasic analysis is recognized, then it is also necessary to recognize theoretically the logical development of the basis of this phasic analysis—the law of segregation according to the plant's vegetative period and vigour.

It seems to me that, after all, the most cogent proof in favour of the proposition we advance is what you saw yesterday: the new varieties bred by the new method as well as the numerous hybrid plots and control experiments conducted for this purpose. You may read about all this in greater detail, comrades, in the papers we have recently published.

I am afraid, however, that not everybody will be able quickly and correctly to understand what is written in those papers, at least judging from some of the remarks made by comrades during our excursion to the fields.

It seems to me that what must urgently be found here is not proof of whether Lysenko is right or wrong, but the best and most practical way of achieving the goal the plant breeder has set himself, namely, to breed the best varieties in the shortest time. Every theoretical proposition that helps practical farming will be more useful and, of course, come nearer the truth than all the theoretical propositions that give neither immediate nor future, neither direct nor indirect, guidance to practical action in our socialist agriculture.

What we are proposing, and not only proposing but employing, gives a new turn to the entire work of plant breeding; it speeds it up, makes it more active and, I would say, more purposeful.

That being the case, I fail to understand many of the objections which, it seems to me, are raised solely for the sake of objecting. It is not always possible to examine the particular facts that are cited by the objecting comrades. I would advise many of those who object to the propositions we advance, for instance, on the question of the length of the vegetative period in heterozygous plants, not only to look for facts (facts must always be looked for, and I myself always look for them), but to try to begin to act—to breed a new variety—on the basis of the propositions we have expounded. This will be an effective and decisive test, which, I am sure, will help the doubters to convince themselves that we are right.

I think that if the propositions we are advancing, particularizing and developing turn out to be fundamentally wrong, it should be regretted not only by me and my staff, but also by all the opponents of these propositions, for we would be deprived of an effective method of breeding new varieties.

I, of course, do not for a moment doubt that the correctness of the laws that govern phenomena is not determined by what we would like them to be. It is not for me to tell you that the laws governing this or that phenomenon in nature operate irrespective of our wishes; but we can understand these laws more or less and, on this basis, bring them under our control.

I shall now close this introduction to my paper, the subject of which is not problems concerning the breeding of varieties, but problems concerning seed growing, problems concerning the way to maintain the varieties produced by plant breeders on the proper level. At any rate, I can only say in conclusion that the best thing to wish for is that the propositions we advance may, after all, turn out to be correct, and that in this controversy over theory and principles, not I and my assistants be proved wrong, but those geneticists who are opposing us. They stand to gain by losing, because, if we prove to be right, it will be a great practical achievement for Soviet plant breeding. In other words, our victory will greatly benefit socialist agriculture, the successes of which will inevitably result in an improvement in the cultural and material conditions of all working people.

The work of plant breeding is crowned with the seeds that are sown on collective- and state-farm fields. I would like at this authoritative assembly to discuss some of the theoretical principles of this important subject.

* * *

More than a year has passed since Comrade Stalin stated at the Seventeenth Congress of our Party that confusion reigns in the sphere of cereal and cotton seed growing. Has the situation changed in this sphere since then? Very little. Seed growing is still one of the most backward sections of socialist agriculture. Agricultural science is largely to blame for this. In many cases, genetics and plant breeding stand aloof from practical seed growing.

The ultimate object of plant breeding is the production of seeds—the sowing material which in the district served by the plant breeder produces the best yield in quality and quantity. Practical farmers are right when they say: "The chief thing for us is not the plant breeder's division of seeds into categories: first, second, third, etc. What interests us primarily is seed of high yield and good quality. That is what we need."

There are different grades of seeds: first, second, third, etc. The seeds are sent to seed-growing collective farms for reproduction and from there to the collective and state farms to be sown for consumption. On the surface it looks as if everything were running smoothly and well. The first grade is the best and is twice as dear as the second grade, the second grade is dearer than the third, etc.

The difference in the prices of the different grades is quite big. Consequently, the dearer the seeds the better they should be; otherwise, why pay a higher price for first-grade seeds? What seeds does the science of seed growing regard as first grade, second grade, etc.? It appears that if, for example, one grain of white wheat is found among 200 grains of red wheat, the latter is classed as half of one per cent adulterated, although in external appearance there is no difference between the red and white wheat, except for the colour of the paleae. One grain in 200 is regarded as an admixture of half of one per cent. And the admixture is a masked one. If in 100 years the testing expert finds 2 spikes that are only slightly different from the rest, the purity of the seeds is classed as 98%, and the expert rejects the seeds and prohibits them from being put in the first grade, irrespective of whether the admixture is pernicious or not. *That is all seed grading amounts to.*

Thus, 2% of harmless admixture in a given variety of wheat, which scarcely affects the quality of the grain, at once reduces the price of the seed material by half.

What is at the bottom of this? A sharp reduction in price in the case of harmless adulteration would be understandable if it were to cause a sharp deterioration in the quality of the grain and a reduction in yield.

A simple calculation, however, shows that a slight harmless adulteration cannot result in anything like a serious reduction in yield and deterioration of quality. For example, a wheat is adulterated 2 or 3%, an admixture that only an expert, but not an ordinary agronomist, could detect. This wheat is put in the third grade; but the yield in this case will not be substantially reduced. Take even 5% adulteration. Let us assume that, when sown pure, this admixture will produce 5 c. less than the basic variety. Even in this improbable case, 5% admixture will reduce the yield by 25 kg. per ha. But nobody in the world has ever taken into account a difference in yield of 25 kg. per ha.!

Pure seeds are certainly needed, and it is absolutely necessary to fight for 100% purity of seeds obtained from self-pollinators. But what is meant by purity? Will a variety always be pure if a crop of red wheat seeds does not contain a single spike of white-kernelled wheat? Will Ukrainka always be pure if not a single grain of rye is found in the whole lot in question?

One thing is certain, namely, that the difference in the price of different grades cannot by any means be attributed entirely to external varietal purity.

The difference in the price paid for different grades of seeds is particularly great in the case of cross-pollinated plants. Before the revolution, sugar refiners paid every year 300-400 rubles per pood for the seeds of mother-stock beets, whereas ordinary sugar-beet seeds cost 1.50-2.00 rubles per pood. It would never enter anybody's mind to attribute the price of 300-400 rubles per pood of sugar-beet seeds solely to their external varietal purity. In the majority of cases, ordinary sugar-beet seeds have no admixtures, not even of a single seed of fodder or table beets; and yet they are hundreds of times cheaper than the seeds of the mother stock.

In prerevolutionary times there was also a big difference in the prices of different grades of seeds of self-pollinating plants, as in the case of cereals, for instance. It is enough to leaf the catalogues of seed-growing firms of that time to see that oats and winter wheat sold by seed nurseries were priced up to 10 rubles per pood, whereas ordinary seeds (pure, of course, unadulterated seeds of the same variety) were sold in the market at 80 kopeks to 1 ruble per pood.

Notwithstanding the apparent dearness of the highest-grade seeds, the landlord and kulak farmers, before the revolution, found it profitable to buy them from seed-growing firms and to sow their fields with them. The peasant farmers—middle and poor—did not buy the highest-grade seeds for the simple reason that they could not afford to do so.

It is profitable, in practical farming, periodically to change low-grade for high-grade seeds because this increases yield; and from this point of view the effort to secure high-quality seeds for our planned socialist agriculture is a matter of supreme importance. However, the results that the Party and the Government are demanding in this sphere cannot be achieved in an extremely short time without the aid of science, without a sound theoretical basis.

Does this seed-growing science, on the basis of which seeds continue to be graded according to whether they have only half of one per cent or three per cent of admixtures, guarantee success in this matter? The grading of seeds solely by their external purity or adulteration is far, very far, from enough.

Take, in particular, the Vinnitsa Region, where in many districts nothing but Ukrainka is sown, and no other variety of winter wheat is used. Ukrainka is a self-pollinator. Hence, there is no need to change the Ukrainka seeds! Nevertheless, the Vinnitsa Region has demanded a change.

It is said that this is too much of a luxury. But it is not a luxury, it is an absolute necessity. In highly efficient and rich farms like our state and collective farms, which have no equal in the world, it is unwise to economize on seeds, it is unbusinesslike to do so.

For the purpose of illustration, an example of vegetable seeds can be quoted. Take cabbage No. 1. Capitalist firms often sold the seeds of this variety of cabbage at 1,000 rubles per kg., and a kg. of other seeds, similar in appearance, at 2 rubles. Both kinds were No. 1, with no admixtures of other varieties. Nobody who knew cabbage No. 1 could tell the difference between the two kinds, either by the seeds or by the young plants. As a rule, however, the 2-ruble seeds remained unsold, whereas the 1,000-ruble kind were sold out. Why? Because no commercial cabbage grower would ever stint a kopek on seeds. We may be told that here it is not a matter of kopeks, but of 1,000 rubles per kg. However, only 50 gr. of seeds are needed for 1 ha., and not a kg. Consequently, for 1 ha. it is necessary to spend not 1,000 rubles, but only 50. A hectare yields about 30,000 heads of cabbage. Hence, on every head grown from high-priced seed there is an extra cost of about 1/6 of a kopek. But that head is sure to have the following advantages: it ripens 10 days earlier, is much more compact, etc. It is a true saying: why waste money on cheap seeds? That is why, on the one hand, our farms must not economize on seeds; and on the other hand, seed-growing science must make the most valuable (highest yield) seeds available. At present, however, seed-growing science is concerned mainly with preventing Ukrainka wheat seeds, for example, from being adulterated with the seeds of other varieties.

Still, it remains unclear why it is necessary to change Ukrainka in a district where no other variety of wheat is grown. The possibility of adulteration is excluded, and yet there can be no doubt that it is necessary to change the seeds from time to time. Why?

Let us approach this question from a remoter angle.

Let us start from the question of "pure lines." One would think that we Soviet scientists ought to remember the elementary fact that all things flow and change; and this applies to pure lines, too. It is a fact, however, that many specialists recognize the mutability of a pure line only in the "historical aspect" and, proceeding from this, they draw the following conclusion: a pure line may not be quite pure, but in our age, during our lifetime and that of our children, a pure line remains almost unchanged; at any rate, practically, as far as the eye can judge.

For two years Johannsen measured the length of beans in the endeavour to prove that the pure line remains unchanged for 20-30-40, etc., years. True, de Vries introduced some amendment in this by his discovery of mutations. We, however, are not strict graders, and if a pure line changes 0.1%, it will not cause a change in the yield.

Nevertheless, it seems to us that the question of the mutability of a so-called "pure line" concerns plant breeders and seed growers much more closely than has been imagined up till now.

We have definitely established the fact that a pure line changes, and does so much more quickly than plant breeders and seed growers believe. This was already established by our great Michurin, but the vast majority of our scientists give little thought to the works of Michurin and are therefore convinced to this day that a pure line is practically immutable.

It is an indisputable fact that the majority of our scientists did not and still do not agree that a pure line can change quickly.

We will leave the question of mutations aside. They cannot influence the variability of a pure line to any considerable degree. Let us pass on to weightier facts and arguments.

What variety of any self-pollinating crop grown from seeds in any part of the globe remains unchanged for 40 or 50 years and more?

I do not know of any variety of self-pollinating crop that has remained unchanged for 30 to 50 years in practice, on millions of hectares. (In a collection, of course, it may remain unchanged for as long as you please.)

I asked a number of specialists whether there is such a variety, and all of them said that they did not know of any, or cited examples of the existence of such a variety on practically negligible areas. That being the case, I doubt very much whether many such varieties exist at all in free nature.

The situation is altogether different with cross-pollinators. Take rye, for example. There are varieties of rye that our great-grandfathers used to grow, and they are grown today on many collective farms. This is a fact. Nor can it be disputed that cross-pollinating varieties remain unchanged far longer than self-pollinating varieties. Why don't the seeds of self-pollinators keep long in the districts, and why do practical farmers demand that they be changed?

Why are old varieties less productive than new ones? Why do new arrivals often oust the old ones?

Here, and here alone, lies the reason why our farmers demand a periodical change of seeds. We have already explained the reason for all this, namely, the fact that a pure line changes quickly, more quickly than scientists have imagined up till now.

Here is an example that proves our case.

Last year we sowed the spring wheat Lutescens 062 on experimental plots. Already at that time we had noted the existence of extremes far apart. On the whole, this variety eared uniformly; but there were certain plants that differed sharply from the rest in time of earing. We selected thirty-four spikes of this variety and this year [1935] sowed them simultaneously in separate families. The following phenomena were observed. An initial plant had eared on June 8 in 1934; this year its family eared on June 12. Another initial plant had eared on June 14, 1934; this year its family eared on June 19. Although they were sown at the same time, there was a difference of 7 days. Thus, the difference in the length of the vegetative period of the initial plants was repeated in their progenies.

Is it not clear from this that if we gather the seeds of these extreme families, reproduce them until we get a kilogram, and then sow them, we will get different varieties? But try to find a morphological difference between them! Twenty years ago the shoots of Lutescens 062 were pubescent, and they will be so for scores of years more. It was an awnless variety, and it will remain so. What does it matter that its external appearance has not changed if its more important characters have changed? Let us assume that, in the bulk, earing is delayed 5 days. What does a delay of 5 days mean? It means a difference amounting to centners of crop per hectare. And this is only a change in length of vegetative period. Perhaps changes have also taken place in the suction power of the roots, and in other of the wheat's properties? How is this rapid change in the 062 line to be explained? By mutational variability? That would be a very superficial and frivolous explanation. Mutations do not occur so often.

It will be easier for us to understand the causes of this phenomenon if we turn to the practical work of our plant-breeding institutions. How is a new variety bred, anyhow?

Two forms of wheat are taken, one, say, from Azerbaijan and the other from Odessa Region. These two forms are crossed. One is a semiwinter, awned and glabrous form; the other is a spring, awnless and pubescent form. What will the hybrid be? It will be a pubescent, awnless, spring form.

In F_2 there will be diversity: awned, pubescent, awnless, glabrous, etc., wheats.

Of course, selection can be applied to F_2, but in selecting plants, one cannot be certain that their progeny will be of a constant form.

What does the plant breeder accept as proof that a given form is constant? The fact that the progeny resembles the initial plant. Resembles, but how? Morphologically. If the initial plant was awnless, pubescent, white-kernelled, of low growth, the progeny will also be awnless, pubescent, of low growth, white-kernelled, etc.

In plant-breeding institutions they have forms, and these forms have 24 headings. If the progeny is described the same under the 24 headings as the initial plant, the plant breeder is convinced that it is a constant variety, especially if the same particulars are entered for a period of 5 years.

Nobody will deny that if a wheat segregate is awned, the crop, with few exceptions, will be awned a hundred years to come. If it was bare, glabrous, for 5 years, it will remain so for 100 years. If it was white-kernelled, the entire progeny will be so. As regards the characters, the homozygous nature of which has been established, the wheat is certain to be constant in the vast majority of cases.

But is there a heading in the form for suction power of roots? Is there one for vernalization phase? Is there one for photo phase? No, nor can there be for an incalculable number of other plant characters and properties.

When segregation takes place, does the hereditary basis segregate only according to awned and awnless nature, pubescent and glabrous nature?

It is clear to us that living organisms differ from each other not in ten, but in a much larger number of characters, properties and qualities. Segregation takes place not only according to the characters that are registered in forms, but according to all the differences actually existent in the parent plants. However closely a child may resemble its father or mother, it will be neither the father nor the mother. However closely it may resemble the father in eyes, and nose, and gait, it will not be the father. It will have its *own* gait, eyes and nose and not the father's. In this case the differences will even be external, let alone internal ones

There are no two organisms in the world with any character absolutely alike. Of course, all organisms resemble each other to some degree; but there are no organisms in the world absolutely alike, even brother and sister, brother and brother, grains from the same ear, etc.

The only question is—how important are these differences for practical purposes? No two peas are absolutely alike, but it does not follow from this that in general no two things in the world are alike. It is all a question of degree of resemblance. If the measure we apply is too general, then all living things in the world may be said to resemble each other in that all organisms of the animal and plant worlds consist of cells. Therefore, it is important for us to ascertain and to concentrate attention upon what agricultural practice demands of us, the practice for which science is working. Practical farming is demanding something more subtle than many theoreticians, especially in the department of genetics, think. The state and collective farms are demanding something much more subtle than the most subtle methods employed by the present-day geneticists of the Morgan school. The latter's methods are so crude that practical farmers cannot employ them.

Indeed, what are geneticists and cytologists (lumping genetics and cytology) engaged in? In counting chromosomes; in altering chromosomes by means of various treatments; breaking them into pieces; transplanting pieces of a chromosome from one end to the other; hooking a piece of one chromosome on to another, etc. Is this work of any use in seeking a solution of the practical problems of agriculture? It is of as much use as the work of the wood chopper to the wood carver. To back my statement, to prevent the analogy I draw between breaking up chromosomes and chopping wood from appearing crude, I will cite an example.

In Michurin's orchard I saw two mountain ashes—an ordinary one, and one bred by Michurin. The one bred by Michurin seemed to differ in no way from the ordinary one. The trees looked alike, and so did the size and shape of the leaves. Fruit hung from both; but those on Michurin's tree were larger than those on the ordinary one.

You would have thought that everything relating to the two trees—size, stems, shape of leaves—was alike, and that only the size of the fruits was different. But when you taste the fruit of the ordinary tree you find that it is sour, whereas that of Michurin's is quite eatable.

This is a difference of immense practical importance. In what way do these two mountain ashes differ from the standpoint of genetics? The chromosome apparatus, the entire hereditary basis of these two trees may be almost alike not only in form but in substance. The whole difference between the fruits of the two trees amounts to 1 or 2% of acid and 5% of sugar. That means that the difference between the two trees may lie in one of the minor, even secondary features of the organism. I am sure that it will be impossible to distinguish between these two trees by their chromosomal apparatuses not only now, but even 10 years from now, for the differences here are infinitesimal.

But these *minute* differences in the hereditary basis, which have nothing in common with breaking chromosomes, hooking one part on to another, etc., are very important for practical workers, for those whose duty it is to create new forms of plant organisms.

The methods employed by the geneticists of the Morgan school are too crude.

The opinion is current that if the progeny of a given hybrid plant resembles the initial mother plant, it is a constant form.

If there really is a resemblance I would say that it is a constant form; but the resemblance is there only because the plant breeder looks at only 10 to 20 conspicuous characters. One man, however, may note 20 to 30 characters while another will note a thousand.

Burbank described in his book *The Harvest of the Years* how he made the first selection for the forms he needed. He would walk quickly through a field where millions of plants were growing. To many specialists these plants would look alike; they would not be able to distinguish one from another. To Burbank, however, they were all different. Burbank did not examine these plants very closely; nevertheless, he distinctly saw that they differed from each other, and that they were not what he wanted. Suddenly he would spy a plant he wanted, throw a piece of cloth over it, and his assistant, who was following behind him, would stop at the plant and begin to work on it, while Burbank would hurry on in search of another plant.

To the uninitiated, all plants look alike, appear to resemble each other; but actually, they are different. Hence, it is wrong to judge whether plants are alike or unlike by their conspicuous features.

Plant breeders are in the habit of judging whether a wheat is a constant form or not from 17, or at most 24, characters.

Let us suppose, for example, that F_3 has been accepted as a constant form (although, as we have ascertained, it is far from being so). The variety is reproduced and the grain sent to be tested by the State Cereal Variety Trial Commission. The state and collective farms begin to sow it in their fields. Here the work of the plant breeder ends. The variety has passed into practical use and further work on it is left to the seed growers.

The seed growers watch this variety, become its guides; but they watch only its "body," not its "soul." They see to it that it is kept clean, that there should not be half of one per cent of white-kernelled wheat among the red and no red wheat among white. *As for its "soul," the genotype, the hereditary basis—this does not interest the seed growers at all. It does not seem to be part of their business.*

A seed grower is deemed to have committed a heinous crime if he has failed to detect in a variety of wheat released by a plant breeder for practical use a 4 or 5% admixture of rye or barley. It is indeed a gross offence, one that can be detected even by a layman, let alone by an expert. Up till now, however, attention has been concentrated *exclusively* on this aspect, but this is far from enough.

We are not opposed to purity of varieties. On the contrary, the adulteration of wheat with rye, red wheat with white, etc., must not be tolerated. An agronomist who does so ought to be dismissed, and so should the field brigade leader who sowed such seed. *It goes without saying that seeds must be 100% pure.* It must be emphasized, however, that this is not the *main* thing the agronomist-seed grower has to see to.

It is the duty of the seed grower not only to see to the "cleanliness of the body," not only to see that white wheat is not adulterated with red, and vice versa, but also to see to the purity of the hereditary basis. But seed growers have paid no attention whatever to the latter.

A hybrid variety which plant breeders call constant, and which, indeed, does not change its jacket for a number of generations, will actually be continuously segregating. This segregation can never end; it will go on infinitely. The more the variety is reproduced the more will it segregate. The rate of segregation is another thing. The older the generation the more homozygous the plants; but the wheat will never become completely homozygous. This, in my opinion, is indisputable.

But we may be asked: what does it matter if the variety never becomes completely homozygous? For practical purposes it is homozygous.

Actually, it is homozygous for practical purposes only as regards those characters for which the plant breeder bred it; but as regards the characters which did not interest the plant breeder it may not be homozygous but heterozygous.

Some may think this is not of practical importance. Let the variety segregate as much as it likes in all succeeding generations, the jacket will keep nevertheless. According to all the headings in the form, the variety will remain unchanged. As for the other characters, invisible to the eye, invisible even to the testing expert, let them segregate. The expert will not reject such a variety.

Is this important for practical farming or not? Here we come to the gist of the matter.

How many years does a self-pollinating variety live in ordinary agricultural practice? Forty to fifty years? We have already established that few cases are known of a self-pollinating variety that occupies a large practical farming area living for more than fifty years. That is why we have expressed the opinion that such a variety is rare in practical farming. We shall now give the explanation of this.

In our opinion, the majority of self-pollinating varieties are not long-lived in practical farming only because they segregate. Even a variety chosen by the plant breeder from hybrid material in the fourth, fifth or sixth generation and, when reproduced and tested, found to produce a yield from 30 to 40% higher than the other varieties in the given district, retains, after 10 or 15 years, only those of its characters that the expert needs to fill up his forms. As for its hereditary basis, it has changed considerably as a result of segregation.

The practical farmer, however, is not interested in the question as to whether the variety changes or not. He says to the seed grower: "Give us seeds that will increase yields, and if you cannot increase yields, at least see to it that the given variety does not deteriorate; and under no circumstances give us seeds that will reduce yield."

The practical farmer is not interested in whether a variety segregates or not. He has one urgent demand—better seeds in place of bad seeds. But what does the seed grower give the collective and state farms? Wheat seeds cleansed of rye and barley, red wheat cleansed of white, awned wheat cleansed of awnless. But the seed grower has no idea that the hereditary basis has changed during these years. *The principal task of the seed grower is to watch the hereditary basis of the seeds of the given variety that has to be preserved in practical farming, but this task has been lost sight of by the seed-growing organizations.*

But perhaps in practical farming a variety always changes for the *better*? Is there not such a thing as natural selection? Natural selection eliminates all the freakish, frail and feeble segregates. Natural selection is not only a sieve that sifts all unviable organisms. *Natural selection is at the same time a builder, a creator.* It creates and sifts. I shall not stop to explain why we associate ourselves with that group of biologists who appraise Darwinian natural selection precisely in this way.

It is obvious that a variety grown in the Odessa district, for example, undergoes other changes than the same variety grown in the Saratov district, etc. A given variety grown in the Odessa and in the Saratov districts will, after a few years, be different in both places. But nature has not pledged herself to work in the way we want her to. A variety does not always change for the better, not always for our benefit. This too is clear.

Who, in such a complex matter as the creation of new varieties, imposed upon nature the duty of working in the way preferred by a given, concrete society? It is for this very reason that Ivan Vladimirovich Michurin launched the slogan: "We cannot wait for favours from Nature; we must wrest them from her." We, however, abandon varieties to their fate. Seed growers, those whose duty it is to be the guardians of a variety's hereditary basis, unwittingly impoverish it. How else can you explain the fact that 50 to 70 years ago, valuable export wheat varieties like Ulka, Girka and Garnovka were extensively grown in the steppes of the Black Sea area, and later, even better lines were obtained from them by means of individual selection (for example, the Odessa-selected Girka 0274 and 0180, and Melanopus 0122), and that now you will not find a single hectare sown to these varieties of spring wheat in this area? This wheat is not grown not because people do not want to grow it, but because it began to grow worse than barley. This is to be explained first of all by the fact that agricultural science has failed to come up to the mark in matters concerning seed growing. Our agricultural science has been unable to advise practical seed growers how to proceed in order not only to prevent the varieties produced by plant breeders from deteriorating, but also to improve them. We must not rest our hopes on the fact that all things in nature are in a state of flux, and that all changes will drift in the direction we want them to. We ourselves must direct this movement into the desired channel.

In the vast majority of cases, the segregation of agricultural field crops reduces the practical value of the hereditary basis. To back this statement we will touch upon the question of inbreeding. Let us discuss the inbreeding of rye, sunflower and maize.

Maize is a cross-pollinator. As long as it remains so, it continues to be a good variety. All the plants are of uniform height, vigorous and produce good yields. If, however, we prevent the egg cells of a maize plant from uniting with the sex cells of other plants and compel them to be fertilized by their own pollen, the hereditary basis of the selfed cross-pollinators will segregate considerably already in the first generation. The yield will drop at once, the plant will lose height and its viability will be reduced. As the result of this operation, we will get a half-dead organism. If we try to fertilize it again with its own pollen we will only do it more harm. Only very few can survive 10 or 11 selfings; the rest perish. Such organisms prove to be unviable, and in such cases, geneticists say: "the lethal genes have got into a homozygous state."

In my opinion, the attributive "lethal" explains nothing. The reason is entirely different. The more homozygous the hereditary basis becomes, the less adapted the organism is to fluctuating conditions. Every organism of a normally cross-pollinated plant consists of the hereditary basis of its father and mother. This hereditary basis contains all the potentialities for adaptation of both the father and the mother. Thus, if the paternal gamete contained the capacity for adaptation to a rainy year and the maternal gamete contained the capacity for adaptation to a drought year, the heterozygous plant will acquire both capacities for adaptation. Whether the spring is dry or rainy, the heterozygous plant will not perish, because it possesses the capacity for adaptation to either of these conditions.

The inbreeding of any cross-pollinator will always lead to the biological impoverishment of the hereditary basis, and thereby to a reduction of the capacity for biological adaptivity. Often, it is enough to inbreed a cross-pollinator, i.e., to make it homozygous, for the best of the inbred plants to be unable, under field conditions in different years, to compete with the worst of the outbred plants.

At our Institute in Odessa we have this year an ordinary variety of sunflower, No. 101, growing side by side with inbred specimens. This No. 101

is not extensively grown in practical farming, but compared with the inbred specimens it is a positive giant.

Is there a single variety in the world, including maize, that has been produced by inbreeding and cultivated on extensive areas in practical farming? And inbreeding has been practised for a long time; in America, the inbreeding of maize has been going on for over twenty years.

Not a single variety of any crop whatever has been produced by inbreeding. Plachek, a plant breeder at the Saratov station, is very good at breeding sunflowers. In a short space of time she bred the variety 169 by ordinary selection of the best heads from the local population. This form has long been regarded as the best standard variety for many districts of the U.S.S.R. But quite some time ago, Plachek took up inbreeding, and to this day she has had enormous achievements in inbreeding to her credit. Nevertheless, not a single variety produced by inbreeding, or by crossing inbred lines, has been released for use in practical farming. People have spent many years and even decades on inbreeding. In my opinion they have wasted their time, because no good can come from the squandering of the hereditary basis that inbreeding involves.

There are enthusiasts of inbreeding who say: "Our aim is not to produce a variety by inbreeding, but only to purge the hereditary basis of undesirable genes." As if there were desirable and undesirable genes "in general," regardless of conditions of development. In the present case, what we mean by conditions of development of individual genes is not only external conditions, but the gene environment as one whole. Indeed, there are bad and good genotypes only under given conditions. (This is dealt with fairly fully in the booklet I have written in collaboration with Professor I. I. Prezent.)[1]

Examples are often cited from the work of the research stations of the State Sugar Trust. There, they have a large number of isolators, glass cases, and so forth. The work was started as far back as 1925; in spite of this, however, their inbred beets are far less viable than the outbred ones, and their yield is lower than that of any ordinary sugar beet.

Let us now go back a little to the field crop variety of hybrid origin which the plant breeder has released for practical farming. As I have already shown, in the present state of plant breeding and seed growing, this variety (in the majority of cases) is gradually losing its yield qualities. This variety, spring wheat for example, is deteriorating because wheat is a self-pollinating plant and, consequently, it naturally *inbreeds*, becomes *homozygous*. Even if it is conceded that homozygosity does not in every case necessarily lead to the economic deterioration of a variety (which is quite possible), the conversion of a variety released by the plant breeder into a homozygote will nevertheless lead to rapid change, i.e., to the disappearance of the old "pure line" and to the creation of a new type of population.

---

[1] In this volume, pp. 65-105.

I may be asked: "How do inbreeding and self-pollination figure here?" The point is that inbreeding is the *artificial* self-pollination of cross-pollinators, whereas self-pollinators undergo *natural inbreeding.*

That cross-pollination is useful for a self-pollinator is a fact of which one can very easily become convinced. I think that our work has fairly convincingly proved that F_1 (which means the product of two crossed parents and not of self-pollination) can never be later in flowering than the earlier parent. F_1 of a self-pollinator, for example, wheat, barley, oats, or any other plant, can never be less vigorous at the time of flowering than the more vigorous parent. The objections that are raised against this proposition spring from the elementary methodological error of comparing parents and segregates which grow in different years, or of failing to take into account the specific development of certain characters.

Thus, as a rule, the cross-pollination of self-pollinators, far from being harmful, can be useful, i.e., can increase the organism's viability. Cross-pollination never reduces the given organism's potentialities of development whereas, as we have ascertained, self-pollination sometimes does not even preserve the previously existing potentialities. On the contrary, self-pollination always reduces these potentialities, because the process of segregation continues not only in the fourth generation (from which the plant breeder already selects the initial forms), but even in the forty-fourth generation. This is how I explain the fact that self-pollinating varieties are not long-lived in practical farming. This does not mean that they cannot live long. They can live long, but they need the help of seed growers. Seed growers have not helped them, plant breeders have also been lacking in knowledge in this matter, and geneticists have never given it a thought.

Up till now, all of us have been in the habit of thinking that it is easier for a self-pollinator to retain its morphology than it is for a cross-pollinator. This view is based on the fact that if a self-pollinating plant is selected and reproduced, all its progeny, year after year, will outwardly resemble the initial plant, and that if we select a cross-pollinator and reproduce it, all its progeny will resemble the initial plant in the first year of reproduction, whereas in the next generation there will be great diversity. This, no doubt, explains the view, widely held among plant breeders, that plant breeding from cross-pollinators is much more difficult than from self-pollinators. No matter where you sow a self-pollinator you will always get plants that will in genotype, at any rate, resemble the maternal form, whereas, as is known, two cross-pollinating varieties must never be sown side by side, because if they are, they will cross-pollinate each other and the progeny will differ from the initial forms.

What is the explanation of this variability in the cross-pollinator in the second year of sowing after the selection of the initial plant, even when there is no alien variety, no alien pollen, in the vicinity?

The explanation is that cross-pollinators are more heterozygous, whereas the isolated progeny is forced to become homozygous.

All this has given plant breeders grounds for saying that it is easier to work with self-pollinators than with cross-pollinators. But plant breeding is an inseparable part of seed growing. From this followed the other proposition, namely, that the seed grower's work with cross-pollinators is very much harder than with self-pollinators.

To many people this was an elementary truth. I held this view myself for many years, but now I have arrived at the opposite view. I assert that it is far easier for seed growers to work with cross-pollinators than with self-pollinators. Practical seed growers have been able to maintain the seeds of cross-pollinating field-crop varieties at the proper level much better than they have maintained the seeds of many varieties of self-pollinators.

Let us examine the following facts. Cross-pollinators, rye, for example, keep in a district for decades without a periodical change of seeds and, in the main, their quality does not deteriorate. Self-pollinators, however, for example, wheat, spring or winter, cannot keep long on farms without a change of seed. We could not easily find, even for the purpose of illustration, a self-pollinating variety that has kept in practical farming for more than fifty years. Hence, the seeds of self-pollinators have been found harder to keep in practical farming than the seeds of cross-pollinators.

If, however, we confined ourselves to this example it might be concluded that this is accidental. I may be told that varieties of wheat disappeared from the scene in a given district only because plant breeders had produced new and better varieties which had rightly ousted the old ones; that the cross-pollinating varieties keep for a long time only because there are none with better yields capable of ousting the old variety of, say, rye.

On the question of whether it is easier for the plant breeder to work with cross- or with self-pollinators, the following argument may be raised against me: Let us say that in the wheat population of a given district there are large-spiked and small-spiked, red-kernelled and white-kernelled forms. We select the white-kernelled forms, reproduce them, sow them, and all the progeny will be white, exactly like the initial, selected plants. If this progeny is worth attention because of its yield and quality, it can be used for the fields. But with a cross-pollinator like rye, for example, the matter is much more complicated. In the second generation it will have a diverse progeny, and not a single plant will be like the selected initial one. Try to breed a cross-pollinating variety under these circumstances! From this my opponents will draw the logical conclusion that cross-pollinating varieties keep in the districts because it is not so easy to breed a new variety and, consequently, there is nothing that will oust the old varieties.

But we, in turn, have a right to put the following question: If a cross-pollinator produces a diverse progeny and not a single plant is like the initial one, how is it that the given variety keeps for decades, and even for half a century, and all the ears are alike? If we examine a rye crop on a hundred hectares of land we will find that all the plants in that crop are alike. More than that, year after year, the plants of one year resemble those of the preceding year. This shows that a variety can sustain itself and can keep constant for a long time. But as soon as we handle this variety unskilfully, as soon as we select 1 or 10 ears from a field of cross-pollinating plants and reproduce the progeny, we get plants that are unlike each other, and not one of the plants is like the one that we took as the initial form.

A cross-pollinator is morphologically more constant than a self-pollinator. Suppose we cross a self-pollinator and also a cross-pollinator and sow the heterozygous progeny in the field each year without performing any further artificial crossing or selection. Which will have the morphologically more uniform crop in 5 or 10 generations, the self-pollinator or the cross-pollinator? Everybody who knows anything about plant breeding will say that the cross-pollinator will have the morphologically more uniform crop. Why? Suppose we take for the crossing of a self-pollinator and of a cross-pollinator a red awnless pubescent parent and a white awned glabrous parent. All the plants in F_1 of the self-pollinator will be alike; the plants in F_1 of the cross-pollinator will not be quite uniform. Now let us see what these plants will be 10 years hence. The self-pollinator will have white-awned, red-awned pubescent, red-awned glabrous, etc., spikes, i.e., all the combinations imaginable of the features in which the parents differed. The cross-pollinator, however, will have only red spikes; all the plants will be awnless, and all will be pubescent. Consequently, which of them is best able to retain its purity?

Up till now the opinion has prevailed in agricultural science that self-pollinators are more constant than cross-pollinators and, consequently, are better able to retain their purity.

One can agree with this only in cases where the cross-pollinators are handled unskilfully: when free cross-pollination is restricted by some means or other, or when the pollen of one variety of cross-pollinator is adulterated with the pollen of another variety. Under normal conditions of free pollination, however, if there is a sufficient number of cross-pollinating plants, the latter will be much more uniform.

We must bear in mind that the segregation of the self-pollinator following reduction division affects not only the potentialities of developing red colour, awnedness and awnlessness, but other characters too.

If a heterozygous wheat, being a self-pollinator (a hybrid variety of a self-pollinator is hardly ever homozygous), segregates in respect to awnedness or awnlessness, then, consequently, it also segregates in respect to all the other properties, characters and qualities. A cross-pollinator, however, given free pollination, can, in the main, retain a uniform jacket year after year, i.e., can also retain all its other properties.

I am sure that if some of our self-pollinators, wheat, for example, were cross-pollinators, we would already have new varieties, better than those we have now. I must confess that I almost went to extremes. I wanted to find wheats in the world collection that flower more openly. Wheat varieties differ very widely in type of flowering. One variety flowers closed, another flowers less closed. I wanted to take a pair that would more or less resemble a cross-pollinator and breed a variety from it. It was only recently that I arrived at the conclusion that any wheat can be given the advantages of a cross-pollinator while all the advantages of the wheat being a self-pollinator can be utilized at the same time. The glumes and paleae of wheat are isolators which prevent the penetration of foreign pollen. At the proper time, before the anthers ripen, the wheat flowers must be castrated and later, when the stigma ripens, the paleta must be pushed aside with tweezers, and pollen gathered from many spikes of the same wheat variety must be inserted.

To pollinate a self-pollinating plant like a cross-pollinator we gather pollen from one hundred to two hundred plants of the same variety (the more the better), mix it, push aside the lodicule of the wheat flower and drop several thousand pollen grains on the stigma. Let the stigma take any gamete it pleases. Having done this, we can leave the field with our minds at rest. We have done our part. We have given the egg cell the opportunity to choose whomever it pleases. Professor Prezent has aptly called this form of pollination "marriage for love." Self-pollination is forced marriage, not marriage for love. However much the given egg cell wants to "marry" the "young man" growing about five inches away from it, it cannot do so because the palet is closed and prevents foreign pollen from entering.

In the case of cross-pollinators, it is indeed "marriage for love." Pollen dust flies about in the air in myriads, and all the gametes are different. Can any gamete satisfy a given egg cell with equal result? According to Morgan, perhaps the first gamete that drops on to the egg cell will satisfy it, but it is not so according to Darwin and Michurin, not by a long shot. There is not a single process, not a single cell, not a single gamete that does not make its own demands upon conditions. As Professor Prezent has formulated it—biological requirements are the reverse side of adaptation. And adaptation is the result of natural selection. Consequently, according to Darwin and Michurin, "marriage for love" is not accidental; it is adaptation, selection on the part of the plant itself. A pollen grain is selected that, under the given conditions, is better adapted for a given egg cell, whereas another pollen grain proves, under the same conditions, to be better adapted to another egg cell. The pollen grains are as diverse as the egg cells. It must not be understood, however, that the air is filled only with "good" and "bad" gametes "in general," and that only the "good" gametes can fertilize an egg cell. There is no such thing as good and bad gametes "in general," there are no such zygotes; and, "in general," there is no bad hereditary basis independent of conditions of development. Therefore, only an ignoramus will attempt to decide whether a given hereditary basis is good or bad without taking the conditions of development into account. There are neither good nor bad hereditary bases independent of conditions of development.

In breeding wheat, for example, plant breeders are often heard to say that they have bred good varieties of spring wheat from F_4, or F_5, etc., but have not yet reproduced them. In appearance, this new variety of theirs does indeed compare favourably with the district standard variety. Since it compares favourably in appearance, you may be sure that its yield will be higher by 3 to 4 c. per ha. On the experimental plot, a difference of one centner per hectare would be imperceptible.

A plant breeder can demonstrate every year really good varieties on small experimental plots. And there ought to be new numbers every year, because the old numbers should already have been reproduced on large areas of land. But it turns out that they have not even reached the state variety-testing stage. At best, they may have reached the State Cereal Variety Trial Commission, only to be rejected there.

On the experimental plot, the variety was a good one, but before it reached the State Cereal Variety Trial Commission it segregated as a consequence of repeated selfing.

How can a self-pollinating variety bred by hybridization be prevented from deteriorating under the conditions of practical farming?

If what is called a constant form of wheat economically valuable to us is found in F_3, F_5 or F_6, I think it is more than possible to prevent it from deteriorating. Let us suppose that an experimental plot of constants was obtained from, say, F_5. Pollen must be taken from as large a number of plants as possible and used to pollinate a few spikes of this same wheat variety. By means of such cross-pollination we not only eliminate the regular annual impoverishment of the hereditary basis, but increase its heterozygosity. Thereby we rejuvenate the variety and send it back to approximately the heterozygous state it was in when the plant breeder demonstrated it on his small experimental plot.

How long will it take to rejuvenate the hereditary basis of the seeds of the best of the old wheat varieties and of the newly released varieties of hybrid origin to be sown on the vast collective- and state-farm fields? This can be calculated without any particular difficulty.

Let us start with 10 crossed spikes, which should give us not less than 100 grains. If we sow these 100 grains in the same year we will get 2,000 grains in the autumn. These we shall sow in a greenhouse in March. In the spring, as soon as the shoots of the spring crops appear in the fields, we shall transplant the young wheat plants to an open experimental plot. In July, the plot will give us no less than 2.5 c. of grain. From our experience in this matter we know that in one year we shall get 2.5 c. of grain from the 10 crossed spikes. In our opinion, that will be the highest and therefore the most valuable category of seed.

This is so simple and at the same time so very important that it ought to be done not only at plant-breeding and seed-growing institutions, but also at collective-farm laboratories.

Collective-farm laboratories can quite easily rejuvenate the seeds of the varieties grown by the given collective farm. All that is needed for this is 5 pots for sowing 100 grains, a pair of tweezers and, perhaps, a brush for pollination.

If this is done, within one year the collective-farm laboratory will have 2.5 c. of rejuvenated seed with which to sow 15 ha. (1 pood to the hectare), and the collective farm will obtain a crop of 15 tons (1 ton per hectare). I am giving the most moderate figures. At the ordinary rate of sowing, the 15 tons of seed should suffice for 150 ha. Thus, in the third year, the old seeds will be ousted by rejuvenated seeds.

Is this possible? It is, provided the collective-farm laboratories receive skilful guidance from the seed-growing and plant-breeding organizations. Seed growers should take this matter in hand; and since seed growing is inseparable from plant breeding, plant breeders must also participate in this work.

It is hard to say how often the hereditary basis of seeds must be rejuvenated. Perhaps it is useful to do so every five or seven years. It depends on the crop, on the district, and on a number of other circumstances. Practical experience must decide this.

It should be stated in conclusion that immense opportunities for improving the work of seed growing present themselves to us. If the assumptions we have made in this paper are correct, we shall be able by infusing "new blood" into cross-pollinators to do the following: firstly, not only maintain the varieties of hybrid origin the plant breeders release for the seed-growing organizations at the level at which they were at the time they left the station, but in many cases improve them;

secondly—a possibility which is not excluded—make a number of the old self-pollinating varieties actually new, more fruitful and more stable than they are now by creating greater heterozygosity (by means of the artificial intravarietal crossing of plants).

A number of self-pollinating varieties are simply begging for the tweezers, brush and scissors. For instance, why not cross different lines of the Humbert tomato? There must now be no less than ten lines of this variety. Why not collect the pollen and cross-pollinate the Humberts? Perhaps this will result in an entirely new and better sort of Humbert.

We must also try to cross peas and beans.

Today we are more than ever confronted with the task of rallying the largest possible number of collective-farm laboratories to fight for the cause of seed growing. In this very important matter the collective-farm laboratories are indispensable. We must not be daunted by the fact that crossing is complicated work. Is it not true that the ordinary working woman at our experimental stations are the best at performing the technical operations of crossing? It is not at all difficult to teach one or two persons on each collective farm to do this work.

It is quite possible that the method of improving the quality of seeds which I propose will considerably increase the yield on collective- and state-farm fields. That being the case, we must not, of course, wait until the state and collective farms are supplied with improved seeds by the seed farms of the plant-breeding stations; but in this great task the plant-breeding stations with their seed farms must certainly play a leading part.

REPLY TO DEBATE

(ABRIDGED)

This session of the Lenin Academy of Agricultural Sciences of the U.S.S.R. has been of great benefit to all its participants, and to me especially. I have learned much that is of use to me and that will benefit the work I am directing from the speeches delivered by the comrades who opposed the fundamental propositions I advanced in my paper and in our recently published works.

An investigator must be persevering and persistent in his work. At the same time he must be able to see beyond the limits of the particular subject he is studying, otherwise he will not see the wood for the trees. Investigators in agronomics who are unable properly to link the subjects they are studying with the other departments of agrobiological science are not fully competent and independent investigators. This does not mean that the work of such people is useless. It is useful, but only within a definite system, when the workers engaged in it are assigned their tasks according to a definite plan.

Many of the speakers here stated that I underrate science, in other words, that I underrate theory. That is not true. I appreciate the value of science no less than any of the comrades in this hall. In the Soviet Union the value of science in general, and of agronomic science in particular, is appreciated far more than in capitalist countries. In our country, scientists, academicians in particular, are judged on their merits. The Academy plays an enormously important role. It must be the leading body, the top rung of agronomic science.

Agricultural science is of such great importance that it must not under any circumstances be trifled with. To underrate the role of theory in our era means not being a Soviet investigator. To talk about seeds, the ordinary seeds that are released for sowing on state- and collective-farm fields, and not to touch upon the fundamental principles of plant breeding and genetics means failure to understand what theory is, means letting practice grope in the dark, divorced from theory. That is why, in discussing questions of distinctly practical importance in my paper, I continuously referred to the theoretical problems of genetics and plant breeding.

Genetics is rightly called the theoretical foundation of plant breeding. Logically, therefore, it should also be the theoretical foundation of seed growing. When speaking of seeds, one must touch upon all the links in the chain of the given department of agronomic science, from top to bottom—from genetics to plant breeding, and then go on to seeds for the state and collective farms.

I shall begin by answering the arguments the geneticists raised against my paper.

The chief, the central argument of the geneticists was levelled against the following propositions that we advanced: 1) segregates from any heterozygous form cannot be earlier ripening than the initial heterozygous form; 2) the first hybrid generation cannot, in the main, be later ripening than the early parent. Dr. Lepin cited facts from his own work which appeared to contradict these propositions. He said that in crossings of Prelyudka wheat (the earliest-ripening wheat) with an early Siberian wheat the length of the vegetative period of the first hybrid generation was intermediate. In proof of this he quoted the number of days from sowing to earing: for one of the parents it was 38 days; for the other—43 days; for the hybrid—40 days. From these facts Lepin drew the conclusion that the first hybrid generation did not take after the early parent, but was intermediate. I think that the explanation is simpler: 38 days and 40 days amount to the same thing. All the hundreds of plants growing on an experimental plot never ear on the same day. In my opinion, the example T. K. Lepin gave here is not an argument against the proposition we advance that F_1 cannot be later ripening than the early parent. In advancing this proposition we have in mind not a gap of 2 days between early and late, but a real gap that exceeds the range of variation.

On the basis of the same facts, T. K. Lepin objected to another proposition we advanced, namely, that the segregates of any heterozygote cannot be earlier ripening than the initial heterozygote. He asserted that in F_3 of the given combination a plant appeared from which a wheat was subsequently raised that was named Extra-prelyud. This variety of wheat is known only to a small group of geneticists and specialists who are working on the world collection of wheats. We have it also in our collection at our Institute.

According to Dr. Lepin, this wheat ripens eight days earlier than the world standard early ripener Prelyudka (one of the parents). Since the vegetative period of the first generation was the same as that of the Prelyudka, the succeeding generations were 8 days earlier in ripening than the first heterozygous generation.

On the face of it, this argument appears to be a weighty one. None of us, including myself, would venture to attribute a gap of 8 days in the ripening of 2 wheats to variation. In the case in question, however, we are evidently dealing with a wrong method of comparing phenological observations, which often happens.

In this case, this applies particularly to the ordinary records of phenological observations made by geneticists and plant breeders who, as far as I know, have never even thought it necessary to compare the lengths of vegetative periods of different hybrid generations of the same combination *when sown simultaneously.* Only now, when arguing against the propositions we advance, have our geneticist opponents turned to literature and to the field records in search of "facts" to confirm the view which has come to be established in formal genetics that a heterozygote can give rise to segregates with any length of vegetative period. As, however, no special experiments on this problem have been made anywhere (except at our Institute), the geneticists are obliged, willy-nilly, to operate with facts that do not fit, such as, for example, the case cited by T. K. Lepin, where lengths of vegetative periods in F_1 and F_3 of the same combination were compared in the same locality, but for *different years.*

Today, during the recess between the morning and evening sittings, Academician N. I. Vavilov inspected, on our hybrid cotton field, plants of over 150 first generation combinations which had been sown simultaneously with their parents. I myself have examined all the records of the phenological observations regarding the appearance of the first flower buds on the plants in this experiment. The vast majority of F_1 plants of all these 150 combinations budded earlier than the early parent. The minority of F_1 plants of all these combinations budded simultaneously with the early parent. And not a single combination contradicts the proposition we advance.

Thus, we have obtained the additional confirmation of the soundness of our proposition, repeatedly tested by experiments, that F_1 cannot, in the main, be later in flowering than the early parent. It can be earlier; and if a phasic analysis is made before crossing, it is possible to foresee the length of the vegetative period of F_1 in comparison with the early parent.

I think that in the case he himself witnessed N. I. Vavilov will agree with me.

Nevertheless, in his speech at this session, N. I. Vavilov stated that on making a special study of publications throughout the world on the problem of dominance in vegetative periods he found the following: in some cases early ripening dominates in the first hybrid generation, in some cases the length of the period of vegetation of the hybrid generation is intermediate, and in all cases of rice, according to the records of a Japanese investigator, early ripening is a recessive character. The first hybrid generation takes after the late-ripening and not the early-ripening parent. Unfortunately, the Japanese records of the hybridization of rice could not include the 10 rice combinations obtained by one of our assistants last year and sown this year together with their parents. Today, each one of you can convince himself by an inspection of the living plants, that the first hybrid generation plants of all the 10 combinations without exception are ahead of, or keeping pace with, the early-ripening parent.

How can we explain the discrepancy between the facts indicated in the literature on rice and the facts shown by plants growing in our experimental field? It is easily explained by the circumstance that the facts cited by investigators in the books they publish frequently are unsupported and disconnected from the actual conditions under which they were obtained. The facts given in scientific literature must not be ignored, of course; but they cannot be used blindly as a basis without first analyzing the specific conditions to which they apply.

You have all seen that the whole of the vast living material at our Institute of Selection and Genetics which has been demonstrated to you, fully confirms the correctness of our theoretical propositions. On the other hand, my opponents have quoted facts gleaned from literature that contradict our theory.

The impression may be gained that our propositions hold only for the conditions prevailing in the Odessa district, that they do not express the general law of development of the hereditary foundation of plant forms. Fortunately, we know of "facts" also in Odessa which seem to contradict our propositions. T. K. Lepin cited these facts in his support. From this one might, of course, draw the conclusion that I withheld these facts from the participants in this session and that they were brought to light by other comrades, workers at our Institute, in private conversation and during the excursions. The sole reason I did not mention these contradictory facts was that I was sure that they "do not exist," just as they never actually did exist except in the head of the experimenter. I have in mind the first hybrid generation of wheat that eared later than the early-ripening parent in an experiment conducted by L. P. Maximchuk.

In July, last year, L. P. Maximchuk sowed the first hybrid generation of wheat in boxes. The parental forms were also sown in boxes. To prevent infection by Hessian and frit flies the boxes with the plants were moved from place to place; they were kept near the greenhouse, in the greenhouse, and in the open. In the autumn it was found that the first generation lagged behind the early-ripening parent in earing. Already at that time I told Maximchuk that this was probably due to the fact that the parents and hybrids were not being grown under equal conditions, and that this could be easily tested in the same combination in the second generation. If earlier forms, or forms as early as the early-ripening parent, were segregated, my explanation of this would be proved correct. Such segregates were actually found in the crop which Maximchuk himself raised.

Plants grown in the greenhouse in the winter also gave rise in F_3 to segregates that were several days earlier ripening than in F_2. But anybody who knows anything about greenhouse conditions in the winter will understand that even a pure-line wheat sown in one box counting 100 plants may prolong its earing for one or two weeks.

To convince T. K. Lepin and the others who jumped at the "facts" communicated to them by L. P. Maximchuk that they are fictions, I must quote another example from Maximchuk's work showing how he himself regards these "facts."

How is it to be explained that Maximchuk, in his own practical plant-breeding work, and also in the work of the postgraduate who is working under his direction, employs 100% the methods we have elaborated and the correctness of which is disputed by many geneticists? One is bound to conclude that Maximchuk is actually guided by methods, the theoretical principles of which he, judging from his remarks, has doubts about.

But we cannot under any circumstances separate theory from practice. We cannot recommend fallacious theoretical propositions for use in practice; on the other hand, propositions that are correct in practical farming cannot be entirely rejected by, divorced from, science.

N. I. Vavilov said in his speech that he was 90% in agreement with our propositions and subscribed to the methods of plant breeding that we suggested. He thoroughly and with convincing logic refuted the theory of inbreeding, but at the same time stated that he accepted only 90% of all this. But it is precisely the remaining 10% that prevented him from thoroughly deciphering the theoretical fallacy of inbreeding. Failing to do this, he cannot accept our proposition concerning the seed growing of self-pollinators. When I say that the theory of inbreeding must be thoroughly deciphered I do not mean that we must cast aside all isolators, dismiss all the people who are engaged in inbreeding and throw away all inbred seeds.

Far from it.

Not long ago, last winter, I insisted that Baransky (a sunflower specialist) should not sow inbred lines. It was already clear to me then that the breeding of sunflower could not be conducted by the inbreeding method. Baransky at that time agreed with me on many points regarding the problem of inbreeding, just as N. I. Vavilov does now. But for all that he insisted that the inbred material which last year occupied (and occupies this year) the major part of the sunflower breeding area (including hybrids of inbred lines) should also be sown in 1935. The weightiest argument he advanced was that in this inbred material there were many lines which (according to the data of past years) were resistant to broom rape. Knowing how enormously important resistance to broom rape is for sunflower, I could not forbid him to sow this material. True, my mind could not grasp why inbred lines, biologically enfeebled, degenerates, should become more resistant to broom rape. Facts, however, cannot always be disproved to the satisfaction of others or even one's own by mere argument. But what did we see yesterday when we inspected the sunflower breeding ground? Not a single inbred line that was "resistant to broom rape" before sowing showed the slightest sign of such resistance after sowing. In the majority of these "resistant lines" the broom rape grew higher than the sunflowers themselves. Next to them are the outbred Zhdanov varieties grown in the North Caucasus. These are powerfully developed and practically resistant to broom rape.

I would like to know why inbred lines that were "resistant" to broom rape before sowing turned out to be absolutely nonresistant after sowing.

I have cited this example only in order to show once again how difficult it is to argue against the bare facts that were brought forward by some of the speakers during the debate on my paper.

During the debate many of the speakers agreed with our proposition that inbreeding was useless as a method of breeding varieties from cross-pollinators. They, however, emphasized that it is possible by inbreeding easily and quickly to breed forms resistant to various diseases, and then to use these resistant although biologically enfeebled forms in hybridization. I still cannot imagine how it is possible by inbreeding to increase resistance to disease. Who has ever heard of a biologically enfeebled organism being more resistant to disease than a healthy, unenfeebled organism?

But even if we picture to ourselves a case where the more a plant is biologically enfeebled the more resistant it is to some diseases, what follows from it? In such a case it will indeed be possible to breed resistant forms by means of inbreeding. Let us suppose that some investigators have already succeeded in breeding such forms. In the given example, however, these forms are resistant only because they are biologically feeble. The question is, what will become of this resistance when the given inbred form is crossed and the viability of the organism is immensely increased? With the increase in the new hybrid organism's viability the resistance linked with the feebleness of the organism will certainly diminish. This defence of the theory of inbreeding is unconvincing.

More convincing is the example cited by N. I. Vavilov of the breeding of maize in America. As you know, America is the home of the theory of inbreeding. In the main, the inbreeding method was worked out on maize. N. I. Vavilov said that the Americans are a practical people and do not waste money. In defence of inbreeding Academician Vavilov said that in America about 5% of the maize area is sown with hybrid varieties obtained by crossing inbred lines.

I cannot see where in this case the practical sense of the Americans lies. Is it in that they employed this "good business," i. e., the theory of inbreeding, in the practical sowing of maize on only 5% of the maize area, or in that they sowed 95% of the maize area with varieties obtained not by inbreeding, not even by individual selection, nor even by linear selection, but by the ordinary method of mass selection, i.e., a method that is the very opposite of inbreeding?

Moreover, I have not found in scientific literature (perhaps I am not acquainted with all of it) confirmation of the statement that 5% represents the maize area in America sown with hybrid varieties obtained by crossing inbred lines. And so, N. I. Vavilov is quite right in saying that the Americans are a practical people.

The statement made by some of those who took part in the debate that it is necessary to employ *all* methods in our work, old and new, and particularly to combine the old with the new, is wrong, in my opinion. To achieve practical aims we must certainly employ *various* methods, but not *every* method. If it is evident to an investigator that the theoretical principles of the method he is employing are wrong, why should he continue to employ that method?

I always advise my colleagues and assistants not to start a single experiment if they have no idea what will come of it, if they do not know the purpose of the experiment.

The debate on my paper, although useful and productive, did not proceed along the line, in the direction that it should have.

The pivotal point of my paper, which should have been subjected to critical discussion, was the question of how the seed grower in his practical work should keep seeds at the level they were on when the plant breeder released them for production purposes.

This problem was only partly touched upon in the speeches of N. I. Vavilov, V. Y. Yuriev and a few others, and then only from the aspect of the longevity of pure lines. Therefore, I will first answer the substance of these speeches.

My thesis is that in practical farming not a single self-pollinating variety keeps for long on large areas. Some of the speakers cited examples showing that there are such varieties in practical farming. For instance, V. Y. Yuriev stated that he knew of a variety that was grown on hundreds and thousands of hectares that had kept for no less than fifty years.

He said that variety No. 120 was grown in practical farming on hundreds of thousands of hectares. But this variety was brought into being, i.e., was bred by V. Y. Yuriev, not fifty years ago, and not even thirty years ago. If you want to call this variety Poltavka, then I don't understand why variety No. 120, which was bred from Poltavka, figures in the achievements of the Kharkov Plant-Breeding Station. It is evident to all the comrades in this hall that a variety bred from Poltavka is not Poltavka itself.

V. Y. Yuriev quotes the example of Kochin's Arnautka, an old local variety. True, it is an old local variety, and a good one, but I don't understand why it now occupies such an insignificant area in practical farming.

N. I. Vavilov cited an example from practical farming abroad of varieties keeping for long periods of time.

All these examples only go to show that different varieties are stable in different degrees—one variety may keep in practical farming for a longer period; another variety may remain unchanged for a shorter period. The majority of self-pollinators, however—this is absolutely clear to me now—cannot keep unchanged in practical farming for a long period of time—the varieties inevitably undergo a change. This is also proved by the examples cited by my opponents who spoke here. It turns out that it is not so easy to quote examples of the longevity of self-pollinating varieties in practical farming, and not so many have been quoted.

I touched upon the transience, the mutability of pure lines only because I wanted to focus your attention upon another problem, less controversial for you, but extremely important for our practical seed-growing work.

All plant breeders release, or intend to release, from their stations varieties of hybrid origin for use in practical farming. A pure line which had undergone self-pollination for scores and, perhaps, hundreds of years before it was singled out by the plant breeder, is found, after one spike has been singled out and reproduced, to have undergone some degree of change in the relatively short period of twenty or thirty years. What happens, and will happen, in ten or fifteen years' time, to a so-called constant variety selected by the plant breeder in the fourth, sixth or seventh hybrid generation, reproduced and put to variety tests which show that the yield of this variety is 20 to 40% higher than that of the old one?

In five, at most ten years' time, this hybrid variety, after leaving the plant-breeding station, will have segregated to such a degree that even if it does resemble the variety released by the plant breeder, the likeness will be only a rough morphological one. In the overwhelming majority of cases, not a trace will be left of the 20 or 40% higher yield, because of which this variety was released for practical farming purposes.

That is the problem which, in my opinion, should have been discussed at this session. This problem involves not only the hybrid varieties created by our method, and which you saw yesterday at our place—although the idea of revising the theoretical principles of seed growing arose in my mind because of my concern for the seeds of the new varieties we have raised. I feared, and still fear, that the fine condition of these varieties revealed on the experimental plots may give way to a bad condition when they are sown on large areas, as has been the case with many other varieties in plant-breeding practice.

V. Y. Yuriev, and other opponents, while arguing against the proposition advanced in my paper concerning the relative transience of pure lines, said not a word about the method we propose for maintaining released hybrid varieties at the proper level.

In my paper I also touched upon the question of the stability of Johannsen's so-called "pure lines," and on the question of inbreeding, and hence, of course, also on the theoretical principles of the present-day science of genetics, for the sole purpose of enlisting the joint efforts of this authoritative assembly to find the theoretical propositions on the basis of which we could in our practical work not only prevent the deterioration of the hybrid variety the plant breeder releases, but also, as far as possible, improve it.

It seems to me that up till now the work of the seed grower has been confined to keeping the variety's jacket clean. Our seed-growing science has paid no attention to the variety's hereditary basis.

I cannot possibly agree that genetics has no relation to seed growing. When speaking on such a distinctly practical subject as seed growing, I could not avoid touching upon the theoretical principles of the Johannsen and Morgan school. I found it impossible to speak about practical seed growing without revealing the theoretical roots of this problem.

THE INTRAVARIETAL CROSSING OF SELF-POLLINATING PLANTS[1]

I READ my first paper on this subject—the intravarietal crossing of self-pollinating field-crop plants—in Odessa, on June 26, 1935, at a session of the Grain Department of the Lenin Academy of Agricultural Sciences of the U.S.S.R. held in that city. A rather vigorous struggle has been raging in science around the problem raised. In this struggle, we, the staff of the Institute of Selection and Genetics, who raised this problem, have been advocating during the past twelve months the extensive organization of experimental tests on state and collective farms. On the other hand, the vast majority of geneticists have been opposing our proposal in rather acrimonious terms.

The history of the question is as follows. About four years ago I set to work at the Institute of Selection and Genetics to breed a variety of spring wheat for the Odessa Region. The methods we chose for breeding this variety of spring wheat were different from those usually employed in the science of genetics and plant breeding. We planned to produce the new variety in three years, counting from the day of sowing the parent forms for crossing, and we said so.

At the present time we have three new varieties of spring wheat, the breeding of which we regard as finished. They have already been reproduced to some extent. In our barns we have no less than fifteen tons of seeds of these three varieties. They have gone through a two-year field variety test. In 1935, they were sown for the variety test on small experimental plots, and this year, 1936, they were sown on experimental plots of normal size for the field variety test at the Institute of Selection and Genetics, as well as for a regional variety test at the Odessa Regional Station (in the Vygoda station district).

The general indices of yield and behaviour of the plants of these varieties have convinced us that we did our work successfully. The varieties are not worse than we anticipated when we began to breed them. Accord-

[1] Revised version of paper read at the session of the Grain Department of the Lenin Academy of Agricultural Sciences of the U.S.S.R., held in Omsk, August 1936.—*Ed.*

ing to the data of the variety test made at the Odessa Regional Station (Vygoda), our varieties have taken first place for yield, and according to the data of the variety test made by the Grain Department of our Institute, they come into the first-class group for yield. It must be borne in mind, however, that the varieties we have bred (1163, 1055, 1160 and especially 1163) developed this year practically without rain, whereas other varieties undergoing the same test enjoyed the most favourable conditions during the last 15 days before wax ripeness. On June 22 heavy rain fell. At that time variety 1163 was already passing to wax ripeness. On July 1 the crop of this variety was already in stacks, whereas the other varieties were reaped 8 to 12 days later. No evidence of blind seed was observed under Odessa conditions this year, but conditions were exceptionally favourable for grain ripening of late varieties. The following fact, for instance, may be taken as proof of this: all the spring sowings of the vernalized wheats of our winter varieties, which had never filled well in previous years under Odessa conditions (owing to their late ripening), yielded 20 and more centners per hectare this year. Even the experimental spring sowings of unvernalized winter varieties like Novokrymka 0204, Kooperatorka and Stepnyachka, which usually do not ear when sown at this date, this year eared normally although late (15 to 20 days later than the spring varieties), and produced good yields.

The data on yield of the 1936 variety test have convinced us still more that the new varieties of spring wheat that we have bred fully warrant being reproduced quickly and included in the spring crop of Odessa Region.

In watching the development of our new varieties, we, already in 1935, were struck by their good behaviour. While still in the initial phases of development they compared favourably with the best of other spring wheats. I myself, as one of the principal authors of this work (of breeding varieties according to plan in an unprecedentedly short space of time), was worried by the question as to whether these varieties would keep, whether they would deteriorate from what they were when I saw them in the variety test plots in 1935. I examined this question from the most diverse aspects. I took the following into particular consideration: nearly every plant breeder has, in small plots in nurseries or in variety test plots, some varieties that compare favourably with the best of the district varieties. And yet, many plant breeders are rarely able to release a variety for production. The necessity of solving this riddle gave me no rest. All the time I was tormented by the thought that this may happen to our new varieties of spring wheat. Everybody knows of many varieties that behaved excellently on the experimental plot but, for unknown reasons, proved useless when released for production.

In June 1935 I solved this problem, at least to my own satisfaction. We found an explanation for the cause of the deterioration of varieties which had shown good indices when on the plant breeder's small experimental plots and had turned out to be bad after reproduction and dispatch to the farms in the districts.

We also discovered the method of eliminating this undesirable phenomenon.

Usually, the discrepancy between the indices of new varieties on the experimental plots at plant-breeding stations and those of the state variety test, and also of sowing for consumption, has been attributed to flaws in the experiment, in other words, to errors committed during the tests on the small experimental plots at plant-breeding stations. I, however, arrived at the conclusion that the deterioration of new varieties after they have left the plant-breeding station is often due to the degeneration of their inherent genetic state.

I can cite again some of the examples I gave in the paper I read last year. How is it to be explained that approximately thirty years ago, in Kherson Gubernia, spring wheat was grown on an area of about 800,000 ha., whereas at the present time, in Odessa Region, which occupies almost the same territory as the former Kherson Gubernia, spring wheat is grown on an area of only 60,000 ha.? Moreover, the varieties Girka, Ulka and Kubanka, which were formerly grown on hundreds of thousands of hectares, are now not grown on a single hectare in this region.

When these varieties are sown for the variety test at experimental stations they show an extremely low yield. It is their low yield that explains their disappearance. Another interesting fact is that usually not in a single district do self-pollinating field crops sown in large areas keep for more than thirty or forty years.

One of the weightiest arguments raised against me on this question last year by many scientists in the field of agronomy, and by geneticists in particular, was that self-pollinating varieties do not keep in the districts for more than forty years for the sole reason that new varieties with higher yields appear. I do not, of course, for a single moment dispute the proposition that a new variety introduced into a district must have a higher yield than the one that is ousted; otherwise the new variety would not be accepted in the district. Neither I nor anybody else will, of course, dispute that. *I only advanced the proposition—and stand by it now—that the only reason why new varieties are often better than the old is that the old varieties of self-pollinators have deteriorated owing to prolonged cultivation, to prolonged self-pollination.*

If we could now have back the seeds of Girka, Ulka and Kubanka possessing the genotypical quality which characterized these varieties thirty or forty years ago, it is doubtful whether many of the present spring varieties would be better than the restored varieties.

Exceptionally rapid is the genotypical deterioration of many young varieties of hybrid origin. The vast majority of these varieties show good results in the first preliminary tests at the station, but they lose them in the succeeding generations. As a result, these varieties do not even reach the state variety test stage.

Everybody who has observed the behaviour of hybrid plants will have seen without much difficulty that in the nursery one may fairly often find among the first hybrid generation combinations that are much better than the variety the plant breeder would like to release for production purposes.

This year, 1936, we observe in the fields of the Odessa Institute of Selection and Genetics that out of 700 combinations of the first hybrid generation of cotton plants, scores of plots (combinations) can be pointed to which are much *better not only than their parents*, but even than the variety our plant breeders, including myself, hope to produce.

Every plant breeder knows that a good first hybrid generation is not in itself a guarantee that a good variety will be produced from this combination. We are not yet able to cultivate hybrids in such a way that their progeny should be exactly like the plants from which the seeds were taken. Usually, the progeny of hybrids are unlike the parent plants. The older the hybrid generation the less the percentage of plants that resemble the first generation of the given combination. Cases often occur when it is quite easy to find plants that are good in all points among the first hybrid generation (of different combinations of parents), but it becomes increasingly difficult to find such in succeeding generations. This partly explains why many plant breeders, after working for decades, have never bred a variety that could be used for production purposes, although they have always had, and now have, excellent material in their fields, and particularly in their plant-breeding nurseries.

On these grounds, and on many others that I have not mentioned here, I arrived at the conclusion that *one of the fundamental problems of plant breeding is the fixation of the genotype*—to prevent the small quantities of the new variety that plant breeders possess from deteriorating in succeeding generations.

What are the causes of the frequent deterioration of self-pollinating varieties after prolonged cultivation? How is it to be explained that the older the generation of a hybrid combination that was good in the first generation, the worse it usually becomes for economic purposes? I think that it is impossible to explain this fact on the accepted genetic grounds. Moreover, to some scientists our proposition that varieties deteriorate as a result of prolonged self-pollination seems absurd. Nevertheless, facts are stubborn things. In many cases self-pollinating varieties vanish after prolonged cultivation only because they have degenerated, because their yield has dropped. This also applies to the fact that the best plants usually appear in F_1 and then deteriorate more and more in each succeeding generation. This phenomenon cannot be explained from the standpoint taken by the geneticists, and consequently, they cannot find ways and means of combating it.

True, it will be nothing new to me if the geneticists, who now dispute the propositions I advance, will, in the future, when they have been confirmed and their value demonstrated in practice, try to explain this phenomenon as being in line with their theory of the corpuscular nature of the hereditary substance. They can do this with exceptional ease in relation to varieties of hybrid origin. They attribute it to quite ordinary segregation. I, however, do not think that it can be attributed to segregation.

In general, the concepts which the geneticists include in the term "segregation" can never serve to explain anything. By segregation they mean merely the mechanical disjunction, separation, of ready-made corpuscles, genes, which always, from the zygote to reduction division, lie opposite each other in homologous chromosomes.

To us, who regard the living organism in its development, i.e., in its changes and transformations, the unsoundness of the concepts which the geneticists include in the term "segregation" is absolutely clear.

I will not dwell on this question any further, but pass on to explain the reasons why, in our opinion, changes, often deterioration, may take place in self-pollinating varieties of both hybrid and nonhybrid origin. We will not touch here upon the question, familiar to everyone, of the deterioration of self-pollinators as a result of mechanical admixtures. Let us begin with the simplest and most comprehensible phenomena.

Every variety of plant requires for its development its own environmental conditions, which in some degree differ from the conditions required by the plants of another variety. Moreover, at different periods of its development the given organism requires different external environmental conditions. At one and the same time, but for the development of different organs of the given plant organism (for example, leaves, roots), different environmental conditions are required. Some scientists, for example, M. M. Zavadovsky and others, dislike our expression: "the organism requires conditions." But the likes and dislikes of this or that investigator do not alter the fact that for its life and development a fish does require water as its environment, and not any kind of water—some fish require river water, others require sea water. Who does not know that for its successful development cotton requires a higher temperature than wheat? Some plants require marshland conditions, others not only do not require such conditions, but cannot stand them and perish if exposed to such conditions.

Everybody also knows why, if plants are to develop from them, cotton seeds require conditions different from those required by the seeds of other plants, wheat, for example. The entire evolutionary process of formation of cotton was different from that of wheat. It goes without saying, therefore, that at the present time cotton requires its own specific conditions, and wheat its specific conditions. It would be surprising indeed if their requirements were the same.

Geneticists, whose concepts to this day do not differ fundamentally from those of Weismann, do not regard sex sells as genuine particles of the organism's body. It is a known fact, however, that the sex cells of plants develop from nonsex cells. There are no sex cells either in the seed germs or in the green grass of wheat. Everybody knows that a plant which has developed from the seed embryos into grass, and then into stems and spikes, develops sex organs and sex cells only in the spikes. Hence, sex cells develop from nonsex cells, i.e., from the organism's body. Thus, at the time of their development, the sex cells are a part of the organism's body.

Every organ, every cell of a given organism, has its own specific features. For example, the cells of a pig's bristles and the same pig's fat or flesh cells are different. Sex cells also have their own specific features, differing from other cells, but this does not contradict the fact that they are genuine particles of the organism's body.

Usually, plants require for their development approximately the same conditions as were required by the preceding generation of plants of the same variety. Seeds gathered from winter wheat will produce winter plants; seeds gathered from spring varieties will produce spring plants. This, with rare exceptions, is always observed everywhere in practical farming. It is so frequent and usual that many laymen, not plant breeders, do not even know that sometimes seeds gathered from spring plants which had flowers of a definite colour may, when sown, produce plants with flowers of a different colour. An infinite number of examples of this kind could be cited. They are to be observed particularly either in unisexual organisms, the male and female sexes of which are represented by separate individuals, or in the hybrid plants of bisexual organisms. In hybrids it is observed that in external appearance and behaviour the progeny, or individual tillers of the progeny, differ from the plant from which the seeds were gathered for sowing. For example, all the hybrid plants obtained from the crossing of winter with spring wheat will be of spring habit in the first generation, whereas the progeny of these spring hybrids will contain both the winter and spring form. In this case, the winter plants obtained from the seeds of the spring hybrid plant will differ from the plant from which the seeds were taken as well as from the sister spring plants.

How did the organism of the spring wheat develop seeds (grains) from which not spring but winter plants developed, and why do these plants require for their development conditions different from those required by the preceding initial plant which produced these seeds?

The development of every new organism always starts from the cells of the preceding plant. Consequently, one would think that the root-cause of the difference in the behaviour of the plants should be sought in the initial cell (and if it is a sex cell, then in that—the zygote). Geneticists, however, represent the numerous properties possessed by the living cell that is capable of development as being different substances, corpuscles, organic molecules. They hold that these particles (genes) of heredity lie in the chromosomes of the cell, and that it is the presence of certain particles (genes) in the chromosomes that explains why some plants turn out to be of winter and others of spring habit, in spite of the fact that all the plants were obtained from seeds that had developed in the same spike of a spring wheat plant.

It cannot be disputed that these seeds differed from one another, for this is proved by the different behaviour of the plants that were grown from them (winter and spring).

There is nothing surprising in the fact that plants in the progeny resemble either the father or the mother. But why are seeds formed from which plants develop that resemble neither the father nor the mother, i.e., the plants from which the seeds were gathered, but resemble their more distant kin?

The explanation that geneticists, the representatives of the Morgan and Weismann trends, give of the case we are examining (when, for example, part of the progeny obtained from the spike of a spring wheat plant is of winter habit and part is of spring habit) is that hidden within the initial spring plant were granules of substance (which they call genes) which facilitate the development of only the property of winter habit, and that, in addition to this substance, there lie in another chromosome granules of another substance, other genes, which facilitate the development in plants of only the property of spring habit. Since experience shows that this is a spring plant, they claim that the granules of substance which determine spring habit repress their partners—the granules of substance which determine winter habit.

Cytologists, who examine the external appearance of the contents of the cell nucleus at different periods of its life cycle, have established that the number of chromosomes that pass into the gamete (sex cell) is only half of that observed in ordinary, nonsex cells. From this the geneticists have drawn the conclusion that the winter-habit genes and spring-habit genes do not lie in the same chromosome, i. e., not along the length of the chromosome, but in different, yet necessarily adjacent, i. e., in homologous, chromosomes. As a rule, these two homologous chromosomes pass singly into different sex cells. From this it was easy for them to arrive at the conclusion that if one chromosome passes into one cell and the other into another cell, the repressed, recessive winter-habit gene is freed from the repression of the dominant spring-habit gene. If such a male sex cell (gamete) has the good fortune to meet a female sex cell which also possesses a winter-habit gene, the fusion of these two cells will result in the formation of a seed, in the cell nuclei of which there will be no repressing dominant spring-habit genes. In that case, the winter-habit genes will be able really to manifest themselves, i.e., will produce winter plants. However, when a gamete possessing winter-habit genes meets another that possesses not winter-habit but spring-habit genes, the fusion of such sex cells will result in the formation of a zygote, and later of a seed in each cell nucleus of which, in definite chromosomes, two antagonists will lie, one opposite the other. One of them, namely, the spring-habit gene, will not prevent the winter-habit gene from living and reproducing, but it will not allow it to assert itself—to influence the development of the wheat plant. This is how the geneticists explain the behaviour of a plant of the spring form. In every cell of this plant, however, in definite chromosomes, there are granules of winter-habit substance, or more correctly, of the substance that controls, influences, the development of the property of winter habit. The geneticists call such plants heterozygotes. The recessive genes receive freedom of action only once, when the sex cells mature. At that time, in the opinion of the geneticists, there takes place a mechanical distribution of immutable genes, i. e., corpuscles of a special hereditary substance that are always present in every cell of an organism.

This is how the geneticists answer the rather difficult and at first sight incomprehensible question as to why a spike of a spring wheat plant (F_1 from a cross of winter and spring forms) yields seeds that produce both spring and winter plants. This is also their explanation of all the other differences among hybrid plants. As we see, their line of argument is a simple one. Since the progeny has such and such a quality, such and such characters or properties, winter habit, for example, and these organisms came from given parents (represented by one plant), therefore this property (winter habit) must have existed in the parents—in the case we are examining, a spring wheat plant. The geneticists are not in the least disturbed by the fact that nobody, not even the geneticist experimenter himself, has detected this property (winter habit) in the spring plant that was investigated. They draw the conclusion that this property of winter habit is present in the form of granules, only it does not manifest itself either externally, to the eye, or by means of any other analysis of this plant. They even say that they know the place in the cell where this special hereditary substance (genes) is situated.

On this question, even I, violently opposed as I am to the Morgan and Weismann corpuscular genetic conception, sincerely sympathize with many of these comrades. These people have got muddled on a rather simple question—the conception of development. They reason as follows: if a given winter plant came from a given seed, then that seed must have been a winter variety; and since it came from a spring plant, then there must have been granules of winter-habit substance (genes) within it (in its "core"), namely, in definite parts of the chromosomes of each cell. This winter-habit substance was repressed by the spring-habit substance and, as a result, it did not manifest itself in the external appearance of the plant. If this is not so, they argue, how is the presence of winter seeds in the spike of a spring plant to be explained? The winter seed was obtained from the fusion of the two sex cells of the given spring plant, consequently, these cells must also have been of the winter form. But these two sex cells of the given spring wheat plant were obtained by division from other cells of the same spring wheat, consequently, they too must have contained the winter-habit substance which, during division, freed itself from the spring-habit substance. Continuing their line of argument they reached the point of saying that the substance of winter habit or of any other property of the organism lies in all the organism's cells.

As we know, the main efforts of "eminent advanced geneticists" are at the present time being directed towards making the granules of this substance which they call genes visible, towards drawing and describing it. We, however, say that we cannot picture the extremely diverse properties and potentialities of the developing plant organism, or of its sex cells, in the form of separate, diverse particles of heredity.

We know that the sex cell is a speck of matter that can be seen and even weighed. It is specific in each individual case. The sex cell of wheat is a wheat sex cell, that of cotton is a cotton sex cell, etc. This living cell possesses the capacity to develop into an adult organism. As it takes nutriment the organism, beginning from the fertilized cell, continuously changes, becomes transformed, develops more and more new characters, organs, properties and qualities. We say "new," because literally only a few days previously the particular individual could not have had these characters, organs and properties in such a form.

The majority of our geneticists fail to understand that development is the means by which the organism builds itself up out of food, i. e., out of nonliving matter. The development of the future organism is always relatively determined by the initial living substance. In sexual reproduction, the fertilized sex cell is the initial substance. From a hen's egg a hen or a cock will develop; from a duck's egg a duck will develop. But since every organism builds itself up out of food, it is easy to conceive that two organisms having identical starting points, for example, two wheat plants obtained from the same variety of seed but developing under different conditions and deriving nourishment from relatively different food, will inevitably be different in external appearance, size and quality. One wheat plant may be dozens of times smaller than the other. There will, of course, be a difference not only in the size and quality of the grain yield, but in all the organs and characters. The contents of any cell of one of these plants will in some way differ from the contents of the analogous cell of another plant.

The differences between these two organisms as a whole and those of the individual cells of the given organisms will not only not resemble the differences in the food they imbibe or in other external conditions, but (and this must be particularly emphasized) will be fundamentally other than the differences in the external environmental conditions that they have utilized.

We must not for a moment forget this aspect of the question. The living organism builds up its body, its organs, out of its food, i.e., by assimilating, qualitatively working up, this food in its own way, characteristic only of the given organism. At the same time we see at every step that two plants of the same type, wheat, for example, may differ very much from each other, depending on the environmental conditions they utilized during their development. As I have already said, the difference will be not only in external appearance, but also in the contents of the cells. Who does not know that pigs of the same breed, if fed on different food, barley or maize, for example, yield different qualities of meat and fat? Young animals reared on different food, under different conditions, develop not only different qualities of meat and fat, but also a different bone structure. Knowing that every organism works up, assimilates and builds up its body in its own particular way out of any food it absorbs we must not forget that within the framework, the limits of this "way," different (in quantity and in quality) cells, organs and characters may develop.

Ever more new characters, properties and qualities, including new cell contents, appear in the developing organism that arises from the zygote. These contents are, as a rule, unlike the substance of the cell from which the new cells developed. The substance of the old cells from which the new ones developed may not even have contained some of the chemical elements of these new substances. These chemical elements appeared in the new cells as a result of the assimilation of food from other cells, and the latter from other cells again, and so forth, down to the suction of mineral substances by the plant's roots and the absorption of carbon derived from the air by its leaves.

From all this it is not difficult to arrive at the conclusion that the organism and, consequently, its individual cells (including the sex cells) do not contain that specific substance which the geneticists call hereditary substance. The organism, or the initial sex cell, is itself the hereditary basis of the future organism; it does not contain a separate heredity substance, it is itself heredity.

Many cells of the plant organism can, under suitable conditions, be transformed by assimilation into a relatively different form, but, of course, not into any form. For example, from a given cell or group of cells a plant may develop different leaves, but certainly not any kind of leaves. If, under unfavourable conditions, a narrow leaf develops, nobody will search in this narrow leaf for the latent form of a broad leaf, although everybody knows that a broad leaf might have developed from the given group of cells, just as a semibroad, a narrow, or other leaf might have developed. Always, only one concrete leaf can develop out of the many possible ones. It is useless to look in the already developed leaf for latent particles, or organic molecules, of other leaves that might have developed. Only that which has developed really exists in the form of matter. The same can, of course, be said about other organs and characters, including sex cells. Any organ of any organism can develop in a relatively different way, and this applies also to the sex cells. But the fact that they can develop does not mean that they already exist. *The main thing—which must not be lost sight of for a moment—is that relatively different, but by no means any kind of new cells, can develop from a given cell.* The entire relative diversity of developmental potentialities is determined by history, by the entire course of preceding life. Preceding development is the basis, the foundation of future development. Neither distant nor immediate future development must be separated from preceding development. But the future development of a given species or variety of plants may be different, depending upon the concrete course of this development.

One of the vital differences between sex cells and all other, ordinary, nonsex cells is that sex cells do not continue the individual life of a plant. They start life anew.

The potentialities of a plant's development from a given fertilized sex cell are limited by the course of development of the preceding generations. The plant organism does not usually exceed these predetermined limits, but in a given period of its development it may closely approach the extremes of its requirement norm, and this may often cause a change in the same direction in the environmental requirements of the succeeding generation.

Knowing the concrete development of, for example, makhorka tobacco of a definite size and quality of leaf, it is possible to direct that development so as to obtain only a definite kind of leaf, although the leaves could have been different had not man watched and intervened in this process.

Among the methods of obtaining definite, well-developed fruit or other organs of plants are the pruning of fruit-tree branches, the pinching of axil shoots from tomatoes, the topping of cotton plants, and other agrotechnical operations. If we also knew the concrete conditions of development of certain of the many possible sex cells, it would be possible to obtain definite and uniform progeny from hybrid plants, without any diversity. It is a mistake to think that F_2 of a pea plant, one of the parents of which bore red flowers and the other white flowers, must unfailingly bear one white flower for every three red flowers. The whole progeny may bear red, white or mixed flowers, i. e., some red and some white.

None of these possibilities is precluded; all three cases are quite possible. They are predetermined by the preceding parent forms. Every cell of an organism that is capable of developing sex cells can, in this case, concretely develop only one of two or more possibles. Everything depends upon the concrete course of development of the sex cells, or of those cells from which sex cells are formed. In this sense, the development of sex cells of one or another, but possible, quality does not differ fundamentally from the examples we have cited of the development of the makhorka leaves, or the development of any other organ or character of a plant. If man does not intervene in the proper way, if the proper conditions are not created at the proper time, the particular qualities of particular organs, and of the sex cells too, will depend upon casual environmental conditions.

Heterogeneity of the developmental potentialities of a plant organism and heterogeneity of environmental conditions cause heterogeneity in the various homologous organs and characters of the given plant, including heterogeneity of its sex cells.

Thus, our argument brings us to the proposition that it is possible, by suitable training, by creating relatively definite conditions at definite periods of development of a heterozygous plant, to obtain sex cells and seeds of a relatively definite and uniform kind. This is one of the most important propositions in present-day genetics, and one in which, in my opinion, great confusion reigns.

It seems to me that the fact that in hybrids, as a rule, different sex cells are formed in one given plant, or even in one boll or one spike, can now be fairly easily explained on the basis of our foregoing argument on plant development. This explanation differs fundamentally from the one accepted by present-day geneticists.

Our explanation contradicts the persistent assumption of geneticists that cells contain some kind of specific hereditary substance separate and apart from the substance of the cell. There are no particles of winter habit in the cells of spring plants; there are no particles of white-flowerness in red-flower plants; there are no particles of narrow-leafness in the cells of broad leaves, etc.

By segregation, geneticists mean the separation, disjunction (at the time of maturation of the sex cells) of the chromosomes containing different particles—organic molecules—which determine the development of definite characters in the organisms. In the opinion of the geneticists, all the cells of hybrid (heterozygous) organisms always contain two sorts of these substances, or genes. For example, in red-flower plants of F_1 (from a crossing of red-flower with white-flower peas) the genes of red-flowerness are always accompanied by particles (genes) of white-flowerness. This, we repeat, is the only way the geneticists can figure out how a red-flower hybrid plant can give rise to white-flower plants in the progeny.

The geneticists who adhere to the corpuscular theory cannot understand that in each concrete case only one of the numerous possibilities of development becomes a reality.

Only those (of all the possible) processes of transformation, of change, take place for which the given conditions of the external environment are most suitable. The developmental needs of each cell, of each organ, of a given organism can be satisfied by relatively different conditions of the external environment, but they must be its own specific conditions. We emphasize that by external conditions we mean all that is assimilated by the given assimilating form, i.e., by the given cell.

When we know what external conditions are required for the concrete development of a given organ or character, the leaf of a makhorka plant, for example, and when we create these conditions by means of the suitable agronomic methods, we can always steer the development of this organ in the direction we desire. We are unable as yet to control these conditions, including the food assimilated at any given moment by those of the organism's cells upon which the development of the sex cells directly or indirectly depends. All these cells, of course, elect from the available chance diversity of conditions only those which are most suitable for them. The different cells out of which the sex cells develop find themselves, however, in relatively different conditions, which explains the diversity of the sex cells of heterozygous plants.

This also explains why, in normal crosses between strains, we usually observe in the second hybrid generation a diversity of forms of individual characters and properties in many (but not in all) cases in the proportion of 3:1.

Indeed, if, when absorbing, assimilating, one kind of food, one set of conditions, the given cells that directly or indirectly take part in the formation—development—of sex cells can turn, change, in the direction of producing sex cells of a given quality, then, when absorbing another kind of food (which they are also able to absorb), they turn, change, in a direction that will produce sex cells of a different quality. If under field conditions one variety of pea produces white flowers and another variety produces red flowers, it shows that this environment contains the conditions required by both these varieties. The organism develops many sex cells, but none of them is the product of another. On the contrary, various nonsex cells take a direct or indirect part in the formation of every sex cell. These cells developed in relatively different ways. Their food was largely determined fortuitously, and was controlled only by the limitations, the elective power, of the organism and these cells themselves.

All this once again confirms our assumption that in all the possible variations which the development of the organs, characters and sex cells of a given organism may assume, the factor that always determines the concrete result of the individual development, including the concrete result of the development of sex cells of relatively different quality, is the external conditions that are assimilated by the organism.

Thus we explain that the production of gametes of a definite quality, from the fusion of which, under given conditions, a white-flower plant will develop, is the result of a concrete course of development. From this it is clear that the production of gametes which give rise to red-flower plants is also due to the relatively different development of the organism's cells from which these sex cells were produced.

The conception which reduces the properties of a living organism to particles, even though lying in the chromosomes, is not only fallacious; it is a hindrance to practical action. This conception prevents us from picturing the changes and transformations of the plant world, i.e., evolution, and hence it prevents us from working out methods of deliberately breeding the plant forms we want.

On the other hand, the propositions we advance on the basis of our theory of development create, in our opinion, real possibilities of directing the individual development of an organism, the producing of definite organs and characters (and not just any kind from among those that are possible) by creating definite environmental conditions for the given plant at definite periods of its development. For example, winter seeds sown in the spring can produce only grass. If, however, we create the necessary conditions (vernalize the sowing material), these plants will develop straw and spikes. It is already well known that by this means we can unfailingly induce any winter plant, if sown in the spring, to alter the course of development customary for it during this season of sowing. Similarly, we think it possible to produce relatively definite sex cells (and not a chance kind from among those that are possible) in any particular heterozygous plant organism. For this it is only necessary to know what conditions are required to produce the best sex cells (of those possible) for our purpose.

This, in our opinion, is the direction in which a real and effective science of genetics ought to proceed. It is not necessary to wait until we have completely mastered a particular subject. It is always not only possible but necessary to improve our practical activities on the basis of the knowledge already acquired. This is a most important source to draw upon for further theoretical elaborations, generalizations and prognoses. One of the forms of practical activity of this kind is the fairly extensive work we are doing in the intravarietal crossing of self-pollinating plants on the grounds of our Institute and on collective and state farms.

Our theory of plant development is still far from perfect. Nevertheless, it is more correct and effective than the knowledge of the laws of plant development possessed by many of those who call themselves geneticists.

Even the opponents of our theory now accept, for example, the method we have proposed of choosing pairs for crossing with the object of creating new forms with a predetermined length of vegetative period under the conditions of a given district. The trouble with the geneticists is that they accept only the method, but refuse to think about, or brush aside, the theory on which this method is based. To this day a number of scientists flatly reject, without advancing any scientific reason, the fundamental law we have discovered governing the development of hybrids as regards length of vegetative period.

Daily observation more and more confirms Michurin's proposition that hybrid plants always develop in the direction for which the environmental conditions are best suited. As a rule, hybrids possess all the potentialities of development of either of the parents, but they scarcely ever take completely after the father or the mother. A hybrid is an integral organism. It contains no division into paternal and maternal potentialities of development. It possesses all these potentialities, but it develops in the directions for which the given environmental conditions are best suited.

An incalculable number of processes, transformations and changes take place within an organism. Some processes are parallel (take place simultaneously); others occur at different times. Some of them are more interconnected, others less. Every process is the result of the interconnection of other processes. All these exceedingly complex interconnections are the realization of the plant organism's possibilities of development from the seed to the ripening of the new seeds of annual seed plants. Hence, the length of the vegetative period of the first hybrid generation will always be determined by those of the given organism's potentialities of development for which the environmental conditions are best suited.

We have, therefore, arrived at the conclusion that:

1. The first hybrid generation, in the main, is earlier flowering than either parent, or as early as the earlier parent.

2. Segregates, i.e., the progeny of heterozygous self-pollinators, are, in the main, not earlier-ripening than the initial heterozygote under the same conditions.

When we discovered this regularity we did not assert that there never have been, nor ever will be, cases of hybrid plants that are later-flowering than not only the early, but also the late-flowering parent. We merely asserted, and assert now, that the general regularity we have revealed exists and operates in all cases. There will, however, be individual cases in which it does not manifest itself outwardly in the form of early flowering, because for some reason or other the reproductive organs do not develop in each individual case or develop late. In all cases when the hybrid is later-ripening than the early-ripening parent, we must look for the cause knowing beforehand that in other cases, or in the same hybrids but under other conditions, the particular cause may not exist. I will mention by way of illustration the controversy I had on this subject with cotton growers over the late ripening of first-generation cotton hybrids obtained from the crossing of early-ripening varieties with perennials, which do not flower in the first year. Several years ago I asserted that such first-generation plants fail to flower, or flower late, in the first year under ordinary field cultivation not because they cannot go through the photo phase like their perennial parents under the long-day conditions of our districts, but because these hybrid plants display excessive vegetative growth. They go through the photo phase like their early-ripening parent, but fail to bud and flower because of this excessive vegetative growth.

This year my assumption was fully confirmed in the fields of our Institute. We have in the fields the first generations of scores of the combinations mentioned. It was enough to top half the number of plants in each combination, and thereby check their rapid growth, for the plants of all these combinations to flower almost simultaneously with the early-ripening parent. The plants that were not topped, and which, consequently, grew too luxuriantly, did not flower at all, or did so 20 to 30 days later than the early-ripening parent.

This one simple example is enough to prove that the regularity we have discovered actually operates. It is a fact that at the plant-breeding stations in Uzbekistan and Transcaucasia considerable funds are spent on breeding such hybrids under artificial short-day conditions. Moreover, the plant breeders there cannot work on a sufficiently large scale. We, however, being guided by the regularity that we have discovered, found it fairly easy, under ordinary Odessa field conditions, to induce the given plants to flower in the normal way.

Without stopping to bring further proof of the correctness of the above regularity relating to the length of the vegetative period of hybrid plants, I can say that it is not worth while continuing the controversy on this question in the form in which our opponents have conducted it up till now. If we do, the controversy will simply degenerate into mere scholastics, and far from being of any use will only mislead the public. Any of our opponents on this question can now come to us in Odessa, to the Institute of Selection and Genetics, and try to point out among the 700 combinations of first-generation hybrid cotton plants combinations that are later-ripening than the early-ripening parent. We ourselves would help them to find a dozen or two of such combinations, but we would at once explain the causes of the late ripening. We would also show that after the removal of these causes this regularity will manifest itself immediately in the plants of these combinations too. To cite examples which it is claimed disprove this regularity in the development of hybrid plants as regards length of vegetative period—a favourite practice of N. I. Vavilov—means misleading the public.

N. I. Vavilov frequently asserts that numerous facts relating to many kinds of crops, including wheat, contradict our theory. Nevertheless, when dealing with this subject in his book *The Theoretical Principles of Wheat Breeding*, he could not cite a single example of wheat, but cited flax. This shows that the author knows of no such examples of wheat, or at any rate, knew of none at the time he wrote his book.

But let us examine this example of flax. The author states that in the second hybrid generation of flax there were plants that flowered 4 to 7 days earlier than the plants of the first hybrid generation. If such cases were observed, then, in our opinion, the cause should have been ascertained. The whole trouble is that, in my opinion, there was no such case, in spite of the fact that N. I. Vavilov said it was his own experiment and that he saw it with his own eyes. I discussed this matter with him in the train on the way to Omsk and was convinced then that he admitted he was wrong in citing this example. Here, however, he again brought up this flax illustration. The case was as follows: there were no more than 20 or 30 first-generation plants in each combination but there were 300 or 400 plants of the second generation. Moreover, N. I. Vavilov compared the date of flowering of the first plants of the second hybrid generation with the date of flowering of the majority of plants of the first generation. It would seem that no "blame" attaches to him for doing so. It is the generally accepted rule to register the flowering of plants of the first hybrid generation as of the date of flowering of the majority of the plants, whereas the date of flowering of the plants of the second hybrid generation is registered individually. It must be obvious to everybody that observations of this kind are unsuitable for the purpose N. I. Vavilov has in view.

To all the comrades who oppose us on this question I can say the following: You have failed to understand the problem. You are not even aware that you are opposing a proposition that we advocate, but which we did not "invent." I have in mind the proposition that if an organism possesses the potentialities of development, and if the external conditions for this exist, then it is certain to develop. We all know that usually the sex cells possess all the potentialities of development that their parents possessed. Hence, possessing both the paternal and maternal potentialities of development, a hybrid, under varying field conditions, will always be no less and often more adapted than its parents for development in any particular period of time.

True, it must not be forgotten (and we have mentioned it many times) that not every adaptation of an organism to development in a given period of time is both economically and biologically useful and expedient for the survival of the particular individual. I have cited the example of the autumn sowing of spring and winter parents and also of their hybrids. In the autumn the spring parent is adapted for going through the vernalization phase; the hybrid (winter form crossed with spring) is also adapted for passing through the vernalization phase. The winter parent, however, is not adapted for this, and therefore it does not go through the vernalization processes in the beginning of the autumn. But the unadaptedness of the winter parent to go through the vernalization phase in the early autumn saves it from freezing in the winter. However, the spring parent plant and the hybrid, which were adapted for going through the vernalization phase in the autumn, come under the action of winter frost in the state (vernalized) in which our cereals are very much less resistant to frost.

Consequently, this example too, in which heterozygosity obviously proved to be even biologically harmful, speaks in favour of and not against the proposition we advocate. Hybrids are richer in potentialities of adaptation to developmental conditions in every particular period of time. It is only necessary to be able to make skilful use of this phenomenon. It will then be possible to make plants more heterozygous when that is to our advantage, and, conversely, to know when and how to make plants homozygous.

It was for these considerations that we not only opposed the inbreeding method employed by many plant breeders in the Soviet Union experimenting with cross-pollinating field crops up to and including 1935, but also proposed that experiments be made in intravarietal crossing of self-pollinating plants.

I must here make the reservation that while opposing close-related pollination as employed in inbreeding, we urge the usefulness of raising thoroughbred animals and pure strains in plants. All we object to is the method employed by our inbreeders.

When plant development is properly understood, it becomes fairly clear how close-related breeding, i. e., inbreeding, can and should be employed.

Geneticists attribute the depression, degeneration, of the progeny of many inbred cross-pollinating plants to the fact that the lethal and semilethal genes, i.e., the immutable particles of the chromosome, become homozygous. But is it possible from the standpoint of the genetic corpuscular theory to explain the numerous cases when the close-related breeding of a given genotype is so harmful that seeds fail to form and, in other cases, in the same genotype, seed formation proceeds normally, and normal, undepressed plants are produced from them?

Today, in N. V. Tsitsin's fields, I saw *Junceum*, a species of couch grass, which, according to Tsitsin, is an inveterate cross-pollinator, and, when self-pollinated, produces no seeds at all. A plant of this couch grass species was planted in the field and during the past two or three years its rhizomes have spread and occupied an area of three to five square metres. All the spikes of the couch grass on this plot are now filled with seeds. From this the conclusion must be drawn that in this case the plants produced from the same seed easily pollinate each other. But the spikes of a couch grass plant having the same, not different, roots, cannot pollinate each other. In this case, the genotypes are apparently the same, but the result of inbreeding is different.

Another example: sugar-beet plants, as a rule, can rarely self-pollinate. If the root of such a plant is cut up into two or three parts and the parts are planted together, isolated from other genotypes, they also rarely produce plants that pollinate each other. If, however, a beet is cut up and the parts are planted separately (preferably under different conditions) and roots are allowed to grow from them, and if this group of roots is planted with the object of obtaining seeds, the plants often pollinate each other as successfully as those grown from other roots. In 1936, D. G. Kornyakov (a specialist at our Institute) on my suggestion performed the following experiment with beets. He cut the root of a sugar beet into two parts. One part he planted in the open field. The other part he planted next to the first, not in the ground, however, but in a pot. The result was fairly good seed formation. The root of another sugar beet was cut in two and the parts were planted under equal conditions, but from these no seeds were obtained.

All these examples show that it is not a matter of lethals, not of genes as discrete particles of the chromosome, but of the identity, similarity in one way or another, of the gametes. To a certain extent, the greater the identity of the sex cells, the less viable, the less adapted to varying conditions is the organism that is obtained from the fusion of such sex cells. By creating different conditions for rearing the initial identical genotypes it is possible to obtain relatively different sex cells, even though these different cells remain of the same genotype. That is how the examples I have cited can, and should, in my opinion, be explained.

On these grounds, the persistent assertions of Darwin, often repeated by Timiryazev, that the prolonged self-pollination of plants is harmful and that at least periodical crossing is beneficial, become clear.

Two or three million plants growing on one hectare from pure Ukrainka wheat seeds will differ from each other in some respect. All of them, severally and jointly, are of the Ukrainka variety, but they all differ from each other in some respect, because the environmental conditions each one utilizes for its development are in some respect different from those utilized by the rest. Although the sex cells of these plants are specific, they are nevertheless particles of the bodies of Ukrainka plants that differ from each other. Hence, the possibility is not precluded that differences may fairly frequently occur also in the sex cells. It must not be forgotten, however, that these are all Ukrainka plants, and consequently, that all the sex cells will be cells of plants of this variety. We are speaking only of intravarietal differences. From this it is evident that the seeds obtained from such a field by the cross-pollination of plants possess great potentialities of adaptation to varying field environmental conditions.

A year has passed since we first urged the need for experiments in intravarietal crossing in the department of seed growing. This year, 1936, we have been able to test our proposition by small field experiments, for we already have in our field the third generation from the intravarietal crossings of spring wheat varieties such as Melanopus 0122, Lutescens 062, Girka 0274, and also our two new wheat varieties 1163 and 1160. The results of the field experiments confirm our anticipations. Hundreds of specialists who have visited our Institute have seen these crops.

The seeds from each intravarietal crossing and the ordinary seeds of each individual variety were sown in the same run of the seed drill. The box of an eleven-row seed drill was partitioned off into two sections. In one the seeds from the third generation of the intravarietal crossing were put; in the other the ordinary seeds of the same variety. In this way the different seeds of each variety were sown in the same run of the seed drill. Not one of the specialists who have visited our Institute was able to point to a single variety in which the plants of the third generation of the intravarietal crossing were inferior to those from the ordinary seeds of the given variety. One could easily see that extremely degenerated old varieties like Girka 0274, and also Melanopus 0122, showed the greatest improvement after intravarietal crossing. Good results were likewise obtained with our new hybrid varieties. The intravarietal crossing of Lutescens 062 also produced economically effective results in yield. The ordinary seeds of this variety showed a yield of 20.1 c. per ha.; the seeds of the third generation of the intravarietal crossing showed a yield of 21.5 c. per ha.; a second replication of ordinary seeds produced a yield of 19.1 c. per ha.; those of the intravarietal crossing showed a yield of 21.5 c. per ha. A third replication of the ordinary seeds produced 17.3 c. per ha., those of the intravarietal crossing 20.6 c. per ha. The ordinary seeds of Girka 0274 yielded 11.7 c., those of the intravarietal crossing 15.1 c. A first replication of the ordinary seeds of Melanopus 0122 yielded 15.8 c., those of the intravarietal crossing 19.9 c.; a second replication of ordinary seeds yielded 14.2 c., and those of the intravarietal crossing 17 9 c.

The ordinary seeds of our new variety 1163 yielded in the first replication 18.9 c. per ha., those of the intravarietal crossing 19.6 c. In the second replication the ordinary seeds yielded 14.5 c. and those of the intravarietal crossing 16.5 c. The ordinary seeds of our other new variety, 1160, yielded 17.8 c. in one replication, and those of the intravarietal crossing 20.2 c.

It must be observed that all those who opposed us last year and argued that our assumptions were useless and even harmful, on seeing or hearing of the results of our field experiment stated almost unanimously that some slight manifestation of plant vigour—"heterosis"—might be expected in the first generation, but that there would be none in the second generation.

These comrades thought that as only a year had passed since this problem arose, we had managed to raise only one generation of wheat. They did not know that this sowing was not of the first generation, but of the generation which, in their opinion, would no longer show any effect. We had a third generation in the field. I cannot say how many generations will show the effect of intravarietal crossing. I do not think that this can be foretold, because different varieties, and also any one variety sown in different districts behave differently. But one thing I do know, namely, that the seeds obtained from intravarietal crossing will be better for sowing than ordinary seeds of the same variety.

I sincerely regret that we were unable, it would be more correct to say that we lacked the determination, last year, to induce the masses of the people on the state and collective farms to start experiments in intravarietal crossing. Had we done so, the practical importance of this problem would have been ever so much more obvious this year.

The masses of the collective-farm people in nearly all the districts of the Soviet Union have enthusiastically responded this year to my article "Make Up for the Lost Year." Several institutes and experiment stations, in particular the Dniepropetrovsk Grain Institute, the Institute for the Socialist Reconstruction of the Azov-Black Sea Regions, and the Moscow Regional Experiment Station, also responded to our appeal.

About two thousand collective farms in different territories and regions of the U.S.S.R. have already this year carried out the intravarietal crossing of different self-pollinating field crops, in the majority of cases spring and winter wheats. We, in conjunction with the collective farmers, quickly surmounted the technical difficulties of the crossing operations. Dolgushin, a specialist at our Institute, devised a way of performing one of the most difficult and responsible operations, namely, pollination. He worked out a method of castrating spikes and leaving them open after the operation for wind pollination with the pollen of other plants. This measure produced excellent results. It can be done five to six times faster than the ordinary way, and the best thing about it is that any collective farmer, man or woman, can learn the operation in two or three hours.

This year over ten thousand men and women collective farmers set to work on intravarietal crossing. The shortage of tweezers made things difficult. The demand for this instrument increased immediately dozens of times over the ordinary demand in the Soviet Union; but the collective farmers overcame the difficulty within a few days—fairly good tweezers began to be made in the collective-farm workshops. That is how the work of intravarietal crossing was conducted in 1936.

In each collective farm an average of 500 to 1,000 gr. of seed were obtained from the intravarietal crossing of winter or spring wheat, and also of other self-pollinating crops. The total amounts to about one and a half tons. This amount of seed obtained directly from the crossing of self-pollinators is, I believe, more than the total obtained by all the people engaged in crossing self-pollinators during the whole period plant-breeding stations have been in existence.

Many people think that a kilogram of seeds, of winter wheat, for example, obtained on a collective farm for experimental purposes, is of no economic value. These people forget that under our conditions not only a kilogram but literally five to ten grams of seeds are of economic value. With skilled scientific guidance a kilogram of seed can be reproductively increased to economically effective quantities on any collective farm within one or two years.

This year we showed the numerous visitors to our Institute, including a group of academicians, a plot of 7 ha. under the ordinary conditions prevailing at our Institute, and also one-hectare plots on a number of collective farms on which were obtained crops of 7 to 8 c. per ha. of spring wheat sown with a seed drill at the rate of 3 kg. per ha. We also showed our visitors a plot of 2 ha. on which no more than a kilogram of spring-wheat seed per ha. had been used, and from which a crop of 8 c. per ha. was obtained. Therefore, if it turns out that the seeds from intravarietal crossing are of better quality, will give rise to more productive plants (and everything goes to show that this will be so in the vast majority of cases, as will be finally demonstrated not later than 1937), there is nothing to prevent a kilogram of seed from being converted into scores of tons on every collective farm. The collective farms which started this experiment have already this year received from our Institute D. A. Dolgushin's pamphlet containing directions for accelerating the reproduction of the seeds from intravarietal crossing.

A kilogram of winter-wheat seed usually contains 30,000-40,000 grains. These 30,000-40,000 grains are sown by hand, one seed at a time, 30 cm. apart in the row, and the rows 70 cm. apart from each other, in a well-tilled and well-manured fallow field. If this hectare is well tended (like an intertilled crop), a yield of no less than 15 to 20 c. should be obtained. Thus, exactly a year later the kilogram of seed will have been converted into 15 to 20 c. This quantity of seed should be sown on well-tilled and well-manured bare fallow at the rate of 50 kg. per ha., i.e., it will take up 30 to 40 ha

The crop should amount to no less than 2 tons per ha. Thus, in exactly 2 years the kilogram of seed should be converted into 60 to 80 tons. If by that time (1938) practice has proved that intravarietal crossing increases yield by at least 1 to 3 c. per ha., then, by the autumn of that year 2,000 collective farms will have enough seed for the whole of their winter crop, or for the whole of their spring crop in 1939.

Hence, our dispute with the geneticists, which seemed to be of an abstract theoretical nature, turns out to be a very important practical question for our socialist agriculture.

The situation is literally the same in many other branches of agricultural science which are necessarily based on some particular genetic conception. This is why we Michurinists lay chief stress on reconstructing the bourgeois theory of genetics on the more effective and therefore truer lines of Darwin, Timiryazev and Michurin.

TWO TRENDS IN GENETICS[1]

COMRADES, the papers that have been read here by breeders who are working in different parts of our great Soviet Union revealed the immense creative breeding work that is being conducted with the most diverse crops and breeds of animals. The success achieved in this work is tremendous. There is no doubt that at this session of our Agricultural Academy only an insignificant part of all the vast work on breeding and genetics carried on in our Soviet Union has been reported.

In our country science is being mastered by extremely wide strata of the working people, from workers in many thousands of collective-farm laboratories to workers in research institutes and academicians. Nobody would dream of comparing our achievements in research work to those of tsarist Russia.

In many departments our agricultural science has already moved to the foremost position in the world.

In view of these unquestionable successes achieved by our science of breeding and genetics, many comrades, including some who are present at this session of the Academy, fail to understand the root-causes of the controversy that is now going on in the columns of the magazines *Sotsialisticheskaya rekonstruktsia selskogo khozyaistva (Socialist Reconstruction of Agriculture) and Yarovizatsia (Vernalization.)* Some of the participants in this controversy write in a rather high-handed style and, in my opinion, often go to extremes and try to twist the facts to suit their arguments. Nobody can accuse me of this. I think that whoever has followed the press must admit that although my articles are written with passion, they are nevertheless dispassionate. (*Applause.*) The articles by Doncho Kostov, Academicians Konstantinov, Lisitsyn, M. M. Zavadovsky and several others, however, although written in cool, calm, measured tones, are, in my opinion, far from being dispassionate.

The controversy that has been and still is going on among us is not merely a conflict of opinion among individual scientists; it affects the most

---

[1] Paper read at the Fourth Session of the Lenin Academy of Agricultural Sciences of the U.S.S.R., December 23, 1936.—*Ed.*

vital interests of research work. This alone, in my opinion, is the reason why this controversy in a seemingly narrow sphere of science,—breeding and genetics—has roused such wide interest among the Soviet public, including collective-farm experimenters. At issue here are not particular minor problems but the main trend of work in agrobiological science. The chief point on which attention is now being concentrated in this controversy is the conception of the evolutionary process in the plant and animal world.

The better we understand the laws of development of plant and animal forms, the more easily and quickly will we be able to create the forms we need in accordance with our wishes and plans.

The materialistic core of Darwin's theory of evolution is at bottom revolutionary and effective. Natural and artificial selection are Darwin's brilliant explanation of the natural purposiveness of the animal and plant world.

Wild plants are distinguished mainly for characters and properties that do not satisfy the requirements of man, but are useful to the plant species or genus itself because of its adaptedness to and greater viability in the conditions under which the given plant is growing. The cultivated varieties of plants are created by man and are therefore adjusted to the requirements of man.

Man selects for seed (for breeding) only those plants that are most suited for the given purpose. Of course, changes which make it impossible for a cultivated plant to live under the given conditions cause it to perish; i.e., in such cases, natural selection also operates among cultivated plants. We know, however, that in addition to selecting for seed such plants as have undergone changes desirable to man, man himself alters the conditions of cultivation, alters agrotechnique, in conformity with the changes in the plant organism. Hence, the higher, the more intensive, the cultivation of the given plant, the greater is the role played by artificial selection in the creation of new forms, and the smaller is the role and significance of natural selection.

Darwin elaborated his theory of evolution on the basis of a generalization of the vast experience amassed by human practice and of his observations of plants and animals in free nature. He showed that the plant and animal worlds change. He discovered the causes of the fitness of organisms for the conditions of their habitat and, in the case of cultivated forms, also their fitness for satisfying man's requirements. Darwin's theory thereby frees man's hands, frees his initiative for action, for creating new forms of plants and animals.

In bourgeois countries Darwin's brilliant teaching did not and could not really flourish nor could it develop further. The best Darwinian scientists in capitalist countries, such as, for example, Burbank in America, and our revolutionary biologists K. A. Timiryazev and I. V. Michurin in tsarist Russia, were lone fighters in the field.

11—742

From the moment it appeared, Darwinism was attacked chiefly for attributing a creative role to both natural and artificial selection, which Darwin's opponents denied.

To save time, I will not deal at length with the first stages of the fight for Darwinism. In capitalist countries Darwinism hewed a road for itself in bitter struggle.

Unable to refute the essence of Darwinism, the critics in the field of biological science—the geneticists—have always tried to falsify it, often on the pretext of correcting Darwin's inexact method of work, or on the plea that in Darwin's time the methods of scientific work were inexact.

The followers of de Vries proclaimed their own theory of mutation in opposition to Darwin's theory of evolution (as if Darwin had not been aware of discontinuous variation!!). The attitude of Bateson and Lotsy towards Darwinism is also known; and so is that of Johannsen, who elaborated the theory of the pure lines of self-pollinators. Johannsen's theory of the pure lines of self-pollinators fundamentally contradicts the central point of Darwin's theory of evolution, namely, the creative role of artificial and natural selection.

A number of passages could be quoted from Johannsen's *Elements of the Precise Theory of Heredity*, where he repudiates the creative role of natural and artificial selection. I shall quote only one passage: "For the theory of heredity, which is a biological discipline mainly of an analytical character, it would be best not to interweave the views of Darwin, and of the other outstanding proponents of the theory of evolution, with current research work."[1]

Thus, Johannsen says, in effect, that Darwin's theory has no relation to the theory of heredity and variability, i.e., to the subject which genetics should deal with. We workers in Soviet agronomic science know very well, however, that all our research work in every department of the study of the plant organism must be permeated with Darwinism. We know very well what was the attitude toward Darwinism of the best selection biologists, who have given the world a vast number of excellent varieties. I will not speak now about I. V. Michurin; I have spoken about him on many occasions. Let us refer to the greatest plant breeder America produced, the late Luther Burbank. In his book, *The Harvest of the Years*, he says that his lifelong adherence to the theory of Charles Darwin was not due to blind faith in Darwin's authority. Owing to his insufficient experience he at first even doubted some of his theories. In the course of time, however, he had more and more occasion to test Darwin's theory in the orchard and field, and the older he grew the more convinced he became that Darwin was the real teacher.

In the same book Burbank tells us what he advised a young man who was interested in the laws of heredity of plant organisms to read in order

[1] W. Johannsen, *Elemente der exakten Erblichkeitslehre*, Jena 1913, S. 219.

to study these laws. He suggested that he start studying Mendel by reading Darwin, then to finish with Mendel and read Darwin again, but more thoroughly.

I mention this only in order to emphasize the great value Burbank attached in his profoundly creative work to Darwin's theory of the evolution of plant forms. He writes that he gave this advice because he saw that the assertions of many well-known scientists were not confirmed and he could do nothing with them practically; whereas he became convinced that Darwin's statements always coincided with the facts and that Darwin never groped in the dark in constant quest of support for some pet theory or preconceived notion, and never took the wrong path.

I don't think anybody can suspect Burbank of not having been a specialist, of not having had to know the laws governing the development of plant organisms.

Burbank repeatedly pointed to the creative role of natural and skilful artificial selection. Johannsen, however, denies the creative role of selection. I may be told: "But Johannsen arrived at these conclusions on the basis of a precise experiment." The whole point is that, in our opinion, Johannsen's experiment is unconvincing. This experiment, the description of which is given in textbook after textbook, was as follows. Beans of a definite variety were divided according to size and planted for seed. The crop showed that the hereditary nature of this variety was not uniform. It consisted of different biotypes. The separate biotypes, or more correctly, their progeny, Johannsen called pure lines.

This part of Johannsen's experiment in no way contradicted Darwin's theory of evolution. It merely proved once again that if big seeds are chosen for sowing, the crop will also have bigger seeds than that grown from smaller seeds under the same conditions.

The originality of Johannsen's conclusion rests on the subsequent experiments he made with these beans. He arrived at the conclusion that in the progeny of self-pollinating plants that had originated from one plant and later were not subjected to cross-pollination, the selection of plants for seeds plays no role. Irrespective of whether the best or worst plants are selected for breeding purposes, when sown under equal conditions the crop will be equal in quality and quantity.

Johannsen arrived at this original conclusion, which contradicted not only Darwin's theory of selection but also ordinary agricultural practice, after experimenting for six years.

Why did the results of Johannsen's six-year experiment in selecting extreme variants (biggest and smallest seeds) for sowing contradict agricultural practice, which has always selected the best specimens for seed (breeding purposes)?

Darwin cited numerous examples of how man has steadily improved the breeds of domestic animals and cultivated plants by means of selection. Perhaps Darwin was wrong; perhaps people in their practical work simply erred, and are mistaken now, in selecting the best plants for seed. But before forming such an opinion of world agricultural practice and suspecting an error in the generalization made by that unexcelled thinker, the biologist Darwin, we must ascertain what material served Johannsen as the foundation for his conclusion that the selection of self-pollinating plants originating from one plant and not subjected later to crossing is useless.

Let us briefly describe Johannsen's experiment. From the progeny of a single bean plant he selected a few of the biggest and a few of the smallest beans and planted them separately. The beans of the new crop obtained both from the big and small beans were, on the average, of equal size. Then he again selected from the total crop of the big beans a few big ones and from the crop grown from the small beans a few small ones, and planted them.

This is all the six-year experiment amounted to. In the sixth year too the experiment in selection produced neither positive nor negative results. The beans of both variants were on the average of equal size.

During the six years of his experiment with bean line No. 1 he obtained from both variants (big and small seeds) a total of 1,525 beans. Thus, during all the six years he obtained from each variant a crop of a little under 800 beans, and the average annual crop was about 150 from each. In other words, every year an average of only 2 to 5 big beans and a similar number of small beans were selected for planting. The selection was not made from the plants which produced, on the average, the biggest or smallest beans compared with other plants developing under the same conditions. Two to five beans were selected from the mixed crop of all the plants obtained from the given variant.

It is a known fact that not only the beans from one plant, but even those in one pod are of unequal size. We likewise know that the difference in the size of seeds does not always indicate a difference in the nature of the embryos in the seeds from which the future plants develop.

That is why, in our opinion, Johannsen, after testing in his own way the creative role of selection established by Darwin, might have arrived at another conclusion, still more disastrous for Darwinism. He might have obtained from small seeds an average crop of bigger seeds than if he had planted big seeds.

In this case, when only two big seeds were selected from the mixed crop, they (or either of them) might accidentally have possessed the nature of the smallest-seed plants, and when selecting two small seeds he might accidentally have chosen seeds of plants which by their nature produce the biggest beans. Such a case is by no means excluded if only two to five seeds are selected, and from a mixed crop of different plants at that, instead of *from plants* (not from seeds) of the extreme variants.

Had Johannsen reproduced the lines of beans he selected in larger quantities, one centner, say, and had he used the method of selecting from extreme variants, taking into consideration the selected plants' conditions of development, or at least had made his selections on a larger scale, then, at any rate, he would not have arrived at the conclusion that selection from self-pollinating plants obtained from the progeny of one seed is useless.

With a larger number of plants there are more chances that individuals will encounter sharply different conditions of development, and this may often cause a variation in their hereditary nature. By skilful (and not any kind of) artificial selection of such crops, man can in his practical work not only preserve good varieties of plants and breeds of animals for a long period, but can also improve them year after year.

The blind, uncritical application of Johannsen's theory of pure lines cannot produce favourable results. All the best plant breeders in the world, even those who appeared to support Johannsen in theory, did not follow him in their practical work.

We must never forget about selection. For breeding purposes, for seeds, always only the best plants must be selected. Not only must Darwin's ideas about selection never be forgotten, they must be inculcated more and more in the minds of our millions of collective farmers. As we know very well from the periodical press, our best Stakhanovite collective farmers not only select the best plants for seeds, but even perform such painstaking work as selecting seeds grain by grain.

I know my opponents pretty well. They will say: "Geneticists nowadays don't think and act like Johannsen; so Lysenko is simply forcing an open door. Geneticists do not deny the creative role of selection." Precisely for this reason I must quote a passage from a book by a present-day geneticist whom we all respect, Thomas Hunt Morgan, *The Scientific Basis of Evolution.* On pages 95-96 of this book we read:

"Darwin's contemporaries seem to have understood that by selection of extreme individuals in any population, the next generation would be moved in the direction of selection. This is true, however, only when different genetic factors are present, and even then the process soon comes to an end—as soon as these factors are sorted out. Nothing really new is accomplished, except that there are more of given kinds of individuals; but the limits of the original population are not transcended."

On page 131 of this book we read: "The argument shows that natural selection does not play the role of a creative principle in evolution." Thus, the geneticists' view of selection, both artificial and natural, sharply contradicts Darwin's theory. *Actually, geneticists deny that selection plays a creative role in evolution.* From their point of view, no deviations in individual development can play any role in phylogenetic changes, i.e., changes in hereditary nature of organisms.

Geneticists regard natural and artificial selection as only a sieve for sifting the hereditary natures of some organisms from those of others. *They refuse to understand that Darwin's conception of natural selection at all times includes heredity, variability and survival.*

Their denial of the creative role of natural and of skilful artificial selection in the process of evolution alone suffices to show that the fundamental theoretical conceptions of genetics are not developing on the plane of Darwin's theory of evolution. *This fundamental question is the pivot of our controversy.*

I, and those who think as I do, stand by Darwin's theory of evolution, by Darwinism in all departments of agrobiological science. Hence, we fundamentally disagree with those views on evolution, on the creation of new forms of plants, that are held by the school of N. I. Vavilov and by many geneticists.

On this question there is a fundamental difference between these two trends in science, which no agreement on minor questions can reconcile.

I am not fond of controversy for the sake of controversy in matters concerning theory. I am an ardent controversialist only when I see that in order to carry out certain practical tasks I must remove the obstacles that stand in the way of my scientific activities. This is what I have done in the course of my work in certain departments of physiology and in certain departments of agrotechnique insofar as it concerned vernalization as an agricultural operation. This is what I have also done in the department of plant breeding. I think that in all these departments the controversy has ended, or nearly ended.

What work is compelling me and Dr. Prezent, and a number of other scientists, to raise the question of revising the premises of genetics? What work gave rise to this controversy? There are two problems. *The first problem is to improve the quality of the seeds of self-pollinating plants by means of intravarietal crossing. The second problem is to alter the nature of plants in the direction we desire by suitable training.* My effort to find a solution of these two problems compelled me to raise the controversy over heredity and variability.

I pass to the first problem. I shall start straight off with the fact that many geneticists deny the possibility of self-pollinating varieties degenerating. This is understandable because many geneticists deny that the genotype varies in the course of many generations; this is also the basis of their denial of the creative role of selection.

We, however, hold different views. After long-continued cultivation, self-pollinating varieties—pure lines—change, and hence often deteriorate, degenerate. Anyone who knows anything about the cultivation of tomatoes can tell you firstly, that they are self-pollinators, and secondly, that if a good variety is cultivated without selecting the best plants for seeds, it degenerates within three to five years. This is very easy to note in tomatoes because these plants easily undergo change. Moreover, people watch this plant very closely, and variations in the shape of the fruit, or in the ripening date, etc., are noted at once.

Darwin made a wide and comprehensive study of the biological harm of self-pollination and the benefits of cross-pollination. He arrived at the conclusion that in practical farming self-pollinating varieties do not stand long cultivation just because they are self-pollinators. They deteriorate, degenerate, and make way for new varieties. I will not quote passages from the works of Darwin. Dr. I. I. Prezent quoted a fairly large number in his comprehensive article in the magazine *Yarovizatsia*, No. 3 (1935).

I shall now pass on to explain the reasons why changes may take place and, as a result, self-pollinating varieties, both of hybrid and nonhybrid origin, often deteriorate. I will not touch here upon the well-known question of the deterioration of self-pollinating varieties as a result of mechanical admixtures. Everybody knows that there must be no rye in a wheat crop, no white wheat in a crop of red wheat. This is not the issue now, and we will not discuss it. It goes without saying that all measures must be taken to secure the utmost purity not only of élite crops, but also of the crops in the collective-farm fields grown for consumption.

Usually fertilized sex cells possess all the potentialities of going through the course of development of their nearest ancestors to a greater extent, and their nearest ancestors are their parents. Therefore, as a rule, progeny possess to the greatest extent the potentiality of going through the same course of development as their parents. To possess the potentialities of development under given environmental conditions means being adjusted to live and develop under the given conditions. Thus, we proceed from Darwin's proposition that sex cells in some measure reflect, accumulate, the course of development of preceding generations, particularly of the nearest ancestors.

In self-pollinating plants, both male and female sex cells develop in the same plant, in the same flower. Therefore, every sex cell, male and female, of self-pollinating plants usually reflects a more identical course of previous development than those of cross-pollinating plants, in which the male and female sex cells of different plants unite during fertilization, and, as a consequence, when fertilization takes place, the zygote represents, reflects, the course of development of not one, but two preceding plants.

I have already said that a plant possesses the potentiality of reflecting, repeating, in some degree, the course of development of its nearest ancestors and not only of its direct ancestors—its parents. Everybody also knows very well that the remoter the ancestors, the less is their course of development reflected in the given generation. The development of succeeding generations obliterates, as it were, the course of development of preceding generations, or rather, does not obliterate, but constantly converts it into a relatively new one.

Hence, it is not difficult to conceive that plants, for example, the winter wheat Krymka or any other winter wheat obtained from seeds after intravarietal crossing, possess the potentiality of repeating the course of development of both the paternal and maternal organism. Therefore, in these plants the potentialities of adaptation to environmental conditions are more diverse than in the father or mother plant taken separately.

The longer plants obtained from cross-pollination are selfed, the more will the difference between the father's and mother's potentiality of development wane.

Thus, in every new generation obtained by selfing, the range of adaptive potentialities of development will become ever narrower. Field conditions, however, are never constant. Therefore, the nonplastic organism, the adaptive potentialities of which have been reduced, will develop less successfully than the organisms whose variability as regards potentialities of development corresponds to the variability of the field conditions. Hence, a variety usually cannot stand long-continued self-pollination. After long-continued cultivation, self-pollinating varieties degenerate, their yield drops, and man replaces them by new, younger varieties with larger and better yields.

The question naturally arises, how is it that in nature species and strains of self-pollinating plants live for thousands of years?

This question was answered by Darwin. After a detailed investigation, Darwin showed (and to this day no anti-Darwinist has been able to disprove it) that there is not a single variety, not a single strain of a cultivated or wild self-pollinating plant, a number of specimens of which have not been subjected from time to time to cross-pollination.

In the case of wild plants, all the seeds of the crop are biologically intended for sowing, but usually the number of plants that survive is about equal to that of the previous generation. Therefore, if only one per cent of a given strain of a wild self-pollinator undergoes cross-pollination, this is quite enough for a fairly frequent rejuvenation of the entire strain, for refreshing its "blood."

In the case of cultivated plants the picture is different. The well-being of cultivated plants largely depends on man. In the case of cultivated plants, wheat for example, not all the seeds obtained in the crop are used for sowing; usually only 5 to 10% is used. The main thing is that by means of agrotechnical measures people create conditions under which nearly all the plants sown survive.

With the improvement of agrotechnique the role of natural selection among cultivated plants steadily diminishes. Hence, one or two per cent of natural cross-pollination is not enough to rejuvenate, to refresh, a variety of cultivated plants. The plants from seeds resulting from natural intravarietal cross-pollination gain an advantage only for individual development; for the rejuvenation of the variety, these plants play a much smaller role than among wild plants. *Therefore, for the rejuvenation of self-pollinating varieties, we propose periodical artificial intravarietal cross-pollination.*

From the vast mass of material that he collected, and also from the careful experiments he himself conducted, Darwin arrived at the categorical conclusion that self-pollination is biologically harmful and that cross-pollination is biologically beneficial. He also showed why self-pollinators

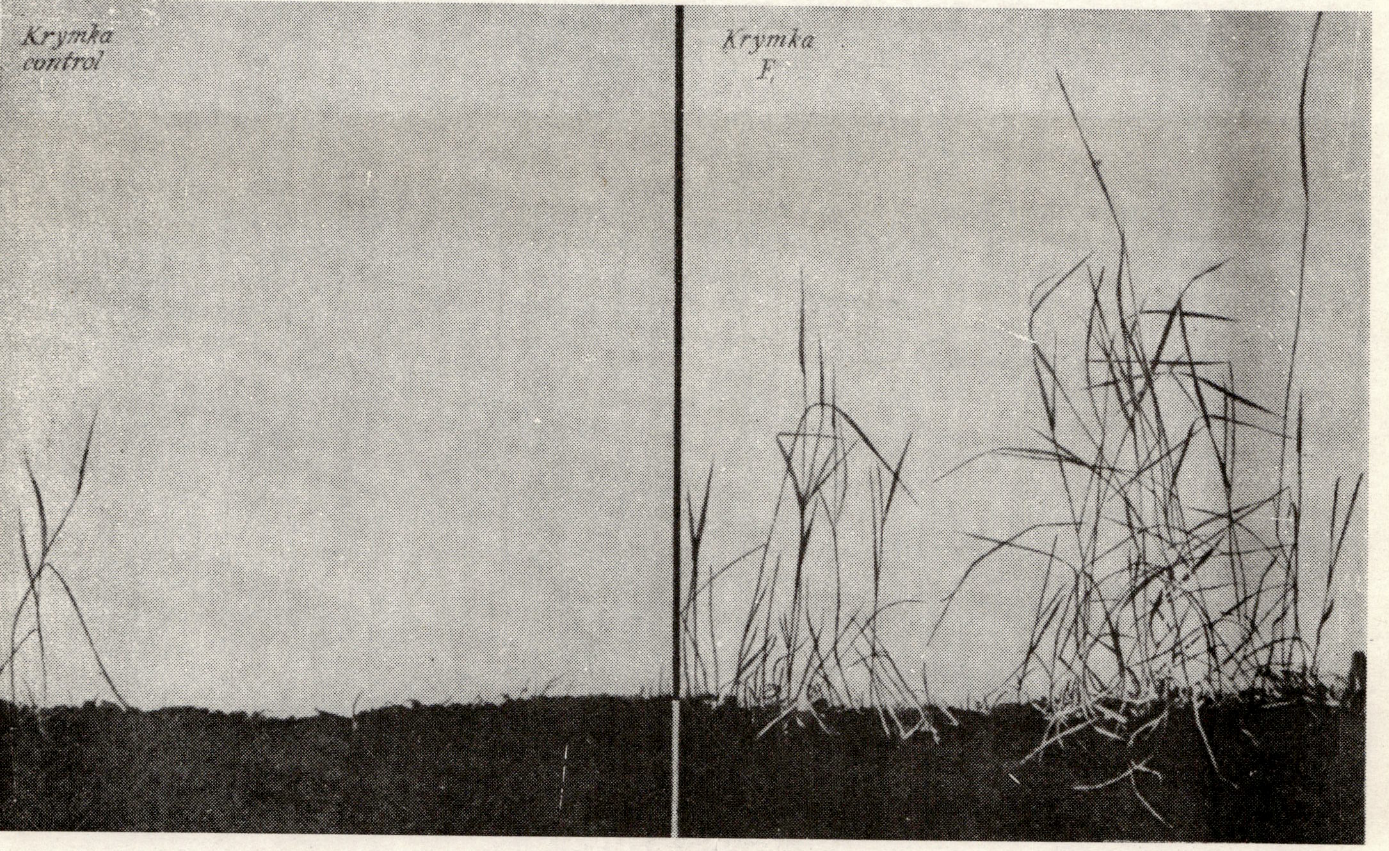

Fig. 42. Winter wheat Krymka

Left—sowing of ordinary seeds (control); right—sowing of seeds from an intravarietal crossing (first generation). When chilled in a refrigerator, the plants from the seeds of the intravarietal crossing showed high frost-hardiness

Fig. 43. Winter wheat Hostianum 0237

Left—sowing of ordinary seeds (control); right—sowing of seeds from an **intravarietal crossing** (first generation). When chilled in a refrigerator, the plants from the seeds of the **intravarietal crossing** showed high frost-hardiness

yere and are being created in nature. In order to survive, to perpetuate the race, it is often more beneficial for the plants to be pollinated with their own pollen rather than remain unpollinated if no pollen from other plants is available, if none has been brought by the wind or by insects. By this Darwin demonstrated the biological usefulness of the ability to self-pollinate.

I must remind you that Darwin's persistent assertion that long-continued self-pollination is biologically harmful and that at least periodical cross-pollination is beneficial was repeatedly confirmed by Darwin's most adept follower—K. A. Timiryazev.

I think it must be obvious to everyone—except, perhaps, those who dogmatically preach bourgeois genetics and maintain that genotypes and certain genes remain unchanged for tens of thousands of generations—that three million plants of the purest variety of wheat growing on one hectare of land cannot be absolutely alike in their nature. If this variety happens to be Ukrainka, then all the plants, severally and collectively, are Ukrainka; nevertheless, they all differ from each other in a greater or lesser degree. This diversity, however, remains within the limits of the term Ukrainka variety.

The apprehensions expressed by our opponents that crossing within a pure-strain variety will create diversity and reduce purity are totally groundless. If the sowing was pure before intravarietal crossing, the variety will, as a rule, be even more uniform after the intravarietal crossing. In the majority of cases, crossing evens out the crop instead of creating diversity. Let me cite a common example: in any family, the children resemble each other more than their mother and father resemble each other. (*Laughter, applause.*)

I have always emphasized that for an experiment in intravarietal crossing, it is essential to take the purest seeds of wheat, barley or other self-pollinating crop. None of us has advised intravarietal crossing with sowing material in which there is white-kernelled wheat among red, in other words, with impure seed material. I have proposed, and propose now, that intravarietal crossings be made only with pure seeds; if the latter are impure, they must first be sorted by mechanical means.

N. I. Vavilov stated in his paper that intravarietal crossing was useless and not worth while. In support of his proposition that self-pollinating varieties do not age, do not degenerate (if admixtures and accidental crossings with other varieties are excluded), Vavilov said that there were numerous varieties of wheat, barley and other self-pollinating field crops that had existed for a century and more. I at once recalled the statement Vavilov made last year (1935) at the Odessa session of the Grain Department of the Academy, where, in the paper I read, I criticized the method of inbreeding cross-pollinating plants. In answer to my assertion that nobody in the world, after working for scores of years, has bred varieties for production by this method, Vavilov said that varieties produced by inbreeding occupied large areas in production. He did not mention definite districts and varieties because he did not then have any such examples at hand.

More than a year has passed since then, and in the paper he read yesterday on this subject Vavilov said something quite different. He has since ascertained that no large areas are as yet occupied by varieties produced by inbreeding, neither maize, nor rye, nor sunflower. Perhaps, after a more detailed investigation, Vavilov will also alter his opinion that self-pollinating varieties do not degenerate even in the course of centuries.

Even if we assume that the number of self-pollinating varieties that have existed for centuries is far greater than N. I. Vavilov mentioned in his paper, these data are no longer required today. They might have played some role, perhaps, in 1935, when I raised the point about the degeneration of varieties due to long-continued self-pollination and proposed intravarietal crossing as a remedy. Fortunately, the geneticists could not cite any such facts in 1935. (*Applause.*) Notwithstanding the strong opposition, counteraction and even ridicule with which the geneticists met our experiments in intravarietal crossing, they could not in 1935 cite any facts confirming their claim that the nature, the genotype, of varieties remains unchanged for a century.[1]

If such facts did exist, it would, of course, not only have been much more difficult for us in 1936 to organize mass testing experiments on approximately two thousand collective farms in different parts of the U.S.S.R., but even at our experimental institute we would have had to conduct these experiments outside of the approved program. From the geneticists' point of view, this question is simply unscientific. In spite of that, I assert even now that varieties do not exist long without undergoing change. And Darwin's assertion that long-continued self-pollination is biologically harmful is as true today as when he made it. Today, however, the solution of this problem has entered a new and higher phase. The problem of whether self- and cross-pollination are beneficial or harmful is now being solved not

---

[1] On the question of the transience of varieties due to long-continued self-pollination I shall point to an example cited by Darwin: "Andrew Knight has observed that the varieties of peas keep very true, because they are not crossed by insects.... But the greater number of varieties have a singularly short existence: thus Loudon remarks that 'sorts which were highly approved in 1821, are now, in 1833, nowhere to be found'; and on comparing the lists of 1833 with those of 1855, I find that nearly all the varieties have changed...." "Nor do I know whether the short existence of almost all the numerous varieties is the result of mere change of fashion, or of their having a weak constitution, *from being the product of long-continued self-fertilization.* It may, however, be noticed that several of Andrew Knight's varieties, which have endured longer than most kinds, were raised towards the close of the last century by artificial crosses; some of them, I believe, were still vigorous in 1860; but now, in 1865, a writer, speaking of Knight's four kinds of marrows, says, *they have acquired a famous history, but their glory has departed.*" (Charles Darwin, *The Variation of Animals and Plants Under Domestication*, London 1885, Vol. I, pp. 348 and 349.)

with the aid of subordinate facts obtained by observing the longevity of self-pollinating varieties, but by performing *decisive experiments.* And the intravarietal crossing made by us and by hundreds of collective-farm laboratories are indeed a resolute experiment.

In my opinion, geneticists should not at the present time set out in search of cases to prove that varieties are immutable, but should ponder over the question of how their "corpuscular hereditary substance" theory can possibly be made to explain, on the one hand, the degeneration of varieties due to long-continued self-pollination, and, on the other, also the beneficial effects of intravarietal crossing.

You really ought to do this, comrades geneticists! What will happen if, in the spring of 1937, hundreds of collective-farm laboratories will be able to prove, by different varieties and in different districts, that winter wheat has become more winter hardy as a result of intravarietal crossing? And if the fairly extensive experiments in artificial chilling made at the Odessa Institute of Selection and Genetics also confirm this? And if, in addition, the variety tests will show a considerable increase in the yield of winter wheat resulting from intravarietal crossing? The field experiments with spring wheat performed at our Institute have already proved this. With us, the problem of intravarietal crossing has already transcended the bounds of theoretical speculation; there was ample time for that before experiments in intravarietal crossing had assumed such large dimensions. Now the problem is in a state in which practice, experiments, will decide and not arguments about varieties existing unchanged for centuries.

Proceeding from the theoretical profundity of Darwin's theory of evolution, we say that there will be no such varieties until man really controls the evolutionary process and, if necessary, compels the nature of plants to change only in the direction we desire.

The geneticists say that intravarietal crossing is useless because in a pure line the hereditary nature of millions of plants is the same. Today, to insure themselves, they concede that in rare cases, the nature of seeds may improve as a result of intravarietal crossing, and they attribute this to heterosis. But if you ask them what this heterosis means in plain language, they answer: well, for example, when a plant is more vigorous or earlier-ripening than its parents growing beside it, that is called heterosis. In other words, the geneticists attribute the increased vigour of a plant resulting from intravarietal crossing to heterosis, i.e., to increased vigour. (*Laughter, applause.*)

This explanation of increased vigour by "heterosis" is far from being the only example in genetics of this method of explaining phenomena. Thus, in genetics, a change in the hereditary basis not due to a cross is called "mutation." When you begin to delve down to the causes of the change in the genotype, the geneticists exclaim in one voice: "This is quite clear, the change takes place because the organism mutates." Translated into plain language, this means that the organism changes hereditarily because a hereditary change takes place.

If the infinite number of genetic terms were translated into plain language, many geneticists would find it much easier to see through the fallacious propositions of their science, which has departed from Darwin's theory of evolution.

What are the results of our experiment in intravarietal crossing?

The results of the field experiment we made in the summer of 1936 in the intravarietal crossing of spring wheat have been published by me several times in the press. The results of the tests of the five varieties of spring wheat with which intravarietal crossings were made are quite convincing to me and a number of comrades who work with me. They have fully confirmed our assumptions. Not only the first and second, but also the third generation even of the new varieties of hybrid origin gave a higher yield after intravarietal crossing. The new varieties gave a higher yield of from 1.5 to 2 c. per ha., and the old varieties from 3 to 4 c. per ha.

Such are the results of the intravarietal crossing of spring wheat.

One of the obstacles to conducting experiments in intravarietal crossing on collective farms was the existing technique of crossing. With the existing method a highly skilled worker at an experiment station could obtain no more than 15 grams of seed (500 to 600 grains) per day. How could we suggest such work to the masses of collective farmers and expect that enough seeds would be collected from the crossing to sow at least half a hectare of seed plot? This was not the main thing for us, however. The main thing was: would this measure have a palpable economic effect? We felt sure that if it did have a favourable effect, a way would be found to overcome the technical difficulty. In 1936, without any moral support from the Agricultural Academy, but with the substantial practical support of the newspaper *Sotsialisticheskoye zemledeliye*, we set out to popularize experimenting in intravarietal crossing at collective-farm laboratories. The local workers—regional and district—were in favour of such work. As a consequence, we succeeded, in 1936, in getting intravarietal crossings performed at nearly two thousand collective farms. On each collective farm, from 5 to 8 persons engaged in this work. This meant that at least 10,000 collective farmers had excellently mastered the technique of crossing.

Why were the collective farmers able to learn the art of crossing so quickly? Because the method of crossing was simplified. And Professor Vakar should not have written in issue No. 12 of the magazine *Sotsialisticheskaya rekonstruktsia selskogo khozyaistva* [Socialist Reconstruction of Agriculture], that the castration method proposed by Dolgushin was useless. "I have tried it myself," wrote Vakar, "and only 1 or 2% developed seeds." But we are not in the least perturbed by the poor results that you, Professor Vakar, obtained by this method. The fact is that on 2,000 collective farms, from 80 to 90% of the castrated flowers developed seeds, and that is more important. Moreover, an inspection on the collective farms of the spikes isolated as controls showed that in the vast majority of cases the castrations had been performed well.

In the autumn of 1936, hundreds of hectares of seed plots on collective farms were already sown for the purpose of reproducing winter-wheat seeds obtained from intravarietal crossing. At three centres variety tests have been carried out: at the Institute of Selection, at the Odessa Regional Station (Vygoda), and at the Moscow Regional Station.

In 1936, as a result of 4 to 5 days' work, 5 to 7 collective farmers each obtained about a kilogram of seeds from the crosses. I think that next year we shall obtain from 5 to 10 kg. (*Applause.*) Some comrades said: "Yes, you succeeded in getting a kilogram of rejuvenated seeds on a collective farm. But what is a kilogram of seeds for a collective farm?" These comrades have forgotten, or perhaps simply do not know, what unlimited possibilities for quickly reproducing seeds exist under present-day collective-farm conditions, with collective-farm laboratories and the unprecedented efforts now being made to secure high-quality seeds. We, in conjunction with 19 collective farms in the Odessa district, succeeded in 1936, under the usual field conditions of a dry summer, in obtaining by the autumn nearly 190 c. of spring-wheat seed from 130 kg. From one kilogram of seeds 50 to 60 tons of them can be obtained in two years. There are no difficulties in the way, and the collective farms can do it easily. Sowings of winter-wheat seeds obtained from intravarietal crosses have been made on hundreds of collective farms in different parts of the Soviet Union. In 1937, *everybody will be able to convince himself that hundreds of tons of rejuvenated seeds will be obtained from the seeds of the intravarietal crosses performed in 1936.*

All the preliminary results obtained from the intravarietal crosses of *winter* wheat also indicate that this will be one of the methods of improving the nature of seeds. That is why I urge academicians, scientific research workers, and the Academy as a whole, to convince themselves as quickly as possible of the usefulness of this measure.

If our assumptions are confirmed, namely, that thanks to the greater vigour of the winter plants that developed from the seeds of intravarietal crosses (and they are indeed considerably more vigorous) they will be more winter hardy, then in the summer of 1937 we must endeavour to get this measure carried out on at least fifty to seventy thousand collective farms.

Every geneticist and other scientific research worker now has the opportunity to convince himself of the efficacy of this measure by what has been achieved on hundreds of hectares on collective farms and at three variety-testing centres. There is plenty of time. We still have the winter and spring before us. If the experiments with winter wheat show good results, then a big organizational task will have to be undertaken.

Take, for example, the supply of scissors.[1] We shall need for this purpose as many as 500,000 pairs. Then, also, there is the training of people. We shall have to train as many as 500,000 collective farmers for the work of intra-varietal crossing. It is the duty of the Academy to undertake this big organizational task.

I shall now pass to the second problem in my paper, that of *altering the nature of plants by training.*

No one will be so bold as to say that external conditions play no role in the evolutionary process of plant forms. Nevertheless, geneticists categorically deny that it is possible to change the hereditary basis of plants in a definite direction by suitably training them for a number of generations.

Any attempt to master this process is at once indiscriminately classed by the geneticists as Lamarckism. They forget that no positive results can be obtained from work conducted from the standpoint of Lamarckism. The very fact that we do succeed in changing the hereditary nature of plants in a definite direction by suitable training shows that we are not Lamarckians, and that we do not take the Lamarckian standpoint.

Surely it is wrong to classify a scientist as a Lamarckian solely because he recognizes the inalienable role of external conditions in the evolutionary process going on in the plant and animal world.

It would be hard to find a bigger enemy of Lamarckism than Dr. Prezent, and yet, as you know, Dr. Prezent not only supports the idea of altering the hereditary nature of plants by suitable training, but he himself is one of the few who conduct fairly extensive experiments along this line.

In general outline it is clear to everybody that external conditions play an enormous part in the infinite process of form-building which plant organisms undergo. As far as I know, however, nobody has yet succeeded in demonstrating and proving experimentally *what* conditions are necessary, and *when*, at *which* moments in a plant's development, they are necessary in order to change the nature of the plants of the succeeding generations in a desired direction.

I think that the level of knowledge of our Soviet scientists concerning plant development is already sufficiently high to enable us to set to work and master the process of directed form-building.

What we know best, most fully and in the greatest detail in the biology of development of plants is that stage in their developmental cycle which we call the vernalization phase. It was natural, therefore, for us to have started our experiments in the directed alteration of the natures of plants by means of suitable training with altering the nature of the processes we had studied, i.e., altering the hereditary nature of plants in respect to the vernalization phase.

---

[1] At the session I mentioned tweezers. Now, owing to the improvement in the technique of castration, scissors are needed.

It is well known by now that the difference between the natures of winter and spring plants is that for one of the periods of individual development known as the vernalization period (phase), winter plants need lower thermal conditions than spring plants. It has been proved up to the hilt that winter and spring varieties of plants (wheat, barley, rye, oats, etc.) are represented by an intergrading series of forms as regards winter habit. Some varieties require low temperatures and a long time for passing through the vernalization phase, others require somewhat higher temperatures, others again require still higher temperatures and can pass through their vernalization phase fairly quickly every year when sown in the spring. That is why the latter are called spring varieties.

Thus, it has been ascertained that different varieties, wheat, for example, are, in different degrees, either winter or spring forms.

The different degrees of spring and winter habit are created and fixed by natural selection in the process of the evolutionary form-building of these plants. The problem of how the property of winter or spring habit becomes fixed has long been clear to agricultural science. Winter varieties, when sown in the spring, cannot reach fruiting; therefore, they drop out, they leave no progeny, and only the spring varieties survive. On the other hand, the spring plants of wheat and of many other crops, if sown in the autumn, freeze in the winter, and only the winter varieties survive. One question remained unclear, namely: how is winter or spring habit itself created?

We have already made a rather thorough study of the conditions that winter plants require for passing through the vernalization phase. Hence, we are able, by presowing vernalization, to induce the winter plants of *any* variety of wheat, rye, vetch and other crops to fruit in *any* district when sown in the spring.

On the basis of this knowledge, we, in 1935, commenced experiments in altering, by training, the nature of the plants' environmental requirements for passing through the vernalization phase.

As is known, presowing vernalization that we have worked out as an agrotechnical operation does not alter the nature of plants, their environmental requirements. By presowing vernalization we merely create the external conditions these plants require for their normal individual development. Now, however, we set out to alter the nature of these plants' requirements.

The theoretical premises underlying these experiments were the following. The environmental conditions that plants require for development, for passing through the vernalization phase, for example, are expressed in every variety by its own range of fluctuation. For example, the winter wheat Kooperatorka requires a temperature ranging approximately from 0° to 15°-20° C.; it can pass through the vernalization phase at 15°-20° C. but very slowly. Thus, at a temperature of 0°-2° C. it requires 40 days for vernalization, but at 15°-20° C. it requires 100-150 days—too long to be of any use for the practical cultivation of winter plants.

Naturally, the passage through the processes of vernalization by plants of the same variety at a temperature of 0°-2° C. and that at a temperature of 15°-20° C. will be relatively different not only in rapidity, but also qualitatively. The cells of the growing point (where the vernalization processes take place) of a wheat plant developing at a temperature of 0°-2° C. and the cells of the growing point of another wheat plant of the same variety developing (being vernalized) at a temperature of 15°-20° C. will resemble each other in that both possess the quality of vernalization. Only this quality (of vernalization) of the cells can undergo further change, can develop in the direction of ear emergence and the formation of reproductive organs. But these cells, vernalized in one plant at 0°-2° C. and in the other at 15°-20° C., will be qualitatively different from each other, because the particular thermal state is a most essential component of the conditions for passing through the vernalization process. Therefore, we assumed that under different thermal conditions relatively different vernalized cells would result, and that this relative biological difference would lie in the fact that the next generation, obtained from seeds which had been vernalized at a high temperature (15°-20° C.), would pass more easily and quickly through its vernalization phase at the same high temperature.

Our line of reasoning was as follows. Vernalized cells develop at the growing point. From them develops the entire subsequent unbroken chain of cells of straw, spikes, and male and female sex cells. Thus, at a certain moment in the plant's development, the vernalized cells are the direct initial cells for the entire further form-building of the organism, which will later produce ripe seeds. Therefore, if the vernalized cells of two plants of the same variety are different because of different thermal conditions during the vernalization period, this difference will be transmitted in some form or other to all the subsequent cells, and will be biologically reflected in the sex cells and hence in the new seeds.

As Dr. Prezent says, in their environmental requirements for passing through their whole cycle of development, plants repeat, and in some degree reflect, the course of development of their ancestors. The nearer the ancestors, the more strong is their course of development biologically impressed upon, accumulated in, the given progeny. Therefore, in spite of the fact that the winter wheat we took for our experiment—the transformation of its nature into that of a spring variety—had for ages, generation after generation, passed through the vernalization process at comparatively low temperatures (much lower than spring wheat), we assumed that it would not take very many new generations vernalized at high temperatures for this wheat to become hereditarily of spring habit.

It must be emphasized that when we say that during a given period of their development (for example, when a winter variety passes through the vernalization phase) plants must be given a high temperature to transform them into spring forms, we do not mean that they must be given the highest possible temperature. If the temperature is higher than that at

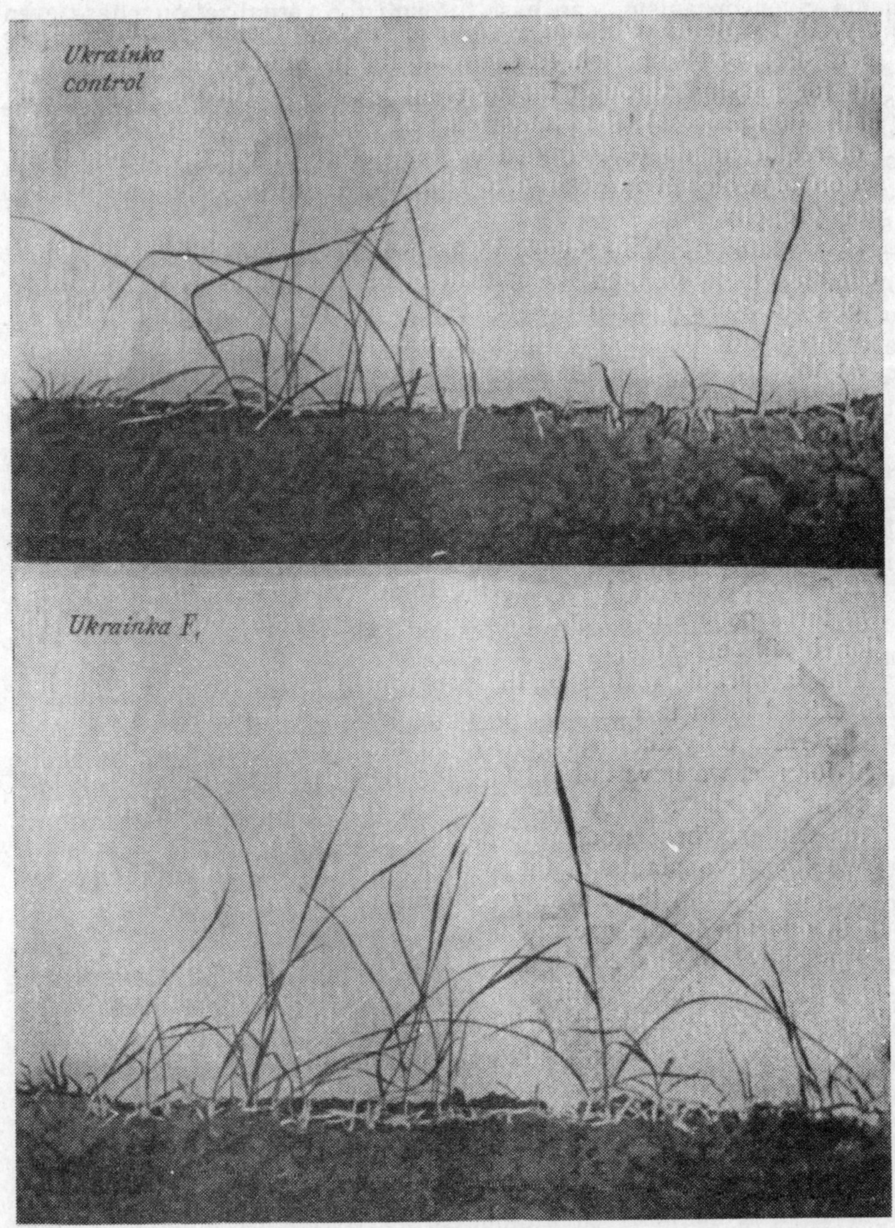

Fig. 44. Winter wheat Ukrainka

Top—sown with ordinary seeds (control); bottom—sown with seeds from an intravarietal cross (first generation). When chilled in a refrigerator, the plants from the seeds of the intravarietal cross proved to be the more frost-hardy

which the given organisms can pass through the vernalization phase even slowly, these plants will not vernalize; they will simply wait for lower temperatures, or else perish. In changing the norm of environmental requirements for passing through the different processes (in our experiment the vernalization process), the plants must be kept at approximately that border line of requirements (determined by the nature of the given plant) in the direction of which it is desired to shift the norm of requirements of the plants' progeny.

Proceeding from the above theoretical premises, and basing ourselves on our knowledge of the development of plant organisms, we, at the Institute of Selection and Genetics (Odessa), are now conducting fairly extensive experiments in altering the nature of plants by suitable training.

At the present time the experiments comprise the following: winter wheat—transforming it in some cases into spring wheat, and in others (and this is most important) into a form possessing greater winter habit; winter rye—transforming it into spring rye; cotton—making it less thermophilic; and several other plants.

Our most prolonged experiments at the present time are those in transforming the winter wheat Kooperatorka into a spring wheat.

These experiments were started in March 1935. Since then we have grown three generations, and in September 1936 we sowed the fourth generation of these plants.

The favourable results of the experiments with Kooperatorka induced us to extend them to a number of other plants, but of these we have managed to raise only the first generation and to sow the second. Therefore, except for rye, we have not yet any results to report on these experiments.

I shall briefly report here the results of our experiments in transforming the winter wheat Kooperatorka into spring wheat.

We took two plants of the winter wheat Kooperatorka and two of Lutescens 329 from the Saratov station, and on March 3, 1935, we sowed them in one pot in the greenhouse.

The object of the sowing was to prevent these winter plants from passing through the vernalization phase and thereby compel them to live as long as possible without earing. The pot with the plants that were sown on March 3 was placed not in a warm, but in a cool greenhouse where from March 3 to the end of April the temperature was often no higher than 10°-15° C. The temperature was raised only in the beginning of May, but the main thing was that it did not drop below 15° C. The two Lutescens 329 plants, being of greater winter habit than Kooperatorka, lived until late in the autumn and perished without earing. By the beginning of August the Kooperatorka plants had tillered considerably and displayed numerous fresh and withered leaves. Individual shoots of these plants had by that time developed straw.

About the middle of August one of the Kooperatorka plants perished because pests had gnawed its roots. Only one plant was left, from which,

Fig. 45. Kooperatorka, Sown September 30, 1936

Left—first generation (control). Right—third generation

Fig. 46. Kooperatorka, Sown September 19, 1936

First six pots on left—plants of first generation (control); in the rest of the pots—plants of the third generation

on September 9, a few first grains were gathered. The fruiting of this plant was very protracted. It lasted right up to January 1936, when the plant, with many of its spikes still green, perished.

On September 9, 1935, the first grains gathered from the Kooperatorka plant were sown. At the same time some ordinary Kooperatorka seeds taken from the store were sown under the same conditions for the purpose of comparison and control. The seeds were sown in the greenhouse where, in November and December, the temperature is usually not below 15°-20° C. About a month after sowing, the experimental plants could already be easily distinguished from the controls. They looked more springy than the controls (leaves more upright, tighter sheaths). In the winter all the plants suffered from the small amount of daylight. Earing commenced at the end of January. The experimental plants eared much better than the controls, and the awns on many of the spikes were so short that they could readily be told apart from the controls. Among the experimental plants there were considerably more sterile spikes than among the controls. In general, it was already evident from this sowing that in their behaviour, their nature, the experimental plants differed from the controls.

On March 28, 1936, we sowed the seeds gathered from the experimental plants. (We called this sowing the third generation.) At the same time we sowed seeds from the former control plants (this sowing we called the second generation), and also ordinary Kooperatorka seeds taken from the store. All these plants were provided with warmer conditions than the first sowing, on March 3, 1935.

The results of the experiment were more striking in the third sowing than they had been in the preceding ones. First of all, the control plants not only failed to ear, but most of them perished by the autumn. The second-generation plants (those from the controls sown on September 9, 1935) were in a better condition than the control plants, but in a much worse condition than the third-generation plants. The third-generation plants began to ear much earlier (30-50 days) than those of the second generation, and what is most important, they eared more uniformly. The second-generation plants had many leaves, but only one or two stems, whereas in the third-generation plants all the tillering shoots eared fully.

In the same sowing we had the second generation of the remaining seeds of the first experimental plant, which was vernalized not in the winter but in the summer of 1935, i.e., under warmer conditions. The behaviour of these plants was more like that of the third generation grown from the seeds of the plants which had vernalized in the winter in the greenhouse. The greatest difference in behaviour was that between the plants of the third generation and the controls.

Beginning with August, all the plants of the third generation eared fully, but of the control plants only two shoots eared, and then only towards the end of September.

Simultaneously with the considerable change in the vernalization phase in the direction of spring habit, the second and, more particularly, the third generation revealed changes in many of the characters of spikes, glumes, length of awns, etc. In the third generation narrow-leaf forms appeared.

In these experiments it was easily observable that the more radically the plant was altered in the vernalization phase, the more disturbed was the harmony of the organism's further development.

We arrived at the conclusion that it was necessary to alter plants more gradually, to put them under less severe conditions. This would disturb the organism less, while no more time would be required for altering the plants, for it would be possible to grow generations more quickly.

In September 1936 we sowed the fourth generation and also the reserve of seeds from the third and second generations and control (ordinary) seeds taken from store. In this sowing the differences between the different generations of Kooperatorka were more striking than ever. On the 50th to 60th day after sowing the older generations already developed stalks (straw). The controls, however, behaved like typical winter plants. Winter rye lends itself still more easily to alteration in the direction of spring habit. In the spring we sowed the Tarashchanskaya variety of winter rye in the field with a seed drill without preliminary vernalization. The plants of this sowing tillered for a long time and in the middle of the summer some of them eared and produced seeds. On August 29, 1936, we sowed part of these seeds simultaneously with ordinary seeds (for control). A month after sowing a difference in the behaviour of these plants could already be easily observed. In the beginning of Novemb r the experimental plants developed straw more uniformly than the control plants; literally, only a few individual plants of the latter eared.

Some comrades have argued that rye, being a cross-pollinator, is inconstant, and therefore, what we have here is not an alteration of the plant's nature, but merely the selection of forms of greater spring habit. We reject this argument. If rye has the spring-habit property in its "blood," and spring habit is a flawless dominant, how has rye been able to winter in our districts for so many years? The spring plants of rye cannot winter in our districts. Moreover, the second-generation plants (of our experiment) show that they are not completely of spring habit.

The conversion of rye is proceeding more quickly and easily than that of wheat only because, in our opinion, rye, as a cross-pollinator, is always a hybrid; and hybrids are always more susceptible to change, their range of adaptive potentialities being wider. Therefore, later on in our work of altering the nature of plants even when experimenting with self-pollinators, such as cotton, for example, we took hybrids (first generation) instead of constant forms.

It may seem at first sight that the experiments I have described are of no practical value, but that is far from being the case. If we are able to transform winter into spring plants by training them when they are passing through the vernalization phase at relatively high temperatures, then we shall be able to alter the nature of plants in the opposite direction too. By training plants when they are passing through the vernalization phase at temperatures lower than those at which they were vernalized under natural conditions, we will steadily increase their winter habit. In other words, we shall steadily increase their winter-hardiness, for although length of vernalization phase is not the only factor of winter-hardiness, it is one of the chief factors.

That it is possible by suitable training to augment the winter habit of wheat from generation to generation is confirmed by a number of facts taken from life. We all know very well that the most winter-hardy varieties in the Soviet Union are the winter wheats bred by the Saratov station. It is known that one of the many services rendered by the Saratov Plant-Breeding Station is its breeding of the most winter-hardy varieties of winter wheat in the world. The least winter-hardy varieties in the European part of the Soviet Union are those bred by the Odessa Plant-Breeding Station (Kooperatorka, Zemka). The varieties bred by the Kharkov station are much more winter-hardy than those bred by the Odessa station and in this respect come close to the Saratov varieties. But the Odessa varieties are more winter-hardy than the winter wheat of southern breed, for example, those of the Kirovabad station (Azerbaijan).

While giving due credit to the plant breeders at all the stations enumerated, we cannot, of course, avoid mentioning that the severer the conditions prevailing in the district in which a given plant-breeding station is situated, the more frost-resistant, as a rule, are the varieties it breeds.

If we recall that the Saratov varieties are of greater winter habit, i.e., have a longer vernalization phase, than those of the Kharkov station and that the Kharkov varieties have a longer vernalization phase than those of Odessa, it is not difficult to arrive at the conclusion that in definite districts certain external conditions must play a part in the creation of plant natures possessing particular lengths of vernalization phase.

Our task is correctly to analyze these conditions. We must learn to create them at any plant-breeding centre in order to be able to breed varieties with the requisite winter-hardiness. These conditions are the field temperature at the time winter plants go through the vernalization phase in a given district.

The field temperature in Saratov and Kharkov at the time winter wheat goes through the vernalization phase is, as a rule, lower from generation to generation than that in Odessa. In our opinion, this is one of the most important local natural conditions which create a given degree of winter habit and therefore in no small degree determine the winter-hardiness of winter varieties.

At a temperature under 1°-2° C. below zero winter wheat does not vernalize (as has been proved by fairly extensive experimental material). Hence, in all districts, vernalization proceeds in the period before constant frost sets in, i.e., at a temperature above zero. It is clear to us now that the nearer the field temperature is to zero at the time wheat passes from generation to generation through the vernalization phase, the greater is the winter habit of the varieties that are created in the given district.

At the same time we see from the table given below that in none of the centres enumerated does wheat always go through vernalization at a temperature of 0° C., or anywhere near it.

AVERAGE MONTHLY TEMPERATURE FOR FIVE YEARS (1905-09)

Centre	September	October	November	December
Saratov	15.4° C	7.8° C	—2.3° C	—5.4° C
Kharkov	14.5° C	7.8° C	—0.1° C	—4.9° C
Odessa	17.3° C	10.8° C	4.0° C	0.1° C

From this table it follows that the vernalization of September-sown winter wheat takes place at a higher temperature under Odessa field conditions than in the districts of Saratov and Kharkov. But even under Saratov and Kharkov conditions the vernalization of winter plants proceeds in September and October at temperatures much higher than 0° C. *The vernalization of winter wheat at a temperature of 0° C. must, however, shift the nature of the plants' progeny in the direction of greater winter habit.*

On this ground we have already started extensive experimental work to increase the winter-hardiness of our wheats. We vernalize seeds under artificial conditions at a temperature of 0° C. and sow them in the field.

Together with a number of other investigators (F. M. Kuperman, M. T. Timofeyeva and E. P. Melnik) we have ascertained that plants become far less frost-resistant after going through the vernalization phase. Therefore, by sowing in the autumn seeds vernalized at 0° C. we doom all the least frost-resistant forms to be frozen. Only the most hardy forms will survive—those which either have a longer vernalization phase, as a consequence of which they are unable to complete the phase in the refrigerator, or, if they completed the vernalization phase, can overwinter thanks to other inherent qualities they possess.

Thus, in our sowing we make a strict selection of the more winter-hardy forms. At the same time the nature of the plants is altered and the winter habit of their progeny is increased, for the parents go through vernalization at considerably lower temperatures than under natural conditions.

We intend to repeat this work year after year. Every year wheat will acquire greater winter habit, i.e., will become more winter hardy. For this purpose, a part of the crop obtained from such sowings will be reproduced for seed, and the other part will be treated to induce a still further change in its vernalization phase.

We have started in the same way to alter the nature of "warmth-loving" plants too. Indeed, if we can induce "cold-loving" plants, for example winter wheat, which cannot tolerate warmth during the vernalization period, to reduce their lower temperature requirements during that period, then it is possible by suitable training to induce a plant like cotton gradually to reduce its higher temperature requirements during the first days of its life.

The same can be said about altering the nature of plants which at given periods of their development need short or long days, etc.

The chief thing here is to understand plant development in the Michurin way, to understand the role and place of external conditions in the evolution of plant forms, and to do deliberately what in nature was and is being done fortuitously.

In nature the finest forms of animals and plants were and are being created by means of variability and natural selection. *By mastering this method, man will be able, firstly, to create such excellent forms in an immeasurably shorter time; and secondly, to create such forms as have not, and could not have appeared in nature even in millions of years.*

The geneticists make no real effort to plumb the depths of Darwin's theory of evolution, which we are effectively mastering, and some parts of which have been further developed and concretized by the best biologists in the world, foremost among them being K. A. Timiryazev and I. V. Michurin.

What I have stated above shows that by developing Darwin's theory of evolution we are experimentally mastering—and in some degree have already mastered—the art of altering the hereditary basis of plant forms in a definite direction.

What I have said in my paper does not, of course, fit into the frame of the theory of genetics. And the geneticists try to shield themselves from our criticism by accusing us of Lamarckism, of denying the material nature of the hereditary basis, and of other mortal sins. This merely shows that the critics do not understand Lamarckism and, as one might expect from autonomist metaphysicians, they call any recognition of the effective role played by external conditions Lamarckism. Nor do they understand materialism any better. Owing to their narrow, mechanistic point of view, they think that by denying that the morphology of the chromosomes plays the *sole and exclusive* role we deny materialism.

The propositions I have formulated do indeed contradict the genetic conceptions. With this part of the accusations levelled against us I fully agree; but far from contradicting the factual material which the geneticists have obtained and are obtaining, what I have said actually explains it from the standpoint of development. Therefore, much of the experimental material obtained by geneticists, which has remained frozen capital in their hands, becomes a source of effective knowledge when dealt with from the standpoint of development.

The fundamental principle on which the theory of genetics is based, and with which we cannot agree, is that the organism, the chromosomes of the cells, contain a "hereditary substance" which is distinct from the organism's body. This special hereditary substance (genotype) exists separately from the organism's body and does not enter the concept "body" for the simple reason that it is not subject to change, to transformation.

The substance of the soma, the body of the organism, develops; hence, it not only can change, but it must change, suffer transformation. In the opinion of the geneticists, the "hereditary substance," which is part of the chromosome, is, as a rule with rare exceptions, not subject to change, and hence is not subject to development, for how can there be development without change, without transformation? The "hereditary substance" (genes) proliferates by self-reproduction, and the most interesting thing is that although it reproduces itself, increases in size—from the zygote to the adult organism—millions of times, it undergoes no change, no transformation.

Thus, the fundamental difference between the "hereditary substance" invented by the geneticists and the substance of the organism's body is that the former undergoes no change in a long series of generations, whereas the latter, from which the organism builds different organs and characters, does change. According to genetic conceptions, the "hereditary substance" is immortal. The thread of this immutable substance runs from generation to generation; however, the substance of which the organism's body consists changes continuously as a result of assimilation and dissimilation and dies when the individual's life ends.

What I have said in my paper also fundamentally contradicts, of course, N. I. Vavilov's "law" of homologous or parallel series of variation. This "law" rests on the genetic theory that the corpuscles of the "hereditary substance," which remain unchanged in a long series of generations, recombine. I feel that I do not possess sufficient knowledge and skill really to smash this "law," which contradicts the facts of the evolutionary process. But in the course of my work I constantly obtain evidence of the unsoundness of this "law." The work itself shows that this "law" cannot be accepted when one sets out to control the evolution of plant forms effectively, in a given direction.

According to the "law" of parallel variation, new forms result not from the development of old forms, but from a reshuffling, a recombination, of already existing hereditary corpuscles. There is exact parallelism in the variation of species, genera, and even families. Moreover, N. I. Vavilov claims that parallelism of variation is a basis for the effective application of this "law" by the experimenter. In the opinion of N. I. Vavilov, it is enough to know the diverse forms of, say, wheat or barley to be able to predict all the different forms of rye, oats and other crops. The diversity of these latter crops must be exactly like that of wheat. If a given form does not already exist in nature then, according to the "law" of homologous series, this form can be created. If wheat could develop, say, a liguleless form, then there is, or can be, such a form of rye. In nature we find apple trees with round fruit, hence there must or can be trees with round pears, cherries, grapes, etc. In general, it is enough to note some character in one species to assume, according to the "law" of homologous series, that other species must or can have the same character.

At first sight the seeming effectiveness of the "law" of homologous series is attractive, especially if one is not, or does not wish to be, familiar with Darwinism. But it can seem so only to people who are ignorant of Darwin's theory of evolution, of the way new forms of plants and animals were and are being created by nature and by man.

A few words must be said about the seeming practical effectiveness of the "law" of homologous series. If there are apple trees in nature that bear round fruit, then round fruit can be borne by pear trees, cherry trees, grape vines, etc. But if there are no wild apple, cherry or plum trees bearing seedless fruit in nature, then, according to the "law" of homologous series, it is impossible to cultivate pear trees with seedless fruit. Plant breeders, however, have created such a variety. The cabbage species have no double flowers, but horticulturists have created a double gilliflower, although no double flowers are observed among the wild crucifers.

It is enough to compare the variability of wild forms with that of cultivated forms for the illusory inevitable parallelism in variation to disappear at once.

According to Darwin's theory of evolution, strains, species and genera of plants are in a continuous process of development and give rise to new strains and species. The theory of evolution excellently explains the common origin and gradual evolution of all the diverse existing living forms and indicates the way to improve the old and create new varieties of plants that we need.

According to the law of homologous series, however, species and genera cannot diverge in their evolution. Divergence is impossible because of the exact parallelism of variation. Since there is no divergence at the present time, then, according to the "law" of homologous series, there was none in the past. It follows, therefore, that all the existing species existed in the past, only in less diverse forms; but every form was richer in potentialities, in its collection of genes. Hence, from the "law" of homologous series we get genogeography, gene centres, gene funds, etc.

This explains our disagreement with the geneticists, and with N. I. Vavilov's school, which proceeds from the "law" of homologous series that contradicts the principles of Darwin's theory of evolution.

Our opponents say that Lysenko rejects genetics, i.e., the science of heredity and variability. This is not true. We are fighting for the science of heredity and variability; we do not reject it.

We are fighting the numerous erroneous tenets of genetics, which are pure inventions. We are fighting for the development of genetics on the basis and on the plane of Darwin's theory of evolution. Genetics is one of the most important departments of agrobiological science and we must master it as quickly and as fully as possible, remould it according to our Soviet pattern and not simply accept the anti-Darwinist standpoint of many of the fundamental propositions of genetics.

None of us denies the brilliant researches in cytology, which have already contributed much to the morphological description of cells, especially of nuclei; we wholeheartedly support the development of this work. Similarly, it would be stupid to deny the extremely useful work of the systematists in describing the morphology of wheat, or of other crops, including wild forms. All these are necessary branches of science which give us knowledge. But we deny that the geneticists and the cytologists will see genes under the microscope. With the aid of the microscope it is possible, and necessary, to see more and more details in the cells, in the nuclei, in individual chromosomes; but these will be particles of the cells, nuclei or chromosomes, and not what the geneticists mean by genes.

The hereditary basis is not some kind of self-reproducing substance *separate* from the body. The hereditary basis is the cell, which develops and is transformed into an organism. In this cell the different organella are of different importance, but not a single particle is free from evolutionary development.

With this I shall conclude. (*Loud and prolonged applause.*)

REPLY TO DEBATE

The results of the discussion on problems of genetics and plant breeding are in my opinion immense. One of them is the fact that the people who are working on the theory of plant breeding and genetics have taken definite sides. The attitude each takes toward the problem of the evolution of plant and animal forms is now clear.

It has also been revealed that the majority of the "pure geneticists" (as Serebrovsky calls them), and especially of the leading lights among the geneticists, are in many cases ignorant in matters of biology. Moreover, these people flaunt their ignorance of everything outside the Drosophila vial. They regard everything outside of that as unscientific. Only they—the Drosophilists—are in possession of real genetics, real science!

Some of these scientists have not read Darwin and Timiryazev, or have read them carelessly; and nearly all of them consider it beneath their dignity to familiarize themselves with the work of their opponents with whom they have entered into controversy. How else can we account for such "authoritative" expressions as the vernalization of fish, or the vernalization of silkworm eggs? These comrades think that in saying such things they are demonstrating their knowledge of Lysenko's and Prezent's views.

It is not surprising after this that many of our opponents among the geneticists who spoke here, beginning with Academician Serebrovsky, asserted that according to Lysenko and Prezent, segregation—heterogeneity of the gametes—is created, or (as Academician Sapegin stated in No. 12 of *Sotsialisticheskaya rekonstruktsia selskogo khozyaistva*) is "determined" by the weather, i.e., by the external environment.

It is an elementary rule never to quote the work of your opponent in your own words; you must always quote the actual words written in the opponent's work.

By stating that Lysenko and Prezent assert that the diversity of the gametes is created by the weather, i.e., by external conditions, our opponents, and Professor Navashin in particular, want to make it appear that we do not recognize, do not take into account, the nature of plants, their genotype. Where did they get the idea that we deny the importance of the nature of plants, the importance of the genotype?!

Somebody mentioned my paper "The Intravarietal Crossing of Self-Pollinating Plants" (p. 138 of this volume.—*Ed.*). Let us see what is stated in that paper on this question. It says the following: "Heterogeneity of the developmental potentialities of a plant organism and heterogeneity of environmental conditions cause heterogeneity in the various homologous organs and characters of the given plant, including heterogeneity of its sex cells."

This passage says: 1) that the organ's potentialities of development are not uniform in type, and 2) that the environmental conditions utilized by the developing organism are always diverse in some degree.

This is our explanation of the diversity that is observed in the progeny of a given hybrid. Moreover, we assert that it is because the developmental potentialities of plant organisms are not one and the same and because the environmental conditions are diverse that natural selection has created different species and genera of plants.

Of all the possible directions of development, the one favoured by the environmental conditions will be the one the organism's development will take.

As I stated in my paper, my chief objection to N. I. Vavilov's "law" of homologous series is precisely that we do not regard the potentialities of a plant organism's development as being uniform in type.

The basis of N. I. Vavilov's "law" above mentioned is precisely the theory that the potentialities of development—of effecting a change in a given gene or group of genes not only of one organism, but also of different species, genera and even families—are of a single type. According to this "law," the variation, evolution, of nearly all plants (a large group of them at any rate) proceeds only in one direction uniform in type and character. The evolution of (change in) wheat is of the same type as that of rye, barley, etc. It is argued that if an apple tree bears round fruit, then pear and plum trees, grape vines, etc., must also bear round fruit, and if they don't, then they can be created, because changes in apple trees, and in a large group of other plants, are uniform in type and character.

It is impossible to picture the evolution of the plant and animal world by starting out from the "law" of homologous series, which is based on the theory that the gene corpuscles are immutable in a long series of generations, and that the development of plant forms is uniform in character. Nor is it possible to picture divergence, i. e., the formation of plant species and genera, as arising out of the accumulation of minor differences.

The genetic conception does not, of course, lead to the conclusion that man can change the nature of plant forms in a definite direction. This was emphasized in the papers and speeches of our opponents. It is this that constitutes our fundamental disagreement with the geneticists and not the argument that diversity of sex cells is created by the weather. The passage I quoted from my paper shows that this is not the case.

Dr. Dubinin ascribes to us the very opposite. In his cleverly constructed speech he proceeded from the false tenets of his conception and then ascribed these tenets to us. By some means or other, he "discovered" in our writings the principle of Weismannism which he himself advocates.

In general, I would ask my opponents, for the common good, to quote Lysenko and Prezent not from memory, which sometimes betrays them, but from our works. This will be a surer way, and nearer to the truth.

Academician Serebrovsky's assertion that I deny the frequently observed segregation of hybrid progeny in the proportion of 3:1 is also untrue. We do not deny this. We deny your proposition that this proportion cannot be controlled. By proceeding from the conception that we are developing, it will be possible (and fairly soon, too) to control segregation.

Academician Serebrovsky is also wrong when he says that Lysenko denies the existence of genes. Neither Lysenko nor Prezent have ever denied the existence of genes. (*Applause.*)

We deny the correctness of your conception of the word "genes," for what you mean by it is particles, corpuscles of heredity. If a man denies the existence of "particles of temperature," denies the existence of a "specific thermal substance," does that mean that he denies the existence of temperature as a property of matter? We deny the existence of corpuscles and molecules of a special "hereditary substance," but at the same time we not only recognize the existence of a hereditary nature, a hereditary basis, of plant forms, but, in our opinion, understand it far better than you geneticists do.

I am grateful to Professor Muller for his splendid paper. His reply to the debate on it today was no less splendid. Muller definitely dotted the i's and crossed the t's. He clearly and distinctly said that genes mutate only in the course of tens and hundreds of thousands of generations; that the phenotype exerts no influence on the genotype. In confirmation, and for comparison, Professor Muller referred to the radio: a man at the microphone (the genotype) influences the listener at the loud-speaker (the phenotype), but the listener at the loud-speaker cannot influence the announcer. Hence, according to Professor Muller, the genotype influences the development of the phenotype, but the phenotype exerts no influence on the genotype.

In general, this means that the hen develops from the egg; the egg, however, develops not from a hen, but directly from a previous egg.

The explanation Professor Muller gave is clear and distinct to us. Professor Muller makes his position as plain as the respected Professor Morgan made his in his latest book *The Scientific Basis of Evolution.*

As for Academician Serebrovsky, he, by his explanation, only tries to confuse others, for he is not confused himself. He firmly holds that genes are immutable, and by genes he means particles, corpuscles.

The principal error the geneticists commit is their contention that genes are immutable in a long line of generations. True, they admit that genes are mutable in the course of tens and hundreds of thousands of generations. Well, we thank them for such mutability!

While recognizing that the genotype changes in the process of the ontogenetic development of plants we at the same time know that it is possible to induce plants not to change for hundreds of generations. To a certain extent we already can, by training, induce the nature of plants to change in a definite direction in every generation.

I am convinced that in the very near future this department of work will spread rapidly throughout the Soviet Union. Already we can, partly, actually control the nature of plants, induce it to change in the direction we want. This is an achievement of our Soviet science.

Academician Sapegin said in his speech that our experiment in transforming winter into spring forms cannot be called directed change. He developed the idea here that in our experiments with the winter wheat Kooperatorka we obtained, in addition to spring forms, such as were of greater winter habit than the initial ones, but they perished. In my opinion, this statement is not only unscientific but frivolous. How could such a miracle have happened that processes which take place in our experiment at the time of vernalization at a relatively high temperature should correspond, be identical, with processes that take place at lower temperatures? What do we mean when we say that a wheat acquires greater winter habit? We mean that the wheat will go through the process of vernalization only at a lower temperature. I would advise our opponents to study more thoroughly the factual material they undertake to criticize. As a matter of fact, neither Academician Sapegin nor the other geneticists have studied the factual material of our experiments with the winter wheat Kooperatorka. Professor Rozanova and several other geneticists who visited our Institute of Selection and Genetics have not quite correctly, and not fully, in my opinion, grasped the details of our experiments. Otherwise they would not have said that in all probability our experiments simply showed the effect of the selection of already existing forms but did not show that the nature of these plants had been altered.

I am sure that as a result of the discussion many comrades have become convinced that it is necessary to make a more detailed study of the factual material concerning the problem under discussion and not simply label us Lamarckians, or come out with such "well-meant" criticism as was expressed in the article by Konstantinov, Lisitsyn and Kostov, in which the authors themselves invent propositions, ascribe them to us, and then criticize them.

I have replied to the substance of this article in fairly considerable detail in the magazines *Yarovizatsia* and *Sotsialisticheskaya rekonstruktsia selskogo khozyaistva.*

Zavadovsky, M. M. How are we to understand you? You say that you love Morgan, and yet you reject his conceptions.

Lysenko. I said that I respect him. I did not say that I "love" him. (*Laughter, applause.*)

I respect Professor Morgan and Professor Muller for their scientific straightforwardness. In his paper Professor Muller said what he thought about the problems of evolution. You, Academicians M. M. Zavadovsky and Serebrovsky, agree with Professor Muller in everything concerning genetics. You agree with him that genes cannot change more often than once in a hundred thousand generations. But if anybody were to ask you whether you held the view that genes do not change in a thousand years, you would at once answer: "Nothing of the kind. It is only that ignoramus Lysenko who thinks that about us."

You, M. M. Zavadovsky, like Academician Serebrovsky, often say things you do not believe yourself. You know very well that everybody in the Soviet Union (except the geneticists, perhaps) agrees that environment influences variability in the plant and animal world, and yet you geneticists label everybody who holds that environment influences the variability of organisms a Lamarckian.

We say that environment influences form-building, but we add that we must know what conditions a plant must be given, and when it must be given them, in order to guide its evolution in a definite direction, i.e., to alter its genotype.

While I respect Professor Morgan and Professor Muller for their scientific straightforwardness, it is also clear to me that one has to be anything but a plant breeder to agree with the principles of their conception, with Muller's assertion that genes remain unchanged for tens of thousands of generations.

It is on this conception of the recombination of the immutable particles (genes) of heredity that the "law" of homologous series and genogeography are based. To accept this theory means abandoning the problem of controlling the nature of plants.

Experience convinces me that it is possible to guide the evolution of the plant world in a definite direction. The experimental data already available are sufficient to enable me to develop work in this direction on an extensive scale.

I am glad that Professor Muller and the other geneticists have clearly expressed their opinions on this problem. Up till now I was accustomed to hearing the geneticists brilliantly explain things retrospectively. They have never dared to make forecasts. Here, for the first time, I have heard them predict that it is impossible to get directed mutations. We shall see. Before a year passes life will show that we are right. (*A voice from the platform:* "They talk about uncontrolled mutations.")

Lysenko. But I am talking about directed variations.

Many theoreticians in the field of genetics argue against making intravarietal crossings of self-pollinators, claiming that it is useless. That this operation should be incomprehensible to you from your standpoint is clear to me, but from the standpoint of Darwinism it is a useful operation. The facts in our possession also convince us of this. Wheat plants from seeds obtained after intravarietal crossing are more vigorous than those from ordinary seeds. By the spring of 1937 we shall have the results of the frost-resistance tests of these plants. The preliminary results show that they will be hardier than the plants from ordinary seeds; and an increase in the winter-hardiness of winter wheat is no unimportant thing.

I fail to understand on what grounds it is stated that crossing within pure lines spoils varieties. It is said that this is difficult work, and that there is even danger of all varieties being killed in this way.

But am I proposing that all other seed-growing measures be dropped and that we should confine ourselves exclusively to intravarietal crossing on collective farms? We have never proposed anything of the kind. We wholeheartedly support all measures that are taken to obtain the purest varieties of seeds for sowing on our collective- and state-farm fields.

At the present time, experimental intravarietal crossing is being performed on about two thousand collective farms. I think that by the spring and the beginning of the summer of 1937 it will be possible partly to test the utility of the measure we propose.

To sum up my paper and my reply to the debate, I would like to cite an example to show the difference in the results of research work obtained, on the one hand, by people who hold that there are such things as immutable corpuscles, a special "hereditary substance" which is independent of external conditions; and, on the other hand, by people who hold that development is connected with every kind of variation, with every metamorphosis of the thing that is developing.

N. I. Vavilov and Kuznetsova have been studying the genetic nature of the properties of winter and spring habit of cereals. Many other geneticists studied this problem before them. As a result, some of them arrived at the conclusion that the property of winter habit depends on one gene, others—that it depends on two, three, or more genes.

N. I. Vavilov and Kuznetsova have been unable to establish the number of genes that determine winter habit, but that is not the main thing. The main thing is that, irrespective of the number of particles (genes) the investigator, whoever he may be, has invented to determine winter habit, knowledge of the nature of winter habit and of its biological peculiarity has not advanced a single step.

Everybody now knows, however, that we have elaborated this problem (winter habit and spring habit) from the standpoint of the theory of plant development so thoroughly that, firstly, anybody can control this property when growing plants; and secondly, after our researches, none of the geneticists has the courage to say now how many genes determine winter habit or spring habit.

This small sector of research work is sufficient to show the difference in the effectiveness and reliability of the respective approaches to the problem. The analysis of the example I have taken does not end here, however.

In the same experiment N. I. Vavilov and Kuznetsova rightly noted that in hybrids spring habit prevails over the winter habit. In the second generation of hybrids obtained from the crossing of a winter with a spring variety, there were ten times more spring forms than winter forms.

The observations made by Vavilov and Kuznetsova are confirmed by our own numerous experiments conducted for the purpose of solving other problems.

Thus these authors have quite rightly established in their experiments that spring habit is dominant. But N. I. Vavilov proceeds from the assumption that genes are immutable in a vast series of generations. Consequently, when in the very same pamphlet in which he describes the above-mentioned experiment and establishes the dominance of spring habit he discusses the question as to which forms came from which (the winter from the spring forms or the spring from the winter), he arrives at the very opposite conclusion. He classes spring habit of plants as a recessive character, and does so only because a number of winter crops (rye, for instance), when sown in the spring, produce a few flowering plants which, however, appear late.

Proceeding from formal genetics, no other conclusion could be drawn than that if spring plants appear among a spring-sown winter variety, they must have been in that variety in a latent form before the sowing. Sown in the autumn for many years, the spring plants did not freeze; and the winter variety did not purge itself of them only because the spring habit genes are recessive.

According to the logic of the recombination theory, these authors are right, but in essence N. I. Vavilov and Kuznetsova allowed a good piece of work to slip from their hands. Had they taken Darwin's theory of evolution as their basis they could easily have arrived at the conclusion that, at a certain period in their lives, and under certain conditions, winter plants can change, can convert their hereditary nature of winter habit into the hereditary nature of spring habit, and vice versa, which is exactly what we are now doing experimentally with fair success.

If N. I. Vavilov accepts our point of view, he too will be able to convert the winter habit of plants into spring habit; and any winter variety can be transformed into a spring variety, irrespective of the number of plants.

Vavilov. You are altering heredity?

Lysenko. Yes, heredity!

Unfortunately the geneticist conception, with its immutable genes in a long line of generations and its denial of the creative role of natural and artificial selection, dominates the minds of many scientists. To understand how this situation in science could have arisen historically, we must turn to the following words uttered by Kliment Arkadievich Timiryazev:

"From 1900 onward, first in Germany and then, louder still, in England, the name of Mendel was extolled and an importance was attached to his work totally out of proportion to its merit. The cause of this unscientific phenomenon should obviously be looked for among circumstances of an unscientific order. The source for this pest, which the future historian will contemplate with perplexity, must be sought in another phenomenon, moving not only parallel but undoubtedly also in unison with it. This phenomenon is the intensification of clerical reaction against Darwinism. In England this reaction was exclusively of clerical origin. When Bateson's own campaign not only against Darwin, but against the theory of evolution in general (*Materials for the Study of Variations*, 1894) had passed unnoticed, he gladly clutched at Mendelism and soon created a whole school. The field for such activities was open to all, since it required neither knowledge, skill, nor even the ability to think logically. The recipe for research was exceedingly simple: perform cross-pollinations (every gardener can do that), then count in the second generation how many take after one parent and how many after the other, and if it comes to about 3:1 the job is done. Then sing praises to Mendel, not failing to take a passing dig at Darwin, and start on another job. In Germany the anti-Darwinist movement did not arise on clerical soil alone. An even firmer prop was provided by the outburst of narrow nationalism, of hatred for everything English and exaltation of everything German. This difference in points of departure even found reflection in the attitude taken towards the very personality of Mendel. Whereas the cleric Bateson was particularly concerned with clearing Mendel of all suspicion of being of Jewish extraction (an attitude that only recently had been inconceivable in an educated Englishman), Mendel was particularly dear to the heart of his German biographer as 'Ein Deutscher von echtem Schrot und Korn.' The future historian of science will probably view with regret this incursion by the clerical and nationalistic element into the brightest field of human endeavour whose sole object is the discovery of truth and its protection from all worthless extraneous matter."[1]

I think, comrades, that we ought to ponder over this passage coming from the pen of Timiryazev.

Can genetics be set up against Darwin's powerful creative theory of evolution? I think that under no circumstances can genetics be set up against Darwin's theory of evolution. Genetics must be developed only from the standpoint of Darwinism and on the plane of Darwinism. Only if that is done will our Soviet genetics be truly effective. (*Loud and prolonged applause.*)

---

[1] К. А. Тимирязев, *Мендель*, энциклопедический словарь «Гранат», т. XXVIII, тр. 454.

COLLECTIVE-FARM LABORATORIES AND AGRONOMIC SCIENCE[1]

THE COLLECTIVE and state farms throughout our boundless country marked the twentieth anniversary of the Great October Socialist Revolution by gathering in an immensely rich harvest. This achievement was by no means accidental. The way for it was prepared by the whole of the preceding development of our great Land of Socialism. Low crops are things of the past and can never recur in a country where agriculture is more highly mechanized and conducted on a larger scale than anywhere else in the world and is based on the principles of Marx, Engels, Lenin and Stalin, on a science whose foundation is dialectical materialism and the splendid practice of socialist agriculture.

Only collective and state farms have full opportunity to make use, and in many ways are already making use, of all the highest achievements of world agronomic science, and the highest achievements of agricultural practice.

It was the endeavour of our socialist agriculturists as rapidly as possible to make the fullest use of all the achievements of agronomic science, both old and new, that created the necessity of establishing numerous research stations and institutes, and also a dense network of collective-farm laboratories. The chief function of this immense network of research institutions, including, of course, the collective-farm laboratories, is concretely to work out and appraise already known and also new agricultural methods in conformity with the conditions prevailing in the particular districts and collective farms.

Our research institutions, stations and institutes will be able promptly to appraise already known agricultural methods and elaborate new ones suitable for the respective districts only if they closely collaborate with the collective-farm laboratories. They will thereby promote the further progress and development of agronomic science. And only in our country is complete unity between theory and practice possible. The aim of all Soviet agrobiological research work should be to raise and improve the quality of the

[1] *Яровизация*, № 5 [14], 1937.

crops of our socialist agriculture. This is the aspect from which the work of the different branches of our agricultural science should be estimated.

On the very eve of our glorious anniversary a big conference of Ukrainian collective-farm experimenters was held under the auspices of our Institute. The conference was attended by experimenters from other Union Republics too, and also by research-institute workers and agronomists; but the majority of those present were managers of collective-farm laboratories.

The characteristic thing about this conference was that nearly all the speakers (and there were scores of them) gauged the work of their respective collective-farm laboratories by the annual increase in the crop yield of their collective-farm fields due to the work of these laboratories. Listening to these speeches, one could not help recalling academic sessions, and also reports of research institutes, in which this correct criterion—gauging the work of the Academy or Institute by the increase in crop yields on collective and state farms—is still only too rarely applied.

In the Soviet Union, the border line between physical and mental labour is being rapidly obliterated. The work of many of our collective-farm laboratories is an excellent illustration of this countryside occurrence. Karpov, the manager of the laboratory at the Kommunar Collective Farm, Voznesensk District, Odessa Region, stated repeatedly in his speech that he never conducted experiments that were not of immediate benefit to the collective farm. If he were to act otherwise, he said, the collective farmers would make a laughingstock of him.

It is greatly to be regretted that this excellent means of keeping scientists from engaging in useless researches is not yet employed in our scientific world. There is still too little genuine criticism among scientists. As soon as collective farmers see that their laboratory is experimenting on something not worth while, they forthwith make a laughingstock of the experimenters. In scientific research institutions people are more polite. At sessions of the Lenin Academy of Agricultural Sciences one may often hear that this or that member of the Academy is just wasting time, but instead of pulling such members up by means of sound criticism, other members get up and make speeches somewhat as follows: "Ivan Petrovich is studying such and such a problem and is arriving at an interesting conclusion, but I must say that I am of a different opinion." And there the matter ends. In most cases, even this "different opinion" differs little from Ivan Petrovich's "interesting conclusion."

Sound criticism guides the work that is in progress. It keeps the investigator from straying from the right road; it guides him along the true road. This is particularly needed in our agricultural science, which today, unfortunately, is still very much entangled by and adulterated with pseudo-scientific ideas.

The work of the collective-farm laboratories is appraised by the only correct criterion—the increase in the crop yields and income of the given collective farm as a result of the measures proposed by its laboratory. When planning a given experiment on his collective farm, every experimenter is responsible for the scientific aspect of the matter; and the scientific correctness of his experiment is at once appraised by life, by collective-farm practice. But what is practice? In July, when the grain crop was being harvested, the collective-farm laboratories conducted experiments in the planting of potatoes. Before that, people had known very well from practical experience that in the South potatoes did not come up well even when planted early, in the spring. Yet the experimenters took it into their heads to plant them in the hottest and driest season of the year! Could the collective farmers help ridiculing such a science?

But the manager of the collective-farm laboratory, who knew the theory underlying the experiment, confidently explained it to the collective farmers, and fully convinced them in the autumn with the aid of tons of fine, large potatoes. And then, of course, the collective farmers laughed at those experimenters who, having failed to understand the theory of summer planting, left their collective farms without potatoes.

If there are still some collective farms in the south of the Ukrainian S.S.R. which lack a sufficient supply of potatoes, it is primarily the managers of their laboratories who feel that they are morally to blame for this. Everybody knows that a method like the summer planting of potatoes is new, and it is primarily the manager of the collective-farm laboratory who is at fault if his collective farm is lagging behind all those which adopted this method, and thereby fully supplied themselves not only with seed potatoes, but also with food potatoes. In other words, it is the duty of the experimenter not only to show that it is possible to grow good potatoes, but, as a member of the collective farm, to see it fully supplied with them.

As a result of all their experiments, collective-farm laboratory managers have already gained high prestige among their fellow collective farmers. Scores and hundreds of collective-farm experimenters can be mentioned at whom nobody will think of laughing when they propose an experiment that at first sight looks fantastic. Nobody will dare laugh at them for the simple reason that the collective farmers have already tested their work and have themselves begun to study the agricultural science that is based on the theory of development of plant organisms.

Collective-farm laboratory managers and active collective-farm experimenters are in extreme need of a lucid and sound theory of agronomic science.

The distinguishing feature of our Soviet agronomic science is its effectiveness. Consequently, Soviet agronomic science must err as little as possible. In the main the old agronomic science held aloof and still holds aloof from life, from practice. Its wrong theories (and there are plenty of them) serve no useful purpose, although the "theoreticians" who invent them often receive scientific degrees even today.

What would happen if the proposals emanating from institutes that are closely connected with collective- and state-farm production were based on wrong theoretical principles? The managers of collective-farm laboratories who took these erroneous proposals on faith would begin to experiment. One after another, these experiments would turn out to be failures. The collective farms would derive no benefit from them. In that case, of course, the collective farmers would laugh to scorn both the collective-farm experimenters and the scientists. And it would serve them right!

If such fallacious proposals were applied directly in production and caused a collective farm harm, its members would kick the laboratory manager who applied them out of the laboratory. In such a case the first to lose prestige would be the research institution that fathered these theoretically wrong propositions.

In general, scientific work for the collective and state farms is an extremely responsible matter; all the more necessary is it therefore to fight for the correctness of our agronomic knowledge. Fighting to purge our agronomic science of pseudoscientific tenets means fighting to make our practical agronomic activities an effective weapon.

I have been informed that the newspaper *Sotsialisticheskoye zemledeliye* enquired of the experimenter Ivanov (Odessa Region) what he needed and that it received the simple answer: "Knowledge." This was a profound and true answer! It expresses the desire of the vast masses of collective-farm laboratory managers. It is hard to imagine that anyone in our country who, having properly set to work to unravel the tangle of biological laws governing the development of plant organism, would not thirst for more knowledge on this subject. That is why competent, talented experimenters like Maltsev, Ivanov, Kruglov, Litvinenko and many others want to get hold of good books on agronomy that correctly expound and reflect the laws of development of plant and animal organisms. These people are perfectly aware that immense work is expected from each one of them, and that from the scientific point of view this work is much more difficult than that of some highly skilled specialists at institutes and academies.

It is enough for a collective-farm laboratory to err once, let alone several times, in making experiments which result in nothing useful for the collective farm, for that laboratory manager to be discharged at once. The thirst for true agrotechnical knowledge is due precisely to the fact that the work of the collective-farm laboratory is tested hourly by life itself. The experimenters must not base their work on false theoretical principles—every mistake here is discovered soon enough. It is to be regretted that the methods employed by many scientists are still far different from those employed at collective-farm laboratories. Some scientists work on a problem for many years, but as the result is not tested by practice they never really know what they may expect to accomplish and of what practical use their work will be. One can easily see that such investigators are unable to delve deeply into the scientific truth of the problems they are studying.

Scientific specialists need true knowledge even more than experimenters do. Unless this true knowledge is acquired, the work of scientific research institutions cannot be really and effectively linked up with that of the collective and state farms; and without the closest contact with practice, theory can make no progress.

I do not know how far I have made clear the role and importance of these laboratories to collective and state farms; but I know and feel that mass experimenting is of tremendous importance both for our socialist agriculture and for our socialist agronomic science.

The work of the collective-farm laboratories is objectively appraised in extra tons of potatoes, centners of wheat, litres of milk and kilograms of wool obtained by the collective farms as a result of implementing the proposals made by the laboratories.

This measure should also be applied in evaluating the work of research stations and institutes. How much has the institution you are at the head of added to the yields of potatoes, wheat, wool and milk obtained on collective and state farms? If it has added much, it shows you have worked well. If it has added little or nothing at all, your work should be rated accordingly.

Every workingman and woman is fully conscious of the beauty of our Soviet reality, of our Soviet system. Indeed, can a man enjoy greater spiritual pleasure than that of feeling and knowing that he has made a contribution to agronomic thought, that he too has laboured usefully in the common cause of making socialist agriculture flourish? Your small theoretical achievements which, as a rule, are simply a generalization of observations in the process of production, yield results in production—when they reach the collective and state farms through definite channels—which you can never attain when working alone, or with a small staff of assistants.

I confess that I myself, when summing up the work of the Institute I head, and consequently, when summing up my own work, notice quite a large number of defects; but for all that, the work gives me profound satisfaction. Can an investigator fail to be inspired by the fact that this year, 1937, the collective farms have harvested 10,000,000 centners of extra crops from the vernalization of grain, that the collective farms have at least 100,000 tons of pure-strain seed potatoes of the best quality obtained from summer planting, and tens, perhaps hundreds of thousands of centners of extra raw cotton obtained in the new cotton-growing districts as a result of topping? The time is not distant (I have hopes of it being in 1938-1939) when tens and hundreds of thousands of collective-farm fields will yield, by way of a mass experiment, one, two and even three extra centners per hectare of grain obtained from seeds rejuvenated by intravarietal crossing. Good varieties of fodder crops have reached the collective-farm fields; and new varieties of early and high-yield cotton are being sent to the collective farms.

All these measures are, in one way or another, connected with the work of the Institute I head. But it should be clear to everyone that this is only a trifle compared with what could be done under our socialist conditions, with productivity of labour ever so much higher than in capitalist society.

It is to be regretted that our economic organizations do not, as a rule, even to this day appraise the effectiveness of scientific work in terms of centners, rubles, etc. This should be the criterion of the correctness, of the success or failure, of a given scientific measure.

Some scientists say and write that the work our Institute is conducting in conjunction with the collective-farm laboratories lacks a theoretical basis. Such statements can be found, for instance, in articles written in controversies over problems of genetics. These scientists admit the practical value of our work, but, they say, there is no theory in it. Is that so?

Let us briefly examine our theoretical views and compare them with those held by these scientists. Let us examine, for the purpose of illustration, such an important practical problem as the growing of seed potatoes under southern conditions, and also of seed growing for other field crops, winter wheat for example.

In both cases we proceed from the Michurin proposition that the conditions under which plant organisms are grown are a factor for the improvement or deterioration of the planting or sowing material.

The problem of combating the degeneration of seed potatoes in the South and Southeast has been studied for decades by entire institutes specially set up for the purpose. Whole societies in countries all over the globe have been trying to solve the potato problem in hot southern districts. None of them have succeeded in solving this problem. Nay, more! It is perfectly evident now that their scientific researches have not brought them anywhere near to a solution of this problem.

Under our socialist conditions this problem has been solved not only quickly but fully. We shall not deal here in detail with the way this problem of combating the degeneration of potato planting material under southern conditions was solved. Most collective-farm laboratory managers are perfectly familiar with this, for they took part in the work. I shall merely mention the kind of potatoes that grew in our southern districts formerly. They were about the size of walnuts. Already in October and November the tubers of these potatoes lost their turgescence. They became soft and sprouted. The yield was usually a ton or ton and a half per hectare.

Figure 47 shows, on the left, tubers of the Ella variety of potatoes which had been grown for seven years in our Odessa district by the ordinary method of spring planting. On the right are specimens of the same variety from the same consignment. For the first two years they were grown in the ordinary way, but in the last five years they were summer planted.

The specimens on the left are small, the yield is no more than a ton per hectare and they are useless either for food or for planting. Those on the right have a yield of 10 to 12 tons per ha., the tubers are large (300 to 600 grams) and are of the best quality both for food and for planting.

The entire problem of making it possible to grow good, healthy, undegenerated potatoes under our southern conditions was reduced to the simple method of planting potatoes in our fields in the beginning of July.

Fig. 47. Potatoes, Ella variety

Left: five tubers from spring-planted crop. The tubers are small and, on the average, weigh 22 gr. each. The seed potatoes of this variant were cultivated in the Odessa district for seven years by the spring-planting method. Yield—1.5 tons per hectare. Right: five tubers from summer-planted crop. The tubers are large, average 470 gr. The first two years the seed potatoes were planted in the spring and the next five years in the summer. Yield—12 tons per hectare

It goes without saying that this simple method was arrived at on the basis of a profound theoretical study of the causes of the degeneration of seed potatoes when cultivated in the South. Simultaneously with ascertaining the causes of the degeneration of potatoes it was necessary to devise methods of removing these causes, and this is what our Institute set about doing in 1933 in conjunction with the thousands of collective-farm laboratories in the south of the Ukrainian S.S.R.

This work at the same time tested and in many ways corrected our preliminary theoretical assumptions, and simultaneously many collective farms in the south of the Ukrainian S.S.R. obtained full supplies of potatoes of their own growing.

Many collective farms in the South already have obtained 100, 200 and even 500 tons of potatoes, and of the best quality. But many South-Ukrainian collective farms unfortunately have no potatoes yet, and for this we and the collective-farm laboratory managers are partly to blame. Profound scientific problems can be solved jointly with the collective-farm laboratories only if the needs of the collective farms are taken into consideration. We must not stop halfway when solving a scientific problem. The work must be carried through to the end, to the practical application of the solution. Take the report made by Rachinsky, for example. He is one of the best collective-farm laboratory managers. He fights hard for a good yield and tests varieties thoroughly. He obtained a yield of 41 c. per ha. of winter wheat on a solid area of 50 ha. But why is it that this year the collective farm where Rachinsky, one of the best collective-farm laboratory managers, is working, has not a single centner of potatoes that he could be proud of, and by which he could show that his collective farm was able to compel potatoes to grow under the conditions of the hot, dry South? Last year he himself stated that he had harvested 900 c. of potatoes from 3 ha. of summer planting.

The collective farm of which Rachinsky is a member has a bad potato harvest this year not because of drought, but because they planted nondescript material and not the material obtained from last year's summer planting. It may also be assumed, and it was no doubt the case, that in many collective farms the seed potatoes from the 1936 summer planting were used up for other purposes and not left for seed only because somebody did not want these collective farms to have planting material, and hence, of course, a good potato crop. It is already possible to flood the South with potatoes, but, unfortunately, we have not yet achieved this. It is our duty to get all our southern collective farms as quickly as possible to grow good crops of the best table varieties. The collective farms in the south of the Ukrainian S.S.R. must not only supply potatoes for the southern towns and industrial centres, but also be the principal purveyors of early young potatoes for our metropolitan centres and thereby show that complex scientific problems are being solved in the Soviet Union.

By the united efforts of our scientific research institutions and of the large network of collective-farm laboratories the problem of combating the degeneration of potatoes in the South has been completely solved. This is confirmed by hundreds of tons of potatoes at thousands of collective farms which grew these crops from their own planting material.

The practical solution of the problem of combating the degeneration of potatoes in the South and the generalization of the mass experience of the collective farms have led to new discoveries, have increased our knowledge of plant life. It has been ascertained that the summer planting of potatoes, as a result of which the new tubers form under the cooler autumn conditions, not only keeps the tubers from degenerating, but alters the *breed* of the potato, not in the direction of degeneration, but *in the direction of acquiring greater vigour*. This has already been tested and proved for three years on large areas under collective-farm conditions. At our Institute we have had the same results for five years with the medium-ripening variety Ella, which A. F. Kotov, a specialist at our Institute, had taken for experimentation.

For four years, year after year, one variant of this potato was planted in the spring and another variant was planted year after year at the end of June and beginning of July. As a result, what seemed to be two distinctly different varieties were obtained. The first was a breed of potato that was good for nothing; the second was a variety that produced a good crop of large tubers. There can be no doubt that an alteration in the breed (genotype) took place here, for how otherwise is it possible to explain the different behaviour of the plants of these two variants when planted in the spring of 1937 under equal conditions? They produced different yields and tubers of different sizes and shapes; and the plants of these two variants bore stems and leaves of different appearance and vigour.

A. F. Kotov conducted a second experiment in altering the breed of potatoes in conformity with the conditions of cultivation. In the autumn of 1935 he took the crop of individual plants of the Early Rose variety, which had been planted in July. The total number of plants selected in 1935 was about three hundred. The crop from each plant, ranging from 5 to 10 tubers, was kept separately. In 1936, half the crop of each plant was planted in the field in spring; the other half was planted at the end of June. In the autumn of the same year the crop was harvested both from the spring- and from the summer-planted tubers. In the spring of 1937 tubers from the spring-planted 1936 crop were planted in the field, and next to them were planted tubers taken from the same plant (1935 selection), but grown from the 1936 summer planting. In the spring and summer of 1937 this experiment presented a magnificent picture. From the appearance of these 300 descendants one could easily tell which of them had come from the tubers grown from the previous year's spring planting, and which from the summer planting. In many cases the crop from plants obtained from the tubers of the 1936 summer planting was two or three times as large as that of plants from the same clone, the seed potatoes of which had been obtained from the 1936 spring planting.

In order to prevent them from degenerating as a result of the spring planting of 1937, Kotov took some of the tubers from the twelve best clones and planted them this year in the summer apart from those he planted in the spring. For the purpose of comparison, tubers from the same clones of last year's spring-planted crop were planted in the summer. Thus, it is possible to compare the crops of four variants obtained from one and the same plant, as is evident from the following table:

AVERAGE CROP FROM ONE POTATO PLANT DEPENDING ON PRECEDING CONDITIONS OF CULTIVATION OF PLANTING MATERIAL

No. of Initial Plant of 1935 Selection	1937 Spring Planting of Tubers from				1937 Summer Planting of Tubers from			
	Summer Planting 1936		Spring Planting 1936		Summer Planting 1936		Spring Planting 1936	
	Crop in Grams	Number of Tubers	Crop in Grams	Number of Tubers	Crop in Grams	Number of Tubers	Crop in Grams	Number of Tubers
59	315	9	153	5	847	8	513	6
82	373	6	247	6	606	9	320	4
105	263	5	126	4	740	7	353	5
181	450	9	205	8	800	6	363	4
203	408	9	45	6	713	9	364	8
213	173	6	82	6	747	7	440	5
221	250	14	87	8	820	10	413	8
224	320	7	208	6	527	6	440	6
232	413	7	83	5	933	11	260	4
244	443	15	212	10	547	7	420	5
265	538	11	42	5	727	5	407	6

From the table it is easy to see that:

1. Summer planting resulted in a much larger crop per plant and in larger tubers than spring planting. This is observed in plantings of tubers from the summer planting and in those of tubers from the spring planting of the preceding year. (Compare sixth column with second and eighth with fourth.)

2. With summer planting we got a more productive breed of seed potato. (Compare second and fourth and also sixth and eighth columns.)

The above results of Comrade Kotov's experiments show how sensitive the varietal nature of potato plants is to altered conditions of cultivation. It is extremely important to know this not only to prevent standard seed potatoes from deteriorating, but also to improve them from year to year, by noting and choosing the best conditions for growing them.

We have already observed that in the southern districts summer planting yields considerably better and healthier seed potatoes than spring planting. It is now necessary to examine this in greater detail. It is necessary experimentally to determine the different dates of summer planting of potatoes for the different districts of the South of the U.S.S.R. and for the different varieties. We must also learn to create such conditions for growing planting material as will produce the best breeds for spring planting in our districts with the object of obtaining young potatoes for food, and for summer planting with the object of obtaining planting material and market potatoes for autumn and winter and early spring consumption.

All that I have said about altering the breed of potatoes in conformity with conditions of cultivation is already known to many collective-farm laboratory managers from the work they themselves have done. Nevertheless, some scientists who espouse the principles of the old agronomics, the principles of the old genetics, are still convinced that conditions of cultivation play no role, or almost no role, in altering the nature of plant organisms. Such views on plant life and development not only made it difficult for these scientists to solve the problem of combating the degeneration of potatoes in the South, but today, when this problem is already solved, prevent some of them from understanding the crux of the matter.

Quite recently a highly-skilled research worker submitted for publication, in a mass edition for the enlightenment of leading collective farmers, a manuscript in which he wrote things about the growing of seed potatoes that cannot be taken seriously today. According to this author, a variety well known to collective farmers, namely, Early Rose, though cultivated for 75 years in different zones of the globe, has not suffered any substantial change. This means that everywhere this variety has remained the same as it was 75 years ago. Something has prevented this author from learning even in 1937 that growing the Early Rose variety under spring or summer conditions in our districts fundamentally alters its breed in literally 3 months, let alone 75 years.

In this same manuscript the author said that varieties of potatoes, chiefly old ones that were bred a long time ago, degenerate for reasons unknown to science. This is not true. Not only are the reasons for the degeneration of potatoes well known to every collective farmer, at any rate in the southern districts of the U.S.S.R., and particularly to collective-farm laboratory managers, but they are already being *eliminated* by means of summer planting.

* * *

Let us take another problem, that of intravarietal crossing.

In 1936, on our suggestion, collective-farm laboratories for the first time began to use tweezers in order to test Darwin's proposition that the cross-pollination of plants subjected to long-continued self-pollination is beneficial. Some scientists hastened to ridicule this proposal. They even cracked jokes about it and said: "Lysenko and Prezent propose arranging 'marriages for love' among plants." But there is no need to be afraid of being laughed at. Ridicule can frighten only those who feel guilty. We, in collaboration with numerous collective-farm laboratories, are performing a useful, scientific task for the collective farms, and we explain the matter to collective farmers and scientific workers. To the scoffers we say: "You don't believe now, but see the results of this experiment in a year's time." Some people laughed at vernalization, at the summer planting of potatoes, and at the topping of cotton plants. But those who laughed maliciously at all this are not laughing now. We are convinced that in a year from now it will be the same with intravarietal crossing.

Agriculture has to do with the rearing of plants and animals. The development of living organisms is, as we say, diverse and varied. This means that if different conditions of life and development are created for plants, they will develop well or ill, yield big or small crops, depending on the conditions created for them.

An unmanured and badly cultivated field will yield a bad crop and poor seeds. Every collective farmer knows this.

It is obvious to us that a seed plot whose crop is to be used for seed must be cultivated as well as possible.

However it goes without saying that we must bear in mind what Rachinsky, the manager of the collective-farm laboratory in the Golovanevsk District, Odessa Region, said in his report. Under steppe conditions Rachinsky obtained an unprecedented crop of winter wheat, 61 c. per ha., not on a small plot but on 10 ha. On an area of, I think, 50 ha. he obtained an average crop of 41 c. per ha. Nevertheless, in spite of this high yield, Rachinsky said that he had not kept his promise; he had pledged to get 70 c. per ha. The most valuable thing about this is that Rachinsky indicated the reason why he fell short of his promise: too much manure was used on a well-cultivated field, which, moreover, had been kept in a good condition in previous years. This shows that a drop in yield may result not only from poor cultivation and inadequate manuring, but also from excessive manuring. On such a field the wheat shot up rapidly and lodged, and the latter had a bad effect on grain ripening.

The examples cited by Rachinsky and other collective-farm laboratory managers prove a simple scientific truth, namely: every agronomic measure that is adopted must always be linked with a whole series of other measures adopted earlier or at the given time, and also with all the succeeding agronomic measures that it is intended to adopt. What I mean is this: every measure that is new for the given field must be concretely linked with the entire set of conditions in that field. A plant's growing conditions in the field are, on the one hand, complex and, on the other hand, inconstant. In some districts there is much precipitation, in others there is little. The composition of the soil in different districts is also different. Moreover, owing to different cultivation and different predecessors, we find in the same district, even on the same collective farm that some plots of land differ from others in soil structure, fertility, etc. All this must be taken into consideration when adopting any given measure.

The complex interconnection of a plant's environmental conditions in the field makes it impossible, so far, to foresee entirely the plant's behaviour when an old agronomic measure is changed or a new one is introduced. Before adopting a new agronomic measure in a given district, or even collective farm, it is necessary, in one way or another, to test it, to ascertain concretely what effect it will have on the crop in combination with all the other agronomic measures employed on the given collective or state farm.

Bearing in mind that "the more the better" is not applicable to all times or to all conditions, it is necessary, in order to get more productive seed, to train plants to thrive well under field conditions. On seed plots it is particularly necessary by means of agronomic measures to create all the conditions that mould the plants' development in the direction of *the utmost biological hardiness and, at the same time, of the highest yield.* If this is done the employment of selection will be most effective. Nothing good can be selected from poorly-grown plants. Before selecting plants for seed growing they must be properly reared.

In view of this, we at our Institute have fundamentally changed our methods and techniques of raising élite and superélite winter wheats. In the main, this means that year after year, after the intravarietal crossing of plants, we sow their seeds one at a time in wide rows (70×25 cm.) on a well-cultivated plot. At the time of earing we select several score of the best plants, castrate them and allow them freely to cross-pollinate with neighbouring plants. In this way we will make the nature of the variety biologically more and more plastic and better cultivated. The seeds from these plants are sown on plots similar to the preceding ones, i.e., in the nursery for the rejuvenation of the variety. In this rejuvenation nursery we select several hundred of the best plants, those that are most typical of the variety, but we do not subject them to a second intravarietal crossing. We sow some of the progeny of these plants in wide rows (70 × 25 cm.) in the seed nursery. In this nursery the variety standard of every offspring is tested and all offspring (families) that deviate from the type of the given variety, as well as the feeble or sickly ones, are rejected. The remaining families are further cultivated in the seed nursery by means of good agronomic measures. The crop from the seed nursery is used for sowing the élite.

This is the new method by which we are creating the most productive élite seeds. Meanwhile, until the new method has been fully tested and approved, we are continuing to grow élite in the ordinary, accepted way.

In less than a year from now the method of producing good élite seeds described above will already have been tested by several score of variety testing centres of the State Cereal Variety Trial Commission, to which seeds grown by our Institute as well as by a number of collective farms in Moscow and Vinnitsa regions have been sent.

Judging by the plants' vigour, we can confidently expect that the seeds we sent for the state test, obtained from intravarietal crossing and, moreover, from plants which had been grown in well-cultivated plots, will be hardier and more productive than all other seeds of the same varieties sent to the same testing centres.

Intravarietal crossing, good agrotechnique and the selection of the best plants will undoubtedly produce winter wheats of great purity, hardiness and yield.

With regard to this problem too I think that we who accept the agrobiological theory of development of Darwin, Michurin and Timiryazev stand on a higher plane than those who see no theory underlying our work, and who fail to see it for the sole reason that their anti-Darwinist blinkers prevent them from seeing it.

It is difficult to escape the thought that in our mass scientific research work in agronomy we are not so much teaching the collective farmers as learning from them.

This year we thrashed at our Institute the crop obtained from the tested seeds of an intravarietal crossing of winter wheats. Krymka wheat gave an additional yield of nearly 8 c. per ha. Most of the other winter varieties gave additional yields of 2 to 4 c. per ha. In general, I became convinced that this is a useful measure. But neither I nor anybody else knew that the intravarietal crossing of winter wheats can improve the quality of the grain, and hence, of the flour and bread. And yet a variety like Yurievka, sent us by the Ilyich Collective Farm in the Kharkov Region, showed, on analysis, 33% of gluten, compared with 29% in ordinary seeds sent us by the same collective farm, and 14% of protein compared with the ordinary

12%. The wheat Dürabl sent us from the Shatsk District, Moscow Region, showed 16.4% of protein as against 12.8%. The variety 2470 from the Moscow station showed 16.3% of protein as against 10.3%. The Zarya variety from Vinnitsa Region showed 20.6% of protein as against 11.4%, etc.

A considerable increase in the protein content of wheat grains makes bread more tasty and nutritious. This is a very important point. I was not aware of all this until I received the specimens from the collective farms. Now, this change in the chemical composition of grain as a result of intravarietal crossing is a problem that I must get to the bottom of, for otherwise what kind of investigator and head of a scientific institute would I be?

I knew that ergot affects the flowers of rye spikes, but I did not know that it can also affect wheat. When conducting experiments in the intravarietal crossing of wheat we, already in August, received alarming news from a number of collective-farm experimenters in Gorky, Chelyabinsk and other regions in the eastern parts of the Soviet Union to the effect that after intravarietal crossing not grains but ergots were formed in some of the wheat spikes. Knowing the biological development of this parasite, which affects cereals only at the time of flowering, it became clear to us why wheat was affected by ergot to a greater extent in experiments in intravarietal crossing than ordinary plants. As a rule wheat flowers are closed; hence the parasite has no access to the ovaries. With intravarietal crossing we artificially create open flowering of the wheat by clipping the palets that cover the flowers. Thereby, access to the stigma is created not only for the pollen in the air, but also for the harmful spores of ergot and, of course, for the spores of stinking rust. The measures to be taken to combat this also became clear to us. Instructions were issued, firstly, to pick off by hand all the ergot from the grain crop obtained immediately after the experiment in intravarietal crossing; and secondly, without fail to disinfect by thermic treatment all the grain obtained from the intravarietal crossing (each collective farm has no more than 2 to 10 kg.). When reproducing these seeds later on, the flowering of the wheat will, of course, proceed in the ordinary way, i.e., it will be closed. Therefore, after the complete elimination of the infection from the small amount of initial seeds, there is no need to fear that in subsequent reproduction the plants obtained from the seeds of intravarietal crossing will be more susceptible to infection by stinking rust or ergot than the ordinary plants of the same variety of wheat.

Agronomic research is very responsible work, being of great importance to our socialist agriculture. Therefore, it is urgently necessary for us to delve more and more deeply into theory, to gain an ever more profound knowledge of the biological laws of plant development.

We know from Darwinism that in nature the entire plant and animal world was and is being created by development, by natural selection. Good breeds of animals and diverse varieties of plants have been created by arti-

Fig. 48. Third generation of Kooperatorka wheat altered by training

Left: two plants (1890 and 1896) from seeds sown on April 3, 1937, in heated greenhouse, the plants eared. Right: two plants (1913 and 1909) from seeds also sown on April 3; the plants were left in the open ground and did not reach earing until late in the autumn

ficial selection in the course of their domestication and cultivation. In view of this we say that before selecting plants for seed, care must be taken to cultivate them in the best possible way, and from these well-cultivated plants to select individuals for breeding purposes. The selected plants must also be cultivated under conditions that will make them most resistant to climatic inclemencies and at the same time give the highest yields.

The nature of organisms is continuously undergoing change and, of course, under cultivation its state of culture improves, whereas the nature of plants deteriorates when they grow wild.

This knowledge can be acquired, and we have already acquired it from our own experiments and from the books of the greatest agrobiologists in the world, such as Darwin, Michurin, Timiryazev, Burbank, Williams and others. Unfortunately, there are as yet no books on agronomic science that tell us concretely how to train cultivated plants in such a way as to make their natures *change quickly and radically* in the direction we want. Rapid and considerable changes in plant organisms are quite often observed in nature. In agronomic science such changes in the nature of organisms are called mutations. *We must gain control over the mutation process, we must learn how fundamentally to change the nature of plants in the direction we want.* Under our socialist conditions of life this not only must, but already can be learnt. We agrobiologists and the entire mass of experimenters must all examine the plant world around us more closely and carefully. We must know the life and development not only of cultivated, but also of wild flora. We must examine and ponder over the question of how, on the basis of what laws, changes in plant organisms take place in nature. By carefully watching plant development in surrounding nature we can learn a great deal that will help us quickly to master the art of directed form-building.

The bourgeois science of genetics, which is divorced from life, comforts itself with the reflection that it will be possible to control the laws governing the mutation process only after many decades. Many geneticists even make efforts to prove that this will never be possible.

I, however, think that Soviet agronomic science will be able to solve this problem within the very next few years; in rough outline it has already solved it. If we now buckle down to work with a will on a wide front, I am profoundly convinced that, under our conditions, immense success in the theory of the subject will be achieved. The solution of this profound theoretical problem must be taken up not only by the scientific research institutes, but also by scores and hundreds of collective-farm laboratories.

At the present time our Institute of Selection and Genetics is working hard to alter the nature of agricultural plants by training. Experimentally, we have already reared a wheat that does not ear when spring-sown in the field because the spring and summer conditions of our warm South are too cold for it. Yet we reared this wheat by training the plants of the winter wheat Kooperatorka. It is well known that spring-sown winter wheats fail to ear because our spring is too hot for them. (Winter plants cannot vernalize because of the high temperature.) The Kooperatorka that we have altered by training cannot vernalize in the field under spring conditions, not, however, because it is too hot, but because this field temperature is too low to enable it to vernalize.

We are convinced of this by the fact that the "sisters" of the plants that failed to ear after being spring-sown in the field do ear if sown on the same spring day as the former, but in a greenhouse, where it is considerably warmer than in the field. Many people, no doubt, know that we did not obtain this wheat by chance; we *deliberately* altered its nature by training. When we started on this experiment we, on many occasions, announced our theoretical expectations at sessions of the Lenin Academy of Agricultural Sciences of the U.S.S.R. and in the press; and today, in spite of the sceptics, our assumptions stand confirmed.

Over two years ago I arrived at the conclusion, from all I knew about the life and development of plants, that if winter-wheat plants, which require cold at one of the stages of their development, namely, the vernalization phase, are trained in that period of their lives under higher temperatures, they can be altered in such a way that the seeds harvested from them will also vernalize at higher temperatures.

The painstaking experiments along this line conducted under my supervision by Laviniot during that period of over two years have been convincing us more and more that in this way it is possible to alter the nature of plants requiring cold at a certain period of their lives into that of thermophilic plants. Moreover, it has become evident to us that it is possible to do this quicker and better than the way we are doing it now. In the process of solving the problem we face, our knowledge in this sphere has been greatly increased. Hence, of course, our greater boldness and confidence in the possibility of the rapid directed alteration of agricultural plants.

The assumptions from which we proceed when starting on these experiments over two years ago were the following:

1. The conditions under which a plant is trained do not play a passive part in the alteration of its nature. Organisms relatively identical at the outset will bring forth increasingly diverse progeny if trained under different conditions. Recall, for example, the growing of seed potatoes by spring and summer planting.

2. Alterations in the nature of plant organisms can and must be made *in a definite direction* if the organism is trained *in a definite direction.* The diverse variations observed in nature round about us take place only because the natures of these organisms and the conditions under which they develop are diverse. It is necessary to ascertain what conditions are required, and for what state of the plant organism they are required, in order to change it in the direction we want.

This explains why in our first experiments in altering the nature of organisms by training we took winter wheat to change its vernalization phase. We know much more about the developmental process of the plant organism called the vernalization phase than we do about any other process. We know when the organism goes through the process of vernalization, and we also know what external conditions are required for this process.

That is why we began to train winter-wheat plants in the period when they pass through the vernalization phase under higher temperatures than are usual for the vernalization conditions of the variety of winter wheat named Kooperatorka.

As a result of these experiments, after two or three generations had been grown under these conditions, the nature of winter wheat entirely lost the ability to vernalize at low temperatures and acquired the ability to vernalize only at high, greenhouse temperatures. The spring temperature in our districts was now too low for the nature of this new wheat. I will state once again that the spring temperature conditions in our districts are usually too hot and not cold enough for the vernalization of winter wheat of all varieties.

The most valuable result of our experiments is that we have made further progress in theory. We have ascertained more concretely the essence of the alteration of the nature of plants and the methods of altering it. We have ascertained that when the nature of the vernalization phase, which requires a low temperature, is being altered the plants must be given a higher temperature *not at the beginning or the middle of their passage through the vernalization phase, but at the end of the process.*

We have now arrived at the conclusion that if it is desired to alter the nature of a plant requiring cold at a certain stage of its life, winter wheat for example, into one that requires cold less, the plant must be given the opportunity to go through this process under optimal, i.e., good conditions, *and to finish this process under conditions in the direction of which it is desired to alter the plant organism.*

If we want to transform a winter into a spring variety, we must provide higher temperatures towards the end of the vernalization process. If, however, the aim is to increase the winter habit of a winter form, then towards the end of the vernalization process it must be given a temperature lower than that under which winter varieties vernalize best.

In addition to the wheat plants and seeds we have obtained experimentally from the winter variety Kooperatorka, which now possess a different nature, we have also obtained real spring plants from the winter variety Novokrymka 204. We also have spring plants from Kooperatorka obtained in the experiments conducted by Shimansky, a postgraduate at our Institute.

The new plants were obtained from the winter variety Novokrymka 204 in the following way. In the spring of 1936, some slightly undervernalized seeds were sown in the field. In the spring of 1937 the crop from these seeds was sown in open ground without preliminary vernalization. Some of the plants from this sowing produced progeny that was entirely of spring habit. Figure 49 (on p. 213) shows these wheat plants which, sown in the greenhouse on September 9 without preliminary vernalization, reached earing on November 2. Hence these plants do not need cold for vernalization.

Fig. 49. Novokrymka 201

Five pots on left: plants transformed into a spring form. Right: ordinary winter plants of the same variety

Our numerous observations of the way in which the Kooperatorka we are altering behaves in the different variants of our experiments, and also of the behaviour of the spring plants obtained from the winter variety Novokrymka 204 and of the plants Shimansky obtained from the winter wheat Kooperatorka, led me to the conclusion that it was necessary to make more precise the method of training plants we employed in our experiments for the purpose of altering their nature.

I deem it necessary to emphasize here once again that our experiments in altering the nature of Kooperatorka have proved beyond doubt that the nature of plants *can* be altered in a definite direction by training: we have already transformed plants that require cold in the vernalization phase into thermophilic plants.

We are now proceeding to make more precise and to simplify the methods of controlling the alteration of the nature of plants. With this object in view we have started a number of experiments based on the reasoning outlined above.

I advise all collective-farm laboratory managers who wish to take part in the testing and elaboration of this problem to start by altering the nature of winter wheat into spring wheat and also by altering the nature of winter wheat in the direction of increased winter habit. For this they must take the variety of winter wheat that is sown on their respective collective farms. From this variety they must make up 40 or 50 packets of seeds, each packet containing 10 grams. At the beginning of February in the southern districts and at the end of February in the midland and northern districts, each day one of these packets should be kept in water for about 12 hours; after that it should be kept at living-room temperature for about 12 hours to allow the seeds to begin germinating. Then the seeds should be placed under conditions where the temperature is 0°-2° C., i.e., the conditions which, as everybody knows, are usual for the vernalization of winter wheat. In this case vernalization can be performed also in the following way: pile up a heap of snow and in it place a tin can. Into this can throw each day a packet of seeds that have been treated in the way described above. Care must be taken to regulate the temperature in the can with a layer of snow and straw. When spring comes all these packets must be taken out and the seeds from each packet sown in a bed of one or two rows, each row being one to one and a half metres long. In this sowing, the seeds will be of the same variety, but their presowing vernalization will have lasted for different lengths of time: from one day to fifty days or more. Part of the undervernalized seeds will complete their vernalization in the field. The seeds that are still more undervernalized will not manage to complete their vernalization during the spring and therefore will not ear. In the summer the seeds must be gathered from all the variants of plants that have eared, keeping those of each variant separately. In the same summer part of the seeds from each variant may be sown in the open or in pots, but the larger part should be kept for sowing in the following spring. Both in summer and the following spring the sowing must be done without preliminary vernalization. I think that from both these sowings—the summer and the following spring sowing—fully spring-habit plants ought certainly to appear from the seeds of one of the variants that have undergone preliminary vernalization in the preceding generation. Which variant will produce spring plants will depend upon the variety and upon the spring conditions under which this experiment is conducted.

For the second experiment, i.e., to increase the winter habit of winter wheat, I recommend the same procedure as described above, but it must be started not later than December 1. This will create the thermal conditions—approximately 1°-3° below zero C.—from the end of January or beginning of February, to enable all these variants to complete their vernalization and then be planted out in the open as early in the spring as possible. The crop from each variant should be gathered separately. The results of the experiment (increased winter habit) should be tested by autumn sowing. The degree of winter habit of these variants can also be tested by presowing vernalization.

I shall also describe how we are now conducting an experiment in transforming a thermophilic plant like cotton into a less thermophilic one. In the case of winter wheat we already knew that the plants had to be treated at the time of vernalization, and now it only remains for us to ascertain *at what particular point* in the vernalization process the winter nature changes *best* into spring nature. In the case of cotton, we propose to conduct the experiment (and are already doing so at our Institute) by means of frequent sowings. (At our Institute this is being done in the greenhouse.) Seeds of the same variety of cotton are sown at intervals of two days, which creates *different states* in plants of the same variety. At any given time we shall have plants of different ages. All the plants will be simultaneously subjected to the influence of the low thermal conditions in the direction of which we want to change their nature. In the spring of 1938 we shall gather the crop of these variants of cotton and sow the seeds obtained in the field. I think that the plants of some of these variants should turn out to be less thermophilic than this variety has been up till now.

I am convinced that in the plant world around us much more rapid and much more profound changes in the nature of organisms take place than geneticists think. These rather profound and rapid changes easily occur in nature only because wild plants of one and the same variety (or sort) are to be found simultaneously in the most diverse states. Next to old plants there are some less old, some quite young, and even germinating and not yet germinating seeds. In nature everything is done through the medium of chance, but since plant organisms of the same variety exist side by side in different states, I think that when the conditions environing the plants change, there is often a group of organisms of the given variety (or sort) which undergoes hereditary change easily and quickly and adapts itself to these conditions. Therefore, it seems to me that in nature plants possess much greater potentialities of hereditary change and, as a result of natural selection, much greater potentialities of adaptation to new conditions, than has been the case up till now in the experiments conducted for this purpose by many scientists in various countries.

It was their unsuccessful experiments that led geneticists to the conviction that environmental conditions, training conditions, play no role in changing the nature of plants. Actually, the nature of the organisms they experimented on remained "unchanged" simply because they conducted their experiments unskilfully, without understanding.

I am convinced that we, jointly with the collective-farm laboratories, will conduct these experiments much more successfully. This is guaranteed by all that we have already accomplished along this line, and by all the unlimited opportunities for conducting real scientific research work that our socialist country offers. If we reflect upon the gist of the results already achieved in altering the nature of plants, it will be easy to draw the conclusion that even more "wonderful things" can be accomplished.

Cotton cannot grow in the central regions of the Soviet Union, Moscow Region, say, because it lacks warmth there, the climate is too cold. But we are now convinced that the nature of the cotton plant can be so altered that the temperature which is at present too low for it will prove to be even too high. We have already begun to experiment on the fundamental alteration of the nature of the cotton plant by means of training.

These experiments open immense prospects for the effective intervention of the people of socialist society with a view to altering the nature of the plant world. Once we learn to control these laws properly we shall soon be able to create varieties of wheat of unprecedented frost-hardiness for the severely cold regions of the Soviet Union. We shall be able to shift many southern plants to the northern latitudes and northern plants to the southern latitudes, and so forth. All we have to do is to set to work to solve this profound scientific problem in a businesslike way. We must not shelve the problem. The entire mass of Darwinist scientists and collective-farm experimenters must, on the basis of what we already know today, unanimously engage in this work that is of such practical use to our collective and state farms.

The collective-farm laboratories must cooperate in solving even the most difficult agrobiological problems. They have already trained a large body of active workers who, shoulder to shoulder with the professional scientific researchers, can take up and solve complex scientific problems. We must more boldly promote capable people from among this body of collective-farm experimenters to our scientific research institutes and stations. In the Ukraine we have eight plant-breeding stations. Some people say that there is a shortage of personnel for plant-breeding work. But even if we had three times eight plant-breeding stations we could find workers for them in the collective-farm laboratories and, as their practical work has shown, they will not be inferior, and in many cases will be even superior, to those scientific workers who are counted as plant breeders, but who, after decades of work on plant breeding, have not by a hair's breadth improved the seeds of the crops they are working on. The research work of a collective-farm laboratory cannot, of course, serve a whole region. For this money and suitable forms of organization are needed. The function of the collective-farm laboratories is to render scientific service to their respective collective farms, and this they are doing. But among the active body of collective-farm laboratory assistants there are comrades who have outgrown collective-farm laboratory work, and these should be boldly promoted to do scientific research work as specialists at institutes and plant-breeding stations.

Soviet science derives its strength precisely from its connection with the masses, who twenty years ago threw off the yoke of exploitation and learned the joy of free labour under the guidance of the supreme genius of Comrade Stalin, the leader of the working people of the whole world, under the guidance of our great Communist Party, which is leading our country from victory to victory under the great, invincible banner of Marx, Engels, Lenin and Stalin.

INTRAVARIETAL CROSSING AND MENDEL'S SO-CALLED "LAW" OF SEGREGATION[1]

THE MAIN thing in the problem of seed growing is knowledge and ability to grow good seeds of the crop with which the particular seed grower is working. The science of seed growing must not be regarded as an appendage of genetics and plant breeding, as many have been and still are inclined to do. Actually seed growing is one of the major departments of agronomic science, and includes all the other branches of science that deal with the life and development of plants.

In the first place, seed growing includes that branch of science which studies the laws of variability and heredity, i.e., genetics. Seed growing also includes agrotechnique, which deals with the conditions that must be created to grow seeds of a definite quality. Also included is plant breeding, which deals with the selection of plants for breeding, and with the time and manner in which they should be selected. In addition to all this, seed growing has something specifically its own. In general, seed growing is one of the major departments of agronomic science and not an appendage of genetics and plant breeding.

Seed growing does not emerge from the tenets of genetics which in many respects are fictitious. On the contrary, the requirements of seed growing have given rise to the genuine branches of science known as genetics, plant breeding and agrotechnique (the creation of prerequisite conditions for cultivating plants for seeds).

The theoretical basis of seed growing, as of every other department of agronomic science, must be the theory of development. In this article I shall deal with only one of the numerous problems concerning the production of seeds, namely, the theoretical basis of the problem of intravarietal crossing.

During the whole period of over two years that the problem of intravarietal crossing has been studied it has always been connected with Dar-

---

[1] Revised stenographic report of a lecture delivered on April 15, 1938, at a seminar on seed growing at the All-Union Institute of Selection and Genetics.—*Ed.*

win's theory. And this problem has been connected with Darwin's theory not only by us, who advanced the problem and urge that the periodical cross-pollination of self-pollinating field crops is beneficial, but also by our opponents in matters of theory, who deny that such crossing is beneficial.

Why do people who hold diametrically opposite views appeal in the present case to the same theory, to Darwinism? The following answer may be given: having unsuccessfully tried to challenge Darwin's principles, the opponents of Darwinism are now trying to lean on Darwin's utterances.

We proceed from the proposition that long-continued self-pollination is biologically harmful because the plants become less fit, less viable, less resistant to unfavourable climatic, soil and other conditions and, following Darwin, we assert that cross-pollination is beneficial.

Our opponents, however, the representatives of the Mendel-Morgan school of genetics, who formerly totally denied the usefulness of even raising this problem, now say: cross-pollination is biologically beneficial only for cross-pollinating plants; self-pollinators benefit from cross-pollination only when the self-pollinator constitutes a population and not an ordinary pure line. Thus, the practical importance of the problem of intravarietal crossing is again reduced to zero.

I repeat—we have proceeded and continue to proceed from the proposition that intravarietal crossing is never harmful and will always be biologically beneficial to some extent. As regards field crops this, as a rule, will coincide with economic utility.

What prevents those who advocate Morganist genetics from appreciating the good of crossing and the harm of self-pollination? This question may be briefly answered as follows: they are prevented by the principle on which the theory they advocate is built; they are prevented by their wrong conception of the nature of living organisms.

The Morganists picture the heredity of organisms as some sort of special substance. This substance, like every other, they divide into discrete particles, or granules. The point is, however, that "hereditary substance" is an invention of the Morganists; it does not exist in nature. As is known, the Morganists have placed this fictitious hereditary substance in the chromosomes of the cell nuclei in a linear, catenulate arrangement. They have endowed these particles of heredity with a property that not a single molecule of the living organism possesses, namely, the property of not developing, of not undergoing transformation. To put it more simply, they have ascribed to these fictitious particles of heredity some sort of miraculous property of growing, of propagating in billions, and at the same time of not undergoing change. But we know that we cannot picture a single living organism, not a single part of an organism, as growing and propagating, and yet not undergoing change and transformation. If there is no change, no transformation, it goes without saying that there is no development.

This is exactly why seed growers and plant breeders who believe in the principles of the Mendel-Morgan theory assert that the hereditary basis of all the organisms which in the past, 10 to 15 years ago, were obtained from one initial seed has remained unchanged. Hence their view that it is useless to cross such plants.

As regards pure-line varieties, in the opinion of geneticists the harm of self-pollination and the good of cross-pollination have been disproved beyond all question.

The inbreeding of cross-pollinating plants, now proved to be a wrong method, is based on the theory of Morganism. Plants obtained by inbreeding are always feeble, of low resistance and, as a rule, less productive. The explanation of this, given by the Morganists in the past and even to this day, is that the failure of inbreeding is due not to consanguineous breeding, as the Darwinists think, but to the fact that closely-related breeding leads to the homozygosity of bad genes (accumulation of bad hereditary particles) which were screened by other, dominant genes, when the organism was in the heterozygous state.

D. F. Jones compares inbreeding to the work of a skilful detective and in effect says the following: Who will condemn the useful work of a detective if he unearths a wicked crime? But inbreeding is condemned by the ignorant who fail to understand that it really ought to be commended. Inbreeding merely reveals to us the bad genes that existed in the given variety but did not manifest themselves. The variety grew splendidly for scores of years, but you did not know that bad genes were hidden in it, and it is only due to inbreeding that the "criminals" were discovered. Were it not for inbreeding you, perhaps, would never know that the variety of plant from which your breed is made contains such evil things as hidden genes.[1]

This is what one can arrive at in the department of seed growing if Morganism is taken seriously for practical purposes, or if it is taken seriously, without a smile, as a science.

We Darwinists hold different views. We proceed from the proposition that the conditions of life, the conditions of training a plant organism, are in some degree reflected in the behaviour of the plant's progeny. No two things in the world are absolutely alike, and no two square metres of field are alike. Hence, environmental conditions are always in some degree different for different plants. For this reason alone—and it is not the only one, of course—it is never possible to picture even two plants of the purest variety as being absolutely alike in morphology or in their nature, i. e., their hereditary basis.

---

[1] "If evil is brought to light, inbreeding is no more to be blamed than the detective who unearths a crime. Instead of being condemned it should be commended." (Edward M. East and Donald F. Jones, *Inbreeding and Outbreeding, Their Genetic and Sociological Significance,* Philadelphia and London 1919, p. 140.)

Plants propagate their kind in the most diverse ways, but principally by means of sexual reproduction.

What is the principal specific feature of sexual reproduction? With sexual reproduction the organism starts life anew. Organisms which, in contradistinction, are not obtained from sex cells, but from cuttings, tubers, bulbs, etc., do not start life anew, but continue it. It is widely known that to sow seeds of, for example, winter wheat in spring it is necessary to vernalize them, i.e., to put them under conditions that will enable them to go through the vernalization phase. If, however, cuttings are taken from the fruiting parts of the plants of the very same wheat, there is no need to put them under conditions required for vernalization. If the plant is already phasically mature, and if cuttings are taken from its top, there is no need to put them under conditions for going through the vernalization phase, or the photo phase.

Sexual reproduction differs fundamentally from every other mode of propagation precisely in the fact that, in the former, the life of the organism starts anew, whereas in vegetative reproduction life continues. This, I think, contains the answer to the question why natural selection created the sexual mode of reproduction, why there are two sexes in animals and in plants. From the Darwinist standpoint this question can be answered fairly easily.

Organisms always possess the property of traversing the same path of development that their ancestors did, but since the environmental conditions of a given plant are never absolutely like those of its ancestors, the heredity of the seeds is never absolutely like that of the seeds of the preceding generations. This is precisely how a new heredity is created by the alteration of the old.

Two sex cells fuse during fertilization and produce one cell (the zygote)—the beginning of the organism. This new, enriched cell produces an organism more adapted for development than it would have been had it developed from each separate unfertilized cell.[1]

It must be emphasized that upon fertilization, i. e., when two cells fuse, a third cell is obtained—the zygote—which is not only better adapted to the conditions of development than either of the sex cells, but is more viable. We will observe, in passing, that greater viability and greater adaptability to the given environmental conditions are not the same thing.

Why, then, does long-continued self-pollination lead to the waning, the enfeeblement of life, and we get what is called degeneration? Up till now we have given only one explanation of this: the shrinking of the range of potentialities of adaptation to environmental conditions. This is true, but it is not the only reason. Probably this is not the chief benefit derived from cross-pollination. If the blood is not renewed, refreshed, by crossing after

[1] It is known that organisms can develop from unfertilized sex cells, but from time to time, these species of organisms also resort to fertilization.

long-continued self-pollination, the viability of the progeny diminishes, wanes. With this, the adaptive potentialities of development possessed by the progeny also diminish.

An opponent of the theory of intravarietal crossing may say, or think: "All this is true, I do not dispute this, but after all, within the limits of a pure variety the heredity of the plants is the same. It may not be the same from the standpoint of dialectical materialism, but in the seed nursery my own eyes convince me that it is the same."[1] And here the opponent of the theory of intravarietal crossing will not fail to mention Darwin, and argue that Darwin himself said that crossing was beneficial only when the organisms taken for crossing differed, if only slightly, in their hereditary natures. Even in Darwin's experiments, he will say, no benefit was derived from crossing when the hereditary natures were the same; and he will quote examples to show that the crossing of peas of one variety performed by Darwin produced no effect.

In citing these examples, however, the opponents of intravarietal crossing juggle with the facts: they misrepresent Darwin's experiments. Darwin proved the very opposite, namely, that if pea plants are taken for the purpose of uniting by cross-pollination and if the preceding generations of these plants had developed even under slightly different conditions, there will be a fairly considerable increase in the viability, the vigour, of the progeny.[2]

Here the following question may at once arise: on a collective farm the plants of a pure variety grow in one field, but Darwin says that cross-pollination is beneficial only when the plants of the preceding generations were grown in different gardens and had, consequently, been exposed to somewhat different conditions. We may be told that the plants of every one of the pure lines of winter wheat (even such as Hostianum 0237, or the varieties bred by L. P. Maximchuk: Od-01, Od-02, Od-03) have the same heredity. It is a fact that in seed nurseries small plots are usually planted

---

[1] True, the point here is not that the eyes of such investigators are imperfect, but that their mode of thinking prevents their eyes from seeing what can and should be seen.

[2] "It should be remembered," wrote Darwin, "that in two of the cases in which highly self-fertile varieties appeared amongst my experimental plants, namely, with Mimulus and Nicotiana, such varieties were greatly benefited by a cross with a fresh stock or with a slightly different variety; and this likewise was the case with the cultivated varieties of *Pisum sativum*" ... "By the term fresh stock I mean a non-related plant, the progenitors of which have been raised during some generations in another garden, and have consequently been exposed to somewhat different conditions."... "If the flowers are not visited by our native insects, or very rarely so, as in the case of the common and sweet pea, and apparently in that of the tobacco when kept in a hot house, any differentiation in the sexual elements caused by intercrosses will tend to disappear. This appears to have occurred with the plants just mentioned, for they were not benefited by being crossed one with another, *though they were greatly benefited by a cross with a fresh stock.*" (My italics.—*T. L.*) (Ch. Darwin, *The Effects of Cross and Self Fertilization in the Vegetable Kingdom*, Second Edition, London 1878, pp. 389, 257, 458.)

with the progeny of separate specimens typical of the variety, and the plants on all these plots are always homogeneous. Only in rare exceptions has it been observed that the plants on one plot deviate somewhat from the type, and this is usually attributed not to a difference between the heredity of these plants and that of the plants on the other plots, but to a difference in the field conditions, to differences in microrelief. Since such an ideal homogeneity is observed in the seed nursery, it would seem to follow, according to Darwin's theory, that it is useless to cross-pollinate plants belonging to pure-line varieties of wheat.

The point, however, is that this reference to seed nurseries is unconvincing as far as we are concerned, if only for the reason that people who talk like this usually have had little to do with examining plants and still less with planting nurseries. They know about them only from books. And in many cases, books on various departments of agronomic science are unfortunately written by people who have no practical knowledge of the subject.

Most of the propositions advanced in agronomic science are of such a nature that not only is it unwise to insist upon them for long, but the investigator should himself try to "undermine" them, even though he himself has advanced them. Such propositions may be relatively correct, effective, and therefore useful; nevertheless, more effective and more correct propositions must be found to replace them. But there are certain propositions which must be understood, always remembered and used as the point of departure in one's work. They must never be deviated from or their correctness doubted. These propositions are the fundamental concepts of the theory of development, the principles of dialectical materialism.

According to dialectics, every identity, every sameness always contains differences. Hence we cannot imagine two offsprings of a plant being in no way different from each other.

We cannot imagine that 200 to 300 plants taken from a large mass of a line, however pure, should produce progeny that are absolutely alike. In some way the offspring will differ from each other.

In most cases, the difficulty in revealing these differences among the progenies of pure lines lies in the lack of experience (the untrained eye, as they say) of many scientists in perceiving differences among the progenies of crops in seed nurseries. This is one reason. The chief difficulty, however, is often the wrong approach to the technique of planting and tending seed nurseries.

How are seed nurseries of winter wheat usually planted? From 20 to 40 seeds are sown from each plant chosen. The space between the plants in a row is 5 cm., and between the rows 15 cm. This mode of planting does not preclude the possibility that among 10,000 offspring of Ukrainka winter wheat the seed grower will fail to see, until earing, even one or two offspring of barley that had got mixed up with the crop. The plants on the plots will spread out like a single carpet, all will look alike, and it will be impossible to distinguish the Ukrainka plants on one plot from the Ukrainka plants on another plot, i. e., from another progeny.

This can be very vividly and practically illustrated on the experimental plots at the Institute of Selection and Genetics.

Plants known to be hybrid (second-generation Krymka × Ukrainka) were taken and their seeds planted in separate families, i.e., seeds from each plant were sown on a separate plot. Technique of planting: grain from grain 5 × 15 cm. From every plant 1,000 grains were planted instead of the usual 20 or 40.

Another nursery was planted with seeds from the parents that had been picked as typical of the varieties Krymka, Hostianum 0237, Od-01, Od-02 and Od-03. The planting was different from that in the first nursery: the space between grains in a row was 25 cm., and between rows 70 cm. Length of plot—100 m. On each plot 1,200 grains were planted.

According to the geneticists there should be a diversity of plants on these nursery plots (second generation Krymka × Ukrainka); but it is difficult and in many cases impossible, at least today, for us to see differences in the plants on the same hybrid plots. On the other hand, on the plots in the second nursery, where, according to the geneticists, there should be no diversity, differences between the progeny of one family and the progeny of other families of the same pure-line variety can easily be seen. Let us note here that for planting this second nursery only spikes that were typical of the variety, i.e., of 100% purity, were taken.

Such planting of a nursery clearly reveals that intravarietal cross-pollination among our pure-line winter wheats does not contradict Darwin's theory, but, on the contrary, is based on it and confirms it; for slight differences in the plants of the pure-line varieties are observed, and this ensures a favourable result of the intravarietal crossing.

When Darwin crossed pea plants growing in one or two pots, in which, of course, the conditions were in very many ways identical, there was no increase in the vigour of the progeny. But when he crossed plants which, although growing in one pot, were of different origin (the seeds had been grown by different owners in different gardens), the results, as is known, were good as regards increase in the vigour of the progeny.

On large collective- or state-farm fields, the plants of pure-line varieties, of course, encounter conditions no less diverse than those obtaining for pea plants of the same variety but grown in different neighbouring gardens. And it was the latter that Darwin used for his experiments, which gave such brilliant results. As regards the environmental conditions of any pot in which plants are grown, they are immeasurably less diverse than those on large collective- or state-farm fields.

Thus, the attempt of the opponents of intravarietal crossing to lean on Darwin in their assertion that intravarietal crossing should not be performed with pure-line varieties fails utterly, because Darwin's experiments with the peas, for instance, and with other self-pollinators point to the beneficial influence of intravarietal crossing. The seeds of pure-line varieties of winter wheat certified by the experts to be 100% pure, are, as a rule, sufficiently heterogeneous in their hereditary natures to renew, refresh, the blood of the progeny as a result of cross-pollination.

In the winter-wheat seed nurseries at the Institute of Selection and Genetics, another extremely important principle that follows from Darwinism may be seen in operation. We have nurseries with progeny from plants of the same varieties of winter wheat that had not been subjected to intravarietal crossing, and progeny (F_2) taken from first-generation plants after intravarietal crossing. In these nurseries it may easily be observed that intravarietal crossing does not increase diversity among the plants, does not disturb the typicalness of the variety. True, it was evident to us without this that crossing does not increase but eliminates diversity in the outer appearance of the population, and this makes the variety more uniform.

But the most important and interesting thing about the phenomena observed in these nurseries at the present time, particularly with regard to the Krymka variety, is that here we find clearly revealed the fundamental error committed by the Mendelists and Morganists on the question of the diversity (segregation) of hybrids in the second generation.

The geneticists assert that the unshakeable foundation of their entire theory is the law, discovered by Mendel, that hybrid progeny must without fail segregate. In the paper he read at the session of the Academy of Agricultural Sciences in December 1936, Academician Serebrovsky said: "The first cardinal fact is that in crossing the progeny segregates according to given characters into groups which are multiples of each other. This law of multiple relation was discovered by Mendel (1865) and the whole edifice of present-day genetics is built on it. . . ."

According to the theory of the geneticists, as everybody knows very well, firstly, the progeny of every hybrid segregates according to the characters of the father or the mother; secondly, Mendel's segregation, to which, in the opinion of the Morganist geneticists, hybrids of peas, wheat, trees, animals, and all other living things on earth in general are subject, requires that diversity of the progeny should always be in the ratio of 3:1. For every three specimens of progeny possessing the characters of the father or the mother, there must be one which possesses the opposite characters.

For the geneticists, this principle is indeed unshakeable, because the entire edifice of Mendelist-Morganist genetics is built on it. But this, of course, does not mean that segregations in the multiple relation of 3:1, or the derivative of three to one $(3:1)^n$, on which genetics is based, is correct. Obviously, it contradicts the fundamental thesis of dialectical materialism. To assume that the hybrid progeny of all varieties of wheat, peas and of various other plants and animals must "segregate" in the same way and in an equal degree means totally ignoring biology, failure to take the environment into account. Can the immense diversity of living nature be forced into the narrow and rigid scheme of 3:1?!

Unfortunately, not only have the Morganist geneticists tried to do this in their own department of science, but they have rammed this idea fairly strongly into the minds of our plant breeders, seed growers, and of all of us agronomists. Actually, I do not think that anybody has seen diversity in the plants of hybrid progeny always being in the ratio of 3:1; that for every three specimens of a particular character there should always be one of the opposite character. Even in Mendel's own experiments not a single hybrid pea plant produced progeny with a diversity of colour of flower, or colour of seed, in the ratio of 3:1. It is enough to study the records of Mendel's experiments to find out without much difficulty that even in the progenies of the 10 hybrid plants of peas given in Mendel's tables, the progeny of one plant had 19 yellow seeds to 20 green ones, while that of another plant had 33 yellows to only one green. In the progenies of different plants of the same hybrid combination, different ratios of types were observed. The possibility is not precluded, of course, that diversity in the progeny of a given plant will be in the ratio of 3:1, but this will be as frequent, or as rare, as the ratios 4:1, 5:1, 50:1, 200:1, and so forth. On the average, of course (not always, by any means), the ratio may be 3:1; but this average ratio of three to one is obtained, and the geneticists obtain it (this they do not deny) from the law of probability, from the law of large numbers. It is common knowledge that the most widespread method of explaining this "biological law" in classes on genetics is the tossing of two coins. Students are asked to assume that the coins are sex cells (of peas, say) and to register, every time the coins are tossed, how many times the two fall tails up, how many times the two fall heads up, and how many times one falls tail up and one head up. They are advised to toss the coins as many times as possible. With a large number of tossings the result is approximately as follows: in 25% of the total number of tossings both coins fall head up, in 25% tail up, and in 50% one head and one tail up, i.e., in the ratio of 1:2:1.

The development of hybrid plants always proceeds in that of all the possible directions for which the conditions of the given field are most suitable. In the development of hybrid organisms, one or other potentiality of the given organism always gains a developmental advantage. The geneticists say that if heads (we will assume that this means red pea flowers) are dominant, i.e., gain the advantage, then all the organisms that come from the fusion of two sex cells, one of which had the potentiality of developing red flowers and the other white, will develop red flowers. According to the "biological" test of coin tossing, the proportion of red-flower plants will be 50% and 25% where both sex cells possessed the potentiality of developing red flowers; total, 75% red flowers and 25% white, i.e., a ratio of 3:1. The geneticists are firmly convinced that this must be the case among all the offspring of hybrids in the whole of living nature, no matter where and how they were crossed and grown. Actually, of course, this is far from inherent in the whole of living nature. It is not even inherent in the hybrid peas from which this "pea law," as I. V. Michurin aptly called it, was deduced.

In short, there is as much in common between biological law and "Mendel's law" as there is between a penny and a pea plant.

After closely observing the behaviour of the plants in winter-wheat seed nurseries, and especially those of Krymka obtained from intravarietal crossing, I boldly assert that nobody has ever seen the hybrid progeny of different plants of the same combination all vary in the same ratio $(3:1)^n$. Such a ratio may be obtained only if coins are tossed a large number of times, or in any other case where only equal probability based on chance plays a role, where necessity is averaged.

A close examination of the Krymka seed nurseries planted in the field where the progeny of seeds from typical Krymka spikes were growing revealed to me that the variety presented such a heterogeneous population that it was literally impossible to point to two offspring that were exactly alike. Of the relative uniformity that people are accustomed to see in this variety there was not even a trace in this case. The wheat plants, still in the grass form (there is no earing yet) differed sharply according to families (lines). At the same time, one could feel confident that after earing, ripening and harvesting, the morphology of the spikes in all these families taken together would constitute 100% pure variety according to experts' standards. After all, the seed-growing experts had chosen only typical spikes for planting in the nursery; nontypical spikes, so-called admixtures (red spikes with tinted awns, or awnless spikes), were not taken when plants were selected for the seed nursery; they were culled.

Seeing this sharp difference between some of the progenies of Krymka and others, and knowing, furthermore, that this great diversity appeared after the variety had been purged of nontypical admixtures, I asked myself: what may one expect, according to the theory of genetics, if several thousand spikes indiscriminately chosen from the Krymka population not purged of admixtures are castrated and left to be freely pollinated by the wind? In the first generation, according to Mendelism, there should be a diversity of plants, since both fathers and mothers differ relatively sharply from each other, and F_1 of many combinations is present.

In 1937 we did observe this diversity in such a first-generation nursery. True, this diversity was less than that of the Krymka which had not been subjected to intravarietal crossing. Every plant of the first generation of the Krymka variety was a hybrid of the parent forms, morphologically differing sharply in vigour, colour of leaf, shape of tuft and other characters.

According to Mendelism and Morganism, in the second generation, the progeny of every such plant should be diverse. That is to say, if intravarietal crossing is performed in the population of a given variety it is impossible, according to genetics, to release uniform seed within the next two or three years. If the slightest faith can be put in the relative truth of Mendelism-Morganism, then, after seeing the seed nursery of our Krymka, where the progeny of separate plants had been planted, it would be a risky thing to perform intravarietal crossing in such a population. Unfortunately, until quite recently I too still believed that hybrids obtained from morphologically different parents must unfailingly show diversity in the succeeding generations, even if not in the ratio of $(3:1)^n$.

But what do we see now on the six hectare Krymka nursery among the second-generation plants obtained from the intravarietal crossing? The diversity among the progenies (we planted 1,200 grains of each line) of the different plants is fairly considerable, but less and not greater than the diversity in the Krymka nursery where there had been no intravarietal crossing. The main thing, and for me the most important, is that for the first time I saw with my own eyes a numerous progeny of plants known to be second-generation hybrids as homogeneous as the progeny of ordinary nonhybrid wheat plants. In this nursery numerous cases can now be seen where there is no segregation in the second hybrid generation of wheat.

It is not difficult to perceive how much we seed growers and plant breeders were hindered by the geneticists' tenet that had been dinned into our ears, that all the progeny of hybrid plants must under all conditions show diversity, i.e., "segregation." On this principle plant breeders, as a rule, mixed the seeds of different plants of the first hybrid generation. The planting of the second hybrid generation was not done with the progeny of each separate plant of the first generation, but a mechanical mixture was made of the different progenies of the same hybrid combination, thus hindering the selection of constant desirable plants in succeeding generations.

Judging by the seed nursery of Krymka obtained from intravarietal crossing, where on different F_2 plots (separate families) there is a varying degree of uniformity, the offspring of the different hybrid plants of the same combination do not vary in the same degree. There are hybrid plants which may not show sharp diversity in any generation, beginning with the first, i.e., will not show what is usually called segregation. On the other hand, there may be hybrid plants of the same combination, the progeny of which show great diversity of forms, i.e., segregate. The importance, however, of finding in the first hybrid generation constant plants that produce relatively uniform progeny resembling the first hybrid generation is very evident to us seed growers and plant breeders. The first hybrid generation of wheat, cotton and other plants fairly often turns out to be so good that if the given variety could be kept in this state in future, plant breeders would, in many cases, be more than pleased with their work. It appears that this can be done. *This is definitely shown by the Krymka winter-wheat nurseries which are planted with seeds from an intravarietal crossing*. Until now, I (and I think all other specialists) did not know this. I was prevented from knowing this by genetics, *which told us, and now still tells us, that in all cases hybrids must segregate in F_2, and that there can be no constant hybrids.*

No small role in the creation of uniformity of hybrid progeny obtained from the crossing of morphologically sharply different forms within the Krymka variety is played by the capacity for elective fertilization. The castrated wheat plants of the Krymka population were allowed much greater freedom to choose pollen grains than is the case in artificial crossing. We know that the more difficult the crossing of two given forms of plants, the more diverse is the progeny of such a cross. It is not for nothing that the term "crazy" segregation has been introduced in genetics to indicate the progeny of plants difficult to cross. When the crossing is easy, for example, one variety of wheat with another, the progeny is less diverse.

It is not difficult to arrive at the conclusion that the more one gamete (sex cell) biologically corresponds to the other during fertilization, the more stable and less diverse will the progeny of this cross become in the succeeding generations.

Geneticists, the advocates of inbreeding (the enforced self-pollination of cross-pollinators), know very well, although they say nothing about it, that the seeds of individual rye plants, or of beetroots obtained by free pollination in the field with the pollen of other plants produce a much more uniform progeny than seeds from the same plants subjected to enforced self-pollination. The whole point, in my opinion, is that the first case provides greater possibilities for the uniting of sex cells that are more fit for each other. That is why a progeny resulting from intravarietal cross-pollination is not only more viable, but also more uniform than offspring resulting from enforced self-pollination.

It is possible, and sometimes necessary, to "violate" the nature of plants, but in doing so it is absolutely necessary to reckon with biological laws.

The more arbitrarily the nature of plants is violated, the less consideration is paid to biological development, the less useful such plants become. Only by agrotechnique, by kindly and skilful treatment of the nature of plants, by showing consideration for their requirements, do people obtain ever-increasing yields from plants. All the more necessary is it, therefore, to treat the nature of plants gently and skilfully if one wants to grow good seeds.

The nursery containing the second generation obtained from the intravarietal crossing of Krymka by means of free wind pollination shows us that it is possible to hybridize in such a way as to obtain comparatively uniform hybrid progeny. In nature there is no such thing as Mendelian segregations invariably in the ratio of $(3:1)^n$. Mendel's so-called law is not a law of biological phenomena, but of averaged, impersonal statistics. As is well known, Mendel himself attached no importance to the deductions from his experiments. This is shown even by the fact that as soon as his leisure was curtailed, when he was promoted from monk to abbot, he gave up his hobby of experimenting with plants. Mendel has nothing whatever to do with biological science. The postulates of Mendelism, developed not by Mendel, but by the Mendelist-Morganists, give us no effective guidance in our practical work of seed growing. As my own experience has convinced me, they are quite a hindrance to the improvement of seeds.

If I have sharply attacked the stronghold and foundations of the science of genetics, if I have made a trenchant assault on Mendel's law, which has been and is now being touched up by the Morganists, it is primarily because this law greatly hinders me in my work, or, to be more specific, hinders the work of improving the seeds of cereals.

THE MENTOR—A POWERFUL MEANS OF PLANT BREEDING[1]

IN THE STRUGGLE for high and stable yields, Soviet agronomic science is inevitably becoming revolutionized. It is being purged of everything that is useless and alien to our socialist country. New germs of genuine and effective agronomic principles are rapidly developing, and new, hitherto unknown, Stakhanovite methods of surmounting the obstacles in the path of continuously increasing yields are being devised.

Real unity between agricultural science and practice is possible only under the collective- and state-farm system of farming. The work of Stakhanovite tractor drivers, harvester-combine operators and skilful growers of big crops shows how the border line between physical and mental labour in agriculture is being erased literally before our very eyes.

Socialist agricultural production is based on the principles of the teachings of Lenin and Stalin—the only correct principles. This kind of agriculture requires the most advanced agronomic science, a continuous improvement in agrotechnique, an improvement in the seeds and plant material of all crops, the breeding of new, high-yield varieties resistant to unfavourable climatic conditions. The theories of Ivan Vladimirovich Michurin, who created hundreds of splendid varieties of apples, pears, cherries (sweet and sour), currants and other crops, being the most advanced theories in agronomic science, open the surest and most effective road to the achievement of that.

No one else in the history of the science of plant breeding and genetics possessed so profound an understanding of the life and development of plants as Ivan Vladimirovich Michurin.

He conducted his numerous experiments not merely to satisfy his curiosity, i.e., not just for the sake of experimenting. His aim was always to surmount the obstacles that stood in the way of creating new varieties and forms of plants. His scientific propositions are not inventions, but

[1] T. D. Lysenko's preface to the complete works of I. V. Michurin, 1948 Russ. edition.—*Ed.*

taken from life. They arose as a result of a long and continuous struggle to master the laws governing the nature of plant organisms.

The works of I. V. Michurin are a synthesis of his long years of splendid and extraordinarily fruitful work for the benefit of the working people. In our struggle for high and stable yields, the generalized experience that Ivan Vladimirovich expounds in his works serves as a most valuable and truly scientific guide. This applies particularly to the seed growing and breeding of different agricultural crops. All his life I. V. Michurin strove to build up genuine, truly scientific genetics and plant breeding. It is our duty to comprehend the Michurin method, to employ it in the Michurin way, to take the teachings of I. V. Michurin not as a dogma divorced from concrete conditions, but as a guide to action.

Experimenting on fruits and small fruits, I. V. Michurin discovered the general laws governing the development of plants. That is why I regard the works of I. V. Michurin as an indispensable and so far unequalled guide not only in the breeding and genetics of fruit and small-fruit plants, but in the seed growing, breeding and genetics of all agricultural crops.

Notwithstanding the fact that Michurin worked mainly on fruit and small-fruit cultures, and that the work I am directing deals, so far, exclusively with annual field crops, I have always found, and find today, that his works are an inexhaustible storehouse of ever new and effective advice and instruction.

It is our duty to develop and apply Michurin's theory, the product of his genius, in practical socialist agriculture. We must always remember the motto of I. V. Michurin's work: "We cannot wait for favours from Nature; we must wrest them from her." In these words he emphasized that the keynote of all our scientific work must always and everywhere be action to satisfy the requirements of socialist construction.

In his works, Michurin developed the materialist core of Darwinism. Therefore, *to be a Darwinist a plant breeder must be a Michurinist.*

Having magnificently grasped the principle that phylogenesis is inseverably connected with ontogenesis, I. V. Michurin deliberately and in a masterly manner controlled the individual development of plants. He skilfully directed the development of trees into a relatively definite channel and, as a result, obtained good varieties. Numerous examples of his work showed that the individual development of an organism affects the variation of hereditary properties (genotype). He brilliantly proved that by dextrously guiding the individual development of an organism in a definite direction we can control evolution, i.e., alter the hereditary properties of the organism in the direction we want.

Guided by the theory of Darwinism, Michurin always found the most diverse methods of creating needed varieties.

Michurin knew perfectly well from his experiments that a needed variety could not be obtained from any parent pair by crossing. When choosing plant forms for crossing he always took into account their historically-formed biological requirements—the result of adaptation—calculating beforehand how the development of the hybrids' hereditary basis would proceed under definite conditions.

He was the first to employ on a considerable scale the hybridization of forms geographically distant from each other and from the place for which the new variety was intended. In doing so he took fully into account the difference in their conditions of existence.

He brought to the foreground remote hybridization and widely employed it, but not merely for the purpose of increasing the diversity of plant forms. His work was not aimless. After choosing a pair for crossing, he at once drew up a plan for further action to create varieties.

Michurin's varieties did not appear of their own accord, not by chance. He created varieties, worked unceasingly on them, and reared them. Of the numerous potentialities of development inherent in the hereditary basis of a hybrid seed, he allowed only those properties and characters to develop that were needed for the creation of the desired varieties. Undesirable ones he did not allow to develop.

I. V. Michurin's works teach us that in addition to having skilfully-created hybrid seeds, one must also be able skilfully to grow plants from these seeds. Unlike the representatives of formal Mendelist-Morganist genetics, he knew perfectly well that the same hybrid seeds, if sown under different conditions, would produce plants with different economic qualities and properties. He altered and directed the individual development of plants by creating definite external conditions at definite times, and thus directed the development of varieties.

Michurin emphasized repeatedly that young plant organisms are highly susceptible to the influence of environmental conditions. That explains his numerous sharp protests against the assertion that it was advisable to graft young hybrid seedlings on to the crowns of old trees. Young plant organisms are not yet firmly established and not yet definitely formed; hence they readily yield to the influence of the stock. In this case, by stock is meant not only the cultivated variety on to the crown of which the young hybrid seedling is grafted, but also the wild variety on to the roots of which the old cultivated variety is nearly always grafted. In his article "Use of Mentors in Training Hybrid Seedlings,"[1] Michurin wrote that while exerting relatively little influence in the direction of altering the nature of the scion—here an old, established, cultivated variety—the roots of the wild variety, acting through the stem and branches of the old, cultivated variety, may very strongly influence a young hybrid seedling grafted on to a branch of the cultivated variety.

As he understood the history of the development of plant organisms from the standpoint of Darwinism, and was aware of the role played by external conditions in the formation of a young organism, Michurin was

---

[1] See I. V. Michurin, *Selected Works*, Eng. ed., Moscow 1950, pp. 105-20.

able to give a correct answer to an extremely important practical question, but one so profoundly theoretical that, in my opinion, it is useless to think of answering it, or even of grasping its profundity, from the standpoint of formal genetics. I have in mind the question why seeds obtained from the good cultivated fruit of apple or pear trees by the natural pollination of flowers or by artificial hybridization produce in most cases, when planted, an extremely high percentage of trees that yield fruit with bad, wild properties. After all, potatoes, for example, which in practice are also not reproduced from seeds but vegetatively (from tubers), produce relatively diverse progeny when planted from seeds; nevertheless the overwhelming majority of them are cultivated forms. Some plum and cherry varieties propagated from root shoots also produce good cultivated trees in the majority of cases, when grown from seeds.

In the above article "Use of Mentors in Training Hybrid Seedlings," I. V. Michurin gives an excellent answer to this question. It appears that the chief cause of bad heredity is the wild stock on to which the cultivated old variety was grafted. The scion itself—the old cultivated variety—is little changed by the action of the roots of the wild variety; but its young organs, i.e., the seeds formed in the fruit, deviate strongly in the direction of the wild stock.

Analyzing this question, Michurin says in his article: "...in other words, what we get is vegetative hybrids of the wild stock with only the most negligible admixture of the properties of the cultivated varieties."[1]

In tsarist Russia the Michurin theory was suppressed. Nor can such a theory really develop abroad, in the capitalist countries. The work of the talented American plant breeder and pomologist, Luther Burbank, will bear this out. The correct theoretical conclusions Burbank arrived at as a result of his work were not developed in capitalist America, either during his lifetime or after his death.

Only in our Soviet Land has Michurin's work been generally recognized and developed: by our Academy and other bodies, right down to the mass of collective farmers.

If Michurin's writings are ably and conscientiously studied, it is always possible to find in them new suggestions which, when applied in practice under our conditions, are at once enormously effective. Take, for example, the above-quoted explanation of the cause underlying the bad heredity of grafted cultivated varieties of apple, pear, and other trees. This explanation must and will certainly be widely made use of in practical and scientific work with the most diverse plants. I shall here point, very roughly, to some of the conclusions that follow from this explanation. (It would be easier for specialists in particular departments to do this than for me, a worker on field crops.) Will not a viticulturist draw from it the

---

[1] I. V. Michurin, *Selected Works*, Eng. ed., Moscow 1950, p. 113.

conclusion that, before crossing, it is necessary to obtain an own-rooted vine of a maternal plant? Will not pomologists draw the conclusion that, before crossing, an effort must be made to obtain own-rooted maternal trees? We who are working on annual plants shall also apply these Michurinian propositions. We shall graft some varieties of potatoes on to others for the vegetative hybridization of tubers, some varieties of cotton on to others in order to endow them by means of vegetative hybridization with certain properties we want, etc., etc.

The principles of Michurin's theory cannot be squeezed into the frame of formal genetics and plant breeding. To test Michurin's propositions, the representatives of this science have taken, and still take them, divorced from the Michurinian conception as a whole, divorced from Darwinism, and, of course, get no results. Michurin repeatedly pointed to the numerous mistakes committed by experimenters who tried to prove, for example, that his proposition concerning the role played by training in breeding new varieties of fruit trees was incorrect. I. V. Michurin elaborated and employed the mentor method as a powerful and effective means of training young plant organisms. In the crown of a young seedling, he grafted a cutting from another variety. The interaction of scion and stock should result, and, as Michurin demonstrated, often did result, in the mutual alteration of the nature of the plant organisms which had been united by grafting.

Michurin attached exceptional importance to the alteration and control of the nature of plants by means of training (including the mentor method). This aspect of Michurin's theory called forth, and unfortunately still calls forth, the largest number of objections on the part of the learned geneticists and plant breeders of the Mendel-Morgan persuasion.

This is understandable, because the substance and significance of the mentor method elaborated by Michurin can be grasped only from the standpoint of the theory of development, and not from the standpoint of formal genetics. To grasp the theoretical profundity and great practical importance of the mentor method means understanding vegetative hybridization, which Morganist genetics also repudiates. All his life Michurin fought the bourgeois science of genetics and plant breeding. Numerous experiments convinced him that as regards "applying Mendel's notorious pea laws in breeding new hybrid varieties of perennial fruiters, only complete ignoramuses in this work can dream of that. Not only are Mendel's conclusions not borne out in crosses of perennial fruiters, but in the case of annuals too. . . ."[1]

To many learned geneticists the idea that it is possible to hybridize by vegetative means seems fantastic at first; for according to the commonly accepted view, hybrids can be obtained only by the fusion of sex cells.

---

[1] I. V. Michurin, *Selected Works*, Eng. ed., Moscow 1950, p. 105.

Only in some of the lower plants does hybridization take place by an asexual process, but even that process is similar to the sexual process, i.e., the fusion of two cells into one.

Nevertheless, many cases have long been known of vegetative hybrids obtained by the grafting of some varieties on to others. Later, all these cases, which fundamentally contradicted the Mendelist-Morganist theory of heredity and variability, were simply proclaimed errors, and all those that could not be classed as errors were called chimeras (plant organisms consisting of the tissues of different breeds). The explanations that were given for the existence of chimeras may in many respects have been correct, and often they are correct in that chimeras are not vegetative hybrids; but it is indisputably true that there can be such things as vegetative hybrids. It is to Michurin's great credit that he learned how to obtain them, and gave us the mentor method, which, if understood, enables every scientist and collective farmer to hybridize many plants by vegetative means.

I will now give a brief preliminary exposition of my understanding of Michurin's mentor method. Every plant cell develops by assimilation and dissimilation, i.e., by absorbing food, and after going through a chain of transformations (intracellular processes) the developing cell divides into two. The zygote, i.e., the fertilized sex cell, which is the basis, the beginning, of the new organism, is formed, however, in a way that differs from the formation of the other cells of the organisms. When the zygote is being formed, two sex cells fuse into one.

The plant organism builds its body out of the food that surrounds it. The organism builds its living body out of nonliving food by properly assimilating it. Assimilation also takes place when two sex cells fuse, although this differs fundamentally from the first case. It may be said that the egg cell assimilates the nucleus of the spermatozoid, but the opposite may also be said: the nucleus of the spermatozoid assimilates the egg cell. More exactly, when two sex cells fuse, they mutually assimilate each other. As a result, neither of the cells remains; a new cell is formed—a zygote, which differs qualitatively from both the egg cell and the spermatozoid.

Such, in my opinion, is one of the aspects of the fertilization process, i. e., the process of the coming into being of the zygote that distinguishes it from the process of development of the somatic (ordinary, asexual) cells.

Further, it is known that, thanks to its heredity, every plant organism possesses the capacity to elect its environmental conditions. We also know that every organ in the organism, every cell of an organ, also possesses this capacity to elect its environmental conditions, including food.

The elective capacity possessed by organisms, organs and cells is the result of the historical adaptation of preceding generations to environmental conditions.

If it comes under the influence of conditions that are not quite optimal for it, a plant, as a result of its development, is found to have become adapted in some degree to these conditions. If it survives in this environment and leaves progeny, many of the factors of this environment which have been assimilated by the cells of the plant organism become, to some extent, essential for the normal development of the succeeding generations.

When two sex cells from two different plants, two varieties of wheat, for example, fuse during fertilization (i.e., mutually assimilate each other) and a new cell is formed, this new cell—the zygote—possesses power of adaptation to, and requires the existence of, the conditions of development of both parents. As a result, a new variety is formed, as a rule. Such, in my opinion, is the way sex hybrids are obtained.

Having ascertained the foregoing we may ask why changes resulting from the mutual influence of stock and scion, for example, when young hybrid seedlings are grafted on to the crowns of other trees, can be called vegetative hybridization. Let us briefly examine this question.

Some plants, potatoes, for example, can be reproduced from tubers, shoots, cuttings, leaves and seeds. (In practice, potatoes are usually reproduced from tubers.) It goes without saying that every cell, or group of cells, of a plant organism (potatoes in this case) from which a plant can be regenerated (grown) usually possesses all the properties, the nature, of the variety from which the initial cells were taken.

The different plastic substances in the leaves, stems and tubers of the potato, of the Epicure variety, let us say, are usually such that when the stolons (the underground shoots on which the tubers develop) of Epicure feed on them, they produce Epicure tubers.

What will happen if we learn how to feed the cells of one variety of plant with (i.e., compel them suitably to assimilate) the ready-made plastic substances of another variety, i.e., merge, as it were, two breeds of plants into one, as happens during the fusion of sex cells? Logically, one would expect the formation of new cells of a new breed—in other words, the formation of a vegetative hybrid possessing in some degree the properties of both the first and the second variety. It seems to me that these hybrids should not differ fundamentally from hybrids obtained sexually.

Understanding the essence of Michurin's mentors precisely in this way, I assumed that if the stolons of one variety of potato are induced to feed on the plastic substances produced by the assimilation of the leaves of another variety of potato, hybrid tubers should result. They will possess, in some degree, the properties of both varieties. To achieve this, two varieties of potatoes must be united in one organism by means of grafting.

Test experiments performed in greenhouses in February-April 1938 by a number of scientific assistants and postgraduates at the Institute of Selection and Genetics (Odessa), and by a group of students at the Timiryazev Agricultural Academy, fully confirmed my assumption. Tubers were obtained in which the properties of both parents were clearly evident. Interesting in this respect are also the results of the experiments performed by the Michurinist experimenter N. V. Brusentsov, near Moscow, in 1937.

The question may arise why vegetative hybrids were not always obtained from potato graftings in the past (and quite a number are known); and why in those rare cases when they were obtained (even Darwin pointed to them in his works), the representatives of the Mendelist-Morganist theory classed them as errors or chimeras. Why have many scientists failed, as a rule, to obtain vegetative hybrids in their test experiments, whereas we, in our experiments, obtained vegetative hybrids in all cases with all varieties (we experimented on as many as fifteen)?

The answer is that it is not merely the grafting that counts, but also the ability skilfully, in the Michurin way, to induce the corresponding cells of one variety to assimilate the food provided, prepared by the other variety. It was necessary to take into account the cells' capacity to choose their food. These cells had to be deprived of the plastic substances which had been specific for them and had to be provided with the substances produced by the plants of another breed.

Indeed, what would have happened if, in our experiments, the stolons of, say, the blue tuber variety of the Odenwälder Blau potato had been allowed to choose between the products of the assimilation of the leaves of its own breed, i.e., of its own variety, and the products of the assimilation of the leaves of another variety, say, Epicure? The stolons would certainly (although, of course, not always) have chosen the food for which they are best adapted, i.e., the food of their own variety. Of course, no vegetative hybrid would have resulted. Usually this is lost sight of in test experiments.

But if no food to which the stolons are best adapted is available, and only the products of assimilation of the leaves of the other potato variety can be had, the stolons will be forced to build tubers out of these plastic substances.

It would be wrong to say that the stolons of the Odenwälder Blau will develop tubers of the Epicure variety from the plastic substances of the complex products of the work of the leaves, of, say, the Epicure variety (white tubers) because the cells of the stolons belong to the Odenwälder Blau and not to the Epicure variety. But it would also be wrong to say that Odenwälder Blau tubers will develop from the cells of the stolons of the Odenwälder Blau variety because the plastic substances out of which the stolons build the tubers belong to the Epicure variety. In such a case, the experiment must result, and all our experiments have resulted, in hybrid tubers.

Therefore, when I. V. Michurin stated that a cutting from a young hybrid seedling of an apple, pear, or other kind of fruit tree, grafted on to the crown of an adult tree, could borrow the properties of the stock, he expressed a new and profound conception of the development of the plant organism. The fact that anti-Michurin scientists have often failed to discern the influence the stock exercises on the scion only exposes the weakness of formal genetic science. We now believe that it is possible in all cases to obtain sharp changes in hybrid character as a result of the interaction of scion and stock. To achieve this, it is necessary, in our opinion, not only to take cuttings from a young plant and graft them on to the crown of the adult tree with which you want to hybridize vegetatively, but also *to remove all the leaves that appear on the scion.* The scion cutting must be induced to build its body entirely out of the food provided by the roots of the stock, and from the plastic substances of the products of assimilation of the leaves of the stock.

In such cases, vegetative hybridization will always infallibly take place. The only question is whether the grafted cutting will be able to take on and live in the crown of the given stock. Thus, the mentor method discovered by Michurin enables us to work out and to make real use of the vegetative hybridization method.

If our conception of Michurin's mentors finds further confirmation (of which I have no doubt), it will be possible, in the very near future, to hybridize on a mass scale many plants that it was impossible to cross sexually. Moreover, vegetatively propagated plants will produce a more definite and more stable progeny than that resulting from sexual hybridization.

In attaching enormous importance to the mentor method we, I believe, guide ourselves entirely by the following remarkable statement by Michurin:

. . . "In a word, when the details of this method and its uses in breeding new varieties of fruit trees have been fully worked out, we shall at last have made a big step towards achieving the long-desired control over the process, without which the results of our work have for the greater part been dependent on various accidental outside factors whose influence we were totally unable to lessen or eliminate, so that we had to content ourselves with only such qualities in our new varieties as fate chose to send us. That being the case, much labour was wasted, and practically 95% of all the hybrid seedlings raised had to be destroyed by reason of one or another shortcoming."[1]

We are now confidently undertaking to hybridize some varieties of potatoes by grafting. I think that when we have mastered this method we shall be able to hybridize potatoes with dahlias, potatoes with Jerusalem artichoke, etc. We shall be able to get vegetative hybrids of tender peaches or apricots and hardy plums or blackthorn. We shall be able to cross lemons, tangerines, oranges and other citrus fruits with *Citrus trifoliata*, which is much more frost-resistant.

In general, to understand the essence of Michurin's theory, to master it completely, means, in our work, in our Soviet land, to do tomorrow what today may seem fantastic.

---

[1] I. V. Michurin, *Selected Works*, Eng. ed., Moscow 1950, p. 109.

I. V. Michurin's works are a daily guide for all Soviet plant-breeding scientists, agronomists and collective- and state-farm experimenters. They must be read and studied over and over again.

His works must become manuals also for our Soviet specialists in genetics, plant breeding and seed growing, and for our student youth. It is time, high time, to base the studies of our students and postgraduates working in the field of genetics, plant breeding and seed growing on the most advanced theory of agrobiology—on Michurin's theory. We must not merely talk about this, however laudatory this talk may be. We must take action, must resolutely eradicate all the pseudoscientific theories that have deeply penetrated agrobiological science, particularly the department that treats of the theory of heredity.

SEED GROWING MUST BE BASED ON MICHURIN'S THEORY[1]

THAT improvement in the natural properties of plants depends on agrotechnical conditions is something the Mendelist-Morganist theory of genetics will not admit. Agrotechnique, the methods of sowing, manuring, etc., exercises influence only insofar as better agrotechnique results in a larger crop of seeds from a given field and worse agrotechnique results in a smaller crop; but the nature of the organisms (heredity) is not altered by the conditions of life, the conditions under which the plants are grown—such is the basic thesis advanced by the Morganist geneticists, which, unfortunately, is consciously or unconsciously used by the majority of the state plant-breeding stations engaged on cereals as the basis of their work. In a number of cases, according to Morganist genetics, the nature of organisms may change (mutate), but the quality, the direction, of the change are in no way connected with the nature of the influence which conditions of life exert upon the organism.

Professor N. N. Grishko, in the recently published *Textbook on Genetics* (Selkhozgiz, 1938), writes on page 175: "Innumerable examples in agricultural practice and thousands of experiments tell us that alterations of characters due to the influence of environment (so-called modifications) are nothing more than the realization of the organism's various potentialities of development. Therefore, modifications are repeated in the progeny only if the environmental conditions under which the changes were observed in the preceding generation are exactly repeated."

According to Grishko, the changes a plant organism undergoes in the process of its individual development are in no degree reflected, are in no way fixed, by the organism's nature. The organism changes, but its nature (genotype) remains unchanged. Thus Professor Grishko fully associates himself with another Morganist geneticist, namely, Professor J. A. Philiptschenko, who, in 1929, wrote: "Let us assume that a high-yield variety of wheat has been bred somewhere. The seed growers procure a quantity of the seeds, sow them in their fields and reproduce them. Some of these seeds come from good plants and others come from inferior, feeble plants, but, as we know

[1] *Яровизация*, 1938 г., № 4-5 [19-20].

very well, this circumstance is of no significance whatever, for the same progeny is obtained from the good and the inferior plants."[1]

True, Grishko says that puny seeds may give rise to feebler plants than those obtained from good, big seeds of the same variety, but he attributes this only to the fact that the puny seeds contain less food for the young plant that is starting its development. The nature of the plants that grow from the puny and from the big seeds will invariably be the same.

What a wonderful "nature" (genotype) that is: to be situated in a changing organism, but not to undergo any change itself! However, the genotype is not "located" somewhere in the organism (as the Morganists assert). The organism itself is the genotype. At the same time the organism is also the phenotype. Only Morganist geneticists can think that the entire organism, down to the last molecule, i.e., the entire phenotype, can change, while the genotype (the nature of the organism) remains entirely unchanged. Or they assume what is still more fantastic, namely, that the organism's phenotype can change in one direction, while its genotype (its nature) undergoes not a corresponding, but an opposite change.

For example, Grishko admits that cases occur of changes (mutations) in the genotype (nature of the organism) in conformity with environmental conditions, but he categorically denies that these changes correspond to the action of environmental conditions upon the phenotype, i.e., upon the plant organism. To put it more simply: although, under specific conditions, plants may grow well, develop luxuriantly, and show a high yield, nevertheless, as a rule, as a law (invented by the Morganist geneticists), the nature (genotype) of the seeds, cuttings or tubers taken from such plants will not be improved. But if it is true that the changes in the phenotype and genotype do not adequately correspond to each other as, in line with all the other Morganists, Professor Grishko and Professor Delone claim in their *Textbook on Genetics*, then there is no need to go to the trouble of raising good livestock and plants for breeding purposes, because improvement in breed would not depend upon improved agrotechnique in cultivating plants for seed, or upon improved zootechnique in raising livestock for breeding purposes. Such are the deductions that follow from the nonsensical theory of Mendelist-Morganist genetics. The leading geneticists do not deny this. Thus, Muller, who is well known among the Morganists, wrote as follows in *Izvestia* of May 24, 1934: "Many examples can be quoted of an enormous amount of energy and money being wasted because of some fallacious genetic theory. Thus, for example, parent plants and animals were placed under the best conditions of development in order that, in conformity with Lamarck's false theory, their progeny should acquire better characters."

In the magazine *Priroda* (*Nature*), No. 6, 1936, Muller went so far as to give utterance to the following, literally howlingly anti-Darwinist statement:

"The results obtained in 1918 and 1919 already showed that . . . the

[1] Ю. А. Филипченко, *Генетика и ее значение для животноводства*, 1931, стр. 46.

stability of the gene is of an order that is good for several thousand years, i.e., in the course of such a period there will be no more than one mutation per gene. We know now that the actual span of time is several tens of thousands of years. In this connection it may be noted that since in every generation every gene forms several score of times a structure completely like itself, the above-quoted frequency of the mutation process in individual loci signifies that even in Drosophila not a single observable false step is taken in the process of forming like structures, with its several million repetitions. In other words, the copy of a copy, repeated several million times, remains practically indistinguishable from the original pattern."

That the Morganist geneticists who call themselves plant breeders do not know how to grow constantly improving seeds and, juggling with "Lamarckism," set seed growers the task of creating only a "copy of a copy," scarcely anybody can doubt (not even the Morganists themselves). But it is indisputable that such profoundly scientific utterances can only serve to discourage those who are striving to produce better and better seeds by good agrotechnique, careful tending, etc., and that under no circumstances can such a science be tolerated.

It is not my intention in this article to deal in greater detail with the *Textbook on Genetics* by Grishko and Delone as a whole. I have merely touched upon the section which directly concerns the subject of the present article.

Being convinced, in conformity with Mendel-Morgan genetic theory, that the conditions of life of plants do not influence the quality of the changes (mutations) in the natural properties of organisms, the vast majority of plant-breeding stations, when growing élites, even now completely ignore questions connected with the training of plants on the seed plots. When growing élites they confine themselves simply to increasing the quantity of seeds by sowing and harvesting the crop, and also to avoiding mechanical adulteration when sowing, harvesting and storing the crop. Their sole concern is that the élites should be of the purest quality and morphologically typical, i.e., should conform to the description of the external appearance of the variety in question given in the testing experts' instructions. The result is that although the élite seeds obtained belong to a definite variety and are of a high degree of purity, their yield potentialities are indefinite, unknown.

In practical agriculture, however, it is well known that some seeds or planting material may be good and some may be inferior, though the whole was of the same variety and purity. It may often be observed that when sown and grown under comparable conditions, some seeds or planting material produces considerably higher yields than other material of the same variety. In other words, the yield, and the economic and biological properties in general, of given seeds or given planting material depend to a large degree upon the growing conditions of the plants of the preceding generations.

One can easily convince oneself of this. For example, it was impossible to grow seed potatoes in the South with the ordinary method of growing potatoes that had long been employed there. When reproduced in hot districts the yield, in the course of one to three generations, dropped to one half or one third of the yield obtained from planting material of the same variety grown in more northern or hilly districts.

By changing the method of cultivation, i.e., by changing the growing conditions for potatoes by means of agrotechnique (suitable cultivation of the field, and planting in July instead of April), we are now obtaining in the South seed potatoes of a higher quality than that of the same variety grown in the best of the old potato districts (Moscow Region, Gorky Region, etc.). This is proved by the tens of thousands of hectares of summer-planted potatoes at the collective and state farms in the south of the Ukrainian S.S.R., and also in the Crimea, in Rostov Region and in Krasnodar Territory. It is also proved by the comparative plantings at the Institute of Selection and Genetics (Odessa) of the same variety of potato tubers grown in the South by the summer-planting method, as well as of tubers obtained from Gorky Region and from the Potato Institute (Moscow Region). In these plantings, the potatoes of the same variety reproduced in the South are now better than those reproduced in the northern regions. Seed potatoes of southern reproduction (summer planted) have been found to thrive better also in Moscow Region, where they were planted experimentally in 1938 (Potato Institute).

We quote this example only in order to emphasize that conditions of cultivation, of raising plants from which seeds or tubers are taken for future planting, play an extremely important role in the creation of the planting qualities of any given material.

It is urgently necessary to work out the agrotechnique for cereals seed plots. We must arrange such conditions for growing sowing material as will continuously improve the seeds of the given variety. Has it not been found possible to obtain the best Soviet seed potatoes in the South by changing the agrotechnique, though no seed potatoes of any kind could be grown there in the past? Why, then, cannot the same thing be done with cereals?

Casting aside Mendelism-Morganism and taking Michurin's theory as the basis for our agrotechnique, we can, by suitable agrotechnical measures on the seed plots, do as we have done with potatoes, namely, radically improve the economic and biological properties of the seeds of the most diverse agricultural plants, including cereals.

To learn to control the conditions required for improving the nature of seeds is the fundamental task of our Soviet, Michurin genetics. This is an extremely important problem for our agriculture. The Party and the Government are devoting great attention to the seed-growing problem. Socialist agriculture has created all the conditions needed to make possible a genuine development of advanced agricultural science. What can be of greater interest to Soviet genetics and plant breeding than studying the problem of growing seeds with ever-increasing yields on our collective and state farms? It is a fact that the seeds of the same variety of, say, winter wheat possessing the same physical qualities, but obtained under different conditions of reproduction, may have higher or lower yields when sown under equal, comparable conditions.

For the purpose of illustration I shall present a table showing the yields of the same varieties of winter wheat in the variety tests made in 1937-38 at the Institute of Selection and Genetics.

Seeds received from:	Weight (in kilograms) of grain per plot of 115 sq. m.			Yield (centner per hectare)
	I replication	II replication	III replication	
Zarya Variety				
1. Vinnitsa Region, Khmelnitsky District	41.1	40.8	35.9	34.2
2. Vinnitsa Region, Komsomolskoye District, Politotdel Collective Farm	43.2	44.7	39.6	37.0
3. Vinnitsa Region, Nemercha Experiment Station	49.0	49.3	42.4	40.8
Ferrugineum 2453 Variety				
4. Moscow Region, N.-Derevensky District, Zarya Collective Farm	30.7	34.3	—	38.3
5. Moscow Region, N.-Derevensky District, Put Krestyanina Collective Farm	32.7	33.8	—	28.9
6. Moscow Region, N.-Derevensky District, Ogorodnik Collective Farm	33.2	36.3	—	30.2
Dürabl Variety				
7. Moscow Region, Moscow Experiment Station	33.5	30.5	—	27.8
8. Moscow Region, Shatsk District, Pobeda Pyatiletki Collective Farm	35.3	33.2	—	29.7
9. Moscow Region, Shatsk District, Karl Marx Collective Farm	38.0	36.6	—	32.4
10. Moscow Region, Ryazhsk District, Nash Put Collective Farm	38.6	36.3	—	32.5

The yield of the Zarya variety of seeds obtained from a collective farm in the Khmelnitsky District, Vinnitsa Region (No. 1 in the table) was 34.2 c. per ha. The seed yield of the same variety, but obtained from the Politotdel Collective Farm, Komsomolskoye District, Vinnitsa Region, was 37.0 c. per ha. And lastly, the yield of the seeds obtained from the Vinnitsa Plant-Breeding Station (Nemercha) was 40.8 c. per ha. According to information received from L. I. Kovalevsky, a plant breeder at the Nemercha station, the Zarya seeds sent us were grown on a record-yield plot.

The example we have cited of different yields (a difference amounting to as much as 6.6 c. per ha.) during the variety tests at the Institute of Selection and Genetics of Zarya variety winter wheat shows for a certainty that the different conditions of cultivation of the preceding generation of plants radically affected the yield properties of the seeds obtained from these plants.

The above table indicates not only the average yield for all test replications, but also the yield at each replication, in order that the reader may convince himself that the differences in the yield of the same variety shown by the tests are not due to differences in the field conditions, but to the relative differences in the nature of the seeds. The seeds of each variety obtained from different farms were planted, one after another, in plots at each replication. In the first replication Zarya seeds from the Nemercha station produced 7.9 kilograms more per plot than the Zarya seeds from the Khmelnitsky District, and the same higher yield of the Zarya seeds from the Nemercha station was observed also in the second and third replications. The same was the case with the other varieties: Ferrugineum 2453 and Dürabl.

There is not a plant whose progeny cannot be improved by suitable conditions of cultivation.

To ascertain, to learn, what these conditions are and then deliberately to create the best of them by means of agrotechnique—this is what we must strive for in seed growing instead of indulging, as Grishko and all the other Morganist geneticists do, in lucubrations about genotypical changes not corresponding to phenotypical changes.

The science of seed growing (which certainly includes plant breeding and genetics) must radically change its approach to its tasks. It is scarcely possible to elaborate an exact science of seed growing from the standpoint of the Mendel-Morgan theory, which denies that conditions of life call forth directed changes in the nature of organisms.

In the first place, the science of seed growing must generalize and develop the best of the practical experience gained in the growing of good seeds. We have quite a number of outstanding practical agriculturists and agricultural scientists, people who know how to grow excellent seeds and planting material.

For example, we can confidently say that the vast majority of cereal seeds from Stakhanovite plots, sown under comparable conditions, will produce higher yields than seeds of the same varieties grown on plots under bad agrotechnical conditions.

The problem of improving the nature of plants by training them on the seed plots at plant-breeding stations is of immense practical importance. Nevertheless, the Lenin Academy of Agricultural Sciences, far from elaborating the theoretical principles of directed training of plants on seed plots, has so far actually hindered the development of such work in our country (cf. report of the Academy's session of December 1936). The All-Union Institute of Plant Industry, the chief plant-breeding institute of the Lenin Academy of Agricultural Sciences, has proceeded and is still proceeding in its theoretical work from a wrong standpoint. This institute bases its work on the principles of Morganism-Mendelism, and yet it has trained and is now training the largest number of postgraduates. Through its published works it has also played a big, but harmful, role in orientating many workers at plant-breeding stations on the formal approach in science, in orientating seed-growing work on the theory of Morganism-Mendelism.

In agricultural science disregard for the question of the influence environment exerts (the influence of agrotechnique) on the changes in the economic and the biologic natural properties of seeds has brought it about that in the practical work of plant-breeding stations nobody *has compared the yields and different types of hardinesses (winter hardiness, resistance to disease, etc.) of the released élites of the different varieties with the pure seeds of the same varieties intended for milling.* Nobody in the Academy of Agricultural Sciences has even thought of making such tests, because, according to Morganist genetics, the genotype of the plants of a given variety is the same, irrespective of whether the seeds were obtained from élite or from ordinary economic plants. That is why, as a rule, workers at plant-breeding stations who grow élite seeds, or the scientists on the basis of whose theories these élite seeds are grown, cannot tell at present whether the yield of the élite seeds released by plant-breeding stations is higher than or the same as that of ordinary seeds of the same variety sown under equal conditions.

The Government's decision of June 29, 1937, on "Measures for Improving the Seeds of Grain Crops" very clearly indicates the measures to be taken not only to prevent the deterioration of seeds, but also continuously to improve them.

Among other instructions contained in this decision we read: "to impose upon the plant-breeding stations the duty of supplying élite seeds for the seed plots of the district seed-growing nurseries in their respective republics, territories and regions, and to hold them responsible for any failure to provide seeds of first-class quality."

Élite seeds are paid for at five times the price of ordinary seeds. One would take it for granted that the sowing of seeds of a higher category should result in a higher economic return.

Nevertheless, up till now the entire work of plant breeding has centred round a single criterion—purity of variety. Everything has been directed *solely* towards reducing to a minimum the number of characters with visible morphological deviations. Responsibility for producing first-class quality élite seeds (meaning by that greater economic yield in the subsequent reproductions of these seeds) seemed to have been forgotten, because, according to the so-called theory of Morganism-Mendelism, it is beyond man's power to improve the nature of seeds or planting material by training (i.e., by agrotechnique).

In order to organize the growing of improved seeds at plant-breeding stations within the shortest time, it is absolutely necessary radically to change the methodological scientific principles of the work at state plant-breeding stations in addition to improving the organizational and economic factors.

The production of seeds must be safeguarded and protected from the pseudoscientific guidance of the Morganist geneticists.

Darwinism and Michurin's theory constitute the surest basis of agrobiology. On this basis it is possible considerably to improve the nature of the most diverse plants, including cereals (and there are sufficient facts to prove this). Michurin's theory and no other must serve as the agrobiological basis for the work of the respective departments of the Academy of Agricultural Sciences, which should be entrusted with the scientific and methodological guidance of the state plant-breeding stations.

Only in our Land of Socialism is the true development of Darwinism possible. The development of Darwinism in agronomic science means, primarily, mastering Michurin's theory. As life has shown, if we guide ourselves by that theory, there is no need to wait (as the Mendel-Morgan theory calls upon us to do) for favours from nature, for mutations that are independent of the conditions of life of plants, and, consequently, independent of man; our task is to wrest them from her.

This is the motto of Michurinian Bolshevik agronomic science.

THE GROWING OF RYE SEEDS

The theory of Morganist genetics is also in a bad way as regards the growing of seeds of the cross-pollinating rye plants. Here, too, all work is directed towards one sole object: to protect the "purity" of the variety from impurities. The difference between the growing of rye seed and the growing of wheat seed, for example, is that in the former all measures are taken to prevent not only mechanical, but also biological admixture, to prevent wind pollination with other varieties. *The regulations of the People's Commissariat of Agriculture lay it down that every variety of rye must be separated from every other variety of rye by an isolation zone not less than a kilometre wide.*

From the theoretical point of view, from the standpoint of Darwinism, that thesis will not bear criticism, and contradicts economic expediency. Not only that; in some cases it is even harmful, not only because it creates considerable difficulty in laying out seed plots on state and collective farms, but also because it deteriorates the yield properties of the rye seeds.

Owing to this demand for a kilometre isolation zone between rye-seed fields of different varieties, and even between them and wheat fields adulterated with rye (and last year even between the same varieties of rye if they were of a different reproduction), tens and hundreds of thousands of centners of good seeds from plots that gave high yields were lost.

It may be confidently asserted that often even the best rye seeds for given districts were rejected in conformity with this theory because the isolation zone regulation had not been adhered to. In a letter they wrote to me, a group of inspectors in Leningrad Region stated that if the regulation of the People's Commissariat of Agriculture were adhered to, it would be necessary to reject about a third of the Vyatka rye seed plantings of 1938 because of their propinquity to plantations of the same rye that had been rejected during the past two or three years (in the variety trials). They stated further that a second group of plantings would have to be rejected because they were situated less than a kilometre from small farm plots of a local variety of rye; and that, lastly, a third group of Vyatka rye seed plantings would have to be rejected because of their propinquity to winter wheat with over 10% adulteration of winter rye, although of the same Vyatka variety. Thus, this regulation of the People's Commissariat of Agriculture creates considerable difficulties for developing rye-seed growing, although *the demand for a kilometre-wide isolation zone has no scientific justification whatsoever.*

In the autumn of 1937, the specialist Avakian planted out at the Institute of Selection and Genetics (Odessa) 11 varieties of rye for a small comparative test. The seeds of every variety were originals—i.e., grown by the plant-breeding station on isolated plots—and also seeds that had been grown at the Kharkov Plant-Breeding Station under variety test conditions, i.e., on 100-metre plots surrounded by plots with other varieties of rye. It was found that none of the 11 varieties obtained from the seeds which in the preceding year had been grown surrounded by various other varieties gave a reduced yield. Nearly all these varieties gave a higher yield—considerably higher, some as much as 7 c. per ha. more—than the same varieties which in the preceding generation had been grown on isolated plots.

The fundamental error committed in seed growing and plant-breeding work with cross-pollinators, particularly with plants like rye, is the denial, in obedience to Mendelist-Morganist genetics, that biological election takes place in the fertilization process. According to Morganist genetics, fertilization takes place only in conformity with the theory of probability, the theory of chance. Fertilization depends solely upon which pollen grain (sex cell) will be the first that the wind brings to the stigma of the rye flower.

This assertion of the Morganists is wrong. Plants have *elective capacity* in the fertilization process, and this elective capacity must be skilfully utilized for improving the quality of seeds.

It has been pointed out more than once in discussing the problem of the intravarietal crossing of self-pollinating cereals that the free cross-pollination even of self-pollinators always results in the progeny having greater biological viability, greater hardiness, and greater adaptability to the conditions of development. This applies in even a larger measure to cross-pollinating plants. It is well known that these plants suffer a strong depression as a result of enforced self-pollination, or even as a result of the presence of too few plants during flowering, as a consequence of which the stigmata have little choice of pollen during fertilization. In virtually all cases of free intravarietal cross-pollination, however, and also of pollination with foreign pollen, when there is free biological choice, the resulting seeds are, as a rule, biologically more hardy, more adapted to the conditions of development.

On these grounds it is perfectly evident that in all cases in which the economic requirements of a given plant coincide with its biological requirements, with the biological adaptability of the organism, the restriction of free cross-pollination, or pollination with foreign pollen, will never be economically useful, and in some cases will even be harmful.

As for rye I do not know at the present time of any economic requirement that diverges from the biological adaptability of this organism to environmental conditions. Therefore, the establishment of kilometre isolation zones between rye-seed plots is simply beneath all criticism from the standpoint of Darwinism. I know of no cases in practice of the economic deterioration of rye seeds resulting from failure to comply with the isolation zone regulation. On the other hand, the entire theory of Darwinism, as far as I understand it, and also the fairly numerous experiments on elective fertilization of plants conducted at the Institute of which I am the director, show the inadvisability of having kilometre isolation zones between rye-seed plots.

Seed growing for cross-pollinating plants, such as, for example, sugar beets, and a number of others is another matter. In certain respects the economic requirements as regards sugar beets sharply diverge from the biological adaptability of the plant. Big roots with a large sugar content are not at all useful for the biology of the sugar-beet seed plant. Therefore, if plantings of sugar beet are allowed freely to choose pollen for fertilization, i.e., if the group of best seed roots (plantings) are not isolated from the other beets, they will forthwith cross-pollinate (cross-breed) with the economically inferior beets. *In this and similar cases isolation zones are essential.*

Therefore, when arguing from the standpoint of Darwinism that kilometre isolation zones are unnecessary and in some cases even harmful for rye, it must under no circumstances be forgotten that for a number of cross-pollinating crops (other than rye and similar crops), isolation zones between different varieties are essential from the same Darwinist standpoint.

This may, perhaps, sound strange to the group of scientists who adhere to the Morganist-Mendelist theory of genetics. It is nevertheless a fact that all the varieties of rye, down to the very last one, that are grown in our collective and state farms, and are classed as standard varieties, were obtained (bred) by plant breeders only because they made their mass selection among plants which had not been isolated from other varieties for several generations, i.e., they had not set up these "scientifically"-backed kilometre isolation zones.

In support of my statement I will quote extracts from the *Handbook on Testing Agricultural Crops* (Selkhozgiz, 1937).

Vyatka—"a variety bred by the Vyatka Plant-Breeding Station from local rye." Obviously, the local rye had not been isolated from other varieties by a kilometre zone.

Lisitsyna (Shatilovskaya)—"a variety bred by the Shatilovskaya Plant-Breeding Station from local rye." I do not think that the local rye had been isolated by a kilometre zone.

Avangard—"a variety bred by the Kazan Plant-Breeding Station in 1920 by family selection from regenerate Alpine rye." By "regenerate" is meant Alpine rye which had crossed naturally with another variety.

Yeliseyevskaya—"it is presumed that Yeliseyevskaya rye was obtained as a result of the hybridization of Alpine rye with the local rye."

Bezenchukskaya Zholtozyornaya—"a variety bred by the Bezenchuk Experiment Station. The initial material was Yeliseyevskaya rye and local rye from the village of Krivoluchye-Ivanovka, former Samara Gubernia." In this case, too, the fact that a kilometre isolation zone was not set up made it possible for the plant breeders to breed the variety.

Tarashchanskaya—"a variety bred by the Verkhnyachka Plant-Breeding Station from local rye that had been affected by the influence of natural hybridization with rye of West-European origin." The appearance of this variety, too, is due to the fact that an isolation zone was not set up.

Nemyshlyanskaya—"a variety bred by the Kharkov Plant-Breeding Station from local rye." The local rye had, of course, not been isolated from other varieties.

Petkusskaya Veselo-Podolyanskaya—"a variety bred by the Veselo-Podolyanskaya Plant-Breeding Station from the original Petkusskaya rye (Lokhov's)."

Petkusskaya Verkhnyachskaya—"a variety bred by the Verkhnyachka Plant-Breeding Station from original Petkusskaya rye (Lokhov's)." I think that it was possible to breed both these varieties of Petkusskaya rye from the original one only because their "originality" was not protected (the originality of Petkusskaya rye lies in that it does not stand our winter well).

These varieties undoubtedly crossed with other varieties of rye, as a result of which the plant breeders, by means of selection, raised the new varieties.

Polesskaya—"a variety bred by the Polesskaya Experiment Station from local rye," which, of course, had never been isolated.

Novozybkovskaya M-4—"a variety bred by the Novozybkov Plant-Breeding Station from a local regenerated Petkusskaya rye." The regenerated rye is a natural hybridization of Petkusskaya rye with another variety.

Pullmana Zholtozyornaya—"a variety bred by the Bogoroditsk Experimental Field from local rye." It goes without saying that the local rye had not been isolated.

Triumph—"bred by the Ramon Experiment Station from local rye." This local rye, too, had not, of course, been isolated.

Mup—"a variety bred by the Steppe Department of the All-Union Institute of Plant Industry from local rye cross-pollinated with a specimen received from Iran." It was possible for this variety to appear only because the workers at the Institute did not protect the collection specimens, and the latter crossed naturally with the local rye.

Omskaya—"a variety bred by the West-Siberian Plant-Breeding Station from the population of the Ivanov, Trostnikov, Shlanshted and local varieties." The story of how Omskaya rye was bred is, to my mind, extremely instructive. In the districts of the Omsk Region the winter is very severe, and it was only because a number of varieties of rye, some of which are by no means very winter-hardy, were able to cross that material was obtained capable of wintering in the Omsk district and of serving as the initial basis for breeding the Omskaya.

Tulunskaya Zelyonozyornaya—"a variety bred by the Tulun Experiment Station from local rye."

I have deliberately quoted all, down to the last one, of the varieties given in the *Handbook on Testing Agricultural Crops* (Selkhozgiz, 1937). In doing so I mean to emphasize that *not a single variety* used in production to this day has been bred by plant breeders isolating the initial material, or by the artificial crossing of pairs.

It is worth recalling the numerous controversies that were initiated by the magazine *Yarovizatsia* over the question of inbreeding rye in plant-breeding experiments. The Morganists asserted that very many valuable varieties of different cross-pollinating plants, including rye, had been raised by inbreeding. Investigation, however, has shown that not a single variety that had been raised by inbreeding, i.e., by self-pollination, is being used in production, and that the plant-breeding stations do not possess even a prospective variety bred by this method.

As is evident from the list of district varieties I have quoted, the plant breeders were able, as a rule, to breed them only because for one reason or another they did not set up kilometre isolation zones.

Where is the logic, the scientific justification, of those men of science who believe that isolation zones are essential for all varieties of all cross-pollinating crops?

The creation of the best agrotechnical conditions for rye-seed plants, i.e., proper training of the seed plants and continuous selection constitutes, to my mind, the basis on which the growing of rye seed should rest.

I believe the People's Commissariat of Agriculture should annul its regulation concerning the testing of rye, should reconsider the regulation calling for a kilometre isolation zone and emphasize as sharply as possible the necessity of singling out for seeds only those plots of wanted varieties from which good and high yields are obtained. Plots which give low yields because of bad agrotechnique cannot be regarded as rye-seed plots.

FUNDAMENTAL PRINCIPLES OF GROWING ÉLITE CEREALS

All the forms of plants that grow wild have been created by natural selection. Man has created cultivated varieties by means of artificial selection. It must be emphasized that the phrase "plants created by means of selection" necessarily includes the growing conditions of plants, the fittest of which survive in free nature, and the best of which are kept for breeding, for seeds, in practical agriculture.

When cultivating plants for the purpose of obtaining from them the best crops at the required date, men have always kept the best of these plants for seeds. In other words, even when not engaged in the special work of seed growing, but simply when cultivating plants for the crop, human practice is constantly improving the breed of plants, improving the existing varieties.

It is an indisputable fact that in countries with a high level of agrotechnique the varieties too are, as a rule, (without exception) superior in quality. This shows that by bare selection alone, without the efficient cultivation of plants, it is rarely possible to create an efficiently cultivated variety, or improve an existing one. But it is only such selection that is advocated by the Morganist geneticists, who assert that this, and this alone, is the essence of the Darwinian theory of form-building. But who does not know that from the general principles of Darwinism it follows that all varieties of cultivated plants have been created by man by means of good, high-level agrotechnique? In other words, good, high-level agrotechnique is the basis for producing higher-grade breeds of plants. In the state of nature, when agrotechnique is bad, not only do wild plants fail to become cultivated, but cultivated plants run wild. The creation of increasingly better conditions of cultivation for the growing of plants (meaning by that the conditions under which the biggest and best crops are obtained) and the selection of the best plants for seeds are the chief means that have served as the basis for breeding the vast majority of varieties of all cultivated plants.

After creating good growing conditions for plants, it is necessary to select the best breeds for seeds, i.e., to select breeds whose seeds will produce better crops than the seeds of other plants grown beside them. Selection must be made according to genotype, as they say, i.e., according to breed. In the case of some crops, cereals, for example, it is necessary to select the best plants and to sow the progeny of each of them separately, but without fail under comparable conditions. The technique of sowing adopted must always be such as will enable the plants in the course of their development sharply to develop those of the organism's properties which will be taken for comparison.

Proceeding from the Darwinian proposition that any two plants, even of the same variety, will differ in some degree with regard to any character or property, the essence of the work of selection amounts solely to the ability to find, see and weigh this difference. In choosing plants for the purpose of obtaining the best seeds, great attention must be paid to this aspect of the technique of the work. Ever new methods of appraising the different properties of plants must be devised. The correctness of the choice—whether they really are the best plants because of their economic qualities and properties typical of the given variety—must be tested by the progeny. For different districts, different crops, and even different varieties, different properties and characters must serve as the criteria of appraisal. In one case most attention will be paid to winter-hardiness, in another case to disease resistance, in a third to drought resistance, etc. And, of course, in all cases attention must be paid to yield and to the quality of the crop.

Once again we emphasize: before selecting plants for seeds at seed-growing farms, and all the more so at plant-breeding stations, the plants must be skilfully trained.

Training plant forms in a definite direction is one of the fundamental propositions of Michurin's theory. Nevertheless, not only is this forgotten by the Mendel-Morgan theory, but the entire logic of Morganism leads to the denial that conditions of life induce changes in the nature of organisms in a definite direction.

After selecting properly- and well-reared plants for seeds, it is necessary, on the one hand, to test the correctness of the selection and, on the other hand, to fix and further strengthen the economically valuable properties for which the selection has been made.

Hence it follows that the work of seed growing does not end with the selection of plants for seeds.

The initial plants chosen from among cereals must have the largest possible yield, so that each plant selected will produce enough seed for the different assessments to be made of the progeny. At the Institute of Selection and Genetics (Odessa) of which I am the director, the winter-wheat plants that were selected in 1937, the Krymka variety, for example, were preliminarily grown in such a way that each plant should yield 2,000, 3,000 and even 5,000 seeds. This quantity is quite sufficient to sow a one- or two-row plot 100 metres long; to sow 300 to 500 grains from each plant under artificial conditions to determine the degree of frost resistance by chilling in the refrigerator; to sow 300 to 500 grains to determine winter-hardiness in districts where winters are known to be severer than in the district served by the given plant-breeding institution; to sow 200 to 300 grains in the field for the purpose of artificially infecting the plants with brand or rust.

On plots sown in families, i.e., progeny from each individual plant (we call such plots "seed nurseries"), it is easy to note the slightest differences in breed, i.e., genotypical differences between the families with respect to morphological characters. Each family is also assessed for different economically important properties (winter-hardiness, disease resistance, etc.). Being in possession at the time of ripening of all these observations and assessments it is not difficult to cull all the families whose nature is qualitatively below the average level of that of the seeds of the given variety released by the station. The plant breeder also singles out the families that are conspicuous for a number of important properties, or even only one (for example, winter-hardiness, or rust resistance), but are typical of the given variety as regards all other properties. All the remaining lines (families) are thrashed separately, the quality of the grain is assessed by external appearance and, on the basis of this, a second culling of the lines is made according to quality of the grain. The remaining seeds are heaped in one group for sowing the plot, the crop from which is called "élite" and is intended for the seed plots of the district seed farms.

As an additional guarantee in testing the correctness of the selection of seed for the élite plot, part of these seeds are sown in the station's comparison plot simultaneously with the sowing of the élite plot. Their yield and other properties and qualities are compared with those of the élite seed of the same variety previously released by the station. *Thus, at the time the élite is sent to the seed plots of the district seed farms a comparison test of the correctness of the plant breeder's work in growing élite seed will already have been made.*

In the comparative tests made in 1937-38 at the Institute of Selection and Genetics (Odessa) mixtures were sown of seeds of the Krymka variety obtained after intravarietal crossing and left over from the selection for the seed nursery. These seeds produced a crop equal to 37.8 c. per ha. Seeds of the same Krymka variety sown as a superélite after a mass selection of ears by the commonly accepted method produced 35.7 c. per ha. The Krymka variety obtained from an intravarietal crossing and preliminary training showed an increase in yield of 1.5-2 c. per ha. also on a number of variety testing plots in Odessa Region.

An increase in yield of 1-1.5 c. per ha. was obtained in Odessa Region in comparative tests of seeds obtained from an intravarietal crossing of Hostianum 0237 and a number of other varieties of winter wheat.

It must be emphasized that in the comparative test, seeds were sown that had not been culled at the seed nursery. The seeds used were a mixture taken from the best plants grown for nursery seed and selected only according to external appearance.

The best families selected at the seed nursery for conspicuous properties (winter-hardiness, rust resistance, etc.) are also subjected to comparative tests and are also at the same time reproduced to a small extent. If the succeeding comparative test confirms the correctness of the selection (i.e., shows that these seeds are more winter-hardy, or are superior in other natural properties), the station will have a new variety differing greatly from the old in particular useful economic qualities.

It is difficult to draw a sharp distinction between the continuous improvement in the quality of the seeds of a given variety and the breeding of a new variety. A plant-breeding station must first of all strive to obtain better and better élites of the varieties grown in the district it serves. If the work is so conducted, the breeding of new varieties will be prompted by the interests of seed growing as such.

On these grounds, the entire work of the majority of plant-breeding stations must be radically changed and focussed on the growing of élite seeds. The more fully plant breeders master the method of creating and training good élites, the better and quicker will they be able to produce new varieties. As a rule, the work of creating élites cannot be divorced from the breeding of new varieties.

It is time that the gulf between plant breeding and seed growing were bridged. It is time it were understood that the work of selection to improve the breed of seeds must accompany the entire course of the work of seed growing.

It is time seed-growing work were resolutely shifted to the track of Michurinian theory.

THE CREATOR OF SOVIET AGROBIOLOGY[1]

(On the Occasion of the Fourth Anniversary of the Death of Ivan Vladimirovich Michurin)

In THE EARLY days of the Soviet regime, Lenin, one of mankind's greatest geniuses, revealed to our country and to the working people of all the world a man who at that time was little known, namely, I. V. Michurin. In tsarist, landlord capitalist Russia, everything scientific and advanced was so widely suppressed that the works of I. V. Michurin, who had lived under that decaying system for sixty-two years, were unknown even to that Bolshevik scientist, the world's most ardent champion of science and Darwinism, K. A. Timiryazev.

Michurin's conception of the laws governing the development of plant organisms has thrived so well only because of the care shown by the Party of Lenin and Stalin and by the Soviet Government, which have created, and are continuing to create, unprecedented opportunities for the development of advanced science.

Michurin's theory represents the Soviet trend in agronomic science; it is Darwinism in agrobiology.

Michurin's theory is generally recognized in our country. True, there are still some scientists, supporters of Mendel-Morgan genetics, who either refuse to accept Michurin's principles because of their ignorance of his theory or, as Ivan Vladimirovich himself put it, "reject with inexplicable fury the very facts involved."

The attempts of certain scientists to clothe Michurin's theory in the garb of Mendelist genetics look ridiculous. All his life Ivan Vladimirovich waged an uncompromising struggle against the theories of formal, bourgeois genetics. It is impossible to fit his theory into the frame of Mendelism-Morganism. It exposes the utter falsity of the fundamental propositions of Morganism.

---

[1] *Яровизация*, 1939 г., № 3 [24].

I. V. Michurin created a new, correct trend in the science of genetics, a trend with which Mendelism-Morganism can bear no comparison, because Michurin's scientific theses are not artificial inventions, but propositions taken from life. They came into being as a result of long and tireless struggle to grasp the laws governing the nature of plant organisms. The scientific propositions of Mendelism-Morganism, however, are divorced from reality. Everybody who is engaged in the practical work of altering the nature of plants in a required direction can easily convince himself of the truth of this.

Indeed, what practical problem can be solved with the help of the theory of Mendelism-Morganism?

Let us take, for example, the breeding of winter-hardy varieties of winter wheat or of winter rye. On this most important problem the decision of the Party and the Government published on January 6, 1939, contains an instruction to agricultural scientists and the Land Departments to produce at the earliest dates winter-hardy varieties of winter wheat and of winter rye biologically adapted to the severe conditions of Siberia. The former is to be produced in 3-5 years, and the latter in 2-3 years.

No assistance in the fulfilment of this task can be obtained from the books written by the Mendelist-Morganists. However, in the writings of Michurin, who all his life worked mainly on fruit and small-fruit plants and elaborated a general biological theory, it is possible to find clear advice, clear theoretical guidance in this matter.

In his article, "Some Interesting Feature of the Influence of Parent Plants on the Characters and Properties of Their Hybrids," Michurin wrote:

"A thing that every plant hybridizer should bear in mind is the following. In natural cross-fertilization, where every maternal plant has the opportunity of, so to speak, freely choosing—out of all the wind- and insect-carried pollen, sometimes from quite a few different varieties—the pollen that is best suited to the structure of its particular fruit-forming organs, the progeny consists of relatively more viable individuals. The latter cannot always be expected in hybrid seedlings, which are obtained by artificial and, of course, enforced crossing. . . ."[1]

In this excerpt one perceives the profound Darwinian theory of the elective capacity in fertilization, and of the biological benefit of cross-pollination.

With this theory as our basis we have outlined a program of scientific research work, to be conducted at plant-breeding stations and by collective-farm experimenters in Siberia with the aim of rapidly breeding winter-hardy forms of wheat and rye. Some of the plants of the best local varieties of winter wheat, surrounded by a collection of other local and bred varieties, are castrated. At the time of flowering the pollen of the different varieties will freely fall on the stigma of the castrated flowers of the local variety. We are

[1] I. V. Michurin, *Selected Works*, Eng. ed., Moscow 1950, p. 92.

quite confident that the seeds obtained from this crossing will possess greater biological resistance to the climatic hardships of wintering.

It is planned to do the same in order to increase the winter-hardiness of varieties of winter rye, with this difference, however, that, being a cross-pollinating plant, it is not necessary to castrate rye.

Not only has it been impossible to find such advice in the theory of the Mendelist-Morganists, but their entire pseudo science runs counter to the path we have chosen in this case. According to Mendelist theory, plants do not as a rule possess biological elective capacity in fertilization; they do so only in rare exceptions.

Further, according to the theory of the Mendelist geneticists, obtaining seeds from two crossed forms is a far cry from breeding a variety. The Mendelists assert that the progeny of these hybrid seeds must, without fail, segregate in a number of generations, i.e., must split up and revert to their original forms; that it is necessary to grow several generations of hybrid plants before one can select so-called constant forms.

It is futile to expect to be able to breed winter-hardy varieties on the basis of this theory by hybridization in the time set by the decision of the Party and the Government. Michurin repeatedly protested against the false laws of segregation laid down by Mendel, calling them "pea laws."

The experiments conducted at the Institute of Selection and Genetics (Odessa) show convincingly that the so-called segregation of hybrid forms obtained from the free, elective crossing of wheat plants is not governed by the laws propounded orally and in textbooks by the Mendelist geneticists. It is particularly with free, i.e., elective, pollination, when sex cells which biologically correspond best to each other unite, that we get hybrids whose offspring vary in different degrees, quite contrary to the laws of Mendel and Morgan. It happens at times that, despite Mendel's laws, the progeny is relatively uniform, i.e., has practically not "segregated."

On these grounds we anticipate that already in the autumn of 1939, after we have obtained hybrid wheat seeds from free, i.e., elective, fertilization, we shall be able to use part of these seeds in districts of Siberia for preliminary winter-hardiness tests. The rest of the seeds will be sown in nurseries with the view to selecting the best of the progeny as regards uniformity and economically valuable qualities and properties.

Experiments commenced in the autumn of 1938 at Gorki Leninskiye (Moscow Region), the Experiment Base of the Lenin Academy of Agricultural Sciences of the U.S.S.R., have brilliantly confirmed the proposition that plants obtained from free cross-pollination have more resistance, are more winter-hardy.

It is known that varieties of rye like Petkusskaya or Tarashchanskaya are not hardy and cannot resist frost. Plots sown with seeds of these varieties grown last year on isolated seed plots suffered rather badly this winter. Neighbouring plots, sown with seeds of the same varieties but gathered from the crop of last year's variety test plots (where different varieties grew near each other and, consequently, were able to cross-pollinate), stood the winter excellently.

To put it briefly, the task the Party and the Government have set agronomic scientists, namely, quickly to breed winter-hardy varieties, can be carried out only on the basis of Michurin's theory.

A number of other examples may be quoted to prove that Michurin's theory will always help to solve a given important practical problem in the sphere of plant breeding. On the other hand, no such assistance can be obtained from the theory of the Mendelist-Morganists. Quite the contrary: as a rule, it leads the investigator into a blind alley.

Often we hear people say: "After all, the world science of Mendelism-Morganism did make some contribution. After all, it has advanced the theory that treats of the life and development of plants."

Again it must be declared that bourgeois Mendelist-Morganist genetics has contributed literally nothing, and can contribute nothing for practical farming. Its principles are fallacious, spurious, purely invented.

During the past few years the Mendelists in our Soviet Union have been talking a lot about their having shown the way to overcome the inability of various species and genera to crossbreed. By treating plants with the extremely strong poison colchicine, and by various other violent processes, they mutilate the plants they experiment upon. The cells cease to divide normally and something in the nature of cancerous growths appear. In such mutilated cells the number of chromosomes is often found to be abnormal. Sometimes there are two or three times the usual number. And the Mendelist geneticists in our Soviet Union call these mutilations directed alteration of the organism's nature! In some cases, these mutilated plants cannot cross with the initial forms, but do so with others with which the initial ones did not cross. And this they call overcoming the inability of various species to cross.

So far, nothing of practical value has been obtained from these experiments and, of course, there is no hope of anything being obtained from them.

However, on the question of how to induce noncrossing species and genera to cross, a wealth of material can be found in the works of I. V. Michurin based on his brilliant practical achievements. For example, in his article "My Experiments in Breeding New Plum Varieties in a Severe Climate" Ivan Vladimirovich wrote that, as far back as the spring of 1889, he hastened to take advantage of the second flowering of his blackthorn, expecting, not without reason, that a young plant, which had not yet had time to develop resistance to fertilization with pollen of a distantly-related variety, would be easier to fertilize as he desired with pollen of Green Reine Claude.[1]

In the same article I. V. Michurin emphasizes that all his attempts to cross blackthorn with the Green Reine Claude plum failed owing to the too-distant kinship between these two plants. Only after he had taken a young blackthorn plant did he obtain twelve fruits out of fifteen pollinated flowers.

---

[1] *Cf.* I. V. Michurin, *Selected Works*, Eng. ed., Moscow 1950, p. 25.

At all events, he achieved a larger percentage of success than is obtained by any treatment with the poison colchicine, or acenaphthene, which are now so much in vogue among our Mendelists.

From these seeds of hybrids of distant forms, which usually fail to cross and, as we know, have different chromosome numbers, Michurin obtained plants that not only were not sterile but could be termed excellent varieties. The Mendelist geneticists, who mutilate plants with poisons and resort to other forms of violent treatment, assert that they are working out a method of making sterile distant hybrids fertile. No, the art of crossing distant species and of obtaining fertile progeny from such crossings must be learnt from the works of I. V. Michurin.

I. V. Michurin was not only able to obtain good varieties himself, but he elaborated a splendid theory which serves as the basis on which numerous Soviet Michurinists are doing brilliant work along the line of creating new forms of the most diverse plants by means of intergeneric and interspecific crossing. I will mention a few such examples.

From the crossing of the wild potato species Akaule with cultivated varieties hybrids rarely result, and when they do they are of wild appearance and produce hardly any tubers. Such hybrids have to be pollinated repeatedly for several generations with the pollen of cultivated varieties, i.e., they have to be impregnated all the time with a cultivated variety. Only after that is done can a variety with cultivated characters be obtained. Of course, some of the wild plant's good properties, for the sake of which it was taken for crossing, are in many cases entirely lost in the process.

A. S. Filippov, a young Soviet scientist at the Potato Institute, tackled the problem of crossing distantly-related forms of potatoes in the Michurin manner. He grafted a plant of the wild potato species on to a cultivated variety (the Michurin method of approximation). When flowers appeared on the wild scion he pollinated them with the pollen of the cultivated variety. In the very first generation (i.e., after a single crossing between the wild and the cultivated species) a cultivated type appeared, at all events, one much more cultivated than the plants growing next to it of the same wild species, which for three consecutive generations had been pollinated with the pollen of cultivated varieties.

Another example. Solodovnikov, a postgraduate at the Potato Institute, obtained vegetative hybrids from a cross between a cultivated potato and a wild potato. These hybrids, grown from the stolons of the wild species Demissum (the stock), already have the appearance of cultivated potatoes.

Again, at the Odessa Young Naturalists' Station, Solovey, a scientific worker, by appropriate training (late autumn sowing), transformed the spring barley Pallidum 032 into a winter variety suitable for the winter conditions of the Odessa Region. More, from the former spring barley he created forms which, planted in the field in the autumn of 1937, produced a crop in the beginning of June 1938. Then these plants sprouted again, lived through the winter and produced another crop in 1939. In other words, spring barley has been transformed if not into a perennial form, since it is too early to speak of that, then at all events into a biennial form, which has fruited twice and has lived through two winters. The initial forms sown next to these hybrids, however, perished entirely in the winter of 1938-39, and even a number of winter varieties of barley which had been sown in the same plot perished too. Many examples may be cited of how Michurin's theory helps to alter the nature of plants. On the other hand, numerous examples may be mentioned of how the false Mendelist-Morganist theory hinders scientists who earnestly want to do useful work. Here is one example.

M. M. Molodozhnikov, a young Soviet scientist who is enthusiastically striving to introduce the quinine tree in our Soviet subtropics but is being guided in his work by people who believe in Mendelism-Morganism, wrote a paper criticizing other authors, in which he stated:

"In other cases the arguments of the naturalizers [of the quinine tree—*T. L.*] were supplemented by groundless expectations of the gradual 'acclimatization' of new generations from the seeds of their more naturalized mother plants."[1]

Now I. V. Michurin always pointed out that in order to grow tender plants in severer regions it is necessary to base oneself precisely on what Molodozhnikov calls "groundless expectations," namely, gradually, from generation to generation, to sow seeds that become more and more adapted to the severer conditions. Thus, M. M. Molodozhnikov, probably unaware of, or failing to understand, Michurin's theory, rejects one of the most effective ways and means of really introducing the quinine tree in our subtropics. And this is so because the Mendelists have stuffed the mind of this young scientist with the notion that environmental conditions do not alter the nature of plant organisms even when these organisms are young, only recently emerged from the seed.

I shall cite another example of how the false, spurious basic principles of Mendelism-Morganism hinder plant breeders in their practical creative work.

According to the Mendelists, the organism, the cell, contains a special "hereditary substance" which consists of granules (genes). Conditions of life change the organism, but its nature, the "hereditary substance" (the gene), remains unchanged. From Mendel's so-called law of the purity of gametes it follows that, beginning with the fertilization of the egg cell, the "hereditary substance" passes through the whole organism without undergoing any change. This conception of heredity, based on the idea of some sort of special "hereditary substance" separate from the body (soma) of the organism, is a very great hindrance to the creation of varieties and breeds in seed growing and livestock breeding. In support of the false law of the purity of gametes, purity of chromosomes and genes, the Mendelist-Morganists usually point to the "segregation" of hybrids, i.e., to the variation

---

[1] Сборник *Хинное дерево в советских субтропиках*, Сухуми, 1938 г., стр. 16.

in the progeny of hybrid organisms. In this variation it is possible, as a rule, to meet with organisms that in their entirety, or in individual properties or characters, resemble either of their parents.

In support of this spurious law seemingly weighty arguments like the following are advanced. The progeny of, say, an awnless hybrid spike (from the crossing of awnless with awned wheat) contains both awned and awnless plants. Pointing to this and analogous examples, the Mendelist geneticists say: "Do you see? You crossed awned with awnless plants and the resulting plants are awnless. But these awnless plants have retained 'hereditary granules' of awnedness which, although not revealing themselves in the appearance of the plants, have remained in the chromosomes in their pure, unaltered, state." In the opinion of the Mendelists, this is "brilliantly" confirmed by the fact that some of the progeny of such awnless plants may be awned.

It must be stated that there are no few cases when the heredity of one of the parents taken for crossing does not manifest itself for a number of generations, or manifests itself only in rare, individual organisms of the progeny. Usually the Mendelist geneticists forget about this, or, simpler still, consider such facts nonexistent.

Many comrades who are working on interspecific and intergeneric hybridization are well aware of cases in which the progeny takes entirely after the mother, or entirely after the father. True, I know of far fewer cases of resulting forms being purely paternal than of their being purely maternal, but such cases do occur. Who does not know that in genetic experiments the Morganists weed out plants with the maternal type of heredity because they regard such plants as the result of unsuccessful crossing (which they usually attribute to faulty castration of the flowers of the mother plant)?

The foregoing facts can be understood only from the standpoint of Michurin's theory, which utterly rejects the metaphysical conception of heredity as a special, separate substance independent of the organism. Michurin also indicated the way to regulate the process of fertilization with a view to producing hybrid progeny that deviates more or less towards the father or the mother, as desired. I have already cited above the instance of how A. S. Filippov, working in the Michurin manner, induced hybrid potatoes to deviate from the wild form Akaule towards the cultivated form.

In 1937, A. A. Avakian, a specialist at the All-Union Institute of Selection and Genetics (Odessa), crossed the awned winter wheat Azerbaijanskaya 2115 (the maternal form) with Lutescens 062, an awnless spring wheat. In the winter of 1938 the hybrid seeds were sown under glass so as to get well-tillered seedlings in the spring. This was done in order to obtain the largest possible quantity of seeds from each first-generation hybrid plant so as to ascertain how the second-generation progeny varies from each first-generation plant.

In the summer, when the hybrid plants were in the field, it was found, after earing, that among scores of hybrid awnless plants (the paternal form was awnless) there were several awned plants which literally in no way differed from the maternal form. According to all the rules of Mendelist-Morganist genetics, such plants should have been weeded out forthwith. The Mendelists would have said that the seeds from which these plants had grown were the result of self-pollination due to faulty or belated castration of the flowers at the time they were crossed with the awnless wheat.

The maternal form of Azerbaijanskaya 2115 is a weak winter variety, but a winter variety for all that, and when sown in the spring it does not ear. The awned plants that we are discussing eared in proper time. But this does not prove in the least that they are of hybrid origin, for they had been planted under glass (for growing seedlings) in winter, at a cool temperature. Under such conditions the seeds of any winter wheat will be fully vernalized by the beginning of the spring and will subsequently ear.

In brief, everything seemed to show that the above-mentioned awned plants were not of hybrid origin, i.e., that the awnless spring variety Lutescens 062 played no part in the creation of these plants.

In 1939, the offspring of these first-generation plants, the "legitimate hybrids" (i.e., the awnless), as well as the offspring of four awned plants, were planted separately at Gorki Leninskiye, the Experiment Base of the Lenin Academy of Agricultural Sciences of the U.S.S.R. I will not deal here with the question of how the progeny of the different first-generation plants varied, because the analysis is not yet complete. Suffice it to say that in the second generation, the progeny of one of the four awned plants (i.e., those that had seemed to be of the purely maternal type) produced several plants with characters which in many ways resembled those of the paternal form Lutescens 062. Of the 180 plants in the progeny of this plant, 172 were found to be of winter habit resembling the maternal form, and 8 of the plants eared; 5 of these were awnless and 3 were awned.

Data of this kind show that the heredity of one parent may, to some extent, be absorbed by the heredity of the other parent. Cases are not rare when crossing, particularly intergeneric and interspecific crossing, results, as it were, in complete absorption, the complete dominance of the heredity of one parent over the heredity of the other parent.

In the beginning of June 1939, at the Michurin Central Genetic Pomological Laboratory, I saw several intergeneric and interspecific hybrid fruit trees and berry bushes. Some of those who are working in this laboratory (P. N. Yakovlev, J. S. Gorshkov, S. F. Chernenko, Kh. K. Yenikeyev, A. Y. Kuzmin and others) stated that they had often observed in their own work (and in that of other experimenters) that after careful castration of the flowers and their pollination with the pollen of another species or genus, seeds are obtained which produce plants of the purely maternal form.

The Mendelist geneticists attribute all such cases to parthenogenesis, i.e., the production of seeds without fertilization. But in a number of experiments, similarly castrated flowers, isolated without pollination, produced no seeds. The Mendelists assert that, notwithstanding this, in cases where plants of the purely maternal form result, the seeds are nevertheless produced without fertilization. In such cases, the Mendelists say, the pollen is needed as a stimulant for the irritation of the stigma.

But even this questionable explanation leaves the Mendelist-Morganists in a difficulty. It is a known fact that the egg cells contain only half the number of chromosomes that will be contained in the cells of the future embryos, including those from which plants of the purely maternal form will develop. The only explanation the cytogeneticists give for the presence of a double set of chromosomes in ordinary cells compared with the number in unfertilized sex cells is that the chromosomes of the spermatozoid nucleus are added to the chromosomes of the egg cell nucleus. But if that is so, how are the facts described above to be explained? According to the Mendelists, no fertilization had taken place (the seeds were the result of parthenogenesis), but the number of chromosomes in the cells of thousands of these plants was normal, i.e., twice the number found in the egg cell. The only "clear" "explanation" that remains is the one resorted to by Rozanova (a specialist at the All-Union Institute of Plant Industry), namely, that in some way the number of chromosomes is doubled.

In Michurinsk, however, and in a number of other places, there have been cases that do not fit into this "doubling" explanation. For example, according to the statement of members of the staff of the Michurinsk laboratory (I. N. Shashkin's), the first generation from the crossing of the strawberries *Fragaria elatior* and *Fragaria grandiflora* in no way differed from the maternal form. The explanation that has always been given for this phenomenon is that, in this case, the strawberry produces seeds parthenogenetically, i.e., without fertilization, and that the pollen was needed only as a stimulant. When, however, the seeds from the first generation were gathered and planted, several of the plants obtained resembled (I was shown these plants) a form which the cytogeneticists do not regard as paternal The facts show that in this case, as in the case of A. A. Avakian's experiment with wheat that I have already referred to, fertilization did take place, only the maternal form had largely absorbed the paternal form.

I shall cite another example from the work of the Michurin Central Genetic Pomological Laboratory related and shown to me by P. N. Yakovlev. The flowers of the cherry *Cerasus Besseyi* were castrated and pollinated with the pollen of a peach. This experiment had been started when Ivan Vladimirovich was still alive. The resulting stones were planted and from them grew plants which in no way differed from the maternal form (i.e., the cherry). The flowers of these plants, which were regarded as nonhybrids, were also castrated and pollinated with peach pollen. Again, the resulting progeny in no way differed from the maternal form.

P. N. Yakovlev continued this operation until the fifth generation, i.e., five generations of hybrids were consecutively castrated and pollinated with peach pollen. The result was that only in the fifth generation two plants were found with the characters of the peach, i.e., of the father among the hundreds of plants that were obtained from the stones of these crosses. This example also shows how, under definite conditions, the heredity of one parent (even in the course of five generations of consecutive fertilization) can entirely absorb the heredity of the other parent.

Many other cases could be instanced, such as the crossing of a currant with a gooseberry, an apple with a pear, etc., where, in the progeny, the influence of one of the parents (usually the male) is often almost entirely absent. To try to explain it as parthenogenesis, i.e., that the seeds form without fertilization, will not do here. It is wrong.

This is confirmed, in particular, by the experiments conducted in 1939 by Kh. K. Yenikeyev at the Central Genetic Pomological Laboratory. In these experiments the branches of several shrubs of *Cerasus Besseyi* were covered with parchment bags. The flowers on these branches were not castrated but pollinated with pollen taken from other flowers on the same shrub. Notwithstanding the presence of their own pollen (which, one would have thought, could have served as the stimulant brought forward by the Mendelists to explain cases of parthenogenesis), not one of the thousands of isolated flowers developed fruit.

That the parthenogenetic "explanation" of cases where the heredity type of one of the parents predominates is invalid becomes particularly evident when the organism resulting from a cross deviates entirely towards the paternal form. In another experiment conducted by Kh. K. Yenikeyev, an interspecific crossing (1937) resulted in a purely paternal form both in habitus and in number of chromosomes. The maternal form was the 16-chromosome American plum Cheresota, and the paternal form was the 48-chromosome Michurin plum Reine Claude Reforma. The plant resulting from this cross has the paternal habitus and the number of chromosomes, as ascertained by T. P. Filosofova, is 48. The parthenogenetic explanation of how the seeds developed is obviously unsound in this case, because the influence of the male sex element is completely manifest here, whereas the influence of the maternal form is, in this case, not apparent at all.

All these examples, and many more that could be adduced are of profound interest to biological science and agronomic practice. They clearly demonstrate the multiformity of the biological process of fertilization, which in no way fits into the artificial Morganist cytogenetical pattern.

The description of the fertilization process as given by the Mendelist-Morganists is utterly wrong. The Mendelists say that with the fusion of the nuclei of the sex cells, the chromosomes of the male gamete pair up with the chromosomes of the female gamete. In the opinion of the Mendelists, the chromosomes contain a "hereditary substance" which is uninfluenced by the conditions of life of the cells. Therefore, according to the cytogeneticists, the chromosomes from the paternal and maternal sex cells retain their individuality throughout the whole life of the organism, i.e., remain unchanged both quantitatively and qualitatively. Hence it follows that the organism will produce exactly the same kind of sex cells as those from which it had sprung. This is the basis of the Mendelist-Morganist principle: that the nature of organisms is immutable, that nothing new is formed.

Actually, however, the process of fertilization, like every other process in the living organism, is subject to the laws of assimilation and dissimilation. The fusion of two sex cells is a process of assimilation, a process of mutual absorption, as a result of which in place of two sex cells (a male and a female) we obtain a third, a new cell, which we call a zygote. The hybrid embryo obtained inclines more or less in the direction of the nature of the sex cell which in its own way assimilates its partner more fully, so to speak, than the latter assimilates it. If, for instance, both sex cells possess equal power of mutual assimilation (absorption) a zygote (fertilized cell) is obtained which produces an organism with an approximately equal distribution of maternal and paternal properties and characters. If the power of assimilation of one of the sexual components predominates, the hybrid obtained exhibits greater inclinations in the direction of this parent, to the point of the complete absorption of the hereditary properties of the other.

On the basis of this explanation, which to my mind follows logically from Michurin's theory, it is possible to control, to facilitate, when crossing plants, the moulding of the nature of hybrid embryos so as to produce a greater or lesser deviation towards either the maternal or the paternal form. I will mention Michurin's direct advice on this question. For example, if it is necessary that the hybrid should obtain from a given form only some of its properties (say, from a little-cultivated local form that is resistant to climatic inclemencies to take only its hardiness) it is better to take the pollen from a young plant that flowers for the first time and the nature of which has not yet become firmly established. On the other hand, the flowers of the other component to which it is desired to give only some of the properties of the first parent (hardiness, let us say), must be chosen from a vigorous tree that has already borne fruit several times and, moreover, from those parts of its branches where they are best ensured of a flow of food. In this way the conditions will be created for the dominance in the progeny of the properties of one (the desired) form and for the considerable grading of the properties of the other.

Michurin also repeatedly gave the following advice. For the purpose of crossing, it is best to choose forms that are distant in place of origin (conditions) not only from each other, but also from the place (conditions) in which the new variety is to be formed. This is particularly necessary in those cases when a cultivated variety of southern origin, bearing good fruit, but unable to winter under severe conditions, is taken as one of the parents. If it is crossed with a frost-hardy local variety, the conditions (climate, food, etc.) will help to augment the absorbing-assimilating power of the sex cells of the local variety and an undesirable hybrid will result. In such a case, Ivan Vladimirovich advised that neither of the parents (hardy and nonhardy) be a local variety, so that the external conditions should to an equal degree facilitate the manifestation of the properties of both components when fertilization takes place. If the plants resulting from such hybrid seed are skilfully trained, it is possible more easily to obtain varieties with fruit of good quality and resistant to unfavourable conditions.

A profounder Michurinian understanding and study of the problem of hybridization, and of the fertilization of plants in general, is extremely important to our practical socialist agriculture. In nature, in the plant world, cross-fertilization, as a rule, predominates over self-fertilization. The elective crossing of some lines (forms) of plants with others results in a progeny that is more viable. At the same time, with regard to its external characters, it often resembles, morphologically, in the mass, the maternal form. Of considerable interest in this respect are the experiments on rye conducted at the All-Union Institute of Selection and Genetics (Odessa) by postgraduate I. E. Glushchenko. In these experiments, the deliberate crossing of a number of varieties of rye with the pollen of other varieties resulted in a progeny that morphologically resembled the initial maternal forms.

By skilfully ascertaining and utilizing the laws that govern the fertilization of plants it will be possible in plant breeding and in practical agriculture to make use of the advantages of the cross-pollination between morphologically different varieties while at the same time preserving the good forms of the old variety. As we see from these examples, the argument advanced by certain scientists—that while there are shortcomings in Mendel-Morgan genetics no other genetics exists—is wrong. *There is Michurinian genetics, the Michurin theory of the life and development of plants.*

In Soviet Land we have every opportunity to learn from I. V. Michurin, from his unexcelled scientific works, the art of crossing both near and distant forms, of producing fertile hybrids and of breeding varieties resistant to different unfavourable climatic conditions. Michurin's works are an inexhaustible fount of knowledge for those whose job it is to alter the nature of plants.

Four years have passed since death took Michurin from us. Year after year the depths of his theory are becoming more and more revealed to agronomic scientists and collective-farm experimenters.

MICHURIN'S THEORY AT THE ALL-UNION AGRICULTURAL EXHIBITION

IN OUR COUNTRY science, theory, is held in high esteem. Agronomists, collective farmers and workers at state farms and machine and tractor stations are not indifferent to the achievements of agrobiological theory, because these achievements help to increase the productivity of their labour. The Party and Government are creating tremendous opportunities for scientific work, for the development of the talents of the masses of the working people.

Only in our country, the land of victorious socialism, can creative Darwinism really develop.

Collective farmers, workers at state farms and specialists in socialist agriculture have a practical interest in the rapid development of agronomic theory. This explains why in our country discussions, controversies and debates on the most profound principles of biological science, on problems concerning the theory of development of plant and animal organisms draw the attention of the broad masses. Prompted by their own observations and experiments, many collective-farm experimenters, agronomists and young scientific workers are joining the ranks of the active fighters for the development of Soviet agrobiology.

The more workers in agronomic science master consistent Darwinism, Michurin's theory, the more evident becomes the creative role this theory plays in the advancement of practice, in improving the varieties of plants and breeds of animals, and the more glaringly and conspicuously the fallacy is revealed of the views held by the Mendelist-Morganist geneticists about the life and development of plant and animal organisms.

The representatives of the metaphysical trend in biological science, the Mendelist-Morganists, say: in capitalist countries, the essence, the core, of Darwinism is not accepted, Michurin's theory is not in use, Mendelism-Morganism is recognized in its entirety, and yet there are good varieties of plants and breeds of animals there. Consequently, they conclude, Mendelism-Morganism is a good and, at bottom, correct theory, and serves as a guide to action.

It pays there scientists to forget that under capitalism theory is entirely divorced from agricultural practice. The breeds of animals and the varieties of plants that men have raised in the course of decades and centuries have no relation whatever to bourgeois biological theory, to genetics. One can be easily convinced of that by tracing the history of the production of any good variety of plant or breed of animal. In the vast majority of cases, this was done not on the basis of theory, but on the basis of personal practical experience. Agronomic practice can achieve, and has achieved, successes without theory, but these successes are slow and small compared with those that must be, and are being, achieved in agriculture in our country, where theory and practice are united.

The role and significance of theory in our country is beyond comparison with the role theory plays in capitalist agriculture. In our country, the slightest theoretical achievement in agronomic science has every opportunity of being quickly put into practice, of easing labour and of increasing crop yields in the collective- and state-farm fields.

The All-Union Agricultural Exhibition mirrors the tremendous achievements of Soviet socialist agriculture and all its departments. It also demonstrates the development of agronomic science on the basis of advanced collective- and state-farm practice. And the first thing that must be pointed out in this field are the exhibits of the great work and immense achievements of the premier agronomist in the Land of Soviets—Academician Vasili Robertovich Williams.

I would like to take a few of the numerous exhibits at the All-Union Agricultural Exhibition as examples in order, very briefly, to show the creative force of Michurin's theory and the impotence of the Morganist-Mendelist trend in science, which is also, to some extent, represented at the Exhibition. The starting point of the irreconcilable disagreements between Michurin's theory of agrobiology and Mendelism-Morganism is their diametrically opposite views on the life and development of plant and animal organisms, their different views on the form-building process. The difference between these views may be briefly summed up as follows.

The Mendelist-Morganists believe that the conditions of life, the conditions of existence, have no effect upon the hereditary, natural, properties of organisms. From this it follows, of course, that it is impossible to control, to alter, breeds of animals and varieties of plants, to improve them by creating definite conditions of life for them. In the opinion of the Morganist geneticists, the breed of plants and animals very rarely undergoes change (mutation), and the quality, the direction, of changes in their breed is not in any degree determined by the conditions of life. Therefore, the Morganist geneticists advise practical plant breeders and seed growers to wait until changes in the nature of organisms take place of their own accord, for reasons unknown to the geneticists, and, moreover, in an unknown direction.

The Michurin theory is the very opposite of Morganism.

The basis of Michurin's theory is that form-building is determined by the conditions of life; the conditions of life of plant organisms influence—and often rather strongly—changes in the breed of the organisms. By mastering the laws that govern changes in the breed of organisms in accordance with conditions of life, with conditions of training, Michurinists can control the development of organisms and thereby create forms and varieties of plants and breeds of animals that are needed by socialist practice.

According to the theory of Morganist genetics, the conditions of life play no role in the alteration of the nature, in the deterioration or improvement, of varieties; and from this it logically follows that it is useless creating good agrotechnical conditions on seed plots for the purpose of preserving the breed of seeds, let alone of improving it.

According to Michurin's theory, however, the conditions of life do play a part in the preservation, and also in the improvement or deterioration, of the breed of organisms. From this it logically follows that bad agrotechnique must not be permitted on seed plots. Under bad conditions of cultivation, the breed of plants deteriorates considerably in the course of one or two generations.

Thus, the pivot of the disagreement between Michurin's theory and that of the Mendelist-Morganist geneticists is that one (Michurin's theory) recognizes that variations and their direction are determined by the conditions of life, whereas the other (Mendelism-Morganism) absolutely denies that the quality, the direction, of variations is determined by the conditions of life, by food, by environmental conditions in general.

A dispassionate and unbiassed study of the wealth of material exhibited at the All-Union Agricultural Exhibition makes it possible for everybody to understand which of the two helps practical socialist agriculture.

In the Michurin orchard at the All-Union Agricultural Exhibition there are splendid exhibits of numerous good varieties of fruit trees, berry bushes and grape vines created by I. V. Michurin and by the scientific workers at the Central Genetic Pomological Laboratory (Michurinsk). Varieties of apple trees, pear trees and berry bushes are growing there that easily stand the severe climate of the central zone of our Soviet Union, and the fruit of which is in no way inferior to the best of the southern varieties of the Crimea and Western Europe, which, as is known, cannot grow in the central zone of the Soviet Union. There are also exhibits of Michurin grapes, which bear an abundance of fruit in the open on the exhibition plot. Some of the varieties of Michurin grapes do not have to be covered in the winter under the severe conditions in Michurinsk, whereas all the other cultivated varieties of grapes that we know of bearing good-quality fruit have to be warmly covered in the winter even in the southern districts of the R.S.F.S.R. The chief thing about the Michurin orchard at the Exhibition, however, is not simply that the varieties exhibited there are good, but that the methods of breeding these varieties are demonstrated. A detailed study of these methods will enable anyone to create the needed varieties himself.

If, for example, it is necessary to cross two distant species that interbreed with difficulty, I. V. Michurin advises the adoption of what he called vegetative approximation. Before crossing, before uniting these two organisms sexually, one of the plants should be grafted on to the other so that a branch of one plant will be fed with the substances produced by the other. As a result of such feeding, the breeds of the plants taken for crossing are approximated, as it were, and after that the flowers of the plant of one breed are more easily and successfully fertilized with the pollen of the other now approximated plant.

At the Exhibition this is demonstrated by the graft of a mountain ash on a pear tree.

In the Michurin orchard there are numerous exhibits that clearly show what must be done when creating varieties of plants in which are combined, for example, the good-quality fruit of southern varieties that cannot stand a severe winter, with another variety that is winter-hardy but bears bad-quality fruit. It is known that if a tender variety of southern origin bearing good-quality fruit is crossed with a local hardy variety that bears inferior fruit, no good variety will, as a rule, result. Michurin attributed this to the fact that the heredity of the local variety that is well adapted to the given conditions predominates when crossed with a southern (imported) variety. The conditions for the development of the sex cells, the conditions for fertilization and also the conditions for the subsequent life of the organisms obtained from such a cross will be much more suitable for developing in the progeny the properties of the local parent variety than of the imported one. Michurin teaches us to choose two varieties for crossing (a hardy variety bearing inferior fruit and a tender variety bearing good fruit), both coming from other districts. Both varieties, both breeds, will thus come under conditions that are not quite habitual for them and, when crossed, the heredity of the one will not greatly predominate over that of the other. In this way it is possible more quickly and easily to create new varieties with increased hardiness and good-quality fruit.

Also demonstrated in the Michurin orchard at the Exhibition are different methods of regulating the development of some properties in the breed that is being created, and of eliminating other, undesirable properties. This is done by creating definite conditions when growing the flowers that are taken for crossing, and also when storing the seeds, planting, and tending the young hybrid plants.

Those who are making a deep study of the theoretical problems of agrobiology can, and ought to, spend weeks studying the Michurin orchard at the Exhibition, learning Michurin's methods of creating the most diverse forms and varieties of plants that have hitherto not existed in nature.

A deep study of the works of I. V. Michurin will make it clear that his theory is not simply the science of growing fruit trees and berry bushes, but is a general biological theory. The thousands of exhibits sent in by the best collective farms in our country, by Yefremov teams,[1] by the Stakhanovites of sugar-beet, flax, cotton and other fields who have achieved unprecedentedly high yields, are the upshot of the employment of a high level of agro-technique, of taking into account the role played by conditions of training, to which I. V. Michurin too attached such enormous, decisive importance.

Many of the exhibits sent in by collective-farm experimenters and scientific workers demonstrate the practical solution of profoundly theoretical problems, which the old science has been unsuccessfully trying to solve for decades, and even for centuries.

There is no need to mention the work, so widely known among the Soviet public, of N. V. Tsitsin and A. I. Derzhavin in creating perennial forms of wheat and rye. I will, however, refer to at least the following example.

A. F. Yudin, a young scientific worker, exhibits naked barley, which, he claims, he has obtained from hulled barley by good feeding and the selection of plants that had accumulated the character of nakedness, i.e., in the way indicated by Darwin.

For several years this Yudin barley has been objected to by the followers of Mendelism-Morganism.

The barley on which Yudin began his work was classed as belonging to the Pallidum variety, but according to the existing mode of classification no variety of barley that possesses naked, and not hulled, seeds, can be classed as belonging to the Pallidum variety. Yudin, however, claims that he obtained a naked barley from a hulled one by the gradual accumulation of nakedness in the course of several generations; he converted one variety into another. The Mendelist-Morganists say that this is impossible; therefore, these scientists regard Yudin's experiment as "illegitimate." They even assert that there is no such thing as a Yudin barley, that this variety of barley was known in this very form before Yudin began his experiment, that is to say, that Yudin did not create, but found and used an ordinary naked barley. Visitors to the Exhibition, however, may have noticed the following on Yudin's barley plots. The straw of many of the plants of this barley has 5, 6 and even 7 nodes, whereas ordinary barley, wheat, and other cereals have only 4 or 5 nodes.

Let the Mendelists say where, when, and which barley had so many plants with such a number of nodes! Of course, it is quite conceivable that in individual cases some plants may be found with such a number of nodes, but not such a large percentage as is seen in Yudin's barley. Yudin's barley has a number of other distinctive characters, such as, for example, a fairly considerable percentage of double-embryo grains, frequent cases of extra stamens in the flowers—more than three, etc.

[1] Followers of Mikhail Yefremov, the Altai collective farmer, renowned for obtaining exceptionally high wheat yields by the use of the most up-to-date methods of farming.—*Tr.*

In the forcing house at the Exhibition one may see numerous exhibits which show that it is enough to grow only one generation under suitably changed conditions for the nature of different varieties of winter wheat to deviate entirely towards the nature of spring wheat, i.e., a wheat that does not require a long period of cool temperature during vernalization for its normal life and development.

In that greenhouse may be seen exhibits that have been created at the Exhibition itself. I am referring to a winter wheat that was planted in the Exhibition greenhouse in the spring of 1938. As was to be expected, the plants of the winter wheat Novokrymka 0204 developed in an unusual way. Their nature required cold for passing through the vernalization process. Nevertheless, late in the autumn they eared and produced seed.

In the spring of 1939 these seeds were planted in the greenhouse. At the same time the seeds of Novokrymka 0204 gathered from plants which had been grown under ordinary field conditions after autumn sowing were also planted in the greenhouse. It now appears that the plants which have grown from the seeds of last year's greenhouse planting and in the development of which low temperatures took no part, do not, this year, require a low temperature when passing through the vernalization process.

These and many other exhibits show not only that conditions of life, environmental conditions, play a part in altering the nature of organisms; they also show that in order to effect a change in the nature of an organism it is not at all essential that the external conditions operate for a great many generations. Such changes can be brought about in the course of one or two generations. These exhibits also show that changes in the nature of an organism (its breed) can be directed adequately, in correspondence with the changes in the organism itself under the influence of external conditions.

This theoretical proposition is of enormous importance to practical socialist agriculture. If changes in the nature of an organism correspond adequately to the changes in the organism itself, then, of course, it is necessary by means of agrotechnique to create on the seed plots of every kind of crop such conditions as will help to obtain from the plants the biggest and best crop within the time limit required by us. It is these conditions that change, incline, the organism's nature too in the same direction, i.e., cultivate the breed of the seeds. Bad agrotechnique, however, not only prevents the improvement of the breed of seeds, but causes the seeds to lose the good properties they possess.

This is clearly demonstrated in the vegetable section at the Exhibition by the living exhibits of potatoes. Although a given variety of potato was planted on different plots at the same time and in the same way, and the plants were subsequently tended in the same way, there is a marked difference in the condition of the plants on the different plots. The yield on one plot is 5 or 6 times as high as that on another plot.

This difference in the yields is due to the difference in the planting material, although, as has been stated, the latter was of the same variety on both plots. The whole point is that the preceding generations of the seed potatoes for each of the plots were grown under different conditions.

In this section it is demonstrated that the "natural calamity" which for centuries prevented the growing of good breeds of seed potatoes in all hot southern districts has been entirely overcome by Soviet agrotechnique, by collective- and state-farm practice.

As is known, in all the hot districts of the South, good potato planting material imported from the North could produce a fairly good crop in the first year. But when tubers grown in the South were planted, the yield, as a rule, dropped heavily even when good agrotechnique was employed. The yields of the second- and third-year seed potatoes grown in the South dropped to one half or one third. Owing to the impossibility of growing good planting material in the South, huge quantities had to be imported from the North. Moreover, the imported varieties were unsuitable for the southern districts, so that, in the end, it was found necessary to import food potatoes too for the southern industrial centres and cities.

The old, formal science, which denied that changes in breed depend on conditions of life, could not, of course, ignore the mass of factual material showing that potato tubers lose their yield qualities even after being grown only one year in the South, but it did not associate the deterioration of the breed of potatoes when grown under hot conditions with the fact that changes in breed depend on the conditions of life. It attributed this change to disease. Until quite recently the view prevailed in world science that potatoes grown in the South suffer from a specifically southern disease, and that this reduces yield.

All that has proved to be sheer imagination. Approaching the phenomenon from the standpoint of creative Darwinism, of Michurin's theory, it was not difficult to reveal the external conditions which contribute to the deterioration of the breed (yield properties) of potato tubers. It was found that even with a slight activation of the eyes (buds) of the potato tubers, under the influence of a high temperature, the breed of the eye cells changes in the direction of a lower yield. In the South, as a rule, at the end of July and in August, the eyes of the tubers begin germination while still in the field, under the plants. The All-Union Institute of Selection and Genetics (Odessa), together with thousands of collective-farm experimenters in the south of the Ukrainian S.S.R., worked out a method of cultivating seed potatoes, as a result of which the eyes (buds) of the tubers, as a rule, do not awaken and, in general, the tubers develop at a time when the temperature is not high. This is the method now widely known in the South as the summer planting of potatoes.

In the south of the Ukrainian S.S.R., potato seed plots are now all planted not in the spring, but in the middle of July. Under these conditions the tubers develop in the autumn. Autumn conditions have proved to be so favourable for the development of the tubers that now, as a rule, they weigh 400-600 gr., and sometimes even one kg. each.

Seed potatoes grown in the South by the summer-planting method can be planted early next spring. When that is done their yield is two to three times as high as that of the same variety of seed potatoes planted in the same field at the same time, but grown by the ordinary method, namely, spring planting.

With this example of potatoes it is easy to demonstrate also the extremely important proposition laid down by Darwin that changes beneficial to the organism accumulate from generation to generation. As practical experiments have shown, every succeeding year of growing seed potatoes by the summer-planting method improves their breed, their yield properties.

On scores of hectares of experimental plots near Moscow belonging to scientific research institutions it is now possible to demonstrate still another interesting phenomenon. The yield of the seed potatoes brought this year from the South, namely, from the Institute of Selection and Genetics (Odessa), when planted in the fields in Moscow Region, far from being inferior to that of the same varieties of seed potatoes grown in previous years in the Moscow Region, has proved to be considerably superior. It is anticipated that the yield of the southern seed potatoes will be about 50% higher than the average yield of the same variety of planting material grown in Moscow Region.

All this shows that in the central zone of the Soviet Union too it is necessary immediately to change, by practical experimentation, the method of growing seed potatoes and to improve their breed year after year by employing the agrotechnique described above.

The most convincing proof that changes in the breed of organisms correspond adequately to the influence conditions of life, conditions of nutrition, exercise upon the changes in the organism itself is provided by the exhibits on vegetative hybridization.

The essence of vegetative hybridization lies in the fact that by causing a coalescence (grafting) of young plant organisms of different breeds (different varieties, or species), a fusion of breed properties takes place in the vegetative and seed progeny similar to that which takes place in sexual crossing.

This alone entirely upsets the basis of the Mendelist-Morganist teaching. The Mendelian geneticists attribute the phenomenon of heredity—the property of organisms to produce progeny more or less like themselves—to a special substance of their own invention, which, they claim, is separate from, and independent of, the body of the organism. According to them this "hereditary substance" consists of granules that are called "genes" and are located in rod-like bodies, chromosomes, which are easily observed in the nuclei of cells at a certain moment of their lives.

As the Mendelist N. K. Koltsov writes, "... chemically, the genoneme with its genes remains unchanged in the course of the entire ovogenesis and is not subject to metabolism—oxidation and reduction processes."[1]

From this it follows, of course, that the conditions of life, conditions of nutrition, the training of plants and animals, can influence only the body of the organism. The "hereditary substance," however, being subject neither to the oxidation nor to the reduction processes, is not, in the opinion of the Morganists, influenced by the conditions of life. Since the chromosomes contain a "hereditary substance" independent of the conditions of life of the cells, then, according to the cytogeneticists, the chromosomes derived from the paternal and maternal sex cells respectively preserve their individuality, i.e., do not change either quantitatively or qualitatively, during the whole life of the organism. From this it follows that on reaching maturity, the organism, irrespective of its conditions of life, will produce sex sells exactly like those from which it itself originated. Changes in the body of the organism due to given influences do not affect its breed—such is the basis of Mendelism-Morganism. They claim that the nature of organisms is immutable. i.e., development proceeds not in a spiral but in a circle.

That is why the theory of the Morganist-Mendelists is directly contradicted by the fact that when the bodies of different young plant organisms were joined by grafting, vegetative and seed progeny have been obtained possessing the properties of the different breeds.

What else can the Morganists do except deny that such cases are possible? But every visitor to the Exhibition can see such cases with his own eyes in considerable number and in diverse plants.

Here is one example. In the Science Hall of the Chief Pavilion at the Exhibition are exhibited two varieties of potatoes that had been taken for an experiment in vegetative hybridization, and the results of this experiment are also shown. The breed of one of these varieties possesses the property of producing blue-skinned tubers and blue flowers. The breed of the other variety possesses the property of producing white-skinned tubers and white flowers. By means of grafting, the blue-tuber-blue-flower variety was induced to feed on the substances produced by the leaves of the white-tuber-white-flower variety. The result was that on the lower part (underground) of the stem of the blue-tuber (by breed) variety white tubers were formed. The plants that were grown from these tubers also produced white flowers. In the same way, the white-tuber-white-flower variety can be transformed into a blue-tuber-blue-flower variety.

In another experiment a variety of tomato was taken which produced ripe yellow fruit. A branch of this breed was induced (by grafting) to feed on the root sap and on the substances produced by the leaves of a red-fruit

---

[1] Н. К. Кольцов, «Структура хромосом и обмен веществ в них», *Биологический журнал*, том 7, выпуск 1, 1938 г., стр. 42.

breed of tomato. The result was that the seeds gathered from the fruits on the branch of the yellow breed produced plants with red fruit.

In the vegetable section of the Exhibition there are living specimens of potato vegetative hybrids grown by the Potato Institute.

Here, postgraduate Solodovnikov demonstrates vegetative (grafted) hybrids obtained from a cultivated and a wild species of potato. These hybrids, grown from the underground shoots of the wild species Demissum (the stock), already have the appearance of a cultivated species of potato.

The representatives of Mendel-Morgan genetics deny that these hybrids have been obtained. They even claim to have found among formerly-known varieties of potatoes plants which resemble these hybrids. But to this day they have failed to find plants "resembling" vegetative hybrids from another wild Demissum species of potato and the cultivated Epicure variety. Such hybrids are exposed at the Exhibition by A. S. Filippov. The plants grown from the shoots of the Demissum, on which the Epicure had been grafted, contain many characters of the wild species and some of the properties of the cultivated variety.

Another thing visitors to the Exhibition can see is how, by grafting, by feeding the stems of tomato plants through the root system of a potato, experimenter Brusentsov succeeded in creating a new dwarf variety of tomato, Karlikovi Brusentsova, the plants of which are thickly covered with fruit.

We have mentioned only a small part of the vast material of profound scientific and immense practical importance shown at the Exhibition.

It goes without saying that only the best is shown at the Exhibition. It provides instruction for the numerous visitors, teaches them how to go about it to raise the productivity of our socialist agriculture to a still higher level.

It must be supposed that the exhibits shown by the scientists who follow the teachings of Mendelism-Morganism also demonstrate the quintessence of the achievements of this trend in science.

N. I. Vavilov has stated more than once that in science he is a follower of Mendelism-Morganism, that he believes in that trend in science. Naturally, he tries to conduct his scientific work from this point of view.

In the Grain Pavilion there is a rather interesting stand. On it there is a map showing the routes of the expeditions sent out by the All-Union Institute of Plant Industry in search of agricultural material for its collection.

At the bottom of the map there is a table (reproduced on pp. 280-81) which, as far as we know, sums up the work of ecologic-geographical classification of agricultural plants, on which a fairly large staff of workers had been engaged. And this is the only exhibit of the All-Union Institute of Plant Industry in this pavilion.

It must be noted that the All-Union Institute of Plant Industry has quite a number of other scientific achievements to its credit. True, they utterly contradict the theory of formal genetics. Perhaps that is why they are not exhibited on this stand. Let us briefly examine the table.

If the gist of this table is approached from the standpoint of formal science it may appear to be valuable. Its heading reads: "Location of Varieties Possessing Economically Valuable Characters." From the columns of this table visitors would seem to be able to learn where to obtain varieties with the properties required. For example, if we take the property of frost resistance, which is rather important for the cultivation of winter cereals in many districts of our Soviet Union, the visitor may learn from this table that high frost resistance is possessed by the plants of the Boreal group, the Middle-European group and the Steppe group. But every agronomist, including the compilers of this table, knows that of all the wheats so far tested, only Volga winter wheats possess high frost resistance, and even they, unfortunately, do not yet possess sufficient frost resistance for the severe conditions that prevail in our trans-Volga districts.

Everybody who is working on winter wheat under Volga and Siberian conditions, i.e., in those districts where high frost resistance is needed, knows perfectly well that none of the existing Middle-European and Boreal groups of wheat (shown in the table as possessing high frost resistance) can stand even a medium winter in these districts.

Nor is all quite well with the classification of the Steppe group as a highly frost resistant group. Like every other group indicated in the table, this group includes a vast number of varieties that have been created by man; and among the varieties which Vavilov classifies as the Steppe group, the different varieties vary in frost resistance to an enormous degree.

The same may be said concerning any of the characters enumerated in this table. Take, for example, rapidity of ripening. The table shows that the Northern group and the Indian group are both early-ripening. The compilers of the table know very well, however, that when the collection was sown in Kirovabad (Azerbaijan S.S.R.), all the varieties were sown at the same time, but the vast majority of the Northern wheats eared and ripened two or three weeks later than the vast majority of the Indian wheats.

Looking at these crops nobody would think of calling the Northern wheats early-ripening under the conditions prevailing in the Azerbaijan S.S.R. When, however, the collection was sown beyond the Arctic Circle, at the Arctic Experiment Station, it could easily be seen that in many cases the Northern wheats that are late ripeners in the South ripened at the same time as the Indian wheats, and there were no few that ripened earlier.

I shall not examine the other characters indicated in the table. I shall merely say that the visitor to the Exhibition can see for himself that this table, which sums up the work of the Mendelist-Morganists, cannot be of any assistance to practical agriculture.

The Mendelist-Morganists cannot understand that every property, for example, frost resistance, drought resistance, etc., every character of a plant, is the result of the plant's development under definite conditions. No property is independent of conditions. Of course, the development of plant organisms proceeds in a relatively different way under different conditions, in

LOCATION OF VARIETIES POSSESSING

Character	Northern	Middle-European	West-European	Steppe	Caucasian Mountains	Azerbaijan-Daghestanian	Transcaucasian moisture loving
Spike	small medium	medium	large	medium	medium	medium large	medium large
Grain	small medium	medium	large	medium	small medium	medium large	medium small
Straw	not firm	lodging medium-firm	firm medium-firm	lodging medium-firm	lodging medium-firm	medium-firm	medium-firm
Ripening	early ripening	medium ripening	late ripening medium ripening	medium ripening early ripening	early ripening medium ripening	late ripening medium ripening	late ripening medium ripening
Drought resistance	not hardy	not hardy	not hardy	hardy	medium hardy	medium hardy	not hardy
Frost resistance	high	high	low medium	high	—	low	low
Grain ripening (rapidity)	medium	medium	slow medium	slow medium	medium	rapid	slow
Resistance to fungous diseases	susceptible	susceptible	nonresistant	susceptible	various	relatively resistant	medium resistant

ECONOMICALLY VALUABLE CHARACTERS

Syrian Palestinian	Mediterranean	Abyssinian	Irano-Turkmenian	Pamiro-Bandakhshansk	Indian	Chinese
small medium	large	small medium large	medium	medium very large	small	small medium
small medium	large	small medium large	medium	small medium large	small medium	small medium
lodging medium-firm	firm medium-firm	thin medium-firm firm	not firm lodging	medium thickness lodging	firm thin	not firm firm
early ripening	medium ripening late ripening	early ripening medium ripening	early ripening medium ripening late ripening	early ripening medium ripening	early ripening	various
hardy	medium hardy	medium hardy not hardy	high	medium hardy hardy	medium hardy	not hardy medium hardy
—	—	—	medium low	medium	medium	medium low
rapid	medium	rapid	rapid	rapid	rapid	rapid
resistant medium resistant	very resistant	susceptible	very susceptible	very susceptible	medium susceptibility	resistant susceptible

different districts. What will be early ripening in one district may often turn out to be late ripening in another district.

The same may happen in the case of many other characters and properties of plants.

It is not clear from the table for which districts varieties or even species are classed as of high or low frost resistance, early or late ripening, etc. No wonder visitors to the Exhibition, and even the guides, simply avoid this stand.

The representatives of Morganism-Mendelism cannot understand that there is only one way to describe the nature of plant organisms, even for the purpose of classifying a large number of varieties, and that is to take the laws of nature as the basis. The nature of different plant organisms requires different environmental conditions for life and development. It is necessary to study the conditions the different varieties require in order that the plants of those varieties may develop particular properties, such as, for example, frost resistance, drought resistance, etc., and only the environmental conditions required by the organism's nature can determine the classification of a variety. or breed.

The above-mentioned table is of no use whatever for theoretical or practical purposes. It can only mislead credulous people who accept agrobiological theories uncritically.

* * *

The All-Union Agricultural Exhibition is a school of advanced, Stakhanovite socialist agriculture. Millions of people engaged in agriculture are learning, and will continue to learn, at this school. One would like the Exhibition to be used as extensively as possible also for the purpose of further developing the theories of our Soviet agronomic science.

A study of the Exhibition will help many to understand that the time has come to eliminate from the curricula of our educational institutions the teaching of the metaphysical pseudo science of Mendelism-Morganism, so that educational training may not run counter to the experience, practice and achievements of socialist agriculture so splendidly demonstrated at the Exhibition.

In answer to the plea of the Mendelists that there is no other science that explains the laws governing heredity and the variability of plant and animal organisms except "classical" Mendelism-Morganism, we can say: "Go to the Exhibition, and, if you wish to do so, you can see with your own eyes that there is Michurinian genetics, which shows how to overcome the obstacles that stand in the way of continuously increasing yield, of creating good varieties and breeds."

WAYS OF CONTROLLING PLANT ORGANISMS[1]

THE FIRST thing that is needed for the practical work of controlling the nature of plant organisms is knowledge of the laws that govern such extremely important properties of organisms as heredity and variability. Different organisms require different conditions for their life and development. The organisms' environmental requirements developed historically, in the course of successive generations. Practical agriculture has been utilizing this property of heredity for thousands of years and by means of agrotechnique has created the conditions required by plant organisms for their development, for producing a crop. By rearing plants well and selecting the best specimens for breeding purposes, men, as Charles Darwin showed, have slowly but surely improved heredity itself.

That alone warrants the conclusion that the general law of heredity is of extreme importance inasmuch as it asserts that the conditions of life play a paramount role in the alteration of the breed of organisms, in the alteration of heredity.

This general law must be looked at from two aspects. On the one hand, an organism of a definite nature, in following the path of its ancestors, requires its own specific conditions, which differ in some degree from the conditions required by other organisms. On the other hand, the conditions of life, in their turn, change the heredity of the organism.

One would have thought that this proposition would be obvious to biological science, particularly after the work of Darwin and the Darwinists.

Nevertheless, not one of the theories of heredity that arose after Darwin (and, as is known, quite a number arose and still exist) deserves to be called a real theory of heredity.

K. A. Timiryazev wrote that "not one of the so-called theories of heredity advanced hitherto satisfies the primary demand made upon it, that it can serve as a general working hypothesis, i.e., as an instrument for directing research toward the discovery of new facts and new generalizations." Timiryazev also gave his reason for writing this. At bottom they are

---

[1] A paper read at the first Annual Timiryazev Reading, April 28, 1940.—*Ed.*

all only variations of one and the same theme: offspring are 'flesh of the flesh and bone of the bone' of their ancestors; but as observation progressed, the more subtle structural features 'cell from cell,' 'plasm, from plasm,' 'nucleus from nucleus,' 'chromosome from chromosome,' etc., are substituted."[1]

Well, today the Mendelist-Morganists tirelessly reiterate: "a gene comes only from a gene."

Soviet scientists know very well that development presupposes the appearance of something new from something old, of certain forms from others. But all the theories of heredity that are built according to the "flesh of the flesh," or "chromosome from the chromosome," or "gene from the gene" pattern lead to the conclusion that nothing new appears in the world, that everything in the world exists now as it was in the beginning. Hence the impotence of such knowledge when it comes to regulating the development of organisms; hence the harmfulness of such knowledge for men engaged in practical work.

These theories of heredity are all based on one fallacious proposition, although it is expounded in different ways. That proposition comes down to this: that the development of organisms is simply increase or decrease, that new properties can *only manifest themselves*, *but cannot appear* in organisms, cannot arise from the old. To this day many biologists continue to assert that in the organism cells can come only from cells, chromosomes only from chromosomes, etc. But everybody knows that every organ of an organism develops from something that is entirely different from that organ: the eyes, for example, do not develop from eyes, a leaf does not develop from a leaf, etc. Why, then, must there be special laws for chromosomes, different from the general laws of development of organisms?

This is what K. A. Timiryazev had in mind when he wrote that to understand the properties of heredity it is necessary first of all "to become imbued with the idea that causes may be potential and not necessarily morphological and, in general, of a character other than the effects they produce."[2]

The laws that govern heredity, the laws of life of organisms, can be understood only from the standpoint of the theory of development. This explains why in bourgeois society biological science is the most backward branch of science. Recognition of the theory of development is detrimental to, is incompatible with, the interests of the decaying capitalist system. But it is impossible to create a science that will lay down the principles for controlling the nature of organisms unless it is approached from the standpoint of the theory of development, from the standpoint of dialectical materialism.

Indeed, take the extremely important problem of the inheritance of so-called "acquired" characters, i.e., characters newly arisen in the organ-

---

[1] К. А. Тимирязев, *Собрание сочинений*, т. VI, стр. 191.
[2] *Ibid.*, p. 193.

ism in the process of its development. The formalist geneticists have confused this question. However, in the light of the theory of development it can be presented and settled in an entirely different way.

There are some scientists (the Mendelist-Morganist geneticists) in our Soviet Union, not to speak of those abroad, who categorically deny that it is possible to inherit "acquired" characters, or to speak more precisely, characters which have newly arisen in the course of the individual's development, i.e., deny that changes in heredity dependent upon the organism's conditions of life are possible.

But K. A. Timiryazev and I. V. Michurin, while developing Darwin's theory, stated repeatedly that controlling the conditions of life of organisms meant controlling their heredity, too. It is common knowledge that by means of agrotechnique and zootechnique, by creating the best conditions for plants and animals, crops and animal produce are being obtained according to set tasks, according to a plan. In the same way it is also possible to control breed and alter it in the direction required.

Though the Mendelist-Morganists have for decades denied the very possibility of inheriting so-called acquired characters, for Soviet agrobiological science this question has been definitely answered in the affirmative. The question was also clear to K. A. Timiryazev. That is why he pointed to changing conditions of life as the most important way of obtaining requisite new characters and properties in organisms.

K. A. Timiryazev wrote: "Physiology is already beginning to reveal the secret of how plant forms are built; it is little by little learning how to control the building of these forms."[1]

In Timiryazev's time, however, science was not yet in possession of facts that proved it possible beyond doubt to alter heredity by changing the conditions of life and that these changes in heredity, though different in different organisms, would in all of them adequately correspond to the organisms' response to the new conditions. True, Michurin had already worked out this problem, but genuine scientific research was so suppressed in tsarist Russia that his work was unknown even to Timiryazev.

A few biologists, the best of them, like Vilmorin, Burbank and Michurin, were very well able to alter the breed of organisms in the direction they wanted. But the high priests of science spurned all these facts and methods, denounced them as unscientific, fallacious, unworthy of being incorporated in commonly accepted, official science. Michurin's splendid theory was not yet known to the broad scientific public. That is why, although the question of the possibility of inheriting so-called acquired characters was already clear to K. A. Timiryazev, the best of the Darwinists, the concrete methods of changing the breeds of organisms in a given direction were not known to official science. In the matter of controlling the properties of heredity and variability, biology began to make real progress, to advance

[1] К. А. Тимирязев, *Собрание сочинений*, т. V, стр. 136.

on a wide front, so to speak, only in the Soviet Union, with the official recognition and development of Michurin's theory.

I. V. Michurin showed that by choosing the training conditions, by suitably feeding plant organisms at definite stages of their development, it is possible to obtain directed changes of heredity, to strengthen useful properties in an organism, or eliminate undesirable properties from its heredity.

What is needed to understand the laws that govern heredity is not a bare, formal, meaningless scheme according to which everything comes from the chromosome and the chromosome itself comes from a similar chromosome, but a general biological theory which embraces all the diverse forms of heredity. In building up such a theory, special importance attaches to the study of vegetative hybridization—a phenomenon that was already noted by Charles Darwin, perfectly understood by K. A. Timiryazev, and for the first time brilliantly proved experimentally in the works of I. V. Michurin. The mentor method worked out by the latter is vegetative hybridization and nothing else.

Concerning cases of the formation of grafted hybrids of individual species and varieties, Charles Darwin wrote: "... if, as I am now convinced, this is possible, it is a most important fact, which will sooner or later change the views held by physiologists with respect to sexual reproduction."[1]

An understanding of the essence of vegetative hybridization is of decisive importance on the one hand, for the correct presentation and solution of the problem of the inheritance of so-called "acquired" characters, and, on the other hand, for a deeper understanding of heredity in general. The wider and more deeply work on vegetative hybridization is developed, the clearer it becomes how right Darwin was in foreseeing the importance of grafted hybrids for studying also sexual hybridization, for creating an effective theory of heredity.

Michurin showed that by grafting, by skilfully feeding the plants of one breed with the plastic substances produced by another, it is possible to obtain not only changes in the heredity of organisms but, as a result, also genuine hybrids. Vegetative hybridization makes it possible to obtain organisms possessing the properties of both (or of several) of the breeds that had been taken for grafting, i.e., what is usually obtained by sexual hybridization.

The experimental material already accumulated clearly shows that *in vegetative hybridization the same forms of heredity may be observed as in sexual hybridization.*

It is not difficult to become convinced of that if the forms of heredity observed in vegetative hybridization are carefully examined and compared with the phenomena of heredity in sexual hybridization.

---

[1] Charles Darwin, *The Variation of Animals and Plants Under Domestication,* 1885, Vol. I, p. 417.

Timiryazev worked out a classification of the different forms of heredity that embraces both asexual and sexual reproduction. He also showed how the different forms of heredity intergrade.

In the light of the data in possession of present-day Soviet science, Darwin's idea, developed by Timiryazev, that there is similarity and intergrading between heredity connected with sexual reproduction and heredity connected with vegetative reproduction stands out much more convincingly today than it did at the time Timiryazev conducted his work.

Classifying the facts of heredity, Timiryazev first of all established two groups: simple heredity and complex heredity.

It is known that plants, grown, for example, from wheat seeds or from potato tubers, or from cuttings, layers, etc., appear to copy the maternal forms in the course of their development. Repetition of the development of the maternal forms is observed with particular frequency in the asexual reproduction of plants. Timiryazev called this form of heredity *simple* heredity, and it is on the utilization of this form that practical agriculture has virtually been based for thousands of years, creating by means of agrotechnique the conditions required by the given natures of plants.

In the sexual process of reproduction, the heredities of two organisms usually unite. Such heredity Timiryazev called *complex*, i.e., dual heredity. It can be divided into several groups according to the forms in which it manifests itself.

For example, there are animals with some patches on their coats the colour of one of their parents and some patches the colour of the other parent. According to Timiryazev's classification, such a heredity is called *mixed*, because in one part of the organism the characters of one parent and in another part the characters of the other parent manifest themselves. These parts, or sections, of the organism may be of different dimensions, from large to microscopically small.

Most frequent are those cases in which the hereditary properties of both parents blend in the progeny (do not each manifest themselves in their pure form), when the progeny acquires new properties. Timiryazev called such heredity *blended* heredity, and to it he attached most importance.

There are cases when parental characters which are the same but expressed oppositely do not mix in the hybrid progeny. For example, when a variety of pea having green seeds is crossed with a yellow variety, these characters do not blend in the progeny. No new or intermediate property is created; the property of one of the parents manifests itself, while the property of the other parent is excluded. This form of heredity Timiryazev called *mutually exclusive.*

In mutually exclusive heredity two categories of facts are observed.

The first category includes cases when the hybrid organisms are uniform in the first and in all succeeding generations. In other words, the hybrid progeny does not vary, does not segregate in succeeding generations; the properties of one parent are entirely absorbed by those of the other parent.

Cases of this kind Timiryazev called *Millardetism*, after the French scientist Millardet, who had made a rather thorough investigation of this category of hybrids.

The second category includes cases which Timiryazev classified as *Mendelism* (although Timiryazev himself pointed out that the few cases which come under this group heading and which occur only under definite conditions were not discovered by Mendel at all). In such cases, beginning usually with the second generation, segregation, diversification, takes place in the hybrids, some forms appearing with the characters of the one parent and other forms with the characters of the other parent.

It can now be said that the same variation of forms of heredity may take place in vegetative hybridization.

In vegetative hybrids it is possible to observe *mixed* heredity, when one part of the organism betrays the properties of one of the breeds, of one of the components, and another part betrays the properties of the other component. Cases of *blended* and of *mutually exclusive* heredity are also met with.

In vegetative hybrids it is also possible to observe either more vigorous development or, on the contrary, reduced viability, i.e., just as is observed in sexual hybridization.

All this does not, of course, mean that there is no difference between vegetative and sexual hybridization. It is important, however, to emphasize that the same forms of heredity manifest themselves in vegetative and in sexual hybrids, to emphasize that these two categories of phenomena are not separated from each other by an impenetrable wall, but are phenomena of the same order.

Soviet science is now in possession of a large number of facts concerning vegetative hybridization.

In experiments conducted at the Institute of Selection and Genetics (A. A. Avakian and M. G. Yastreb), a yellow-fruit tomato Albino was grafted in 1939 on to a Mexican 353 stock with small red fruit. The yellow-fruit Albino scion developed fruit of different colours, including red.

The seeds taken from an absolutely red fruit that had developed on the Albino scion, originally yellow-fruit by breed, were planted in the greenhouse. In the spring of this year some of the plants from these seeds developed bright red fruit, others developed raspberry-coloured fruit, others again developed pale yellow fruit like that of the Albino, and lastly, a fourth group developed fruit of a bright yellow colour, unlike either of the parent forms, the scion or the stock.

In the same experiment, progeny was grown from the seeds not of the red fruit, but of a yellow fruit with red stripes, also obtained from the Albino which had been grafted on to the Mexican 353. The progeny from that graft too was diverse: there were plants with fruit which had lost their red stripes, and plants with fruit which had numerous pink stripes.

In general, in this experiment the seed progeny of the vegetative hybrid behaved as sexual hybrids often, do, i.e., on the one hand, the characters

Fig. 50. A Grafted Plant

A cutting of the white-fruited Albino tomato variety was grafted on a red-fruited tomato No. 353 from Mexico. All leaves on the Albino scion were replaced (by the graft) by tomato 353 leaves. On the twig of the naturally white-fruited Albino scion a red fruit developed (the big fruit shown slightly above the pot)

Fig. 51. The pot on the left contains a No. K 1014 tomato plant from Mexico. The pot on the right, a Rosso Grosso variety tomato plant. The middle pot has a plant of the first seed generation of the vegetative hybrid obtained from grafting Rosso Grosso (the scion) on a No. K 1014 tomato (stock)

The seeds for the sowing were taken from the fruit that developed on a shoot of the No. K 1014 stock. The picture shows that the fruits of the vegetative hybrid are considerably bigger than those of the No. K 1014 tomato. In shape some of the fruits of the hybrid resemble the fruits of No. K 1014; others are almost like the Rosso Grosso fruits

segregated into paternal and maternal, and on the other hand, there was new form-building, the appearance of characters which the parents had not possessed.

We have on numerous occasions also described cases of the transmission of the colour of potato tubers from the scion to the stock. In experiments conducted by various scientific workers, the grafting of a white-tuber scion on to a blue-tuber stock produced white tubers, and, on the contrary, a blue-tuber scion coloured the tubers of a white-tuber stock.

In an experiment conducted by A. A. Avakian two years ago, the following happened. He grafted a cutting from an Odenwälder Blau blue-tuber potato on to the stock of a plant of the Ella variety. The Ella stock developed white tubers, which is usual for this variety (for the breed of the stock).

When, however, the buds of these tubers sprouted, it was found that instead of white shoots, which are characteristic of the Ella variety, there were pale-violet shoots (characteristic of the former scion). Now, two vegetative generations later, we have (without grafting) tubers on which a blue patch is distinctly visible, similar in colour to that of the tubers of the former scion (Odenwälder Blau).

Thus, the bluish tuber property (of the Odenwälder Blau scion) did not manifest itself in the white-tuber stock for two vegetative generations; it did so only in the third generation.

All these facts are analogous to a number of cases of heredity in sexual hybridization.

In experiments conducted by E. P. Khazina, a postgraduate at the Institute of Selection and Genetics (Odessa), a cutting from a young Humbert tomato plant was grafted on to a nightshade. From the fruit of these tomatoes seeds were taken and planted, and cuttings from the resulting plants were also grafted on to the nightshade. The seeds from each fruit of the second Humbert grafting on the nightshade were planted separately. It was found that some of the plants from one and the same fruit produced fruit of a shape that differed sharply from the Humbert. They were not elongated but round, i.e., the shape of the nightshade stock.

We could describe many more experiments in vegetative hybridization conducted by scientific workers and postgraduates at the Institute of Selection and Genetics (Avakian, Yastreb, Khazina, Bassarskaya, Kovalevskaya and a number of others) and also at other scientific institutions.[1]

To scientists abroad (and there are admirers and uncritical mouthpieces of this science among our scientists too), it seems utterly impossible to obtain hybrids except by sexual crossing. For Michurinists, however, vegetative hybrids are now no longer a rarity.

During the past few years a fairly large number of vegetative hybrids of the most diverse plants have been obtained in different parts of our Soviet Union.

[1] A number of experiments in vegetative hybridization are described in *Yarovizatsia*, Nos. 3, 4-5 and 6, 1938; Nos. 1, 3 and 4-5, 1939, and No. 1, 1940.

The cases of vegetative hybrids that have been obtained utterly refute the views of the Mendelist-Morganists concerning the phenomena of heredity. The Morganists entirely, or almost entirely, associate the phenomena of heredity with the chromosomes, or with particles or corpuscles of chromosomes—genes—and they assert that the transmission of different hereditary properties from one organism to another is impossible without the transmission of chromosomes (or of their particles).

In vegetative hybridization the scion and the stock do not exchange chromosomes. That is why the Morganists cannot, from their standpoint, concede the existence of vegetative hybrids. That is why the cases of vegetative hybrids that were already known to Darwin, for example, *Cytisus Adami* or the vegetative hybrid of a hawthorn and a medlar, were not recognized, were rejected by bourgeois science. What simply was impossible to reject was classed among the incomprehensible and inexplicable phenomena called chimeras, but not among hybrids. Certain scientists claim that chimeras are organisms whose tissues are mechanically built up of the tissues of two breeds.

Actually, however, the so-called chimeras may be regarded as manifestations of mixed heredity, when one part of the organisms bears the properties of one of the components and another part those of the other, i.e., cases analogous to a piebald or dappled cow, some of the patches on which have the colour of the maternal organism and some the colour of the paternal organism. Who would ever think of calling a piebald cow a chimera?

The facts at the disposal of Soviet agrobiological science create the basis for building up an integral, effective theory of heredity that will fully satisfy the requirement of serving as a "general working hypothesis, i.e., as an instrument for directing investigation towards the discovery of new facts, new generalizations."

In the final analysis, both sexual and vegetative hybridization may be regarded as a metabolic process, as an assimilation-dissimilation process.

In vegetative hybridization each of the components is fed by the other; there is interchange of substances. As a result of this mutual influence of the two breeds upon each other a new organism is formed, which contains in some measure (depending upon conditions) the heredity of both components.

In my opinion, sexual hybridization can be looked upon in the same light, for in it, too, there is a metabolic process between the fusing components (cells) of the cross.

If vegetative and sexual hybridization are phenomena of the same order, it follows that the two should have a common basis. This common basis is the fact that both vegetative and sexual hybridization are processes of the mutual assimilative activity of the components of hybridization as a result of which a new hybrid is produced.

I. V. Michurin contributed much to a correct understanding of the sexual process in plants. He showed that by suitably training the organism, by suitable feeding, it is possible to induce forms otherwise biologically incompatible to cross. Michurin devised a method of overcoming the inability to cross by the mutual feeding of the components of the cross with each other's nutritive products. This method is called preliminary vegetative approximation. Michurin also showed that choice of the conditions of life, choice of diet, makes it possible to alter, direct the sexual process by creating the prerequisites for the absorption of the heredity of one of the components by that of the other. Michurin showed also that the hereditary properties of hybrid trees continue to form throughout the course of their individual lives right up to the first fruiting years. And the way the hybrid is fed will determine whether given properties will incline towards the one or the other component of the cross.

From all this follow the interconnection and intergrading that exist between vegetative and sexual hybridization on the one hand, and between vegetative hybridization and the influence of environmental conditions on the other.

In this connection, mention must be made of the result of experiments conducted by A. A. Avakian at the Institute of Selection and Genetics (Odessa) and later in the greenhouse at Gorki Leninskiye, the Experiment Base of the Lenin Academy of Agricultural Sciences of the U.S.S.R. This result is of theoretical interest to the student of general biology.

Several years ago, at the Institute of Selection and Genetics, Avakian observed the following phenomenon repeating itself over and over again in experiments. When crossing the winter wheat Hostianum 0237 with the spring wheats 1160 or 1163 (the two latter are sisters) the resulting seeds are normal. From these seeds there at first develop shoots that outwardly appear normal. But as soon as the third leaf appears on the shoots, the first leaf withers; as soon as the fourth leaf appears, the second leaf withers, i.e., all the time only the two latest leaves remain alive. In the end, the plants perish. In short, we have here a phenomenon which the Morganists call the action of lethal genes. Although they have proposed a new term for this phenomenon, the Morganists have not been able to suggest any means of combating it. They have declared it to be fatal, insurmountable, and argue that there is only one way out in such cases, namely, not to interbreed plant or animal organisms that carry lethal genes.

In the experiments repeated by Avakian, there were at different times thousands of plants of this kind, but not one of them lived even to earing; they all perished. At the present time there are in the greenhouse at Gorki Leninskiye hundreds of hybrid wheat plants of the combination indicated and they are all on the verge of utter extinction.

Meanwhile, a crossing of the same combination of Hostianum 0237 and 1160 resulted in hybrids which are vegetating excellently in the same greenhouse, and developing into viable plants. The whole point here is that one of the components (the paternal form—1160) was grown for two generations before crossing not from a spring, but from an autumn sowing (1160 is a spring variety). This proved to be sufficient to produce a viable progeny after crossing this 1160 with Hostianum 0237. The changed conditions of cultivation changed the sex cells of the 1160 wheat plants; hence the difference in the hybridization results.

Here is another interesting phenomenon belonging to the same category of cases.

In experiments conducted by Avakian at Gorki Leninskiye, castrated plants of Hostianum 0237 wheat were pollinated with the pollen of 1160 (as we have already said, this combination usually results in nonviable progeny) mixed with the pollen of the maternal form Hostianum 0237. Some of the plants grown from the resulting seeds were obviously hybrids.[1] These plants proved to be viable, they did not perish. Thus, the presence of the pollen from Hostianum 0237 influenced the process and result of fertilization with the pollen from 1160, as a consequence of which a viable instead of a lethal, nonviable, progeny was obtained.

This shows that exchange of substances can take place between different varieties of pollen if a mixture is placed on the stigma, or between the pollen of different varieties and the egg cells of the mother plant. The physiology of these processes has not been investigated, but at all events it is an indisputable fact that the result of pollination with mixed pollen is different from that of pollination with the pollen of 1160 only. Michurin pointed to the advisability of mixing pollens. In this way he succeeded in crossing species and genera that could not interbreed otherwise.

I think this fact too shows that the sexual process, fertilization, is a peculiar process of assimilation, a metabolic process, just as it is in cases of vegetative hybridization.

This conception of the sexual process is also borne out by the category of phenomena connected with cross-pollination. As was shown by Darwin and confirmed by Timiryazev, cross-pollination, as a rule, is beneficial to plants. The progeny of seeds obtained by cross-pollination is more viable. Darwin explained this in the following way. Different organs, developing under relatively different conditions, build themselves out of the surrounding food in different ways. The result is relatively different organisms and hence, also, different sex cells. The union of such sex cells, differing somewhat in heredity, produces more viable organisms.

The intravarietal crossing conducted at plant-breeding stations is based on this Darwinian proposition. And this year, too, in the fields of the Institute of Selection and Genetics, the sowings of the fourth generation from an intravarietal cross of the Krymka winter wheat passed the winter comparative tests much better than ordinary Krymka seeds (which had not been intracrossed). Already, judging by the condition of the plants after passing through the winter, it can be confidently anticipated that the difference in the

[1] This was proved, for instance, by the following: the shoots of the maternal form Hostianum 0237 were not pubescent, whereas the shoots of the paternal form 1160 were pubescent. By obviously hybrid forms we mean those plants that are conspicuously pubescent.

Fig. 52. Diversity of plants in seed generation of vegetative hybrid

Left: tomato plant of the Mexican variety No. K 1014. The other pots contain the first seed generation of vegetative hybrids obtained from the graftings of a Rosso Grosso (scion) on to a No. K 1014 (stock). All five plants were grown from seeds taken from one fruit that developed on a shoot of the stock No. K 1014

crop will be no less than five centners per hectare in favour of the intravarietal sowings. It should be mentioned that for two years already our Institute has been supplying district seed nurseries with élite Krymka seeds obtained from an intravarietal cross.

As is known, intravarietal crossing is based on elective fertilization.

Every organism requires for its life and development relatively definite conditions in conformity with its nature, its heredity. Usually, the organism does not take inferior food elements from its environment if better conditions are present in an accessible form.

Herein lies the historically developed adaptability of organisms. Every process that takes place in an organism possesses relative elective capacity. The sexual process also possesses this elective capacity, and the assertion of the Mendelist-Morganists that fertilization is purely accidental, that it takes place only according to the law of probability, is, of course, totally inacceptable to anybody who knows anything at all about biology.

The study of elective capacity in the fertilization of plants is of great practical and theoretical importance for an understanding of the laws governing heredity.

In the Grain Department of the Institute of Selection and Genetics, D A. Dolgushin conducted the following experiment. In 1938, on the winter-wheat variety testing plots on which over twenty different varieties had been sown, several score of plants of each variety were castrated and thus allowed to pollinate with the pollen of any variety. It can be confidently asserted that the quantity of alien pollen available for every castrated flower of a given variety of wheat far outweighed the pollen of the uncastrated plants of its own variety. The plants of each variety occupied a plot one metre wide and a hundred metres long, so that the total area of all the other plots far exceeded the area of any one plot.

The first-generation plants from the seeds of the castrated spikes were only a little more viable, more vigorous, than the maternal forms growing next to them. Except for a small percentage, none of these plants differed morphologically from the maternal forms, notwithstanding the fact that the characters of some of the maternal forms were recessive (for example, awnedness, white-earedness, etc.). As a rule, plants from seeds of castrated spikes resemble pure-variety maternal forms.

In the autumn of 1939 the seeds of the second generation of these intervarietal hybrids were sown with a seed drill. Next to them the maternal forms were sown. When these plots were inspected on April 17, 1940, one could see at once that the second-generation plants of the free, elective intervarietal crossing in all cases (no less than twenty varieties were sown) passed through the unfavourable winter of 1939-40 better than the maternal forms. Not one of the varieties obtained from free elective crossing showed reduced winter-hardiness. And yet, in this experiment there were varieties like Lutescens 0329, which, according to the Morganist view, had no means of acquiring increased hardiness when pollinated with other varieties (all the other varieties were less winter-hardy). Another interesting point is that not one of these low winter-hardy varieties, Kooperatorka, for example, increased its hardiness to any considerable degree. As is known, Kooperatorka, when artificially (forcibly) crossed with more frost-hardy varieties, produces hybrids that are much more frost-hardy than it itself. In D. A. Dolgushin's experiment, all the varieties improved their winter-hardiness after free elective intervarietal crossing, but not to any considerable extent.

This experiment, and a number of others like it, show that in elective fertilization, that which is biologically most suitable for the heredity of the maternal plants is elected. Our observations have shown that, as a rule, seeds result which produce plants differing little from the maternal type, provided, of course, real freedom of choice had been ensured, i.e., that there was something from which to choose. On the other hand, such seeds, as a rule, produce plants which are invariably better, more viable, more resistant to climatic inclemencies, though they may be so only to a small extent.

I shall mention another such fact. In the autumn of 1938, A. A. Avakian sowed spring rye in plots half a metre wide and fifty metres long, in between winter varieties sown in plots of the same size. The total area on which this experiment was conducted was about a fourth of a hectare. In that sowing all the varieties flowered at the same time.

At a distance of three or four metres from these experimental plots the winter rye Pullman was sown on a plot five metres wide. The seeds obtained from that plot were planted in the greenhouse and only 1 to 1.5% of the resulting plants were of the spring variety; and yet, quite a large amount of pollen from spring plants had floated over the Pullman winter-rye plot. The preservation of the maternal plant forms in the progeny, as was the case, for example, in this experiment with the Pullman variety, is by no means due solely to elective capacity; it is also due *to the property of one heredity, in this case the maternal, to absorb* (completely assimilate) *the other.*

No few cases are known of the pollination of castrated flowers with the pollen of deliberately chosen alien forms resulting in seed from which seemingly pure maternal plants grow, which in their turn, in succeeding generations, also produce seemingly pure maternal forms. I have already described elsewhere the experiment conducted by P. N. Yakovlev (at the Michurin Central Selection and Genetics Laboratory) in crossing the *Cerasus Besseyi* cherry with a peach. In that experiment the *Besseyi* was pollinated for five consecutive generations with the pollen of the peach; nevertheless, the progeny was purely maternal. Interesting also are the experiments conducted by I. E. Glushchenko (at the Institute of Selection and Genetics) in sowing a collection of cross-pollinating rye plants in small plots. In this collection there were varieties that differed sharply from each other morphologically. In spite of this, the majority of those varieties have retained their forms for three generations and differ from the original, i.e., pure, seeds only in that they are somewhat more viable and slightly more resistant to winter inclemencies.

All this cannot be attributed solely to the fact that plants choose the pollen of their own variety. Undoubtedly in these cases the property of one heredity to absorb the other also manifests itself.

Cases could also be cited of the complete absorption of the maternal by the paternal heredity.

All these and similar experiments show that it is possible, gradually, but surely and unmistakably, to improve the biological hardiness of plants, to increase their viability, by means of intravarietal and intervarietal, free elective fertilization, just as it is possible by means of good, skilful agrotechnique to improve the breed of plants from generation to generation.

It goes without saying that both in the elective pollination of field crops, cereals for example, and in improving breeds by means of agrotechnique it is always necessary to select the best plants from each generation for seed.

For the practical breeding of field cereals, free intervarietal elective pollination is, in our opinion, the surest method of continuously improving the heredity of plants, of strengthening their resistance to climatic hardships, and also of improving the quality of the grain and flour.

As regards the improvement of the breed of plants by means of good agrotechnique, by means of good feeding conditions, this can be most clearly demonstrated by the example of summer-planting potatoes in the South. This method was worked out by the Institute of Selection and Genetics in cooperation with the collective and state farms. The summer planting of potatoes creates conditions under which it is possible to obtain tubers weighing 300 to 500 gr. This shows that summer planting creates good conditions for the development of the tubers; and it explains why the nature of the tubers improves from generation to generation. We know now that when tubers taken from a summer-planted crop in the South are planted in the spring in any district of the Soviet Union, they produce a far better crop than tubers of the same variety planted next to them, but taken from a spring-planted crop of southern reproduction. Tubers of summer reproduction produce crops two and three times as big as those of spring reproduction, and sometimes even bigger.

At the All-Union Agricultural Exhibition hundreds of thousands of visitors were able, in 1939, to see comparative plantings of the same varieties of tubers obtained in the South, at the Institute of Selection and Genetics, from spring and summer reproductions. The difference in the crops obtained was very marked. For example, the Early Rose variety, planted at the Exhibition with fourth-year spring reproduction tubers, produced a crop equal to 144 c. per ha.; but the same variety, planted with tubers of summer reproduction, yielded a crop equal to 693 c. per ha. The same contrasts were to be seen in a number of other varieties.

With the summer planting of potatoes in the South the breed of potatoes is improving year after year and their yield is steadily increasing. For example, in experiments conducted by A. M. Favorov at the Institute of Selection and Genetics, different plots of tubers of the Lorkh variety, after simultaneous planting under equal agrotechnical conditions, produced, in 1939, crops of different size, depending upon the number of years of previous reproduction from summer plantings. Tubers of second-year summer planting produced a crop equal to 103.7 c. per ha., those of third-year—111.1 c. per ha., and those of fourth-year—126.8 c. per ha.

Facts of the following nature may be cited here. In the first year of summer planting the largest tubers found in the crop weighed 300 to 500 gr.; in the second year, i.e., after the second year of summer planted reproduction, tubers were found weighing 500 to 600 gr.; in 1937, I had tubers weighing 800 to 900 gr., and in 1938 I had some weighing 1,000 to 1,470 gr.

These facts show that with summer planting the breed of potatoes improves from generation to generation.

Putting it broadly we can say that when plants are grown under good agrotechnical conditions and the best of them are selected for seed, there is a gradual but sure improvement in breed, in literally the same way as when the seeds (of wheat, for example) obtained from free elective intravarietal or intervarietal crossing are sown.

But, as a rule, neither the selection of plants grown under good agrotechnical conditions nor elective fertilization causes a radical change in heredity. To obtain such a change it is necessary to intervene drastically in the development of the plants. For this purpose it is necessary to resort to "violence," but to "violence" tempered with wisdom, as they say. This fully fits in with the "training of plants" as I. V. Michurin understood it. Therefore, while advantage can be taken in practical plant-breeding work of the organism's elective capacity in respect to conditions of life and pollen for fertilization when a radical change in heredity is to be effected, it is possible and necessary, at the same time, to produce such a change by inducing plants to cross, to be fertilized with pollen which they usually would not choose; or by inducing them to assimilate food to which they are not accustomed, to live under unaccustomed conditions.

Cases of radical changes in heredity as a result of artificial, enforced, fertilization are widely known, and therefore, I will not deal with them here. I will pass on to cases of radical changes in the nature of plant organisms brought about by changing the conditions of life. These cases most convincingly confirm the proposition advanced by K. A. Timiryazev and I. V. Michurin that it is possible to control the hereditary variability of plant organisms through the influence of the environmental conditions.

We already possess methods for converting hereditary winter wheats into hereditary spring wheats. We already have hereditary spring forms for many widely known varieties of winter wheat, for example, Kooperatorka, Ukrainka, Stepnyachka, Novokrymka 0204 and Krymka.

In the spring of this year, Avakian handed over for the variety test at the Institute of Selection and Genetics (Odessa) a spring form obtained from the winter wheat Novokrymka 0204 for the purpose of ascertaining its fitness as a *spring variety* for Odessa conditions.

The properties of winter habit and spring habit are stable hereditary properties. Winter forms and spring forms of wheat, say, have been such for hundreds of years. Only by studying these properties in the way indicated by Timiryazev and Michurin, only by studying the environmental conditions that take part in the formation of the hereditary properties of winter habit and spring habit, did scientists acquire the ability to change them in an indicated direction. The heredities of winter and spring forms differ in that each has different requirements, different responses to conditions, primarily temperature, for passing through the processes that are called vernalization. In the vernalization phase winter forms require lower temperatures than spring forms.

The properties of winter habit and spring habit in plant organisms are, of course, adaptive properties. But this in itself does not explain how these properties arose. Proceeding from the Darwinian conception of the laws governing the development of organisms, a conception further developed by Timiryazev, we arrived at the conclusion that *in the creation of the hereditary properties of an organism an absolutely essential part was played by the same environmental conditions that the organism requires in order that these properties may manifest themselves in its progeny.* It has now been proved by experiment that in the creation of the hereditary property of winter habit, for example, an absolutely essential part is played by lower thermal conditions, and that in the creation of spring habit, an absolutely essential part is played by higher thermal conditions.

It can now be confidently asserted that there is not a single variety of winter wheat from which it is impossible to obtain whole kilograms of hereditarily stable seeds of spring forms in the course of two or three generations by the suitable training of the plants. The way to do this is to effect a change in the conditions of life, specifically to alter those which take part in the process of going through the vernalization phase. A generalization of the experiments conducted by a number of scientific workers at the Institute of Selection and Genetics brought us to the conclusion that the greatest part in changing the hereditary property of winter habit is played by the completion of the vernalization process. To convert the heredity of winter wheat into that of spring wheat it is necessary to bring higher temperatures to bear upon the plants precisely towards the end of their passage through the vernalization phase.

At the present time we also have cases of hereditary spring forms that have been converted into hereditary winter forms.

An abundance of experimental material on the conversion of winter habit into spring habit, for example, shows that in the period of abrupt conversion of some hereditary properties (environmental requirements) into others, heredity becomes extremely unstable.

Cases of unstable, destabilized, heredity were also pointed to by the best of biologists—Vilmorin, Burbank and Michurin.

Fig. 53. Conversion of the hereditary nature of the winter wheat Stepnyachka into a spring form

Left pot: The winter wheat Stepnyachka (the plant did not ear). The other two pots: Stepnyachka converted into the spring form (fifth generation). All three pots were sown on November 19, 1939. The plants were grown in the greenhouse of the Institute of Selection and Genetics (Odessa)

Fig. 54. Conversion of the hereditary nature of the winter wheat Ukrainka into a spring form

Left pot: The winter wheat Ukrainka. Two middle pots: Ukrainka converted into a spring form (third generation). Right pot: spring variety of wheat Lutescens 062. All four pots were sown on November 27, 1939. The plants were grown in the greenhouse of the Institute of Selection and Genetics (Odessa)

The practical value of organisms with a destabilized heredity is obvious. Such organisms become particularly susceptible to change and serve as excellent plastic material for creating forms of plants with the needed hereditary properties.

An analysis of the results of the experiments in converting spring wheats and barleys into winter varieties (conducted by Solovey, Shimansky and other scientific workers at the Institute of Selection and Genetics) clearly indicated to us the method of speedily creating exceptionally frost-resistant forms. At the present time interesting facts are to be observed in the fields of the Institute of Selection and Genetics (Odessa), where winter wheats are undergoing variety tests.

The winter of 1939-40 was rather severe in the Institute's fields. The plots that were sown with the seeds of the spring wheat Erythrospermum 1160, converted into a winter form, stand out to advantage when compared with winter varieties like Ukrainka, and the local Krymka (without intravarietal crossing).

The history of this wheat is briefly as follows. Beginning with 1935, A. F. Kotov sowed three consecutive generations of the spring wheat Erythrospermum 1160 as a winter crop. In the first two years the seeds were planted late in the autumn, so the wheat shoots did not appear above the ground, and in this state they passed through the winter. Later, work on this wheat was continued by N. K. Shimansky. In addition to the conversion of this wheat into a winter variety, new formations of a most diverse kind appeared in it, such as, for example, awnless forms, red-spiked forms, and a number of others. In brief, instead of the spring wheat Erythrospermum 1160, a population of winter wheat appeared.

In the autumn of 1939 Shimansky planted in a nursery plot the progeny of a few plants of the winter wheat variety obtained from the spring wheat 1160. It may be anticipated that from this nursery it will be possible to select the best families, which, in winter hardiness and yield, will be superior to the most winter-hardy variety grown under Odessa conditions.

A similar example may be found among the experiments conducted by Solovey, a scientific worker at the Institute of Selection and Genetics. He started these experiments at the Young Naturalists' Station (in Odessa) with spring wheat and barley. The spring wheat Lutescens 1163 was planted late in the autumn of 1937, passed through the winter and in the summer of 1938 produced a crop. The plants turned out to be awned instead of awnless. (Even in ordinary sowings of Lutescens 1163 the appearance of awned spikes is observed.) In the autumn some of these plants, the heredity of which had already been destabilized, sprouted again and passed through a second winter (1938-39).

The seeds gathered from these plants in 1939 were sown on August 10 and 17 at the Institute of Selection and Genetics. A small percentage of the plants of the August 10 sowing eared. The rest behaved like typical winter wheats. As is known, winter wheat sown in August, which is too early for the Odessa district, is less frost-resistant than winter wheat sown normally, in September.

Nevertheless, the plants from both sowings (August 10 and 17) successfully passed through the rather severe winter of 1939-40.

Solovey's experiments in converting the spring barley Pallidum 032 into a winter variety presents a no less interesting picture. Sown late in the autumn of 1937, the spring barley produced a crop which was harvested in the summer of 1938.

Towards the autumn, under ordinary field conditions, some of these plants sprouted again and passed through a second winter. In the summer of 1939 a second crop was harvested from these plants, although the remnants of the stalks of old straw still remained.

In the autumn of 1939 the seeds obtained were sown in the fields of the Institute of Selection and Genetics (Odessa) at the usual date for winter crops. The condition of the plants after passing through the winter shows that among all the standard varieties of winter barley sown in this experiment, none is more winter-hardy than the barley which Solovey converted from a spring into a winter form, although even this one is not yet sufficiently winter-hardy to stand up well to a relatively severe winter such as we had in the fields of the Institute of Selection and Genetics this year. Some of the barley plants in Solovey's sowing passed through the winter of 1939-40 without any harmful effects, in a green state (as usually happens with rye plants), and with strong and well-developed rosettes.

It is interesting to compare the winter-hardiness of the winter wheat converted from the spring wheat Erythrospermum 1160 and the barley converted into a winter form from the spring variety Pallidum 032 with the winter-hardiness of the standard winter varieties of wheat and barley. The wheat converted from a spring into a winter form has proved to be no less winter-hardy than the ordinary winter wheats of steppe origin, such as Banatka, Ukrainka, and Krymka, i.e., wheats whose winter-hardiness was formed under conditions similar to those under which the winter-hardiness heredity was created during the conversion of the spring wheat Erythrospermum 1160 into a winter wheat. But the new wheat proved to be less frost-hardy than Lutescens 0329 or Hostianum 0237, which originated in the Volga region (Saratov), a district where wintering conditions are more severe.

In the experiments conducted in Odessa the barley that was converted into a winter form from the spring barley Pallidum 032 proved to be not inferior in frost resistance to any of the standard varieties of winter barley used in the experiments.

All the standard varieties of winter barley were formed in districts where wintering conditions are milder than in the Odessa district. I mention this only in order to emphasize once again the role and significance of external environment in the formation of an organism's hereditary properties.

The rapid conversion of spring into winter forms and winter into spring forms—in two or three generations—shows that knowledge of the

Fig. 55. Conversion of the hereditary nature of the winter wheat Novokrymka 0204 into a spring form

Extreme left pot: winter wheat Novokrymka 0204; second pot from left: Novokrymka 0204 plant grown from seed vernalized 45 days; third pot from the left: spring form of wheat obtained from the winter form of Novokrymka 0204. Extreme right pot: a plant of the Lutescens 062 spring wheat variety. All 4 pots were sown at the same time in the spring of 1940 and grown in the greenhouse of the Gorki Leninskiye Experiment Base of the Lenin Academy of Agricultural Sciences of the U.S.S.R. (Photographed on July 27, 1940)

laws governing the development of plant organisms makes it possible to create within a very short space of time better, more hardy, forms than the ordinary ones that were created in free nature in the course of a long period of time.

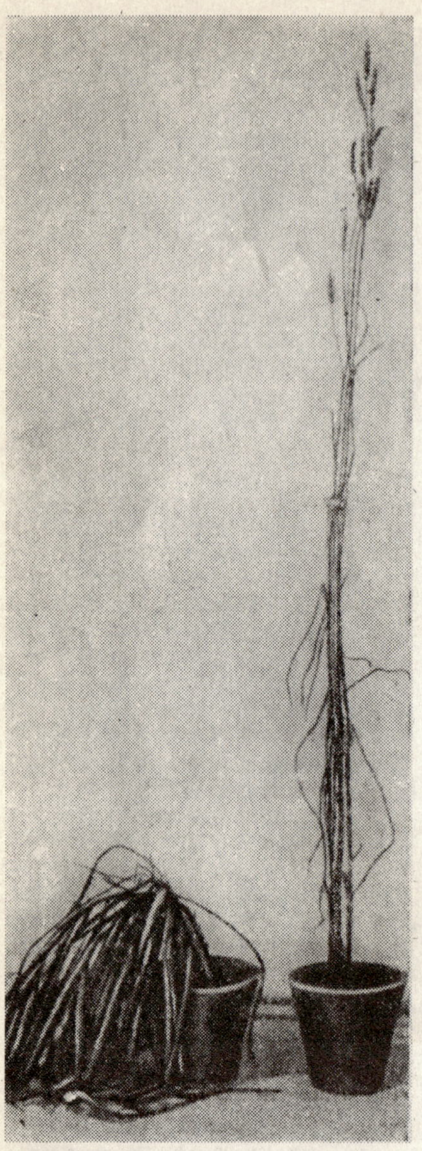

Fig. 56. Conversion of the hereditary nature of winter wheat Kooperatorka into a spring form

Left: the winter wheat Kooperatorka. Right: fifth-generation Kooperatorka plant converted into a spring form. Both planted March 17, 1940; grown in greenhouse at Gorki Leninskiye Experiment Base of the Lenin Academy of Agricultural Sciences of the U.S.S.R. (Photographed July 27, 1940)

It would be wrong, of course, to think that the most winter-hardy varieties of winter crops can be obtained only by the conversion of spring crops. The examples we have quoted show only one thing, namely, that with a correct, Darwinian approach to the heredity of plant organisms and a conception of heredity that accords with the conception taught by K. A. Timiryazev and I. V. Michurin, it is possible to obtain frost-resistant forms in two or three generations even from totally nonfrost-resistant forms of cereals, as a result of the suitable action of environmental conditions.

It goes without saying that in this way it is possible to convert the existing varieties of winter wheat into still more frost-resistant varieties. Therefore, for the purpose of creating winter varieties capable of standing the severe conditions of the trans-Volga regions and Siberia, work on intravarietal and intervarietal crossing of winter varieties is now being supplemented with work on altering, strengthening frost resistance by bringing suitable agrotechnical influences to bear upon the hereditary nature of winter plants.

At the Gorki Leninskiye Experiment Base, the spring experimental sowings of different variants of winter wheats which had not eared, or had not fully eared, by the autumn, were left standing in the winter of 1939-40. These plants started the winter with a winter heredity that had been destabilized by high summer temperatures. By analogy with the conversion of spring into winter forms, we anticipate that these winter plants with a destabilized heredity will change sharply under the influence of low autumn and winter temperatures. In the winter of 1939-40, these sowings were covered with a thick layer of snow and they passed through the winter well.

The seeds from these plants (which are fairly numerous, covering an area of about one-fourth of a hectare) will be sown in the autumn of 1940, separately, according to progeny, and also in mixtures, in the fields at the Gorki Leninskiye Experiment Base, and under the severe wintering conditions of the trans-Volga regions and Siberia. There is every reason to believe that not only will natural selection of the most winter-hardy forms take place in those sowings, but further changes in the plants in the direction of greater frost resistance will also ensue. In other words we are of the opinion that in the course of one or two generations these plants will become localized, as it were, in districts with severe wintering conditions. They will become as well adapted as wild forms.

There is no doubt that *the entire process of development, including the development of the properties of heredity and variability, depends upon the source of life, namely, nutriment.* Living substance, which once upon a time originated from nonliving substance, is now too rooted in the nonliving, building itself up out of the latter. Without food, without metabolism, living substance cannot develop.

Assimilation, metabolism, which is the essence of life, is also the basis of such extremely important properties of organisms as heredity and variability.

Heredity and all its forms, obtained both with and without hybridization, can be controlled by providing the necessary organic and inorganic conditions for assimilative activity.

By choosing the conditions that, so to speak, please plants most (elective fertilization, better agrotechnique, etc.), the breed properties of plants can be slowly, gradually but continuously improved, perfected.

By choosing conditions of cultivation that tear a plant out of its historically-built system of adaptability, by "destabilizing" its heredity (enforced fertilization, including distant crossing and abrupt changes in conditions of cultivation) and by selecting training conditions, it is possible in succeeding generations quickly to create new requirements in plants, to create new breeds and varieties that will sharply differ from the originals.

NEW ACHIEVEMENTS IN CONTROLLING THE NATURE OF PLANTS[1]

I INTEND to deal with only one problem of the big subject of controlling the nature of plants, namely, the role of environmental conditions in the development of plants. To the present day this problem has evoked great and heated discussion in the scientific world. While some scientists recognize the role of environmental conditions in moulding the nature of organisms, others consider that environmental conditions, the conditions of life, cause no alteration in the nature of organisms, or that, if they do, the alteration is in any case not a qualitative one. The former are Darwinist-Michurinists, the latter Morganist-Mendelists.

Let us analyze the fundamental postulates of each of these trends in biological science.

As you know, living matter at one time originated from nonliving matter. Although we do not know the conditions under which the first living beings appeared, or the time of their appearance, for Marxists it remains an incontrovertible fact that the living arose from the nonliving. Further you also know that all organisms (whether they be animal or plant) build themselves up only out of nonliving matter. In other words, every living organism builds itself up out of food (in the broad sense of the word). Plant organisms absorb minerals from the soil and gaseous substances from the surrounding air. These inorganic substances are converted into organic matter by the plant organism, which from the products thus converted builds all the organs and all the parts of its body. Simultaneously with the process of assimilation a process of dissimilation takes place in the organism. The two processes are inseparable, integral. All this is common knowledge, and is undisputed in science.

Further. The various plant organisms differ from one another. Anyone of you can observe the great diversity of plant organisms in nature or in practical agriculture. One comes across scores of different plants in any

[1] Lecture delivered on July 6, 1940, at a conference of heads of Marxism-Leninism departments at higher educational institutions.—*Ed.*

small patch of grass or woodland. Each one of these plants possesses its own specific properties. In a small kitchen garden bed for example, both pepper and tomato may be growing. One of these plants bears sweet fruits, the other bitter. Moreover, there are different varieties of pepper; some varieties produce sweet, others bitter fruit. Nor are these differences limited to sweetness or bitterness. Wild and cultivated plants or their fruits differ in form, colour, size, etc. Two plants may grow literally side by side; they may have the same environment—the same soil, the same air, the same light, and yet they develop differently, form different bodies. Both plants build themselves up out of the substances that occur in one and the same environment, but the results are different. The organisms produced are not the same and can easily be distinguished.

Why is this so?

To this question the answer of plant breeders is unanimous: it is so because different organisms, as, for instance, the pepper and the tomato, have different natures. Nor is there any disagreement among scientists on this score. Each organism builds itself up out of food that surrounds it; but each organism does so in its own way, for each of them has its own nature in accordance with which it lives. Each organism absorbs substances from its environment that correspond with its nature, and these substances and the conditions elected from the environment by the particular organism are then modified and converted by it in its own peculiar way.

The difference in the behaviour of organisms in one and the same environment is explained by the difference in their natures, or, as it is expressed in modern biological science, by the difference in their genotypes.

It would be only natural for you to ask: but what about the breed, the genotype, itself; is it built up or not, does it alter or not? If the breed, the genotype, builds itself up, develops, changes, then the question arises: how, from what, and under the influence of what forces does this take place?

It is on these questions that the opinion of scientists is divided. A dispute has long been raging around these questions among biologists. This dispute has become particularly acute in our country of late.

In our country, where agriculture is being built up on strictly scientific principles, great practical interest attaches to the question of how to control the nature of organisms, of how we can and should improve the nature (genotype) of organisms according to a definite, assigned plan. Obviously, before replying to this question the biologists must decide as to whether the nature (i. e., the genotype) of organisms is mutable or not. Perhaps the genotype is not subject to change at all?

I am confident that the answer to this last question is clear to everyone here. To you the variability of the nature of organisms is axiomatic. But the variability of the nature of organisms, this indisputable law of life, is far from being an axiom to everyone. Both in 1938 and in 1939 a number of periodicals in the Soviet Union published articles by scientists contradicting what is self-evident to all of you, and to me in particular.

There are still quite a number of scientists in our country who consider the nature of organisms immutable. These scientists argue that the breed of an organism, the genotype, is a special substance consisting of particles, a "hereditary substance," intrinsically different from the ordinary substance, from the body of the organism. The "hereditary substance" (called by Academician Koltsov "genoneme") is allegedly not subject to change, does not undergo transformation in the life process of the organism. Hence it appears that the general law of life—the process of assimilation and dissimilation—is not applicable to the "hereditary substance." Thus, for instance, not so long ago one such scientist, the above-mentioned Academician Koltsov, wrote:

"Chemically, the genoneme with its genes remains unchanged in the course of the entire ovogenesis and is not subject to metabolism—oxidation and reduction processes."

It is not often that one comes across such undisguised metaphysical assertions in journals published in the Soviet Union. After all, not every editor by far will let such things go through. However, we quoted Academician Koltsov from a journal edited by Academician Koltsov himself. This explains why lines which nohow hang together with the generally recognized conceptions of the laws of life came to be printed.

But there are not a few scientists in our country who, although they are essentially in agreement with Academician Koltsov, make their contention that the nature of the "hereditary substance" is immutable in a disguised, veiled manner. They do not assert (but neither do they deny!) that the "genoneme" is not subject either to oxidation or reduction processes. Such a contention would, indeed, be too glaring a contradiction of all the conclusions of science. These geneticists verbally admit the mutability of the genotype, but at the same time they say: We do not know how this change takes place in the genotype, but we do know that it is not through assimilation and dissimilation.

Such scientists actually differ in nowise from the scientists who altogether refuse to admit variability, since both in their journals, works and textbooks they maintain that the quality of the genotype's variability does not depend on environment, on the organism's conditions of life.

Let us take wheat, for example. The scientists whose views we are now analyzing say the following: irrespective of whether wheat grows in cold or in warmth, the nature of the plant is not altered in any way, and even if a change does take place, the quality of this change does not depend on cold or warmth. In other words, according to this point of view, the very same changes can take place in the nature of an organism in either cold or warmth if the state of the organism is the same.

Such is the contention of the Morganist geneticists. The Darwinist-Michurinists maintain the opposite: changes in the genotype, i. e., in the nature of a living body, correspond to changes in the body—the soma—occurring under the influence of environmental conditions. It is this proposition that at present constitutes the pivotal point of the controversy about the variability of the genotype and the ways of controlling this variability.

It is necessary to examine the question under discussion from all angles. Perhaps changes in the nature of organisms and the quality of these changes actually do not depend on external conditions?

Perhaps environmental conditions are only a stimulus with regard to changes in the breed of organisms, a spark, so to speak, falling into a powder magazine? The powder explodes for inherent reasons, and the spark merely sets it off, simply raises the temperature where it comes into contact with the powder.

The Morganists who propound such theories frantically attempt to prove that their theories follow directly from Darwinism. They strive to demonstrate that their contention that the quality of genotypic variations is independent of the quality of the environmental changes derives from Darwin's teaching. Hence, they declare, we are one-hundred-per-cent Darwinists. In their opinion, the people who claim that environment is not a matter of indifference as regards the quality of an alteration in the nature of organisms are nothing but Lamarckians.

In this connection, let us remark that it is idle for the Morganists to try to terrify people by accusing them of being Lamarckians. Lamarck was a wise man. But his teachings, of course, cannot be placed on the same level with Darwinism. There are serious errors in Lamarck's theory. But Lamarck was the most progressive biologist of his day. It is wrong to make a bogey of Lamarck. Men of science who know their business have no reason to be afraid of Lamarck. They take what is good from his theory and discard what is false.

But that is not the point. The Morganists do not understand and do not admit Michurin's thesis that variations in the breed of an organism are bound up with the conditions of life of the given organism. They endeavour to label this thesis Lamarckism, although it is an organic part of Darwinism.

The basis of Darwinism is the theory of natural and artificial selection. Recognition of the theory of natural selection is conducive to a correct understanding of the formation of the genera, species and varieties of animals and plants. Artificial selection, on the other hand, makes it possible for man consciously and deliberately to create better races, better varieties. It is only by natural selection that one can explain the amazing harmony, so to speak, in nature, the adjustment of organisms to their environment—to the season, to the soil; the adjustment of the organs to one another within the organism, etc. Look around you in a forest or field—no sooner does a caterpillar appear than right beside it you can see a leaf already beginning to unfold. The leaf is board and lodging for the caterpillar. It seems to have developed especially for the latter. One might wonder: how is it that this caterpillar did not emerge two weeks earlier? Simply because if it had been born before the leaf appeared that would have been the end of it.

In nature, of course, only relative adjustment, harmony, cooperation, exist. At the same time a struggle is going on, mutual destruction, some organisms surviving by devouring others, etc.

The principles of Darwinism are known to you. Therefore I shall not dwell on this question in detail. I shall only point out that Kliment Arkadievich Timiryazev, who did so much for the further development of Darwinism, often said and wrote that Darwin's expression "natural selection" must be understood metaphorically, figuratively. Darwinian selection, wrote Timiryazev, involves three phenomena: *variability*, *heredity*, and *survival*.

In order to select anything (in artificial selection), or for anything to be selected in nature, what is necessary above all else is variability of organisms. Variation provides the material for selection. The variations themselves may be beneficial, injurious or immaterial to the organism. However, for selection to lead unfailingly to the perfection of organisms, not only variability of organisms is necessary, but also the preservation, fixation and accumulation of these variations in the offspring. It is this that is called heredity. Variation creates a diversity of forms, but it is heredity that fixes these new properties of the organisms, while the entire environmental complex of the given organism and the interaction of the latter with the external conditions determine the fate of the organism in nature, i.e., decide whether it will survive or perish, whether it will have offspring or not.

We agrobiologists cannot regard selection as a mere sieve, cannot refrain from taking an interest in how and why a given variation occurs, how and in what conditions it is fixed and becomes transmissible. In studying selection, we are obliged at the same time to study the phenomena of variability and heredity. If we fail to study the laws governing these phenomena, we shall be able to understand the development of organic forms from less perfect to more perfect ones only in its most general aspect. But we shall not be able to discover the concrete laws of development of organisms.

In that event progressive development in nature will, of course, take place independently of us, just as was the case hitherto. In practical farming, however, things are altogether different. If we are to bear practical ends in mind it is impermissible, intolerable to reduce Darwinism simply to the selection of ready forms. Unless we know concretely how to produce such variations of organisms as we require and how to make them hereditary, we agrobiologists cannot become proficient in our work.

We cannot, and have not the right, to sit and wait for the time when, for instance, in a field sown to wheat a single spike, out of the tens of thousands of other spikes there, will appear, possessed of a new character of importance to farming. We cannot wait passively for a spontaneous variation, and at that a variation we desire. Moreover, we should be unable to tell whether this variation would be fixed, whether it would be preserved in the offspring. For lack of knowledge, we would be unable to help matters. Such passivity and such ignorance are incompatible with, contradictory to, the revolutionary spirit of Darwinism.

Practice and science that want to make use of Darwinism as a guide to action stand to gain very little from passive waiting and dependence solely on the selection of ready, random forms that appear independently of man. It is incumbent upon us to learn to alter and direct the nature of organisms in accordance with the requirements of man. At the same time, we must be able to fix the variations already produced, to make them heritable, and to select for further breeding the best of these changed organisms, those most fully answering our desires and requirements.

The branch of science that treats of the control of the variability of heredity has been very little developed to this day; the causes of variation in the nature of organisms have been little studied as yet. Darwin himself did not do much to elaborate this branch of science as applied to practical farming, and therefore contemporary Darwinism is inconceivable without the work of Michurin, Timiryazev and Burbank, and without taking into account the enormous amount of scientific data and facts that have been accumulated as a result of the work being done in this field in the Soviet Union.

I shall endeavour as I go along to tell you in brief how we Darwinist-Michurinists understand the development of plants, the role of environmental conditions in the development of plants, how we understand the role of environment in the creation of an organism's body and in the creation of this organism's nature.

There is no need to dwell on the role of environment in creating the body of the organism. Everyone knows that the better the conditions provided for a plant, the higher will be its yield. Yield now depends on the collective farmers who know the rules of scientific farming. The Stakhanovite collective farmers, the followers of Yefremov, fully understand that the plant builds itself up out of the food that surrounds it. It all depends on man whether more or less food is given the plant and whether this food is of a better or worse quality. Agrotechnique teaches us how to provide plants with more and better food so as to obtain big yields. I shall not deal here with this branch of science.

Let me now pass on to the role of external conditions in creating the nature of the organism, the genotype.

It is common knowledge that every organism possesses its own nature, or, to use the scientific term, its own genotype. In essence, breed, nature, genotype and heredity are all synonymous. Scientists say genotype; the collective farmers say breed, nature. Actually, one and the same thing is meant.

Every organism has its own nature; rice has its own, and wheat has its own. It is characteristic of the nature of rice that it requires relatively definite external conditions, peculiar to itself. Wheat, in its turn, also requires specific environmental conditions. For instance, rice requires the field in which it is grown to be inundated with 5-7 inches of water. Such conditions, which are favourable and even indispensable for rice, would prove fatal to wheat. Not only does wheat not require the same conditions as rice but it cannot even bear such conditions.

The nature (genotype) of each of these crops was formed historically and is usually extremely conservative. People have been engaged in agriculture for hundreds and thousands of years, and still rice invariably requires a surface covering of water, while wheat, on the contrary, cannot stand it. Two different organisms can live and develop in the same environment, yet they will build themselves up in different ways because they absorb different substances from the environment and absorb them in different ways, transforming or modifying these substances differently.

It is this that constitutes the heredity, or nature, of an organism, as the property of the organism, of the living being, to draw on the environment for just those substances, those conditions' of life, that are suitable, specific for it, and to absorb, assimilate them.

The property of heredity is the capacity of the organism to take from the environment only what agrees with the nature of the given organism and not to take whatever does not agree with it, even when the conditions that the nature of the given organism requires are lacking. This, in my opinion, is what constitutes the property of heredity.

The property of heredity is conservative. The conservatism of heredity expresses itself in the fact that if the conditions which the nature of the organism demands do not exist, the organism will not accept, will not assimilate, different conditions, conditions that do not agree with its heredity, its genotype. It frequently happens that an organism does not find conditions suitable to its heredity, and since it will not assimilate the existing conditions, which are different and unsuitable, it perishes. However, if the organism lacked this conservatism in its choice of conditions of life, the comparative order which we observe all about us in nature would not exist.

Allow me to illustrate this by an example. I assume that everyone here has either heard or read something about vernalization, about the phase of vernalization. Most likely, you have also heard that this phase of development in winter plants necessarily requires cold, in addition to a number of other conditions. If cold is lacking, winter plants will grow, develop roots and leaves, but will not pass through the stage of vernalization. And until winter plants traverse this phase, they can form neither stalks nor spikes even though the conditions of the external environment are suitable for the development of these organs. At a definite stage the nature, the heredity, of winter plants requires cold.

That is why ordinary seeds of winter wheat which are sown in the spring, when there is no prolonged period of cold, produce plants that will grow and tiller until the autumn, but will not grow stalks or spikes. Today, however, we already know how artificially to force winter plants to bear fruit even if sown in the spring. At the end of the winter, before the seeds of the winter plant are sown, they are moistened with a specified quantity of water. The embryo now begins to grow. The nutriment it needs is contained in the seed itself. The required low temperature (approximately 0° C.) is obtained by regulating the thickness of the layer of moistened seeds. Exact experiments have proved that the slightly-grown embryo within the seed, even before it breaks through the seed coat, is already capable of passing through the phase of vernalization. After passing through this phase before being sown, winter plants can bear fruit even if they are sown in the spring.

You see before you two tufts of Novokrymka 0204 winter wheat. They were both sown in the spring of this year in Gorki Leninskiye, at the Experiment Base of the Lenin Academy of Agricultural Sciences of the U.S.S.R. One of these wheat plants looks like a tuft of grass. This grass can continue to grow until the autumn, but it will not ear. The other wheat plant is of the same breed, the same variety. It was sown at the same time and, as you see, has already formed spikes and will soon blossom. In somewhat over a month it will yield ripe seeds. The seeds of this wheat were vernalized before being sown. In other words, in the second case seeds were sown whose natural, genotypic demands for the conditions requisite to pass through the process of vernalization had been satisfied. Thanks to this the development of the plant proceeded normally. In the first case, on the other hand, seeds of the same variety were sown, but they had not passed through the phase of vernalization. We did not satisfy the requirement of this breed for vernalization conditions, did not provide it with cold, hence the plant failed to form stems.

In order to obtain a crop from plants, it is necessary to comply with their nature, to satisfy the demands of their heredity with regard to the conditions of development of the given plant as a whole, and in particular of those organs which produce the harvest. The better and more fully we satisfy the requirements made by the nature of the plant, the greater the harvest we shall be able to gather.

The two plants of winter wheat which I have shown you are different in appearance, are different organisms. But their natures, i.e., their heredities, are comparatively the same. The organisms are different, their appearances differ, because the natural requirements of one of these plants were satisfied (it was vernalized), while the natural requirements of the other plant as regards the phase of vernalization were left unsatisfied.

But chilling the suitably prepared (soaked) seeds and thus satisfying their hereditary requirements does not change their hereditary character of being winter plants. That is why we say that the nature of these two different organisms is comparatively the same. If vernalized winter wheat is sown in the spring, we can harvest seeds in the summer. Upon sowing the seeds thus obtained, they will again require cold in order to vernalize, just like all winter plants in general. They will not be satisfied with warmth and will not pass through the phase of vernalization in warmth.

Here we have a very definite manifestation of the conservatism of heredity.

Everyone knows that in addition to winter wheat there is also spring wheat. This wheat is sown in the spring; it does not require the low temperatures that winter wheat requires. In spring wheat the process of vernalization takes place at a higher temperature. This hereditary property of spring wheat is likewise conservative.

What would be the result if the heredity of winter plants, for example, were not conservative?

The seeds of a wild winter plant would ripen in June and then shatter. After a rainfall the seeds would begin to sprout. At this time of the year it is warm, there is no cold. If heredity were not conservative, the plant would not require cold and would easily begin to pass through the process of vernalization in the warmth. After this the stalks, the straw would form. But we know that if cereal grasses evince even the faintest sign of straw formation, the plants are no longer capable of withstanding severe frosts, i.e., of living through the winter. Cereal grasses that have passed through the phase of vernalization and begun to form stalks (straw) are unable to develop resistance (hardiness) to frost.

For this reason if wild winter cereal grasses did not possess the conservative hereditary property of winter habit they would be unable to exist.

What would happen to winter plants in agriculture, in practical farming, if their heredity were not conservative? We should quite simply fail to obtain a harvest.

For hundreds and thousands of years winter plants have been sown on millions of hectares in August and September, when it is still warm. It is only due to the fact that plants have a strongly conservative heredity that winter plants sown in the early autumn, when it is still warm, develop roots and leaves but do not pass through the stage of vernalization. The winter plants do not have suitable conditions for vernalization since it is not cold. That is why these plants are able to live through the winter. In the late autumn and winter, cold sets in and the plants become vernalized, and in the spring they begin to ear.

When we harvest the seeds of winter plants we may be quite sure that the plants grown from these seeds will also be winter plants, and the offspring of seeds from spring plants will be spring plants. The same is true of all the other inheritable characters and properties of plants. Thus, for example, the offspring of awned wheat will be awned, of red wheat—red. From these simple examples we may pass to more complex ones, and all of them alike will testify to the conservatism of heredity. The favourable aspect of the conservatism of heredity is that this property makes it possible to have definite varieties in agriculture, while in nature it ensures the preservation of the ability developed by organisms to adapt themselves to their external conditions.

But the conservatism of heredity also has its unfavourable aspects. The stable and conservative property of heredity obliges man to humour plants in every way and to suit conditions to the plant with the help of agrotechnique. This is not always possible or convenient. Hence, the question naturally arises: Cannot the conservatism of heredity be broken? Is it impossible, for example, to force winter wheat to require for its vernalization the warmth that prevails in our fields in the spring, instead of cold?

In order to answer this question we must have a clear conception of how the various heredities are formed and under the influence of what forces they are modified.

The property of heredity inheres in living matter alone. All living matter builds itself up out of food, out of its environmental conditions, by means of assimilation and dissimilation. More than that, the very first living matter arose from nonliving matter. But if life originated from nonliving matter and every plant organism builds up its body from such matter, from food, it naturally follows that all the inherent properties of the living body—including the property of requiring specific conditions of development, i.e., the property of heredity—develop, build themselves up and change simultaneously with and inseparably from the development of the very body of the organism.

We already have at our disposal a vast amount of factual material, both experimental and practical, which shows that not only the body of the organism but its heredity as well builds itself up in the process of development, i.e., in the process of absorption, of assimilating the environmental conditions of the organism.

Diverse hereditary characters are conservative in diverse degrees, but all of them are to a certain extent conservative. Such a hereditary property as the winter habit of grain is one of the most conservative properties. For thousands of years people have been obtaining winter crops from the winter plants they sowed. From the seeds sown in the autumn grass would sprout, in the spring stems would form, and then spikes and grain. During the autumn, winter, spring and summer thousands of changes would occur in the plant, every day would bring some new transformation, and at the end of the ripening apparently the same seed, with the very same hereditary properties, would be obtained as was originally sown.

But if one examines matters a little more closely, it is not difficult to see that the nature of the organism does not remain unchanged from generation to generation, but also changes. These alterations are of different degrees, from those that are scarcely noticeable to quite considerable ones. We have in view here changes that occur when we humour the plant, when we meet the requirements of its nature.

But what would happen if the plant were not provided with the conditions it requires and given other conditions? What would be the result of this? You could reply: "The plant will not accept conditions that do not suit it, will not assimilate them, and as a result will perish." That is true. However, it is not always so.

If one approaches the plant from the standpoint of the Michurin theory, if one treats the plant organism properly, it is possible not only to increase the yield by humouring the nature of the organism but also to remould its very nature, its very heredity, in accordance with the conditions prevailing in the particular bed, field or garden. In other words, the heredity may be changed deliberately in the direction we desire. This can be done by means of skilful training of the plant.

What, then, does this skilful training of plant organisms consist in?

It consists not only in humouring the plant's nature but also in opposing it with a view to inducing new demands in the given plant.

Winter habit is one of the stable hereditary properties of cereals. In farming it is absolutely necessary to cater to the requirements of winter plants with the help of agrotechnique and to furnish them with cold, otherwise there will be no harvest. But what would be the result if we "mistreated" winter wheat Novokrymka 0204 and did not provide it with the conditions essential for vernalization? The result would be no crop. A study of the biology of winter wheat has shown that the different varieties require different periods of chilling for vernalization. Novokrymka 0204 requires a temperature of approximately 0° C. for a period of 35 days in order to go through the entire process of vernalization. If the temperature is 3°-5° C. the process of vernalization will require 40 days. If the temperature is 15°-20° C., the process of vernalization will not take place altogether, or, if it does, will take a considerably longer period of time.

But what will be the result if we chill soaked seeds of this same Novokrymka 0204 for only 25-30 days? Vernalization will proceed normally. After 30 days we cut the cold treatment off. Thus, the seeds will still be 5 days short of the normal period for completing the process of vernalization. From numerous experiments we know that if the period this variety normally requires for vernalization is artificially shortened by even one or two days, the other processes, which follow vernalization, cannot take place. In every phase of the development of the plant organism, including the phase of vernalization, a qualitative change takes place in its demands on environmental conditions. But for this change to take place certain definite external conditions quantitative in character must be available. After the organism has been provided with these conditions and has assimilated them, it suffers a qualitative change, development enters a new phase and the demands of the organism on its environment also change. For example, the demand of winter plants for the cold which they need in order to pass through the process of vernalization is replaced by a demand for warmth.

For the phases following vernalization, for the subsequent processes, warmth is now indispensable.

Thus, the soaked Novokrymka 0204 seeds which we chilled for 30 days (we vernalize them that long) are then sown in the field in the spring. It is not very warm in the spring, but neither is it cold. And now, instead of completing the process of vernalization in 5 days, which this sort of winter wheat would do at a temperature of about 0° C., the plant begins to experience distress, figuratively speaking. It "languishes" in this way for 15-20 days, but in the end it completes its vernalization. But once vernalization has actually taken place, despite the abnormality, despite the "distress," further development proceeds at a very rapid pace. The field conditions for this are good: the days are long, there is plenty of light, it is warm and there is sufficient nutriment.

In the summer the seeds of the plants ripen. The question arises: will these seeds be normal, common seeds possessing the normal heredity of the winter variety? It turns out that they will not. In these seeds the old, long-established, conservative heredity of the winter habit has been broken down. In this generation the heredity has not been reproduced as it was in the numerous preceding generations. Cold was not given the plant at a particular, critical moment, towards the end of vernalization.

In our experiment the vernalization was completed in warmth and not in cold. What does this mean with regard to the plant? It is impossible to conceive that one and the same process can take place in a living organism both in cold and in warmth in absolutely the same way. If we take seeds of any variety of winter wheat, divide them into two lots and let the vernalization of one lot be completed in cold, and of the other lot in warmth, then, of course, the processes of vernalization in these two lots of seeds will differ qualitatively. The heredity as regards the phase of vernalization will also be different in the organisms grown from these seeds.

Suppose we sow the seeds of the plants that completed vernalization in warmth. The new organisms, as experiments have shown, no longer stand in particular need of vernalization chilling. They no longer have that urgent need of cold which exists in ordinary winter plants. The old, conservative hereditary property of winter habit has been destroyed by us in 10-15 days, i.e., during the time when the plants of the preceding generation completed the process of vernalization in conditions of spring temperature.

I must state that this is not only easily said but easily done. All that is needed to destroy any conservative hereditary trait prejudicial to the definite end we pursue is a knowledge of the conditions with which the organism should be provided, and when. It is necessary to know when to stop humouring the old, strong heredity, when to remove the conditions which the old heredity requires and to substitute for them conditions for which we wish to create a demand in the organism. I must stress that in this interesting task everything depends on the skill and knowledge of the experimenter.

We tried giving the winter plant warmth at the beginning of the vernalization process instead of at the end. In this experiment, after very prolonged treatment, the heredity changed, but we obtained deformed organisms.

In order to change the nature of the vernalization phase, it is necessary to modify the conditions under which vernalization takes place at the end of the process.

How did we arrive at this conclusion?

Here it would be in place to mention the absolute necessity of coordination between the experimental work of scientists and practical life. Whatever the experiment we undertake, we must always keep an eye on what is happening around us in life. We must examine life closely and try to understand the facts we observe in it and to link them up with the experiment we are working on.

In practical farming, people have been sowing winter wheat for thousands of years in the early autumn, in August, when it is warm (cold sets in much later), and the winter habit of this variety of grain has never changed because of this. Why then provide warmth at the beginning of vernalization? The process of vernalization simply will not take place in this case; the organism will wait for the onset of cold. In order to convert winter wheat hereditarily into spring wheat, warmth must be supplied only at a strictly definite time, namely, before the end of the process of vernalization.

Several researchers attempted giving the plant warmth after earing, when not a breath, so to speak, of the process of vernalization was left; vernalization had long been completed in the given plant. In such experiments these researchers wanted to change a property which the organism had possessed in the past, but which it had already outlived, a property which could reappear only in the succeeding generation. Thus, these people tried to change something in the organism which it no longer possessed at the time of the experiment. They explain this mode of experimentation by asserting that an organism possesses a specific hereditary substance—genes—which in its normal state is not subject to the processes of either oxidation or reduction. As they conceive it, the genes of winter habit are always present in the form of particles in the cells of the winter varieties of plants. Therefore, according to their view, the time of treatment and the state of the organism at that time are of no importance. And despite their obviously incompetent experiments these scientists say that it is impossible to change winter varieties into spring varieties by bringing external conditions to bear upon them.

When we provided warmth for winter plants at the beginning of vernalization, no good results by way of transforming the nature of the plant were obtained. But when we began to provide warmth at the end of the process of vernalization, the results were considerably better—the old heredity was often broken down immediately. On the basis of these experiments we arrived at the following conclusion: *the conditions in accordance with which we wish to create a new heredity must be furnished at the end of the process whose nature we are changing.*

Today there is not a single hereditary winter variety of wheat that cannot be converted into hereditary spring wheat in two or three generations. From wheats preferring cold at the vernalization stage it is possible, by a proper growth of wheat varieties, to obtain spring forms which do not require cold.

It may be objected: very well, granted that you already know how to break down the conservative hereditary winter habit and to create in its place a hereditary spring habit. This does not yet prove that heredity, as a property of living bodies, is created from and depends on the same nutrition, the same conditions, from which the organism's body is created by means of assimilation.

And indeed, how can we prove, how can we obtain experimental evidence that heredity really depends on nutrition, that it is built up in the process of the formation of the organism's body itself, in full conformity with this process?

Let us turn to experimental examples. In agriculture there are various sorts of tomatoes having diverse heredities. There is, for instance, the Mexican (wild) tomato. The fruits of this variety are small, round, and, when ripe, red. Then there is the cultivated tomato known as the Albino. The ripe fruits of this sort are not small but of normal size, and not red but yellow.

It is known that if a cutting of a tomato plant is grafted on to another tomato plant, the graft, after union of stock and scion, begins to develop. In experiments carried on by A. A. Avakian and M. G. Yastreb (Odessa Institute of Selection and Genetics) a cutting of a young yellow-fruited Albino tomato plant was grafted on a Mexican red-fruited tomato plant.

What substances are produced by the roots and leaves of the given red-fruited stock? It is quite clear that they elaborate those substances that are characteristic of the given variety, that are indispensable for the formation of red fruits. Hence the grafted cutting of the yellow-fruited Albino plant is forced to use nutrients of the red-fruited variety instead of its habitual food. As a result of such nutrition the yellow-fruited variety graft frequently bears fruit that are not yellow but red, or reddish, or yellow with red streaks. From this it is perfectly evident that here the nutrition has exerted a direct influence on the quality (in the given case, the colour) of the fruit.

To this, the Morganist-Mendelists say: true, the quality of the fruit has changed. But that is of no particular significance. All agree that the body of an organism changes under the influence of external conditions, including nutrition. But the heredity, they continue, undergoes no change, i.e., remains that of the Albino variety.

On the other hand, we who adhere to the Michurin theory hold that if we were to take the seeds from a red fruit that had grown on the grafted yellow cutting and were to sow them, the new plants would bear not only yellow fruit but red ones as well. The Morganists were and still are wholly unwilling to admit this, as was evidenced by their speeches at the special conference on questions of genetics called by the editors of the journal *Pod znamenem Marxisma* in October 1939.

When I took the floor to speak at that conference, I had in my hand only the fruit of the grafted plant. On demonstrating this fruit I said approximately the following: if experimenters succeeded in raising red fruit on a yellow-fruited cutting, how much easier must it be to obtain red plants from red fruit. After all, in practice people as a rule obtain red fruit from the seeds of red tomatoes and yellow fruit from the seeds of yellow tomatoes.

The Morganists, however, retorted: you will not get them—the body of the fruit has changed but its heredity has not.

This controversy took place in October 1939. In the winter of that same year, the seeds of the red fruit were sown in a greenhouse in Odessa, at the Institute of Selection and Genetics. The requisite time elapsed, the plants grew up, and some of them produced red fruit, others yellow. There were also plants bearing fruits of intermediate colouring ranging from red to yellow.

The same results were achieved by Avakian in Gorki Leninskiye, at the Experiment Base of the Lenin Academy of Agricultural Sciences of the U.S.S.R.

Of particular importance to us is the fact that in addition to the plants bearing red fruit we also obtained plants bearing yellow fruit.

What conclusion can be drawn from the given experiment?

It is possible to draw the direct conclusion that under the influence of nutrition the heredity of the developed graft not only changed in general, but changed in the direction of the heredity of the stock, and this modified heredity was transmitted to the seed offspring. It follows that heredity may be shaped by the agency of nutrition, by the interchange of substances (in this case, between the scion and the stock). By changing the nature of the metabolic process we can achieve the directed alteration of the nature of the organism.

The Morganists seek to prove that heredity is located (in the literal sense of the word) in the chromosomes, in the form of particles. They even draw diagrams showing the position of these particles and the linear arrangement of the genes (the heredity corpuscles). The assumption that these particles are transmitted from one organism to another is precisely what the Morganist chromosome theory of heredity is based on.

But we know that chromosomes are not transmitted to the scion from the root of the stock. The protoplasm was also not transmitted; yet the heredity of the indicated graft underwent a directed change, was "transmitted" from the stock to the scion and vice versa.

In the light of experiments with vegetative hybrids, nothing remains of the Morganist chromosome theory of heredity. Chromosomes, of course, still remain in the cells, and of course they possess the property of heredity just as do other reproductive elements of the body. But the chromosome theory of heredity, i. e., the theory maintaining that the chromosomes differ intrinsically from the rest of the body with regard to heredity, must be thrown overboard in its entirety.

But perhaps what I have related here is an isolated case that involves only the colour property of the fruit?

No, it is far from being the only case of vegetative hybridization. There have been fairly numerous cases where hybrids have been obtained, in other words, where the hereditary properties of two organisms united in a third, new organism; moreover these hereditary properties combined without any fusion of the nuclei of the cells and without fusion of protoplasm.

No few examples of changes in heredity by means of grafting could be cited.

These experiments were performed by various people, all of whom, however, hold the same views, basing themselves in biology on the teachings of Michurin. The possibility of vegetative hybridization has been unerringly proved in theory, and as for practice, numerous vegetative hybrids are already being raised in our country.

There is a variety of tomato called Humbert. The fruits of this variety are oblong in shape. They are widely used for canning. The Humbert was grafted on to an early variety having round fruit. The seeds were taken from the graft (the Humbert scion) and sown. The offspring obtained were diverse: there were fruits resembling the Humbert, round fruits, and fruits which were round on top and Humbertian, oblong, below.

This diversity in the shape of the fruits obtained as a result of grafting is extremely gratifying to us, but most depressing to the Mendelist-Morganists.

Why is that? you will ask. It is because these facts disclose to us a number of interesting regularities.

For one thing, these facts show that the heredity of the stock and the heredity of the scion *build one another* through the medium of nutrition, through the metabolic processes; what is obtained is, so to speak, a union, a fusion of two breeds.

For another thing, and this is most noteworthy, we see that the old breed can be preserved; in other words, the result is analogous to the result obtained in sexual hybridization.

In sexual hybridization the fusion of two sex cells results in a new organism. This new organism usually possesses a dual heredity—both paternal and maternal. This dual heredity separates, as it were, in the individual cells, for instance, in the sex cells. The result is what is known in genetics as "segregation." The Morganists explain this separation of the dual characters, this "segregation," as a dissociation of the homologous chromosomes, in which, they assert, the hereditary particles (the genes) are located. One chromosome contains the hereditary properties of one parental form, another chromosome those of the other.

But experiments with vegetative hybrids, I repeat, patently proved that chromosomes are not transmitted from the scion to the stock (or vice versa), whereas hereditary properties may be and are transmitted, and that these hereditary properties of two varieties may combine in one, and may then vegetatively, or through the medium of seeds, again separate, so to speak.

All these facts oblige us to regard the sexual process in an entirely different way from the way in which it has been regarded in biology hitherto. I have not the time to deal with this question in detail. I should merely like to point out that Darwin himself had already foreseen a change in the scientific views of sexual reproduction. Recognizing that vegetative hybridization is possible, he wrote the following:

"... if, as I am now convinced, this is possible, it is a most important fact, which will sooner or later change the views held by physiologists with respect to sexual reproduction."[1]

The example of vegetative hybrids constitutes splendid proof of the fact that heredity is not only changed under the influence of conditions of life, but is also *built up*, is *created* under the influence of these conditions, under the influence of nutrition, of metabolism. In other words, heredity, as a property of living matter, is created under the influence of the *very same* things as go to create the body of the organism. Vegetative hybrids are excellent confirmation of this proposition.

To us a correct understanding of heredity and its alteration is of the utmost interest. This question is of extreme importance not only in the realm of theory but equally so in actual practice. It is my custom to tackle any question of theory only from the standpoint of practice, to solve problems practically. Nor has this become a habit with me because I do not care for theory. Quite the contrary: divorce from practice leads not only to fruitless but also to erroneous work in the sphere of theory. This may be confirmed by dozens of examples taken from works dealing with theoretical problems in the field of genetics—the science of heredity and variability.

The Michurin theory of heredity and of the means of controlling it furnishes the scientific worker with simple and at the same time very effective methods of remoulding the nature of plants.

In 1935 I did not know of a single instance of spring wheat having been produced from winter wheat. It was not known how to change the hereditary winter habit into a spring habit.

At present, however, anyone who tackles this work can fairly easily convert hereditary winter forms into hereditary spring forms. At the same time we have also learned how to transform spring forms into winter forms. Thus, for example, G. T. Solovey, formerly with the Odessa Young Naturalists' Station, and F. A. Kotov and N. K. Shimansky, both research workers of the Odessa Institute of Selection and Genetics, set themselves the task

---

[1] Charles Darwin, *The Variation of Animals and Plants Under Domestication*, **London** 1885, Vol. I, p. 417.

of procuring winter varieties from spring varieties. And they succeeded in doing so.

In order to convert hereditary winter plants into spring plants it is necessary to bring a higher temperature to bear on the organisms at a definite moment in their lives, in other words, it is necessary to apply warmth to the vernalizing winter seed at a definite moment. Solovey and Kotov did the contrary, i.e., they treated spring varieties with cold. By their nature, spring varieties do not require chilling for vernalization, but nonetheless they were sown in the field in the late autumn and subjected to cold for a prolonged period of time. In this way they forced the spring varieties to change their heredity in the course of two generations and become winter varieties. Now we sow the seeds of these altered plants in the spring, and they no longer behave like spring varieties but like winter varieties; they do not develop stalks, nor do they ear, and they require cold for vernalization.

Such sowings are demonstrated at the All-Union Agricultural Exhibition. Erythrospermum 1160 wheat is growing on one of the plots there. One half of the plot is sown to the spring variety, the other half to the winter variety that has been obtained from the latter. Both varieties were sown at the same time, in May. At the end of June the spring form has already eared, while the winter form is still in the rosette stage, tillering and showing no signs of coming into ear, since it requires chilling, and there are no prolonged periods of cold.

The endeavours to convert winter varieties into spring varieties, and spring varieties into winter varieties are of interest not only from the theoretical standpoint—for studying the basis of heredity and its alteration. These endeavours open up a new stage in the breeding and seed growing of our cereals.

Not only were the results of the experiments in transforming winter plants into spring ones and spring plants into winter ones such as were expected, but in addition a number of other very interesting and important results were obtained.

Some of the wheat seeds that Shimansky converted from Erythrospermum 1160 spring wheat into winter wheat in the autumn of 1939 were subjected to variety tests at the Institute of Selection and Genetics.

In the spring of 1940 it was evident that this wheat had wintered like a good typical winter wheat. Moreover, when the conversion experiments were made—transforming the spring variety into a winter one—we did not suppose that this winter wheat would be able to withstand the winter cold better than all the varieties of winter wheat coming from the steppes of the Ukraine. But this was just the case.

The winter of 1939-40 was not a favourable one for winter wheat in the fields of the Odessa Institute. Winter wheat like the Ukrainka (standard) wheat, for instance, was quite badly damaged by frosts.

In these variety tests, the winter wheat developed from Erythrospermum 1160 spring wheat was sown side by side with Ukrainka. A survey of the plots in the spring, after the wintering, revealed that the new winter wheat, developed from spring wheat that was totally nonresistant to frost, was in no way inferior, and even superior, to the hardiness evinced by the standard, Ukrainka winter wheat.

The very same was the case with the *winter barley* obtained by Solovey from Pallidum 032 spring barley. In variety tests at the Institute of Selection and Genetics all the standard varieties of winter barley almost completely perished in the winter of 1939-40. The winter barley which Solovey had obtained from spring barley also suffered greatly. But, nevertheless, it withstood the winter incomparably better than all the other, standard winter varieties of barley—its winter hardiness was greater.

Today it has become clearer to me how to produce a winter variety of wheat such as will be able to resist the severest frosts that occur in regions of the harshest climate, as, for instance, the open Siberian steppes.

This is an absolutely real possibility.

In the two instances I mentioned, the agriculturists had not set out with the purpose of developing frost resistance in plants. We did not try to make these varieties frost-resistant, they became so of themselves. Now we have a pretty good idea of why this took place.

So long as the heredity of the organism remains conservative from generation to generation, it is very difficult to do anything with the organism; it does not lend itself to rapid improvement. The new winter varieties obtained from spring varieties, however, no longer have the old heredity. It has been destroyed. And the new heredity is just forming in the organism, is no more than a propensity, so to speak. Hence, such heredity is not yet conservative, is unstable. It may be termed destabilized heredity, to use the expression of Michurin and Vilmorin. And organisms with a destabilized heredity are excellent material for the plant breeder. Given skill, remarkable things can be done with this material.

An organism possessing an unstable, destabilized heredity requires conditions that are suitable to it no less than an organism whose heredity is fixed and conservative. But in contradistinction to the latter, which would perish in the absence of such conditions, the organism with an unstable heredity does not wait if these conditions are not forthcoming, but begins to assimilate many others from its environment.

The distinguished biologist and remarkable plant breeder Vilmorin wrote in his time that it was of the utmost importance to induce variations in plants. It is of importance to change plants even if it is in the direction opposite to the one desired, so long as a change is effected. After this has been achieved it is easy to force the plant with such a destabilized heredity to do what we want it to do. The very same idea has been expressed by the outstanding American biologist Burbank, only in different words. And this idea has been elaborated more profoundly on a scientific theoretical basis by the Soviet scientist Ivan Vladimirovich Michurin. Michurin not only pointed out the definite scientific paths along which to proceed in the question of altering and destabilizing the heredity of plants in general, but he likewise showed how to induce directed changes. The followers of Michurin have really mastered the technique of making systematic directed changes in the nature of organisms. Today we understand this question much more fully and profoundly than we did yesterday.

Today we not only know how to destabilize and destroy the old heredity, but we also know why the new heredity is unstable, and what we must do in order to fix it in the desired direction.

If seeds of plants possessing a young and as yet not very stable heredity were to be put into the hands of the Morganist-Mendelists, they would simply ruin the whole thing because of their disregard for the scientific Michurin approach to plants. And having done so, the Morganists would say: We took your seeds for testing and failed to obtain good results. Consequently, your claim that directed alteration of heredity can be induced is just idle talk.

It is necessary to know not only how to destabilize heredity but also how to fix the young heredity in the desired direction. For this it is necessary to know how heredity is constituted and then to provide the conditions favouring development in the requisite direction. Once we know this, the seeds of the second generation will possess a more stable heredity, the seeds of the third generation an even more stable one, and so on. And seeds of the fourth and fifth generations may even be handed over to the Morganists, who, desire it as they may, will no longer be able to spoil our work, to reverse the plant's course of development.

Research work in the field of variation by the numerous followers of Michurin, coupled with a study of heredity, has brought us to the following, what we think important, conclusion: *the environmental conditions that heredity demands for the development in the organism of a given property or character are the very conditions that absolutely must have taken part in creating the heredity itself, in creating the heredity of the given property or character.*

By basing ourselves on this conclusion, we are in a position to perform interesting experiments of use in practical farming. For instance, we very much need a frost-resistant winter wheat for Siberia. To develop such a variety it is absolutely necessary to utilize the severe conditions of the open Siberian steppes. These conditions must be made to act on an organism having a destabilized, nonconservative heredity. We are now able to produce such heredity. Hence we are confident that in a short period of time we shall be able to grow winter wheat capable of resisting the severe conditions of the open Siberian steppes.

Why did the winter wheat that Kotov and Shimansky produced from Erythrospermum 1160 spring wheat prove more winter-hardy than any other winter wheat coming from the Ukrainian steppes that possesses a conservative heredity? Because fairly severe winter conditions helped to shape the heredity of this wheat. These conditions acted on the destabilized heredity of the new wheat in such a way as to incline it to greater frost resistance. If this wheat withstood the severe Odessa winter of 1939-40, its seeds will undoubtedly be even hardier than those sown in the autumn of 1939.

If we sow this wheat (and analogous kinds which we now have) under the severe conditions prevailing in the trans-Volga area or Siberia, furnish it with inclement conditions (but, of course, not so inclement as to kill the wheat altogether!), we shall thereby force the young heredity in the direction of still greater frost resistance.

The experimental work of the followers of Michurin makes it fully possible to convert the inordinately conservative heredity of an organism into an extremely unstable heredity—and this cannot but make us happy.

A person who does not know what heredity is built up from and how it is built up cannot do anything with an organism possessing an unstable heredity. But for people who know what heredity is built up from and how it is built up, organisms with a shaky, unestablished heredity constitute veritable gold mines. We shall subject these organisms to harsher and harsher conditions from generation to generation, shall subject them to severe cold (but, I repeat, only to the extent that it does not kill the plants!), and after two or three years of such training, the wheat we obtain will be in no way different from the local plant forms with respect to hardiness.

You might object that on the open Siberian steppes there is no native winter wheat and hence that there is nothing with which the new varieties of wheat could be compared. It is true that in a number of districts there are no local forms of winter wheat. But there are local forms of weeds which developed under Siberian climatic conditions and fear no frost whatever. Once we know what heredity is built up from and how it is built up, we can make use of this knowledge to train wheat in such a way as to develop in it a winter-hardiness in no way inferior to that of the local weeds.

ORGANISM AND ENVIRONMENT[1]

Distinction Between Living and Nonliving Matter In Their Interrelations with the Environment

As we know, organisms are closely connected with the conditions of the external environment, and not only connected but adjusted, in a definite way, to the environment in which they live. This purposiveness, as it were, of the structure and mode of life of organisms, this harmony of living nature, was splendidly explained by Darwin in his theory on natural and artificial selection. Natural selection explains why beings are perfected under natural conditions. Artificial selection explains why cultivated plants and domesticated animals, as a rule, well satisfy man's requirements.

In general, both living and nonliving bodies exist in certain relations to the environment that surrounds them. However, the interrelations between organisms and their environment differ in principle from the interrelations between nonliving bodies and their environment. The main difference is that the interaction between nonliving bodies and their environment is not a condition of their preservation but, on the contrary, is a condition of their destruction as such. For instance, the more isolated a nonliving body is from the action of oxygen, moisture, temperature, etc., the longer it remains what it is.

But if, on the contrary, a living organism is isolated from the environmental conditions it needs it ceases to be an organism, living matter ceases to live. Living matter is inseparably connected with its environment, with the conditions of a continuous exchange of substances.

Thus, essentiality of interrelations with environment is, for living beings, an indispensable condition of their existence, nourishment and development in the broad sense of these words, i.e., in the sense of forming the hereditary properties of organisms.

The discovery of the laws governing the mutual relations between organisms and their environmental conditions is the principal task of

---

[1] Stenographic report of a lecture delivered in the Polytechnical Museum on January 11, 1941.—*Ed.*

agrobiology. This problem was and still is of paramount importance also to practical farming.

The better we understand the interconnections between organisms and environmental conditions the better organisms can be governed, by making use of the possibilities of regulating and creating environmental conditions.

THE IMPORTANCE OF SELECTING THE OBJECT OF RESEARCH

The general laws of development of organisms can, of course, be demonstrated on any animal or plant object. But if the best theoretical and practical results are to be obtained it is not a matter of indifference what object is chosen for that purpose.

During the last few decades the Drosophila fly has become one of the "classical" objects of investigation for the discovery of the laws of variation and heredity. Many scientists, not only abroad but also here in the Soviet Union, picked (and some even now pick) it as their object of research aimed at discovering the laws of heredity of plant and animal organisms, at discovering the laws governing the interrelations between organisms and external conditions.

The choice of the object largely determines the success or failure of theoretical work.

If the object is of economic importance the work is carried on with a will and, I would say, of necessity. There are any number of interesting problems, but one must be able to select what is of greatest significance among all the material that has research value. A research object of practical importance compels one to decide the most profound questions of theory from the point of view of practice, and this is the most important thing in deciding questions of theory, since practice is the criterion by which the veracity of theory is gauged.

It seems to me that choosing Drosophila as an object of research renders genuine investigation more difficult. Of course, this object too is governed by the general natural laws inherent in any living body, whether animal or plant. But what can be ascertained with the aid of the Drosophila, what can be expected of it from the practical point of view? What problems can receive a practical solution with the Drosophila as the object of research?

It requires particular genius and erudition to discover anything of real practical importance even in an object that has no bearing on practical life.

As a rule we choose objects of investigation that are of practical importance.

I shall dwell on one such object, the potato.

The various problems of agrobiological theory are brought to light and solved more easily if the potato or some other item of practical importance rather than the Drosophila is taken as the research object. The results obtained in working on potatoes have emboldened me to agree to the much-promising title of my lecture, "Organism and Environment," although the lecture will be simply on the potato and the best way of cultivating it. The laws discovered with regard to this object can in some degree or other be applied also to other plants.

CAUSES OF THE DEGENERATION OF POTATO PLANTING MATERIAL IN THE SOUTH

Under the conditions prevailing in the South where the summers are hot the potato fares badly. It quickly loses its nature, its fine hereditary qualities.

The first year of planting, with good agrotechnique, a fairly good crop (10-15 tons per ha.) will be obtained from sowing material brought from the North. But if you take tubers from this crop and plant them the next year again in the same southern area, the new crop, as a rule, will be considerably smaller than the preceding one. If you plant material grown twice in the South (in two generations) the crop will drop to half or a third. When early varieties of potatoes were reproduced three or four years in the South (and in the South alone could they be cultivated) things reached such a pass that from a ton or somewhat over a ton of seed potatoes per hectare used in spring a harvest of somewhat less than a ton would be received, i.e., the crop would not even return the planting material expended. In practical farming this behaviour of potatoes in areas where the summers are hot was termed degeneration.

In order to maintain potato growing in the South, planting material had to be brought in annually from the more northern or the upland districts.

Before the revolution planned deliveries of planting material to the South were not feasible. For that reason the potato was almost unknown in the steppes of the Ukraine and the Northern Caucasus as well as in the cotton regions of Central Asia and the Transcaucasian republics. The urban food supply of potatoes was brought in from the North or the upland districts.

In our time, after the October Revolution, seed potatoes began to be delivered to the southern regions every year. But it was rather difficult to develop potato growing to any considerable extent in these regions by the use of outside seed material.

The rapid decline of southern potato crops when planting tubers bred under southern conditions was attributed by the old scientists to disease and not to a change in the nature (heredity) of the potato. For to admit a change in the nature of the potato when cultivated in the South would mean to admit a change in the heredity of an organism due to conditions of life, and it was not to the advantage of the bourgeois science of heredity to admit such a dependence. It does not recognize it to this day.

The swift and readily detected change in the behaviour of potato varieties when reproduced in the South was called by scientists a specific southern potato disease. But since the pathogenic, i.e., the contagious, element could not be discovered, this disease was included among the diseases caused by filterable viruses, and virus diseases are known to be infectious. In this connection it was the established rule that in the South seed potatoes were not to be planted at a distance of less than two or three kilometres from virus-infected potatoes, regardless of the percentage of infection.

Every year carloads of good planting material were brought to seed-potato farms in the South from the northern regions. It was planted in isolation, far away from degenerate potato fields. But there was not a single case of a seed-potato farm in the South producing even one ton of good, sound planting material for farms growing potatoes for consumption. Sound potatoes brought in from the North degenerated in the course of one or two years during the process of reproduction.

Little success attended also the efforts of plant breeders in the southern regions who for decades tried to produce a variety of potato that would resist degeneration, resist the so-called virus diseases.

True, by sowing seeds instead of tubers, plant breeders quite easily succeeded in obtaining tubers which when sown the following year produced a crop qualitatively and quantitatively superior to that of any variety brought in from the North. But three or four years later, when this new variety (new strain) had increased to, say, 20-30 c., it usually proved to be already degenerating and yielding small crops.

Today it can be said with certainty that people in the South suffered reverses in their efforts to grow potatoes because the accepted "scientific" explanation of the degeneration of potatoes in the South was incorrect, was at variance with the facts. *Scientists did not approach the potato from the angle of the interrelation of the organism and its environment. This was the basic mistake of science in explaining the question of potato degeneration in the South.*

When in our research work we decide particular questions of agrobiology we begin, as a rule, with an analysis of the interrelations of the given organisms with the environment that surrounds them. Such an analysis must, of course, not fail to take into account the various diseases and the conditions under which they develop.

Let us try, as far as possible, to recall and briefly recapitulate the course of our argumentation in 1933 when at the Odessa Institute of Selection and Genetics we for the first time undertook to solve the problem of combating potato degeneration in the South.

How are we to explain the fact that potatoes shipped south, for instance, from Gorky Region to Odessa Region, yield a good crop the first year but a considerably smaller crop the second year? Why, from what external causes does planting material deteriorate after reproduction in the South? After all, the preceding year, when it was brought south from the more northern regions, its breed qualities, i.e., its seed qualities, were good. It follows that here in the South, certain conditions must exist which differ from conditions elsewhere and which rapidly deteriorate the breed qualities of the potato, its seed qualities.

At first the idea may occur to one's mind that the length of the day is to blame. For in summer the days are shorter in our southern districts than in the northern districts, and scientists know that the length of the day frequently plays no small part in the development of plant organisms. One need only compare the results obtained by growing, for instance, certain varieties of millet under conditions of different lengths of the day for different plant behaviours to be easily observed. Plants which at the inception of their development were given even only 10 short days of 10-12 hours each (and correspondingly long nights) sharply reduced their vegetative period.

But the supposition that the length of the day has any bearing on the change in the property of the potato we are discussing must be rejected. In any district of the South the potato develops splendidly and does not deteriorate in breed if planted on foothills or in the uplands. Contrariwise, potatoes degenerate in the South if planted in valleys where the day lasts about as long as in the mountains. Hence, the length of the day is not a cause of the potato's degeneration.

Again one might assume that the soil was the most important factor in the degeneration phenomenon. We know that in the South the potato thrives better in vegetable gardens near brooks or on sandy soils, and there holds out somewhat longer, i.e., its degeneration is slighter than in the ordinary soils of fields. Yet in the more northern and upland regions with their diversity of soils the potato does not deteriorate as it does in hot regions. Hence it is clear that neither the soil nor, consequently, the nutriment of the plants in the direct sense was the chief cause of the degeneration of the potato.[1]

Only environmental differences in temperature explained, and that quite easily, why in the extreme South, for instance, in the Azerbaijan Soviet Socialist Republic, the potato could not hold out more than a year in the valleys, whereas 30-40 km. away, in the foothills and upland districts, it grew better than in the more northern districts, for instance, in the Ukraine (before the introduction of summer planting there).

It would be incorrect to assume that a high temperature has a bad effect on the potato plant during the entire process of its development. High temperatures occur alike in Moscow Region and in the southern highlands

[1] However, generally speaking, in a plant organism everything depends on food, in its wider sense, on metabolism, on the interconnection with the environment.

during certain periods of the potato's vegetation, yet, as is known, the potato grows well in these localities. Consequently, a high temperature does not run counter to the numerous biological requirements of the potato. It is therefore not at all a question of the action of high temperatures on all processes of development of the potato plant but a question of the action of high temperatures at particular moments in the development of its separate organs. The fairly large number of experiments performed by that time in the domain of what is known as the phasic development of plant organisms helped us to find our bearings in this question.

At that time we already knew that if you take, for instance, a seed-grown annual plant which can be propagated by cuttings, cut up its stem lengthwise and root the cuttings under identical conditions, it will be readily observed that plants derived from cuttings taken from the crown and a little below it will, as a rule, produce flowers earlier than plants derived from cuttings taken from the part of the stem near the roots. The latter cuttings will flower almost simultaneously with plants obtained from seeds sown when the cuttings were rooted. Plants obtained from cuttings near the stem's roots prove to be as young, as near to the commencement of individual life and as far from its end as organisms which start life from the seed. Conversely, cuttings taken from the top, i.e., above the part of the stem where there have been buds or flowers, yield buds and flowers early. Annual plants grown from such cuttings are at the very beginning near the end of their individual lives, near the formation of seeds.

Hence, we arrived at the conclusion that the cell tissue along the stem of a plant organism is qualitatively heterogeneous in the sense of phasic development, in the sense of the stages of individual life of the organism.

The older in growth a tissue is (the lower down the stem, i.e., the nearer the root it is), the younger is it in development, the more ways remain for the various complicated and largely still unstudied transformations and modifications through which plant organisms normally pass before completing their lives.

Conversely, the nearer to the top of the plant the cell tissue is—it usually is the youngest tissue—the fewer possibilities there are for vital transformations. The individual life, the general cycle of development, approaches its normal end, old age. When a plant starts from the seed, life begins anew (with a repetition of the phasic stages). In cases of propagation by cutting, tuber, etc., the life of the organism does not begin de novo. It is in direct form a continuation of the life of the previous organism, or, to be more exact, of the cell tissue which was taken for propagation.

This gave rise to the thought that the rapid degeneration of the potato in the South is, basically, a result of the phasic aging of the potato as a flowering seed plant. But how can this be proved?

In 1933 E. P. Melnik, a researcher of the Institute of Selection and Genetics, at our suggestion prepared cuttings from Ella variety potato stems. These cuttings produced plants which yielded tubers. The tubers were picked and kept until the spring of 1934, when they were planted. It was found that the tubers which in the experiment came from upper-stem cuttings produced approximately half as big a crop as the tubers derived from cuttings taken at the base of the stem, i.e., from the place where tubers usually form under the ground. The former yielded an average crop of 120 gr. per plant, the latter—250 gr. Tubers from top cuttings produced potato plants similar in appearance to those which in the South are called degenerated, while tubers from cuttings taken from the lowest parts of the stalk produced considerably better plants.

We now began to understand why the potato, a flowering annual plant, propagated year after year not by seeds but by tubers, in many localities neither degenerates nor ages in decades. The explanation is that year after year the tubers get their start from the phasically youngest cells, from cells of the lower, underground part of the stem. By that time it had already been ascertained that phasic changes in seed plants take place in the cells of the growing points of stems.

It was assumed that if high temperatures act sufficiently upon the germinating eyes of the tubers, i.e., on the growing points of the future stems, the tissue of these eyes will become decrepit and the tubers having such eyes will, when planted, yield degenerated plants. The stems will originate from decrepit eye tissue. The new tubers will likewise develop from decrepit tissue and therefore will possess little vitality, will be degenerate.

This phenomenon was tested experimentally in the following manner. Absolutely healthy, undegenerated tubers were taken and cut in half. Some of these halves were placed in a basement, the others in a thermostat at a temperature of 30°-40° C. and kept under these conditions for 25-30 days. Then they planted all the halves (from both the basement and the thermostat). The potato plants obtained from the basement halves were healthy while the thermostat halves were suffering from the southern disease.

When early-maturing varieties of potatoes are cultivated in hot southern districts the eyes on newly-formed tubers frequently begin to sprout even before the crop is picked. On the basis of all this it was assumed that the action of high temperatures on even slightly stimulated eyes, eyes which had just about started to sprout, was the principal cause of the rapid mass degeneration of potato planting material in the South.

Such, briefly, was the line of reasoning and the minor experimentation which in 1933 and 1934 enabled us to ascertain why potatoes quickly degenerate in the South and to map out the course to be taken in order to combat this phenomenon.

The question arose of what to do to prevent high temperatures from affecting the eyes of tubers of one year's crop intended to be used as planting material for the following year's crop.

Let the high temperatures act on the stems and foliage of the potato plant. The higher (to a certain extent) the temperature the faster is the progress of the life of the potato plant, the formation of the crop, the normal aging of the stem and leaves. All this constitutes the normal course of life of the potato plant. Only do not let the decrepit, aged parts be used the following spring as the basis for the new offspring, for planting purposes. High temperatures must not be allowed to affect for any great length of time the stimulated eyes of tubers from which the new plants will begin to grow after the planting.

Thus originated the idea, in 1933, of trying to plant seed potatoes in summer instead of in spring, so that the tubers might develop in the cooler autumn weather. On the other hand, since the young tubers obtained from such a crop would not be ripe by harvest time and would therefore fail for a long time to stimulate the sprouting of their eyes, they would not, in winter, be subject to aging or decrepitude under the rather warm conditions of the South. It has already been ascertained that as long as the eyes of the tubers are not stimulated to sprout they cannot be affected by high temperatures in the sense of changing the nature of the potato.

SUMMER PLANTING TESTS AT THE INSTITUTE OF SELECTION AND GENETICS AND ON COLLECTIVE FARMS

On July 6, 1933, at the Odessa Institute of Selection and Genetics, 0.25 ha. of land were planted with potatoes of the Ella variety after a crop of young potatoes had been gathered in that field early in the spring. The summer was rainy that year. The Institute therefore succeeded in obtaining potato sprouts on a section where one crop had already been raised, which in ordinary years cannot be done under southern conditions on unirrigated plots.

In the autumn the July planting of 0.25 ha. yielded approximately 2.5 tons of potatoes. The tubers of this crop weighed the usual 100-150 gr. on the average.

In the spring of 1934, 1.5 ha. were planted with tubers from the summer-planted potatoes. Part of the tubers were left for a second planting in the summer of 1934. Alongside of the potatoes planted in spring from the crop yielded by the summer planting were planted tubers of the Ella variety obtained from normal, spring-planted crop. It was difficult to detect any difference between these two plantations; nothing in particular distinguished the behaviour of the plants obtained from the tubers of summer-planted crop. When I recall the behaviour in 1934 of the plants obtained from the tubers of the first summer planting in 1933 I believe hardly anybody could have anticipated how effective, in the sense of improving the planting material, summer plantings would subsequently be under southern conditions.

One might think the behaviour of the plants in the above experiment of planting in the spring of 1934 tubers obtained from Ella plantations of the preceding summer and spring did not bear out our above-expounded assumption that summer plantings would stop the deterioration of potato planting material when reproduced in the South. It is difficult to say today whether in view of the results of this first experiment we would have ventured to propose that the collective farms plant potatoes in summer. I think we would have. What actually happened was the following. In the winter of 1933-34, i.e., before the above experiment was performed for the purpose of testing the seed quality of the planting material obtained from the 1933 summer plantation, A. D. Rodionov, a specialist of the Institute of Selection and Genetics (now its director), ordered a carload each of potatoes of the Epicure and the Early Rose varieties to be shipped from Gorky Region to 16 Odessa district collective farms for the purpose of testing the idea of planting potatoes in summer.

To plant potatoes in summer under the conditions of the arid south of the Ukrainian S.S.R. on unirrigated land necessitates the careful retention of the winter and spring moisture in the soil. To this end it was suggested to the collective farms that they leave the autumn-ploughed sections untilled and in early spring begin to till them in the way good bare fallow is tilled under the conditions obtaining in arid districts. The planting was done at the end of June, i.e., when the grain was being harvested. The summer of 1934 (particularly the first half) was hot and exceedingly dry. Many collective-farm members, men and women, as well as a number of agronomists, looked upon the proposal to sow potatoes in summer with distrust. That was quite a normal reaction. In these regions potatoes, as we have already stated, had been faring rather poorly even when planted early in spring. To plant them at the driest and hottest time of the year, when gathering in the grain, seemed of course impractical and uncalled for. Owing to the distrustful attitude to this proposition the collective-farm experimenters, who had undertaken to summer-plant 1-2 ha. of potatoes on each collective farm, encountered great difficulty in securing assistance for the proper preparation of the fields and for the planting. In the autumn, when picking the crop, all the doubts of the collective-farm members were dissipated. Despite the dry summer the potato tubers obtained from the summer-planted crop were on the whole of a size beyond all precedent in the South—400-500 gr. each. The tubers were considerably bigger than in our first summer-planting experiment, in 1933, which had not been performed on land left fallow since spring, but on a plot on which an early-spring potato crop had already been picked. I shall dwell below on the causes responsible for the considerably greater size of the tubers obtained in 1934 as compared with 1933. I shall only note now that we had not expected an increase in the size of the tubers as a result of the summer planting of potatoes.

Tubers which the collective-farm experimenters obtained from the summer plantings were planted in the spring of 1935 on the same collective farms. Alongside were planted potatoes obtained from the ordinary spring plantings.

The collective farmers performed these experiments not on small plots but on whole hectares, for the practical purpose of obtaining potato crops.

After the appearance of shoots a sharp difference was observed between the plants of the fields on which summer-planted tubers and ordinary spring-planted tubers had respectively been used. The early variety plants (Epicure and Early Rose) obtained from summer-planted tubers of southern reproduction were in no way inferior in vigour and healthiness to the plants of the same varieties obtained from tubers brought in from northern districts in the year they were planted, i.e., in the spring of 1935. Conversely, the plants of these same varieties obtained from tubers that had gone through one year's spring reproduction in the South were weak, degenerate. The fields under summer-planted tubers yielded crops about twice as big as the fields with spring-planted tubers of southern reproduction.

In the spring of the same year (1935) Ella variety potatoes were planted at the Institute of Selection and Genetics to compare the results of the new second-year summer reproduction with those of the ordinary spring reproduction. In 1935 the experiment with the Ella variety, after two years of summer reproduction, also showed good results for material obtained from summer-planted crops. It was found that only 8.8% of the plants obtained from the previous summer planting was degenerate. But the potatoes obtained from the former, spring-planted tubers contained 58.9% of degenerate plants.

How are we to explain the fact that the potatoes from the 1933 summer planting produced no difference in the quality of the planting material when planted in the spring of 1934 and that a difference was found only in 1935?

Today it is quite easy to explain this.

The Ella variety, being a plant of medium maturity, degenerates more slowly when reproduced in the South than early-maturing varieties. The material taken for the experiment was brought to the Institute from Kiev Region in 1931 and in 1933 did not yet have time to degenerate to any considerable extent. The summer of 1933 was cool and rainy. Therefore the spring plantings of that year likewise deteriorated the planting material only a little. As a result, in the spring planting of 1934 the plants derived from former summer- and spring-planted tubers proved to be almost alike.

At that time we suggested the summer-planting method for the sole purpose of stopping the degeneration of potato planting material in the South. We had as yet no idea of what is now well known to us, namely, that seed potatoes can be improved by summer planting.

Different results were obtained in experimenting with summer planting in 1934. It was a hot and dry year, particularly, as has already been stated, during the first half of the summer. Under these conditions the potato crop obtained from the spring planting showed great degeneration. But when planted in summer the potatoes not only did not degenerate or deteriorate but exhibited considerable improvement compared with what they had been when brought from the North.

In order to show how much better summer-planted seed potatoes are than those spring-planted under southern conditions I shall refer to an experiment performed by A. F. Kotov, a specialist of the Odessa Institute of Selection and Genetics.

In the autumn of 1935 he selected 300 of the best plants from the Early Rose summer-planted potato section. He did not mix the tubers taken from these plants but stored them separate in a basement till the spring of 1936 when half of the tubers obtained from each plant were planted in the field and half were left in the basement till the end of June. The latter were then planted in the field in beds alongside of the early-spring plantations. The beds were so arranged that the progeny of the tubers of each summer-planted plant, picked out in the autumn of 1935, grew next to the spring-planted progeny of the same plant. The total crop was harvested in the autumn, simultaneously. Until the summer planting the field had been kept in a clean friable state.

It could readily be noted that in the progeny of every plant selected in the autumn of 1935 the crop from the summer planting was bigger than from the spring planting. The tubers of the summer-planted part of the crop were several times as big as the tubers of the spring planting. The crop picked from each clone (the progeny of the separate plants) of the summer as well as the spring plantings was kept separately in a basement. In the spring of 1937 about half the tubers from the crop of the preceding year, both of the spring and the summer plantings of each clone, were planted for the purpose of comparing the quality of the seed potatoes obtained from the former spring and the former summer planting. The other half of the tubers from each clone was left for planting in summer. On the whole 4 variants were obtained from the progeny of each plant selected in 1935:

1. Potatoes spring-planted in 1936 and 1937.
2. Potatoes summer-planted in 1936 and spring-planted in 1937.
3. Potatoes spring-planted in 1936 and summer-planted in 1937.
4. Potatoes summer-planted in 1936 and 1937.

On examining the spring plantings of this experiment it became strikingly apparent that the plots of all the clones planted with tubers from the previous summer plantations were superior to the plots planted with tubers obtained from the previous spring plantings. I have already published the results of this experiment ("Collective-Farm Laboratories and Agronomic Science," in the journal *Yarovizatsia*, No. 5, 1937).[1] I shall therefore not deal with them in detail. I shall only point out that the crops of plants derived from tubers of prior summer planting are in many cases two or three times as big, and even more, as the crops picked from plants obtained from prior spring plantings. Thus, for instance, clone No. 59 from tubers

[1] See present volume, pp. 202-04.—*Ed.*

spring-planted in 1936 on being spring-planted in 1937 produced an average crop of 153 gr. per plant. Tubers of the same clone planted in the spring of 1937 alongside of the tubers mentioned above but which in 1936 had been obtained from a crop planted in summer and not in spring produced an average crop of 315 gr. per plant instead of 153 gr. Clone No. 232 from tubers spring-planted in 1936 on being planted in the spring of 1937 yielded 80 gr. per plant on the average, whereas the same clone derived from summer-planted tubers produced an average crop of 413 gr. per plant. There was not a single case of a crop of any one of the 300 different clones obtained by planting tubers reproduced in summer being smaller or only equal to the crop from plants derived from tubers of a prior spring planting.

The same results were observed in the variant of the 1937 experiment with summer instead of spring planting, the only difference being that in the case of summer planting a considerably bigger crop was obtained regardless of whether the tubers were planted in spring or in summer the preceding year.

From 1935 on an increasing number of collective farms (at first hundreds, then thousands and tens of thousands) in the southern regions of the Ukrainian S.S.R., and later also in the southern districts of the Russian Federation, began to participate in the practical farming experiment of planting potatoes in summer for the purpose of stopping potato degeneration under southern conditions.

The spring planting of potatoes on collective farms with tubers from crops planted the previous summer strongly influenced the South in favour of planting seed potatoes in summer.

After extensive practical and experimental testing it can be said that our above-mentioned assumption relative to the causes of potato degeneration under the conditions of heat prevailing in the South seemed to have been splendidly confirmed. I say "seemed."

Although the above assumption as to what causes potatoes to degenerate in the South is relatively true (it can be tested experimentally any year) the practical results of summer planting have shown this explanation to be narrow and one-sided. As a matter of fact, it has been palpably plain for several years now that the summer-planting of potatoes in the South is a method *not only of stopping the degeneration of potato planting material brought from northern districts for reproduction under the conditions prevailing in the southern districts but a method of considerably improving this planting material.* This, after all, does not follow from the above explanation. The entire explanation was built on a divorcement of the individual development (ontogenesis) of plants from the development of the breed (phylogenesis). This mistake, of major importance to science, was easily and quickly discovered in practical farming. The grain of truth in the assumption was made use of by the practical farmers, while practice supplied theory with a splendid series of facts that demonstrated the unity of ontogenesis and phylogenesis.

It was noticed that in the mass, with each new reproduction brought about by summer planting, the tubers increased in size. I shall point out at least the following general observation we made. Whereas in the first year (1934) that potatoes were summer-planted in collective farms tubers weighing 300-500 gr. could be found, in subsequent years some tubers weighed as much as 1 kg. and later 1.5 kg. There were rare cases of their weighing even 2 kg. This alone was sufficient to warrant the conclusion that if potatoes shipped to the South from northern districts are planted in summer, they not only do not deteriorate, do not degenerate, but may even improve the breed, i.e., their seed properties, from year to year.

It turned out that in summer planting the seed qualities of potatoes are as liable to change as in spring planting, however with this essential difference that when planted in spring in the South the potatoes deteriorated, degenerated with each reproduction, but when planted in summer their quality on the contrary improved with each reproduction.

It used to be generally accepted that if planting material of, for instance, the Early Rose variety obtained from Moscow Region crops and planting material of the same variety obtained from Odessa Region crops are planted under comparable conditions, the yield of the Moscow Region planting material will almost always be considerably greater than that of the Odessa Region planting material. Now a multiplicity of experimental data points the other way. Last year, too, in 1940, I. E. Glushchenko (a researcher of the Institute of Genetics, Academy of Sciences of the U.S.S.R.) in his experiments performed on a plot of land near Moscow obtained a potato crop of the Early Rose variety from southern tubers reproduced in summer at the Odessa Institute of Selection and Genetics amounting to 480.5 c. per ha., while under the same conditions the same variety of local origin (the Potato Institute, Moscow Region) produced a crop of 219.5 c. per ha.

All this argues in favour of the proposition that in the South planting potatoes in summer is a method of improving potato breeds and not merely of stopping their degeneration.

While finding a solution for the special and practically important problem we have discussed—the problem of combating the degeneration of seed potatoes under southern conditions—it seems to me there have been clearly revealed a number of regularities which in some measure or other have a bearing on numerous plant and, I think, also animal organisms. Thus, for instance, unimpeachable experimental proof was supplied on small and big plots of summer-planted potatoes in support of one of the most important and fundamental propositions of agrobiological science, which holds, in substance, that *the environmental conditions which are conducive to good development of particular organs of plants, for instance potato tubers, also improve the breed of plants in the same direction.*

Practice has shown that in the South seed potatoes obtained from summer plantings are considerably better in point of breeding qualities than those secured from spring plantings. That this breed improvement takes place as the result of good environmental conditions during the period of development of summer-planted tubers is borne out by the fact that tubers from summer-planted crops are on the whole considerably bigger than tubers of the same variety derived from plants planted in spring. And it goes without saying that big tubers are obtained as a result of good environmental conditions at the time they develop.

It must also be stressed that the fact that tubers of various sizes are taken for comparative plantings—bigger ones from summer planting and smaller ones from spring planting—is not the reason why the yield of seed potatoes of summer reproduction is considerably higher than that of planting material derived from spring reproduction. In the South, at the Odessa Institute of Selection and Genetics, numerous experiments were made in which the smallest tubers, weighing 10-20 gr.—tubers which fail to develop by the time the autumn frosts set in—were taken from a summer-planted crop to compare breed qualities. These tubers were planted alongside of the considerably bigger ones (50-100 gr.) derived from spring-planted southern crops. Nevertheless the yield of plants from tubers of summer-planted parentage was, as a rule (and in the case of early varieties in the South without exception), considerably higher than the yield of plants from tubers of spring-planted parentage. This goes to show that the breed of potato tubers derived from summer plantings is changing. If planted in summer good conditions are established for the development of potato tubers. Owing to this the tubers obtained are big and at the same time the potato breed is improved, changed in the direction of large-sized tuber production.

The proposition that efficient agrotechnique and zootechnique improve the quality of plant varieties and animal breeds, and that inefficient agrotechnique and zootechnique cause even the existing good plant varieties and animal breeds to deteriorate—a proposition generally recognized in practice and, contrarily, rejected by bourgeois genetics—is thus easily corroborated experimentally. This conception of the breed (heredity) of organisms naturally leads to the following practical conclusion: seed plots must be spared bad agrotechnique since bad agrotechnique deteriorates the breed of plants. Good agrotechnique must be taken to mean the creation of conditions that will facilitate the development of desired qualities and quantities of seed or planting material.

Why do good environmental conditions result in the South if potatoes are planted in summer; why do these conditions cause big potato tubers to develop and at the same time the potato breed to improve? In my opinion this phenomenon is to be explained as follows.

In the arid districts of the South it is absolutely necessary, in order to obtain potato shoots when planting between the end of June and the middle of July, to take a section of land deep-ploughed in autumn, to keep this land from early spring up to the time of planting like the best bare fallow, i.e., clear of weeds and with a loose upper layer of mulch. Under these conditions the supply of winter and spring moisture is well retained. Simultaneously, about the middle of summer, as is shown by the extensive data in possession of experimental stations, a great quantity of soluble nutritive substances accumulates on this land. The accumulation of nutrient nitrates alone amounts to 50-60 poods a ha.

When summer-planted in the South potatoes in a good and properly tilled field enjoy what I should call exceedingly good food conditions. During the latter half of summer—in the beginning of autumn when the temperature is no longer too high and a large quantity of easily assimilated nutritive substances is available—stems and leaves develop luxuriantly and in two or three weeks large tubers begin to develop.

In general we assert even today that the principal cause of potato degeneration in the South under spring planting is the action of high temperatures on the potato eyes that are being stimulated to sprout at the time they are under the vines, or after the harvest while in storage. *On the other hand, improvement of planting material derived from summer planting is caused by good food conditions, good conditions for the development of potato tubers.*

This explanation makes it intelligible why in the first year of our experiments with the summer planting of potatoes, in 1933, no improvement in the planting material was achieved. I have already said that in comparable plantings in the spring of 1934 we discovered no difference in the behaviour of the plants, whether they were obtained from tubers of the preceding year's spring planting or summer planting.

The medium-maturing Ella variety, when planted in the spring of the cool, rainy year 1933, did not degenerate, nor did summer planting improve its breed, because it was produced on a plot poor in assimilable nutrient substances, on land from which a potato crop planted in early spring had been harvested. This explains the lack of difference in the case adduced between seed potatoes obtained from spring and from summer planting.

The practice of planting potatoes in summer has brought up a question of profound interest to agrobiology: not only the body of an organism but also its nature, its heredity builds itself from food. Many Soviet investigators have substantially confirmed this by experiments, first, with potatoes and then with many other plants. In the given case I have in mind experiments in vegetative hybridization.

In his splendid teaching on mentors I. V. Michurin showed how the nature of plants can be changed by means of grafting. Until quite recently the high priests of science did not want to recognize this. Now when vegetative hybrids have been obtained literally in masses this can no longer be denied.

ENGELS AND CERTAIN PROBLEMS OF DARWINISM[1]

THE FUNDAMENTAL thing in Darwinism is the theory of natural and artificial selection. The essence of natural selection is that organisms which become adapted to life in a given environment survive, whereas those that do not, either perish or leave no offspring. Darwin attributed natural selection mainly to the struggle for existence arising from overpopulation.

No one will deny the fact that in the plant and animal kingdoms, as a rule, more young are produced than there is room for their life and development. Hence it is clear that a struggle for existence as a result of overpopulation does take place in nature. But it is not here that one must look for the principal motive forces of development in the organic world.

The founders of Marxism-Leninism attached very great value to Darwin's teaching. They frequently referred to the tremendous importance of Darwinism for science in general, and primarily for the materialist conception of living nature. At the same time it was Engels who quite rightly pointed out that it is impossible to sum up the multiformity of historical development and the complexity of life in the one formula: "struggle for existence."

Development in the organic world can be explained, as Engels showed, without the struggle for existence, although such struggle is, of course, frequently to be observed in nature.

In his *Dialectics of Nature*, Engels wrote the following noteworthy lines on this question:

"... even in regard to nature, it is not permissible one-sidedly to inscribe only 'struggle' on one's banner. But it is absolutely childish to desire to sum up the whole manifold wealth of historical evolution and complexity in the meagre and one-sided phrase of 'struggle for existence.' That says less than nothing."[2]

---

[1] Lecture delivered on November 28, 1940, at a meeting of the Department of History and Philosophy of the Academy of Sciences of the U.S.S.R. in commemoration of the hundred and twentieth anniversary of the birth of Frederick Engels.—*Ed.*

[2] F. Engels, *Dialectics of Nature*, New York 1940, p. 208.

Engels gave an objective appraisal of the part played by the struggle for existence in the evolution of the organic world. First of all he limited this struggle strictly to the struggle resulting from overpopulation. Struggle consequent on overpopulation does in fact occur at definite stages in the development of the plant and animal world. But at the same time he pointed out that "species alter, old ones die out and newly evolved ones take their place, *without* this overpopulation: *e.g.*, on the migration of animals and plants into new regions where new conditions of climate, soil, etc., are responsible for the alteration."[1]

Thus, Engels has in view here the variation of species where overpopulation does not exist; moreover, he points to the importance of *adaptational variation* in the evolution of organisms.

"If there the individuals which become adapted survive and develop into a new species by continually increasing adaptation, while the other, more stable individuals die away and finally die out, and with them the imperfect intermediate stages, then this can and does proceed *without any Malthusianism*, and if the latter should occur at all it makes no change in the process, at most it can accelerate it."[2] Alterations in the climatic, soil, etc., conditions can cause alterations of organisms. Organisms capable of change, of becoming adapted to altered conditions, survive and leave offspring. Those that are incapable of changing to the necessary extent in the process of adaptation die out.

We are of the opinion that these remarks of Engels' are of enormous importance to all who are fighting for creative Darwinism, for the Michurin theory in agrobiology.

Darwin's theory of evolution affords an excellent explanation of the way in which new organic forms arise—through natural selection in nature, and artificial selection in practical agriculture. He lucidly explained how changing organisms perfect themselves, how in nature the development of the organic world from a few original forms leads to a multitude of forms. But he himself dealt very little with the question of the *direct* causes of variation in organisms.

Engels wrote:

"... Darwin, when considering natural selection, leaves out of account the *causes* which have produced the variations in separate individuals, and deals in the first place with the way in which such individual variations gradually become the characteristics of a race, variety or species. To Darwin it was of less immediate importance to discover these causes—which up to the present are in part absolutely unknown, and in part can only be stated in quite general terms—than to establish a rational form according to which their effects are preserved and acquire permanent significance."

[1] *Ibid.*, p. 235.
[2] *Ibid.*

"...But once again, the man who gave the impetus to science to investigate how exactly these variations and differences arise is no other than Darwin."[1]

Darwin determined the direction in which agrobiologists must research, must assemble data in the domain of variation and discover the causes of this phenomenon. But in his day it was difficult to reveal the concrete causes of the variation of organisms. Science as yet lacked sufficient data for this. It had not yet reached the stage where it could solve such a problem.

It is of importance to agronomists and agriculturists that the concrete causes of variation in organisms be found. If this is done Darwinism will become so much the more effective in the hands of the agrobiologists.

For example. It is common knowledge that in agriculture the best plants and the best animals are kept for breeding purposes. In this way people are able to improve the varieties and breeds. This is, of course, the proper thing to do. But should one simply sit and wait for the random occasion when variations of advantage to man arise spontaneously in order to seize on them, to pick them out? Were this done, one would often have to wait a long and tedious time. Revolutionary creative Darwinism cannot reconcile itself to such passivity.

Ivan Vladimirovich Michurin, that great transformer of nature, not only realized the necessity of studying the causes of variation in organisms more than any other biologist, but also elaborated an excellent theory on this problem. He showed concretely what causes the variation of organisms and how Darwinism can be made an effective instrument in the hands of agrobiologists.

The ideas guiding the further elaboration of the question of the causes of variation in organisms, a question of such importance to agrobiological science, come from Engels. Moreover, Engels' works not only contain the general guiding ideas for a study of the variability of heredity, but they also give direct and concrete indications of the causes of alterations, of how they arise in organisms. For us biologists these suggestions are of the utmost value.

The question of the causes of variation in organisms has long been a disputed one among scientists. In order to solve this problem, which is of such importance both for theory and for practice, one must turn to the works of Engels.

One of the chapters of Engels' *Anti-Dühring* contains this passage:

"From the exchange of matter which takes place through nutrition and excretion as the essential function of albumen, and from its peculiar plasticity, proceed also all the other most simple characteristics of life..."[2]

---

[1] F. Engels, *Herr Eugen Dühring's Revolution in Science* (*Anti-Dühring*), Moscow 1947, p. 107.

[2] *Ibid*, p. 124.

He then goes on to enumerate these factors of life, mentioning irritability, which consists in the interaction between the protein and its nutriment, contractibility, and the capacity for growth, which, so he indicates, in the lowest forms includes propagation by fission.

From what Engels points out here, it follows that with an alteration in the process of assimilation and dissimilation the very properties of the organism change, among them heredity.

But many agrobiologists, particularly the Morganist geneticists, hold that such an assertion contradicts the actual facts that can be observed daily. Related organisms, as for instance plants, may live in different regions and varying environments for many generations. In these dissimilar environments their nutrition is different, but upon investigation it appears that their heredity remains the same. Despite differences in nutrition the nature of the organisms does not change.

Such facts are not isolated ones. Engels also was fully aware of them For instance, in *Anti-Dühring* he wrote:

"Species of grain change extremely slowly, and so the barley of today is almost the same as it was a century ago."[1]

At the same time it is clear that barley, whether it grows in different fields for only one generation or in the same field for a long succession of annual generations, will encounter a variety of external conditions in the course of its development. Nevertheless, despite the divergency and variability of its surrounding external conditions the heredity of barley has changed comparatively little.

But does this mean that organisms whose conditions of life call forth changes in their metabolism do not change?

No, it does not. The fairly slow change in species of grain indicated by Engels does not contradict the assertion that with a change in metabolism, i.e., with a change in the process of assimilation and dissimilation, organisms, their nature and heredity, change, too. This becomes quite clear in the light of the teaching of Michurin. I shall return to this extremely interesting question later, but for the time being let us continue the above-quoted passage from Engels. After pointing out the slight variability of species of grain, and in particular of barley, Engels goes on to say:

"But if we take an artificially cultivated ornamental plant, for example a dahlia or an orchid; if we treat the seed and the plant which grows from it as a gardener does, we get as the result of this negation of the negation not only more seeds, but also qualitatively better seeds, which produce more beautiful flowers, and each fresh repetition of this process, each repeated negation of the negation increases this improvement."[2]

In analyzing the problem of variation, Engels takes two groups of plants as illustrations: cereals and decorative plants (the dahlia and the

[1] *Ibid.*, p. 202.
[2] *Ibid.*

orchid). Each of these groups has its own specific features. Cereals, barley for instance, remain approximately the same as they were for nearly a hundred years, whereas decorative plants, being plastic, change incomparably more rapidly.

Would it be correct to hold that metabolism is not always the basic cause of variation, that it is not such for all plants, but only for some of them as, for instance, for the plastic, decorative plants? I believe that it would not. Michurinists now know full well that it is possible through experiment to make cereal, too, plastic, amenable to change. If cereal grain is nevertheless practically the same as it was a hundred years ago, this means that its metabolism is practically the same today as it was in the past century.

If you succeed in changing its metabolism, you will at the same time have changed its nature, its heredity; the cereals will have become plastic.

Engels' statement that all the other most simple factors of life, which include, of course, the variability of heredity, are derived from metabolism, has been brilliantly borne out by the teaching of I. V. Michurin with regard to mentors, to vegetative hybrids.

In uniting young plants of two different breeds by means of grafting, there takes place as it were a transmission of heritable properties from one component to the other. If the seeds of such grafts are planted, the offspring obtained will be exactly the same as are usually obtained from sexual hybridization. The result of vegetative hybridization is a blending, so to speak, of the heredity of both breeds.

There are hundreds of examples today where the grafting of two plant organisms of different varieties has resulted in the formation of a third, new, hybrid organism. In this case the hybrid is formed through the exchange of matter between the components of the grafting. The organism, the result of the coalescence of components belonging to two breeds, is obliged to obtain its nutriment through the exchange of plastic substances by the two breeds. Such a change in nutrition, in metabolism, leads to a change in the hereditary properties, as well.

There have already been a number of cases where through the exchange of plastic matter by means of grafting different varieties of tomatoes, the colour of the fruits, the structure of the racemes, the shape of the leaves and vines, the property of early or late maturity and many other properties and characters have been transmitted. In the seed progeny of such vegetative hybrids one frequently observes what in sexual hybridization is called segregation, i.e., heterogeneous offspring.

I have here before me live fruits obtained from tomatoes of the second seed generation of a vegetative hybrid. The plants that bore them were on display at the All-Union Agricultural Exhibition. These fruits are contained in three boxes.

A cutting of an Albino yellow-fruited tomato plant grafted on to a red-fruited tomato plant from Mexico bore red fruit instead of yellow. The seeds of this fruit were sown, and produced the first generation of hybrids. Continuing the experiment further, one fruit each—red, crimson and yellow—was taken from three tomato plants of this first generation. In one of the boxes I am showing you are the fruits of the plants obtained from the red tomato, in the second, the fruits of the crimson tomato, and in the third, of the yellow.

As you see, the majority of the offspring of the red tomato are red, the minority yellow and white. The offspring of the crimson tomato are mostly crimson and red, the minority yellow. The progeny of the yellow tomato are yellow and white, but some of the fruits, as you can see, have a reddish tint. We have taken the seeds of such reddish fruits and intend to continue the experiment further. It may be assumed that from these seeds we shall obtain plants some of which will bear red fruit.

Numerous experiments with vegetative hybrids fully confirm Engels' statement about the role of metabolism in the variability of the heredity of plant organisms.

In investigating the causes of variation, the agrobiologists must proceed from Engels' assessment of the role of metabolism in the development of organisms. Only if this is done is it possible to count on good scientific results in our work. By learning to understand the process of metabolism or, in other words, by learning to understand the details, the subtleties of the nutrition of organisms (taking nutrition in the broad sense of the word), we will be able to control plant organisms better and better.

The hereditary properties, i.e., the nature of organisms, can be changed only by changing the process of metabolism. For agrobiologists this is an extremely important thing, since it is imperative that they have a clear understanding of the laws governing the variation of organisms.

Every organism has its own heredity. The property of heredity is the property of every organism to require for its development relatively definite conditions of life.

If the organism is not afforded the conditions its breed requires, the conditions its heredity demands, it will not develop. At best it will remain quiescent, if it is able to live at all in these conditions so prejudicial to its development, and will not develop further. For example, take the wheat seeds stored in a granary. They are alive, capable of germination, and have a corresponding heredity. But the seeds do not have the necessary amount of moisture (there is usually sufficient warmth), and hence do not germinate.

The entire branch of agricultural science known as agrotechnique has been built up on the basis of satisfying the requirements of organisms under various conditions of development. Through experiment and observation people ascertain the conditions necessary for the nature of a given organism, i.e., for its heredity, in order that it may develop and that in doing so those organs, those of its parts which we harvest may develop to the greatest degree. We know that the cycle of individual development of an organism proceeds by stages. This may be confirmed if only by the fact that at different moments of its life a plant requires different conditions. Furthermore, at one and the same moment of the development of one and the same organism, but where different processes are taking place, where different organs of the plant are developing, different conditions are likewise required.

All this is well known today, at least in its general features, and, by skilfully controlling external conditions, it is possible to control the nature of plants. By altering the conditions, it is possible to alter the metabolism of the organism and thereby to change its nature.

Heredity, as a rule, is a conservative property. The conservatism of heredity manifests itself in the fact that the organism does not accept conditions that are not peculiar to its nature, that it waits for the appearance of conditions suitable to its nature. The conservatism of heredity explains the fact that the barley of today, as Engels pointed out, has changed very little from the barley of a century ago. The conservatism of heredity in plants and animals should not be regarded as a "bad" property or a "good" one. Both in nature and in agricultural practice the conservatism of heredity in organisms is a necessary property.

I have already given a number of examples illustrating this necessity. In August or at the beginning of September millions of hectares are sown to winter wheat. At this time the fields are warm and the winter wheat grows well, but shooting does not take place. In order that it may do so winter wheat must pass through the phases of vernalization. But this phase, through which the plant passes at the very beginning of its development, requires a low temperature, in accordance with the heredity of winter wheat. There is no such temperature in the fields in August and September. Therefore the process of vernalization does not take place. During this time all that can develop in winter wheat (the roots, the leaves) does so, but the process of vernalization does not. In the given case the organism waits for the conditions which it requires, in accordance with its nature. A month, a month and a half, or two months will pass (it varies with each particular case) until the autumn cold sets in, whereupon the process of vernalization commences. In this instance we see that the capacity to wait for the necessary conditions (which is present only by reason of the conservatism of heredity) is an indispensable property.

What would happen to the winter wheat if it did not possess this conservative heredity and failed to wait for the onset of cold weather for vernalization? It would pass through the phase of vernalization when it was still warm. But we know that winter wheat which has traversed the phase of vernalization and begun to develop straw perishes at the very first frost.

The conservative nature of heredity in organisms is what makes it possible in practical farming to have relatively stable varieties of plants and breeds of animals. In nature, too, it is thanks to the conservatism of heredity that we observe relatively constant plant and animal organisms.

But, as Engels pointed out, there are also organisms, like the dahlia and the orchid, for instance, as well as many other cultivated plants, which are pliable, plastic. If we subject these plants to artificial treatment, cultivate them better, the result will be not only a greater quantity of seeds but at the same time qualitatively improved seeds, which produce new plants with better flowers, etc. In these cases the breed of the plant has been altered by the treatment to such an extent that the effected change is actually visible. Such organisms possess what Michurin termed a destabilized heredity.

In recent years we have learned experimentally to produce cereals with destabilized heredity. If we take Michurin's teaching as a guide it is not difficult to learn how to destroy the conservatism of the property of heredity in plants and to produce plants such as Engels called plastic, plants that lend themselves to various changes.

How do plants possessing a destabilized nature, a destabilized heredity, behave?

Usually the organism with a conservative heredity does not accept, does not assimilate conditions that are not native to it. That is why it is difficult for it to change, difficult for it to adapt itself to new conditions. Organisms with a destabilized heredity, however, act differently. They have not yet elaborated a fully developed stability, they lack conservatism with regard to electing conditions for assimilation. All that they have is a tendency, a preference for assimilating certain conditions. If these conditions are lacking in the given environment, the organism having a destabilized heredity does not resist long, does not persist; the conditions which are not inherently assimilable by this organism "get in of themselves," as it were. The organism with a destabilized heredity assimilates the conditions which environ it with less discrimination, so to say, with greater appetite. A skilful experimenter can literally mould such organisms like clay and produce good, new and desirable races.

But how is it possible to produce organisms possessing a destabilized heredity; moreover, destabilized not in general but only with regard to some particular character? You all know that various processes take place in an organism simultaneously. Each of these processes requires its own particular conditions. That is why we cannot speak of heredity in general, but must speak of the heredity of a given property, of a given character.

Having set out to do away with the conservatism of heredity, one must first of all provide the organism with the conditions which are required by its heredity. In other words, it is necessary to begin by satisfying the organism.

But it is a well-known fact that if we furnish the organism with the conditions its heredity requires, then in the next generation (when the cycle is repeated) the organism will require the very same conditions. Hence the heredity will not be changed.

But it is likewise a well-known fact that if the conditions required by the heredity for a given process are not furnished, the process will not take place. It is this that constitutes the conservatism of heredity. The result is a sort of vicious circle. From this the formal geneticists arrive at the conclusion that the nature of organisms does not change under the influence of the conditions of life. Actually, however, there are moments when a developing organism is in such a state that, as it develops one or another process, it quite easily assimilates conditions that are not altogether native to it, even if it has a conservative heredity.

Experiments show that the organism must be deprived of the conditions required by its old heredity before the process which we wish to alter comes to an end. Before the end is reached (it is impossible to say how many days before—this must be established experimentally) the conditions should be altered; the old conditions should be removed and new ones furnished according to the tendency we wish the organism to develop. Here great skill is required on the part of the experimenter. At a definite moment he must remove the old conditions and substitute for them conditions that will promote the formation in the organism of the new and desired heredity.

After the conditions required by the old heredity are removed, and different, new conditions established, the process of assimilation can no longer culminate normally, in the usual way. Having been deprived of the old conditions, the organism at first seems to refuse to accept the new ones. But since the process has almost reached its culmination it usually continues, though slowly, to the end even under the new conditions.

We arrive at this conclusion on the basis of various experiments in converting winter cereals into spring cereals.

In these experiments, towards the end of the vernalization process, cold is replaced by warmth. Given a temperature of 0° C. another three or four days would have enabled the winter grain to complete its vernalization normally. But three or four days before the completion of vernalization, the winter cereals are treated with above-zero temperatures, namely, with the usual spring temperature. The organism begins to experience distress, as we say, because the new conditions do not suit it. The process is delayed, and for its completion not three to four days but ten or fifteen are now required.

Nevertheless it is completed. And as soon as the process of vernalization in the winter plant is completed, the plant will begin to change rapidly (literally under one's very eyes) in the spring field conditions. Instead of spreading on the ground, it will erect its leaves, shooting will begin, it will change colour, etc.

As a result of the treatment we have described here, the conservative character of winter grain is done away with. Once vernalization is completed under conditions not specific for winter grain (instead of the lower temperature prevailing in autumn and winter—at a higher, spring temperature), the next generation of such organisms will no longer require cold to go through this process.

But does this mean that the form produced has become a spring form, i.e., a form that now requires warmth for the process of vernalization to take place? Does it mean that the vernalization process can take place in the new form only when it is warm and that wherever it is sown it will remain a spring form? No, by far not. The propensity of these organisms to pass through the phase of vernalization at spring temperature exists, but this is only a *propensity* and nothing more. In the spring the temperature varies, not only in different years, but also on different days. Even in the course of a single day in spring, the temperature fluctuates: in the morning there is one temperature, at noon another, and in the evening a third. Under such changing conditions an organism with a conservative heredity selects, literally seizes on, the conditions it needs, rejecting the ones not required by it. But the organism that has been rid of the old heredity, and in which the new heredity has not yet taken firm hold, is incapable of waiting. In this organism the process can proceed under at least considerably more varied conditions, if not under all conditions. If such organisms are left to their own fate, to chance, many of them usually turn out to be deformed, abnormal in structure. The training of organisms with a destabilized heredity is of the utmost importance, and success here depends on the skill of the experimenter.

By proceeding along the path described, we have not only learned how to break down the old, conservative heredity and to obtain plastic organisms, but also how to give the organism a new, strong heredity.

A. A. Avakian and a number of other experimenters have already produced spring forms from many varieties of winter wheat, and vice versa, winter forms from spring varieties. These experiments have fully proved in our opinion that cereals can also be made plastic, that they, too, can be forced to change just as orchids do, which Engels pointed out.

The experiments in altering heredity, in remoulding the nature of plants, are extremely interesting from the standpoint of theory, for they irrefutably prove the absolute correctness of Engels' assertion that all the simplest manifestations of an organism's activity must be derived from the metabolic process. By altering the process of metabolism, all the properties of the organism are altered, including its heredity. And this fact is of importance to us from the standpoint of practice, as well.

Allow me to relate briefly just how we want to verify the above-mentioned thesis that organisms possessing a so-called destabilized heredity are actually of great value for practical agriculture, for our work with cereals. Since organisms with destabilized heredity can easily assimilate diverse conditions of life, then by creating the necessary conditions we can mould, create new requisite forms from these organisms.

Taking as our guide the above-mentioned theoretical propositions on the role of metabolism in the life of organisms, theoretical propositions based on Engels' statements, we have undertaken to produce in short order new forms of winter wheat that will not fear the Siberian frosts.

In my opinion this is a realistic vision. It is based on the scientific theses of which I have already spoken. After all, in the open, snowless Siberian steppes with their bitter frosts there are plants, even if they be weeds or wild plants, that are capable of withstanding the severe cold. These organisms developed in the past, develop now and will develop in the future spontaneously. But if such organisms can be created spontaneously why should it not be possible to create them at will?

To achieve this only one thing is necessary, namely, to understand how this takes place spontaneously in nature. And this can be understood only on the basis of Engels' teaching on the dialectics of nature.

The achievements of our contemporary Soviet biological science and Michurin's teaching have made it possible for us to produce also pliant, plastic cereal grains. But once we are able to produce pliable, destabilized organisms, it becomes possible to treat wheat in the same way as people have treated, say, orchids for hundreds of years. In the latter case, if the organism was raised under good conditions, then after each generation a better variety than the preceding was obtained, a variety capable of producing more beautiful flowers. The very same laws operate for any other group of plants.

A. A. Avakian has now sown in the experimental fields of the Lenin Academy of Agricultural Sciences in Gorki Leninskiye scores of varieties of wheat with a destabilized heredity, including hundreds of variants of each variety, the destabilization being quite thorough. Destabilization was provoked during the phase of vernalization, but it was not only the property of vernalization that was affected by this process. At the same time the colour of the leaves, the shape of the tufts, etc., were also altered.

In pure-line strains, the spikes all resemble one another, and irrespective of where one sows Ukrainka wheat, for instance, this variety can always be distinguished from any other. When, however, a number of spikes were taken after the heredity had been destabilized, and the seeds sown in a single row, the progeny varied even more than in the second generation of sexual hybrids. The cause of this marked variation in most of the offspring of these spikes is quite clear to us, namely, the fact that organisms with a destabilized heredity are incapable of waiting for some definite condition, and assimilate the conditions existing at the given moment.

There are hopes that, given a skilful approach to this work, the forms we need may be produced from organisms possessing a destabilized heredity. Nor is it so difficult to display skill in treating organisms with a destabilized heredity. All that is necessary is to create such conditions for the organisms that they will assimilate not just anything that comes their way, but primarily whatever we want the organism to have a propensity for.

We have sown plants with a destabilized heredity in various environments, in Gorki Leninskiye, not far from Moscow, and in localities with a more rigorous climate—the trans-Volga region and various places in Siberia. Undoubtedly errors are possible in our work. But we believe that these errors will be noticed in the course of the work and eliminated. We are confident of a favourable result. We have already received information from a Barnaul plant breeder, Comrade Kondratenko, that winter wheat obtained in Odessa from spring wheat Erythrospermum 1160 has this year already withstood a frost of —29° C. without snow. I do not know what further results this wheat will show, but this fact is in itself most noteworthy. Spring wheat of the Erythrospermum 1160 variety is unable to resist frosts of 5°-10° C. below zero, while after this grain has been converted into winter wheat we see that it can withstand 29° C. below zero. Of course I do not know what will happen to the wheat in December and January. As the temperature in Barnaul can drop to 50° C. below zero it remains to be seen how our wheat will resist such cold. But leaving the unknown ultimate result aside for the time being, we can already see that destabilization of heredity plus plant training yield remarkable results.

We shall be more successful in controlling organisms if we abandon the idea that natural selection is based on the struggle for existence resulting from overpopulation. It is not the struggle for existence that underlies natural selection. Natural selection is based on alteration of metabolism, on the process of adaptation.

Nor should the process of adaptation going on within the organism be confused with the usefulness, the purposiveness of the given adaptation for the harmonious functioning of the organism as a whole, as well as of the interconnection of the given organism with its environment. The harmony of organisms and their purposiveness are created solely by natural selection in nature, and by artificial selection if we are dealing with the harmony of cultivated organisms, and their fitness to satisfy our requirements.

No one, I believe, will deny the fact that agrobiology has made considerable progress in the Soviet Union. We can justly pride ourselves on Michurin's theory. Academician V. R. Williams contributed a great deal to the theory of agrobiology in the sphere of agronomy, as has Academician M. F. Ivanov in the field of animal husbandry. Science has spread to the masses; many young pupils of Michurin are growing up or have already attained maturity. Nevertheless, agrobiology is still, unfortunately, one of the most backward branches, one of the most backward departments of natural science in its entirety.

On the other hand, it is not mere boasting when I say, as I do so often, that in this field of biology there is no need for us to look up to the West and America. In this field of science we are in the lead. Those countries do not have the Michurin theory, nor is this so for lack of gifted scientists. There have been and there are outstanding people there, too, but neither the West nor America has the conditions which we possess for bringing out talent and developing it. In America, for instance, there was Burbank, a biologist of genius, yet there is no Burbank theory although there might have been.

But for all this, agrobiology's achievements in our country still fall short of what they ought to be. I am confident that the time is not far off when we agrobiologists will no longer have cause to say or be told that our branch of science is one of the most backward of all natural science. I am confident that in the near future agrobiology will overtake the other branches of natural science, overtake them despite the fact that these other branches are likewise progressing rapidly in our country.

We have all the prerequisites for the closest bond between theory and practice. And this bond is the prime and indispensable prerequisite for the successful solution of theoretical problems. Not only is it possible to combine theory with practice in the Soviet Union, but practice in our country impels science to do so.

There is no need to talk of the material aspect of the matter. Pavlov long ago stated that in our country science is very generously endowed. But of most importance is the fact that we have an advanced revolutionary theory, Marxism-Leninism. We have the opportunity for constant study, the opportunity to master this science of sciences.

WHAT IS MICHURIN GENETICS?[1]

A FULL exposition of Michurin genetics would require more than one lecture—a whole series of them. I shall therefore confine myself to the fundamental principles of Michurin genetics, of the Michurin conception of the variability of heredity in plant organisms. At the same time I shall point out the difference between Michurin genetics and anti-Michurin, i.e., Morganist-Mendelist, genetics.

Michurin's teaching proceeds from the premise that the conditions of life of organisms influence the quality of changes in their breed, the quality of changes in the genotype.

Morganist genetics asserts the opposite. It asserts that the conditions of life, the conditions of the environment, do not influence the quality of changes in the breed, in the nature of the organism. The conditions of life, the Morganists claim, influence only changes in the body (soma).

Both Michurin and Morganist genetics cite facts and experiments in support of their conclusions.

Take, for instance, the well-known fact that the body of an organism may sharply deviate from normal as a result of the influence of particular conditions of life, but that the offspring of this organism, if provided with the ordinary conditions, grow up normally. From such facts the Morganists arrive at the conclusion that a change in the body of an organism resulting from the action of the external conditions does not effect a change in its breed, i.e., its genotype. The genotype, the breed, cannot be changed, they say, by changing the body of the organism.

The Michurin theory, as has already been stated, takes the opposite stand.

Any change in the breed, i.e., the genotype, is always due to the action of the environmental conditions upon the organism's body. A change in the genotype is adequate, corresponds, to the action of the environmental conditions. Such are the conclusions drawn by the Michurinists.

[1] Revised stenographic report of a public lecture delivered on October 15, 1940, at the Leningrad State University.—*Ed.*

23—742

To support their conclusions the Michurinists produce facts, facts which show that breeds depend on the conditions of life, that breeds change under the influence of environmental conditions, and changes in a manner adequate to the action of these conditions.

Consequently, both Michurinists and Morganists take facts as their point of departure yet arrive at opposite conclusions from which two opposite, mutually exclusive trends in science have come into existence.

How is that?

The trouble is that certain facts only seem to be such.

Indeed, let us, to begin with, see what facts the Morganists adduce in their attempt to prove that the action of environmental conditions upon the organism finds no reflection in a change of its genotype, and that if there is such a reflection it is not adequate to, not in correspondence with, such action.

Everybody knows that an ordinary mutilation, such as results from the loss of an organ or part of it, is not inherited. The removal of separate parts or of organs of plants or animals does not affect the progeny of the operated organisms. The Morganists make use of these facts as cogent proof of the absence of any connection between the organism's conditions of life and changes in the genotype, i.e., in heredity. They make frequent reference to experiments performed by Weismann. Mice had their tails cut off for several generations but the operated animals invariably reproduced mice with tails. This led to the conclusion that mutilations are not transmissible. Is this true? Yes. The Michurinists never disputed such facts. It is general knowledge that mice's tails take no part in the procreation and development of the progeny. The relation of parents' tails to their offspring is a very, very distant one.

Thus, the examples given of mechanical injuries or mutilations have no bearing on the question of whether heredity, or the genotype, as it is called, changes when a change occurs in the living body due to the action of the conditions of life.

Examples like the one analyzed merely go to show that the removal of a number of organs or parts of an organism does not deprive the rest of the living body of its principal property, the reproductive capacity, the possibility of restoring the whole organism from part of it. After all, it is well known that in many cases whole organisms may easily be obtained or restored from the cuttings of plants. In certain species of animals it is also possible to restore whole organisms from separate organs. This disposes of one series of facts adduced by the Morganists for the purpose of proving that the action of the external conditions on the body of an organism does not change the heredity of the organism.

Let us now pass on to another series of facts utilized by the Morganists.

They take two like plants or animals and make them different by different feeding.

Depending upon the conditions of life some plant organisms are able to develop ten and even a hundred times as intensively as others, though at the outset, before the experiment started, they were alike. Thus, for instance, we had a tuft of millet, grown under bad conditions. Its roots, stem, foliage and panicle together with the seeds weighed less than a gram while another tuft, produced from the seeds of the same millet but under different, good conditions weighed over 900 gr., i.e., about a thousand times as much as the first. Both tufts differed, of course, not only in weight, i.e., quantitatively, but also in respect to quality. Incidentally, the fact that the quality as well as the quantity of a crop depends on the life conditions of the plant is generally known and requires no proof. The same may be said of the productivity of animal organisms.

But what offspring are obtained from two organisms grown under different conditions if the seeds from the two organisms are grown under equal conditions? The progeny is almost identical.

These generally known facts the Mendelist-Morganists undertook to ascertain by "exact" experimentation, whereupon they concluded that the conditions of life change only the body of the organism but not its breed, its genotype. There now logically follows a second conclusion: the organism consists of the ordinary body (the soma), which is dependent upon the conditions of life, plus something else, some other substance, which is independent of the conditions of life. The latter "substance" the Morganists call hereditary substance.

The conclusions set forth were arrived at by these scientists on the basis of what are claimed to be "exact" experiments, on the basis of "facts."

They take, for instance, two seeds from one spike of wheat. One seed they grow under conditions of good nutrition, the other under conditions of bad nutrition. The seeds of the sharply differing plants obtained are then likewise sown but under equal conditions. These seeds produce identical, indistinguishable plants. After the experiment the above conclusions were drawn.

The performance of such experiments requires no scientific knowledge, but to be able to understand them one must at least have common sense. Before they asserted, in reliance upon the experiments described, that the conditions of life change the organism's body but leave its breed untouched, it certainly should have dawned on them that such a conclusion contradicts all that we see in nature round about us. Who does not know that proper conditions improve the breed of plants and animals? Good strains are obtained only with good agrotechnique. Under bad conditions even the existing good strains are turned into inferior, bad strains; they run wild.

However, instead of correctly interpreting the data obtained in their experiments the Morganists draw the general conclusion that heredity is independent of changes in the body of the organism.

Likewise the Morganists bolster up their claim that bodily changes do not influence changes in the breed, i.e., in the genotype, by referring to experiments in the use and disuse of particular organs of animals. Granted that exercise promotes the development of the muscles of an organ and that its muscles become stronger. But these changes are not preserved in the progeny, just as in the experiments described above the changes that took place in the body of the organism induced by good nutrition were not preserved.

On the basis of such and analogous experiments the Morganists drew the conclusion that no change in the body is transmitted by inheritance. But since each property and each character is acquired by the organism in the process of its development, there followed the general conclusion that "acquired" characters, i.e., such as the organism becomes possessed of in the process of its development, are not reflected in the heredity, are not transmitted hereditarily.

What the Morganists noticed with regard to changes in the chromosomes was different. These changes, as a rule, are transmitted to the daughter cells, i.e., are transmitted by inheritance. The reservation must here be made that so far investigators have been able to observe in the cell nucleus only the morphological changes visible under the microscope, analogous to the changes, visible to the naked eye, in the external form of the organism or of its parts. The facts obtained by the Morganists in their experiments indicate that the morphological changes which take place in the chromosome of the cell are transmitted by heredity.

Relying upon such facts and comparing them with what has been set forth above some scientists have reached the conclusion that the organism as a whole and each of its cells consists of two fundamentally different substances: the body (soma) and a "hereditary substance"—chromosomes and genes. A change in the former, i.e., the body (soma), does not influence heredity; however a change in the "hereditary substance," i.e., in the chromosomes, alters the heredity. This is the basis on which the so-called chromosome theory of heredity rests.

In order to show that what I have set forth is not an exaggeration I shall here quote a few excerpts from Sinnott and Dunn's textbook which our institutions of higher learning mainly go by.

On page 29 we read the following:

"The point at issue is whether the particular response made by an ancestor (as, for example, the greater growth reaction of plants to good soil or of animals to food, heat, or light) has any determining effect on the responses to be made by its descendants. Does the fact that the parent has responded in a way which leads to a visible variation predetermine the response of the offspring, or make it any easier for the offspring to develop this same character in the absence of the same stimulus?"[1]

---

1 Edmund W. Sinnott and L. Dunn, *Principles of Genetics*, 3rd ed., New York and London 1939, p. 29.

The argument is thus introduced by a question which is followed by an exposition of rather numerous experimental data meant to prove that only bodily changes depend upon the conditions of life and that the quality of changes in the breed does not depend on the conditions of life. The authors then draw the following conclusion:

"Does the response of the parent to a particular set of stimuli predispose the descendants to make the same or a similar response? In the language of Weismann, are somatic acquirements transferred as such to the germ cells? This is the question, 'Are acquired characters inherited?' in its narrowest sense.... It is evident, however, from the representative experiments and observations cited that the probabilities are decidedly against the inheritance of such somatic acquirements as disease, mutilations, the effects of poisons, malnutrition, variation in food, light, and temperature, and the changes produced by use, disuse, and training."[1]

Such is the fundamental tenet from which Morganist genetics proceeds.

A diametrically opposite tenet was advanced by Michurin: No changes in the genotype can be independent of the conditions of life. Literally nothing can occur in an organism out of connection with its environment.

After all, what distinguishes living from nonliving matter? The fact that living matter always necessarily requires—this is intrinsic in living substances—relatively definite conditions of life. The better a nonliving body is preserved from the influences of its environment the longer it remains as it is. But a living body is unable to continue living if even for a fraction of a second it is isolated from the environmental conditions which it needs for existence, for metabolism. Hence it is inconceivable that changes in an organism should take place independently of the conditions of life.

But what about those experiments from which the conclusion was drawn that not only mutilations but even sharp differences in the nutrition of parental forms fail to produce changes in the progeny while even morphological changes in chromosomes are, as a rule, transmitted to the progeny? Is this not proof, say the Morganists, that the chromosomes are a substance sui generis, differing in its essence from the substance of the ordinary body (the soma)?

And they go on to say that inasmuch as material continuity from one generation to another pertains solely to the hereditary substance the following conclusion must be drawn: What (they opine) would otherwise be inexplicable facts, such as obtaining, from a given pair of organisms or from a single plant (if it is self-pollinating), diversified offspring with the characters and properties of the ancestors that had been taken for the cross, can only be explained as separation of the chromosomes at a certain moment in the development of the cell. It is well known that in hybrid progeny the characters frequently begin to dissociate, so to say. Such a dissociation

---

[1] W. Sinnott and L. Dunn, *Principles of Genetics*, 1925, p. 300.

of characters, the Morganists say, can only be explained by segregation of the chromosomes.

Heredity, they reason further, is transmitted from generation to generation through the chromosomes, and it is only by way of exception that separate properties of characters can be transmitted through cell plasma.

All these ratiocinations seem to be logical throughout, but their logic is formal. By making use of such logic investigators involuntarily arrive at the conclusion that nothing new is formed in the world. Flesh has its origin in like flesh, blood is derived from like blood, cells from like cells, chromosomes from like chromosomes, chromosome granules (genes) from like granules.

According to this logic the transmission of hereditary properties is conceived of as a simple transfer of certain "hereditary substance" granules in the chromosomes from ancestor to progeny.

With regard to alterations of the "hereditary substance," usually called mutations, scientists of the Morgan trend frankly say (both in their textbooks and lectures) that they do not know what causes mutations, why the "hereditary substance" changes.

It would not be half so bad if after this admission the Morganists went no further. However they do not stop here but declare: Though we do not know what mutations are caused by we do know that mutations, i.e., changes in the "hereditary substance," are not brought about by assimilation or dissimilation; we know that a change in nutrition, in the conditions of the organism's life, will not change its nature.

This makes it altogether bad. For on the basis of this proposition the conclusion is drawn (drawn openly a few years ago, at present in veiled form) that only illiterate, blind practical farmers can instal good zootechnique or agrotechnique on stockbreeding farms and seed-growing fields. People, however, armed with the science of genetics ought to realize that this will never do. The function of the animals bred on pedigree stock farms or of the plants grown for seed is not to produce milk, wool, fat, flax, grain, etc., but to transmit to their offspring their genotype, their breed. And a breed, say the Morganists, does not change—neither improves nor deteriorates—because of good or bad conditions of keeping. If alterations (mutations) do occur, they do not correspond to the action of the conditions of life. Therefore, what is to be gained by establishing a system of good tending of herds on the pedigree stock farms and good cultivation of seed-growing fields?

Under the conditions prevailing in our country such conclusions sound simply preposterous. Hundreds of thousands of Stakhanovites would laugh their proponents out of court. Our Morganist disciples do not therefore write in this fashion nowadays. But only a few years ago such conclusions were voiced openly. They follow inevitably from the foundations of Morganism, which is preached to this day to our students from university chairs.

Here we must return to our starting point.

The conclusions drawn by the Morganist "scientists" are contrary to common sense and good practical farming, but the Morganist geneticists may retort: we proceed from facts, from experiments, don't you see. After all, we have demonstrated experimentally that changes which take place in an organism due to the conditions of life are not reflected in the heredity while changes in the chromosomes are hereditarily transmitted to the progeny of the cells. Such is the testimony of facts.

That the Morganists are in possession of facts is beyond dispute. But when these facts are analyzed from the standpoint of Michurin genetics they inevitably acquire a new meaning. An analysis, made in the Michurin way, of the facts adduced by the Morganists is bound to lead us to diametrically opposite conclusions, which argue in favour of Michurin genetics and against the principles of Morganism.

We shall not analyze here "mutilations" and various mechanical dissections of plants, for instance, for purposes of propagation since they have no bearing on the question of the "transmission of acquired characters" by inheritance, as has been shown above.

But how about facts concerning the influence of conditions of life (nutriment in the broad sense of the word) upon changes in heredity? Let us take, for instance, the example referred to above in which, owing to different nutritive conditions, one tuft of millet was a thousand times as big as another of the same variety while the heredities of these quite different tufts were equalized by growing a new generation of these plants from their seeds.

Mention has already been made of how the Morganists interpret such cases. They say that the body (soma) of one organism may differ greatly from the body of another organism while their heredities remain unchanged. It is the same for both plants. The conditions of life do not change the heredity.

But this argument loses sight of the fact that the plants differed greatly from each other while their seeds did not. The seeds of these plants ordinarily differ far less than the plants themselves. And the embryos in the seeds differ still less. Hence the plants obtained from such embryos usually show little difference. When small plots are sown for experimental purposes, and so much the more when plants are grown in pots or small beds, these differences may not be noticed, because they are screened by the fact that the experiments are performed under varying conditions.

The experimenting geneticist who has raised a small number of plants arrives at the conclusion that the heredity of a big, well-nourished plant and of a small plant puny from malnutrition remains the same, unchanged. As a matter of fact he has observed a big difference between the parent tufts but for his test has taken those of their parts (in the case in question, their seeds) which differ very little. On the basis of this the Morganist experimenter asserts that heredity does not change with a change in the body.

I may be asked: What then, if not the seeds, is to be taken to elucidate the question of whether the heredity has changed in consequence of a change in the body of the organism?

To this question the following may be said in reply: True enough, in practical farming millet is only seed-sown. But in experiments performed for the purpose of elucidating the general question of whether heredity changes when the body changes, the altered body of the organism must be taken and its heredity ascertained. In the case in question corresponding rooted cuttings should have been taken from the millet tufts and grown under identical conditions. I believe that in such an experiment the difference between the heredity of the cuttings from the one tuft and the other would be more telling than if seed-grown progenies were compared. For the Morganists know perfectly well that different parts of one and the same organism and even cells lying next to each other may have different heredities (somatic mutations).

We consider experiments in which the least altered parts of the organism are taken to elucidate the question of whether the heredity changes when the body is altered by the conditions of life to be improperly conducted, unscientific. No wonder the conclusions drawn from them are absolutely wrong and are contrary to practical experience. To practical gardeners it is a well-known fact that it is not immaterial what cuttings, from apple or pear trees, for example, are taken for propagation purposes. It is not a matter of indifference whether for bud grafting you take buds from shoots of excessive vegetative growth or from normal growing shoots (and not all growing shoots by far are of equal value for propagation purposes even if they all grew on the same tree).

We know that fruit trees obtained, for instance, by bud grafting from shoots of the mother tree that manifest excessive vegetative growth are also relatively vegetating and for a long time do not bear fruit.

Reference may also be made to the example of changes in potato breeds (improvement and deterioration) depending upon whether in the South the planting material is planted in spring or summer. Many are aware of the fact that if in the southern districts tubers of an early variety of potatoes from a summer-planted crop and tubers from a spring-planted crop are planted under like conditions the crop obtained from the former will be 2-3 times the size of the crop from the latter.

Can it be said in these cases that the heredity did not change upon a change in the body due to altered conditions of life? Obviously not. We likewise know that if in southern districts potatoes are planted in summer the conditions created for the development of the tubers are considerably better than when they are planted in spring. Accordingly, the breed, the heredity, of the tubers becomes better, more productive, if they are taken from summer-planted crops than the heredity of tubers from crops planted in spring in the same districts.

All these illustrations go to prove the one thing, that heredity is a property of the living body and that this property changes only when there is a change in the body.

How is it, then, that in organisms which sharply differ from each other because of different conditions of life (such as nutriment) the heredity of the seeds rather frequently exhibits but minor differences?

The heredity of seeds gathered from sharply differing millet tufts differs little because the seeds themselves, the bodies of the embryos of these seeds, likewise differ little. In the example we have cited the millet plants which sharply differ (qualitatively and quantitatively) because of different conditions of life produced embryos in their seeds that differed little because in a scantly nourished vegetable organism not all organs and not all cells of the organs starve in equal measure. Whenever a plant starves its nutriment is directed primarily toward building those cells from which in the final analysis the sex cells and later also the embryo are formed.

This explains why plants sharply differing from each other on account of their conditions of life frequently produce seeds that differ little in heredity. This does not mean, however, that when there is a change in the body the heredity does not change. But it does mean that different cells of one and the same organism depart from the norm in different ways in their development.

Usually everything in an organism tends toward a minimum deviation in the development of the bodies of those cells, those organs, which are meant for reproduction, for the propagation of the race.

In the light of Michurin's general biological doctrine the instances, cited in the textbooks of the Morganist geneticists, of the noninheritance of acquired characters, as they are called, are thus given a different interpretation. In substance all these facts have no bearing on the Morganist conclusion, drawn in reliance upon them, that a change in the living body (soma) does not influence the alteration of heredity.

The Morganist cytogeneticists' treatment of the morphological variation of chromosomes observed by them is likewise wrong, unscientific. These facts, the simplest and easiest to ascertain, are unimpeachable evidence that simultaneously with a change in the body, with its deviation from the norm in the process of development, the heredity changes, and in the same direction.

The Morganists, who proceed from these facts, have declared that chromosomes cannot be classified as parts of the ordinary body, that they consist of a special substance, special with regard to their hereditary properties, and that this substance is fundamentally different from that of the organism's body. Hence, the Morganists arrived at the conclusion, as has already been stated, that the organism and every one of its cells consist of an ordinary body (soma) and of a hereditary substance, the chromosomes. According to the Michurin's theory, on the other hand, the organism consists solely of body, and of nothing else besides this living body with all its properties. No special hereditary substance exists, either in the organism or in its cells.

What every living body and every particle of it does possess is breed, heredity. The chromosomes are not special hereditary substance but ordinary body, part of the cell, performing some biological function but, of course, not the function of an organ of heredity. An organism can and does have various organs, including organs of reproduction, but it cannot and does not have an organ of heredity. To look for an organ of heredity in an organism is tantamount to looking in it for an organ of life.

All facts relating to changes in heredity and connected with changes in the chromosomes militate not for but against the chromosome theory of heredity, which avers that a change in the living body (soma) does not entail changes in the hereditary properties.

Indeed, how many facts have been accumulated (and not by just anybody but by the Morganists themselves) which go to show that due to the action of environmental conditions any morphological changes in an organ or organella of the body, namely, the chromosomes, are hereditarily transmitted, and rather exactly, too. Chromosome changes acquired in the process of the individual development of a cell or organism are, as a rule, always heritably transmitted to the daughter cells. Does this not betoken the transmission of acquired characters by inheritance? Does this not betoken change in the breed corresponding to the action of the external environment on the preceding generation? Are not variation of the chromosomes and the heritable transmission of these changes a refutation of the chromosome theory of heredity?

In brief, the Morganists are in possession of facts; however they are not pro-Morganist but anti-Morganist facts.

It must be emphasized that the possibility of inheriting properties that appear in the process of an organism's development is far from being confined to morphological changes. To convince oneself that with regard to inheritability the chromosomes are ordinary living body (and not special organs of heredity or a special hereditary substance) one must turn to Michurin's teaching on mentors, on vegetative hybrids.

Vegetative hybridization is positive proof that the heritable properties of two organisms can combine in one without a transmission of chromosomes, as well as without a transmission of the protoplasms of the cells. The heritable properties of two plant organisms can be united in a third organism experimentally, by a skilful approach, by grafting. The union is here effected by means of an exchange of substances. And such an exchange does not necessarily involve an exchange of chromosomes or an exchange of protoplasms as such. The theoretic aspect of this question was splendidly developed by I. V. Michurin. When the metabolism of a developing new young organism changes, its breed changes too. And when you graft a cutting from one breed on to the crown of another breed, metabolic relations are established between the two components after they coalesce. As a result, an obviously hybrid organism is often obtained in which both breeds of the initial forms, correspondingly remoulded, are united. And all this takes place solely through mutual nutrition, the organism of the one breed being fed substances produced by the organism of the other.

Changes in breed consequent upon such metabolism as is conducive to the formation of a new breed differ in nowise from breed changes obtained in the hybridization of two organisms by sexual means, i.e., by crossbreeding.

Like the offspring of sexual hybrids the offspring of vegetative hybrids are not infrequently heterogeneous, to a greater or smaller extent, or are homogeneous.

As in sexual so in vegetative hybridization hybrids are obtained possessing properties like the one or the other parent or intermediate between them. Initially dominant characters may become recessive and vice versa. In other words, all the forms of behaviour of offspring that occur in sexual hybridization are encountered also in vegetative hybridization.

To illustrate what I have said I shall analyze a case in which a vegetative hybrid was obtained by a cross between the white- and large-fruited Albino variety of tomato and a red- and small-fruited Mexican tomato No. 353. I already referred to this case repeatedly and refer to it today once more not because there are no other cases of vegetative hybridization but because I exercised special control over this experiment, this combination. After all, an experimenter performs experiments primarily for himself but draws conclusions for the public. It is therefore necessary that he pick out those experiments which convince him beyond peradventure of the correctness of his conclusions, experiments which dissipate all doubts in his mind as to the possibility of error.

I have taken personal charge of some specimens of hybrids of the combination indicated. In this experiment I did everything myself, including even the picking of the seeds from the fruits and sowing them in the field, entrusting the technique of this business to no one in order to make sure that there would be no confusion.

I demonstrated this combination at last year's conference on questions of genetics organized by the editors of the journal *Pod znamenem Marxisma*, and every three or four months it furnishes me with additional proof of the correctness of Michurin's understanding of the laws of genetics.

What was the gist of this experiment? A cutting of a white-fruited variety of the Albino was grafted on a red-fruited Mexican form of tomato. The grafted Albino cutting fed on the juices of the red-fruited breed. As a result the graft did not produce white but red fruit, which I displayed at the above conference.

Most of the plants grown from the seeds of this red fruit bore red and crimson fruits; a minority bore yellow or whitish-yellow fruits.

For the subsequent experiments I took seeds from one red, one crimson and one whitish-yellow fruit. In the spring of 1940 the seeds of these fruits were sown in the greenhouse of the Institute of Genetics of the Academy of Sciences of the U.S.S.R. Most of the plants of this crop were shipped to the All-Union Agricultural Exhibition for display. The majority of the offspring of the red fruit were found to be red, the minority yellow. The majority of the crimson fruit offspring were crimson or red and the minority yellow. The most interesting fact was that while the offspring of the whitish-yellow fruit, i.e., the fruit having a recessive colouring, were mostly white or yellow, a few plants bore fruits almost red in colour, i.e., a reversion apparently to red-fruitedness directly on the graft. The seeds of these fruits have not yet been sown. I believe it will be possible to obtain a few plants with red fruits from them.

This fact, as well as all other facts of vegetative hybridization, denotes that hereditary properties (in the case in question, colour) can be transmitted not only without a transmission of chromosomes but even without transmitting protoplasm. After all, the latter did not diffuse, did not penetrate, from the stock to the scion.

Thus, hereditary properties can be transmitted by a metabolic exchange of matter between two organisms of different breeds. This is the first conclusion. The second conclusion that can be drawn is that in vegetative hybrids seed offspring may possess a great variety of properties with apparently relative reversion to former parental forms, i.e., "segregation," as it is called, may take place, which, in the opinion of the Morganists, can only be explained by a dissociation of the chromosomes which the hybrid received from the parental forms when they were crossed.

From the experimental data accumulated hitherto it is clear that vegetative hybrids can be obtained not only with respect to the property of colour. There is no hereditary property which under definite conditions cannot be transmitted by vegetative means from one plant to another. For example, the experiments with tomatoes conducted by A. A. Avakian and his assistants have already yielded results which show that vegetative hybridization can produce hereditary changes in size, form of fruit, shape of leaves, form of cluster, number of chambers in fruits, etc.; association of properties, called linkage by the Morganists, may be observed.

Guided by his own teaching I. V. Michurin literally moulded new forms of plants that man needed. When a variety of apples, pears, sour cherries, etc., deviated from the plan he had mapped out he engrafted upon that variety the desirable properties it lacked. Thus, for instance, he determined to produce a fruit of a variety that would stand lengthy winter storage and at the same time possess a delicate aroma, fine flavour, etc. He often failed to obtain such a combination of properties in the varieties he bred. He would then graft upon the crown of a young variety cuttings from other varieties possessing the properties which it itself lacked, and the newly-bred variety would thus acquire new and economically useful properties. Necessary characters were added to the new variety and unnecessary characters eliminated.

Michurin explained theoretically why and under what conditions vegetative hybrids could be obtained. His theory has equipped the Darwinist agrobiologists for the practical work of changing the nature of plants. His writings have helped many of his followers during the last two or three years to obtain hundreds of vegetative hybrids of the most diverse plants and varieties. Grafting has been practised for thousands of years. Yet why were vegetative hybrids not obtained before Michurin? How did science happen to skip them? For it is a fact that only such geniuses in the field of biology as Darwin, Timiryazev and a few others of prominence recognized that it was possible for vegetative hybrids to exist and be produced.

In the recent past vegetative hybrids were still a rare and ununderstood phenomenon. The theory of Morganism does not permit of the possibility of obtaining vegetative hybrids. Hence, most of the facts relative to vegetative hybridization were classified as errors or were even decried as jugglery or imposition. Such of the facts as could not possibly be described as juggled (as, for instance, those pertaining to *Cytisus Adami,* a vegetative hybrid well known to Darwin) were denominated "chimeras." This designation alone suffices to show that vegetative hybrids were regarded as something unnatural, abnormal, monstrous.

It took Michurin to end the ignorance prevailing in this field of biology. The theory he evolved on this question is profound, veracious and at the same time intelligible to all. It can be grasped by every person having anything to do not only with books but also with living nature, with plants.

Vegetative hybridization shows plainly that metabolism, the change in nutrition that takes place in grafting, results in a change of not only the organism's body but also its breed, its genotype. Moreover, the breed does not only change. What takes place, as a rule, is a transmission of hereditary properties from one organism to another. Frequently that type of heredity is obtained which Timiryazev, in application to hybrids produced by sexual conjugation, called dual heredity, when an organism is formed possessing the heredity of both breeds. We have already seen how under the influence of a red-fruited stock red fruit grows on a yellow-fruited cutting. The seeds of this fruit yield plants bearing red and yellow fruits. In the second seed generation we again have segregation, diversity. This bears out the dual heredity of hybrid organisms. The properties of this heredity may, as it were, dissociate, segregate.

Similarity in the behaviour of offspring of sexual and vegetative hybrids compels us to revise the old notions concerning the essence of the sexual process. The behaviour of seed offspring of vegetative hybrids denotes that hereditary properties inhere not only in chromosomes and protoplasm but also in plastic substances.

By altering the metabolism, by altering the nutrition of organisms, changes may be produced not only in their bodies but also in their hereditary properties. Such is the conclusion to be drawn from Michurin's teaching on vegetative hybrids and from the fact that such hybrids are obtained by Michurin and the Michurinists.

This conclusion arms the experimenting biologist. Guided by this conclusion we can be bolder in our approach to the work of changing the hereditary properties of plants in the direction we desire. After all, if a change in nourishment by means of grafting can lead to an alteration of the breed, the properties of heredity may be altered also by different methods of changing nutrition, of changing the conditions of life.

And indeed, quite a number of experiments made along this line show that the action of environmental conditions may effect major or minor changes in the breed of an organism. Besides, a change in the nature of a living body is always adequate to the action of the external conditions upon this nature. However, in order to obtain the requisite effect from the said action one must know in what state the organism should be at the time it is subjected to treatment, and when and to what conditions it should be subjected. Before acting upon the organism one must find out as much as possible about the organism's biology and ascertain its specific features.

At the same time it should be remembered that while success in altering breeds by conditions of training depends upon skill, skill depends on desire. If an experimenter is imbued with a strong desire not to produce the proper effect he will find a way of satisfying his desire. That is exactly what has been happening in the case of a number of Mendelist-Morganists in their experiments on vegetative hybridization and the directed alteration of breed properties by bringing conditions of life to bear on them. Yet experience has shown that where there's a will there's a way, and a not very difficult one, of finding a state of the organism and effective means that will render it possible to make directed changes in heredity through the conditions of training, to change the properties of the breed.

I shall now pass on to an exposition of a few pertinent experiments with which I am well acquainted.

As we know, there is winter wheat and spring wheat, winter rye and spring rye, and other winter and spring plants. They all possess the hereditary property of the winter habit or the spring habit.

I don't think anybody will dispute the conservative nature of these hereditary properties. In agriculture winter varieties are sown for a great number of years and they remain winter plants; spring varieties are also cultivated for a great many years and hereditarily they retain their spring habit.

But, as experiments have shown, when the plant is approached skilfully, in the Michurin way, its conservative hereditary property of the winter habit or the spring habit can be eliminated in its entirety by a change in the organism's conditions of life. All that this requires is a knowledge of what alterations are to be made in the organism's conditions of life and in what state the organism must be.

If plants of the winter variety, for example wheat, are vernalized before they are sown, when they are still in the form of embryos that have only just started to germinate, i.e., if the organisms are supplied with the vernalization conditions demanded by their genotype which include a low temperature (approximately 0° C.), then upon sowing in spring they will bear the same summer. But the seeds of these plants will retain the winter habit of the preceding generation. This will be repeated in the following generations. They will demand similar conditions, i.e., their heredity with regard to the winter habit will remain practically unchanged. Under these conditions each organism produces a generation of like individuals because the organism possesses a conservative heredity. And the property of conservative heredity consists in this: that an organism requires definite conditions of life, actively takes from its environment such conditions as are peculiar to its breed and does not take such as its breed does not require, even when the conditions that fit its racial requirements are lacking.

It is well known, for instance, that the vernalization of winter plants requires cold. Winter plants sown in August may develop well, produce leaves and roots but they will not go through the phase of vernalization and will not develop stems and spikes. Winter plants wait, so to say, to be chilled and the process of development in the direction of vernalization will set in only when they are chilled.

The more capable an organism is of holding out, of not picking conditions unsuited to its breed, the stronger and more conservative its heredity.

The conservatism inherent in heredity should not be underestimated. It may be observed quite frequently that a plant will refuse to accept conditions unsuited for it and will perish without having completed its development.

All agrotechnique is based on ways and means of best and most completely "humouring" the conservative nature of the heredity of plant organisms. Such "humouring" is necessary in order that organisms may develop in accordance with their natures, and particularly in order that that character, that organ, may best develop which is most important to us in the plant in question. In vain certain Mendelist-Morganists are beginning to demonstrate, under pressure of the criticism the Michurinists level at them, that heredity after all is not so stable, that mutations are frequent, etc. What kind of heredity is it if it has no conservatism? What sort of variety and what sort of breed is it if it tends to develop whichever way the wind blows? Michurinists recognize the heredity of organisms such as it is in nature: strong, conservative, unpliant.

This conservatism of the organisms' heredity is a frequent impediment to practical farming, which demands that this heredity be changed, be adapted to particular conditions.

Methods of effecting such changes in the nature of organisms have been worked out and pointed out by Michurin. Michurin changed the breed of organisms by good and proper training. Here good, skilful training must be taken to mean not only "patting on the back." Sometimes you have to go "against the grain." If a plant is only "humoured" from generation to generation its breed may improve, to be sure, but only a little at a time. It will get better, change in the direction given by the training, the selection. You know full well that Darwin evolved his theory precisely out of a host of similar facts with which practical farmers had long been acquainted, and offered a splendid explanation of the evolution of organisms in nature by natural selection.

"Humouring" an organism in order to develop organs useful to us is a sure method of improving breed, but it is a slow, gradual improvement.

We have now gained knowledge of methods of changing breeds more rapidly, by means of training. When you know at what moment in the development of an organism it must not be humoured but, on the contrary, be given conditions alien to its nature, the old hereditary properties may be shattered. Sometimes they are wholly eliminated. The organism will no longer possess the hereditary properties it formerly possessed. The heredity established in the preceding generations will be destroyed. The further task will then arise of selecting the conditions of training, thereby causing the plant to deviate more and more in the direction intended and thus inducing new requirements, a new heredity, in a few generations.

I shall now pass on to an exposition of the results of a number of similar experiments in the rebuilding of heredity.

With a view to altering its hereditary winter property, winter wheat undergoing vernalization was given at the end of the process (5-6 days before its completion) the temperature that prevails in the fields in spring instead of the cold which it requires. What happened? Under normal conditions, in cold, the process of vernalization would have been completed in 5-6 days. Under the changed conditions the process was delayed and ended only in 10, 20 or 25 days. But all the same it went on to the end, as could easily be seen by the external appearance of the plant. As soon as the vernalization process is completed the plant begins to develop further, changing accordingly, inasmuch as the conditions required for the development of the subsequent processes usually exist in the field in spring.

The seeds of these plants are frequently found to have already lost their conservative hereditary winter property and if skilfully trained will possess the hereditary spring property in the subsequent generations.

Today there is no winter variety of wheat, rye or barley that in two or three generations could not be made to produce kilograms of seeds possessing spring heredity. Likewise there is no spring variety of cereal that cannot be converted into a winter variety.

The principal object of experimenting in the alteration of winter plants into spring plants and vice versa was to test the assumption that the nature of a plant organism can be altered by environmental action during its development. It was shown that the breed changes responsively, in accordance with the action. This means that if the plant is treated with cold the breed will change in the direction of requiring cold. If the plant is acted upon by warmth the breed will change in the direction of requiring warmth.

In these conversion experiments (winter habit to spring habit and vice versa) the heredity of the organism was counteracted, was not given what its old nature required. But, as has already been pointed out, an organism does not usually include in its development conditions that are not congruous with its nature. It was therefore necessary to probe experimentally for the moment when it would be easiest for the organisms to begin to include such conditions. This moment arrives at the end of vernalization. It is then that the organism assimilates what before it did not ordinarily assimilate.

There are grounds for asserting that this law applies to any hereditary property. If an experimenter wants to change the heredity of any particular property or character of an organism he must study the conditions required by the old heredity and ensure the existence of these conditions; but at the end of the process in question he must change these conditions and substitute those in accordance with which he wants to create a new hereditary property.

But here a fact was discovered which is of great interest to the experimenter, to science (and at the same time of great importance to the practical farmer). We say it is exceedingly easy to alter the old heredity and even to eliminate it. One need only, as was said above, provide a winter variety with warmth instead of with cold at the conclusion of the process of vernalization and there will be an end to its hereditary winter habit. But this does not mean that spring forms are obtained forthwith. The old heredity can be swiftly eliminated when the organism is not furnished the conditions out of which the body of the organism built itself by means of assimilation and dissimilation in past generations. But once other conditions are provided and in consequence another body is built up, hereditary properties that existed previously are no longer there. The old heredity has, so to speak, been exploded.

Instead of requiring cold such an organism develops a tendency to require warmth. It is only a tendency, not yet a want. The old heredity was strong, conservative. Even if the necessary conditions do not exist the organism in consequence of its conservatism will not accept the conditions it does not want. The new heredity, still young, behaves quite differently in such cases. If in the concrete case we are analyzing the temperature which the organism has a tendency to require is not forthcoming, the latter will not hold out. Its heredity will deviate in the direction in which the environmental conditions existing at that particular moment will incline it.

If an organism possessing a still unstable, young heredity is approached skilfully, is provided with the conditions which it has a propensity to assimilate, we shall have a situation of which Darwin said: If conditions favour the development of the given tendency it will develop, grow strong and become fixed; if the conditions are contrary, it will not develop.

The old, strong heredity compels the organism to hold out, but a heredity which has not yet become strong does not possess this quality. Organisms possessing heredities not firmly established are able to develop also under conditions for which they have no propensity. After all, they do not yet have a real, strong heredity or fixed requirements. This, as Michurin said, is a destabilized heredity.

Some scientists, on observing the inviability of mutations under experimental conditions, reached the conclusion that it was very unlikely for evolution to proceed by way of new formations. But frequently the trouble is caused by inability to manage an organism with a destabilized, unestablished heredity.

Yet such organisms are of great value in plant-breeding work. They are amenable to changes in their heredities; skilful training will make them yield fine varieties.

What, for instance, were the results of the effort to remake spring wheat into winter wheat and of the experiments to convert spring barley into winter barley?

It was found that winter wheat and winter barley could be obtained from the spring forms. This for one thing; for another thing, that the winter forms obtained from the spring forms proved quite frost-resistant.

In a varietal trial at the Odessa Institute of Selection and Genetics held in the spring of 1940 it was found that the winter form of wheat obtained from a spring form was more frost-resistant than any other wheat originating from the Ukrainian steppes. Although the winter barley obtained from spring Pallidum 032 suffered severely in the fields of the Institute from the harsh winter of 1939/40, it was damaged less than the standard winter varieties of barley.

The explanation for this is that the new winter forms of wheat and barley had not yet consolidated their heredity and during the relatively severe winter changed and tended toward greater frost resistance.

In the autumn of 1940 we shipped small quantities of the winter wheat obtained from the spring form to some plant-breeding stations in Siberia for sowing. We hope that the cold Siberian winter will bring out the hereditary frost-resistant property of this wheat still more. The wheat will increase its frost resistance and we assume that we shall succeed in finding among these forms those needed to cope with the severe wintering conditions in Siberia.

Cereals with destabilized, unestablished heredity must be sown from generation to generation under increasingly severe conditions in respect of cold. These organisms will then become frost-resistant in ever-increasing measure. At the same time their heredity will become fixed. The more years, the more generations pass before the heredity becomes fixed, the better. In these cases it will be possible to shift the heredity gradually toward greater frost resistance.

If one proceeds from Michurin's teaching it is not difficult to obtain plant organisms with destabilized heredity. They can be produced both by skilful training and by crossing suitable forms selected in the Michurin way. After obtaining such forms observation of their behaviour will make it easy to see in the phenomenon of form-building much that is new and of interest and that will not in the least fit into the framework of formal Morganist genetics.

Permit me to refer to one of the experiments performed by A. A. Avakian at Gorki Leninskiye (Experiment Base of the Lenin Academy of Agricultural Sciences).

In the spring of 1938 several varieties of winter wheat incompletely vernalized in varying degrees were sown at the Odessa Institute of Selection and Genetics for the purpose of converting winter forms into spring forms. The seeds of the eared variants were sown in the spring of 1939 in the fields of the Academy's Experiment Base near Moscow without presowing vernalization.

The plants of a number of variants were found to be spring forms and came into ear. Almost all the plants in the beds survived and were green in autumn. The plants which had eared also continued to grow. Knowing what destabilized heredity meant, that it meant switching a plant organism off its track, as they say, we decided not to replough the plot in the autumn of 1939 but to let the plants winter. We assumed that winter would set the plant organisms with destabilized heredity going in the other direction, toward winter-hardiness. The grass that had fully grown in autumn passed the winter splendidly. In the spring of 1940 it was found that some variants, especially of the Krymka variety, had in many respects lost their cultivated appearance. They had a dark blue coating and thin spikes. Diversity could readily be observed among plants of such varieties as Kooperatorka, Hostianum 0237 and Krymka. There were awnless spikes among awned varieties. Some varieties had plants whose spikes differed in colour. Thus, for instance, in Ferrugineum 1316/2 (which originated at the Kirovabad station, Azerbaijan) almost all the plants on the plot had changed from red spikes to spikes of another colour, i. e., had been converted into other varieties, most of them into Erythrospermum.

In the autumn of 1940 A. A. Avakian sowed seeds of the winter wheats indicated. Where in the experiment seeds of individual spikes were sown, it was easy to observe even in the sprouts the differences between the offspring of the individual spikes within the particular variety. Some of the seeds of the different variants of winter-wheat varieties harvested from this experiment were shipped for sowing in the autumn of 1940 to several Siberian stations.

A few more facts, to which attention was paid only thanks to Michurin's teaching. We now have in our possession numerous data from various sources which indicate that winter forms are obtained from spring forms quite rapidly and easily in nature. These facts compel us to approach investigation work in a new way; for instance, the search for the winter-hardy forms of wheat we need for various districts of Siberia. Let me cite an example.

In the wastelands of Siberia and on its roads, even far from any wheat field, isolated wheat plants may be found. One will readily arrive at the conclusion that this wheat originates from spring-wheat windfalls, for in many of these districts winter wheat is not sown.

M. I. Sekisov, a member of the Michurin Collective Farm, in Barnaul District, accidentally came across such windfall wheat in the summer of 1938. He was gathering wild-growing fodder plant seeds in a waste field and discovered wheat. He must have been convinced that what he had found was spring wheat, for where was winter wheat to come from here? He therefore did not sow the seeds in autumn 1938 but in spring 1939. The majority of the plants eared, though they were behind time. Some failed to ear and behaved like winter plants.

Michurin's theory made it possible to understand and explain this fact.

I found out about the wheat Sekisov had encountered from plant breeder Kondratenko of the Barnaul Plant-Breeding Station. It was sown in spring and not in autumn. But it had to be sown without fail in autumn, otherwise the good nature had done in the course of several years might easily have been spoiled. In the autumn of 1939, when I went into this matter thoroughly, it became clear to me that the wheat Sekisov had found was derived from spring-grain windfalls. The question then arises of why a great many of the plants in Sekisov's spring-planted crop were found to be of the spring variety and only a few of the winter variety.

Sekisov obtained his wheat from windfalls of spring wheat. The seeds of the windfalls had to winter in the field. It is rather difficult to suppose that spring wheat seeds could lie in the ground during the winter without swelling. The plants that grew from the seeds of the windfall produced seeds in their turn. After shattering, in autumn, they produced shoots which also passed the winter. This went on for several years in succession. As far as I could ascertain this stretch of land had not been ploughed for about nine years. In general it may be assumed that in this wheat the old hereditary spring habit was broken but the new winter habit had not yet established itself. Therefore Sekisov's crop produced a diversity of plants: a majority of spring forms, i.e., forms that eared when sown in spring, and a minority of winter forms.

I keep constant track of Sekisov's wheat. I am certain that it will produce a good, frost-resistant variety for the Barnaul steppes.

In the summer of 1940, on our way to Omsk, we also picked up on a country road quite a number of windfall spring-wheat plants. Seeds of this wheat and a small number of Sekisov wheat seeds were sown in the autumn of 1940 at several places.

* * *

At the beginning of my address I gave warning that it was impossible to expound even the principal propositions of the Michurin theory, or of Michurin genetics, in the course of two or three hours.

I. V. Michurin produced an integral theory. He dealt with literally every question of biology in its application to agronomy, to practical agriculture. The Michurin theory is creative Darwinism in agrobiology.

The more one studies this science the more one understands, appreciates and cherishes it. Each day greater sections of collective farmers, agronomists and men of science come under the influence of Michurin's teaching. And what does it signify that ever broader strata are mastering a particular science? It signifies that that science is marching onward with gigantic strides.

Michurin's teaching—creative Darwinism in agrobiology—is making tremendous progress in our country. Nowhere else in the world have biological problems been solved as swiftly and as profoundly as in the Soviet Union.

K. A. TIMIRYAZEV AND THE TASKS OF OUR AGROBIOLOGY[1]

UNDER the difficult conditions of a backward, despotic tsarist Russia, Kliment Arkadievich Timiryazev paved the way for the theory of development of the organic world, the theory of Darwinism. No one made so valuable a contribution to the further elaboration and popularization of Darwin's teaching as Timiryazev. Love of learning, the aspiration to reach the summits of knowledge, a correct generalization of facts and the elucidation of natural laws on the basis of which much becomes "possible and foreseeable," coupled with undisguised hatred of all that is reactionary, a principled struggle against everything that impedes the advancement of science, abhorrence of pseudo science—these are the qualities that characterized Timiryazev the scientist.

Timiryazev's works serve the Soviet agrobiologist as a manual of thorough universal instruction in his search of the most effective methods of developing theory as an aid to producing bigger crops and increasing the productivity of labour in collective and state farms.

Unity of theory and practice is the true road taken by Soviet science. To work for science and write for the people, i.e., in popular style, was Timiryazev's motto and aim in life. These profound words reveal the innermost secret of real scientific work. They are the compass, the lodestar, by which one must be guided in choosing the object of one's scientific thought and work.

K. A. Timiryazev displayed great perseverance in his struggle to democratize science, to enlist large sections of the working people in scientific work. Here in the Soviet Union his ideas have been carried into life. Nowhere else do such wide strata of the population participate in scientific endeavour. Our science is a mass science. A single thread leads from the research laboratories, studies and greenhouses to the experimental fields and the collective- and state-farm experimenters. The Party and the Government have created all the conditions necessary for the development of science. Science and scientists are held in high esteem by us and given the

[1] Address delivered at the celebration meeting of the Academy of Sciences of the U.S.S.R. held on June 3, 1943, at the Moscow House of Scientists in honour of the 100th anniversary of K. A. Timiryazev's birth.—*Ed.*

utmost care and protection, so as to enable them to devote themselves to fruitful scientific activity.

In our country the pursuit of science is not a private but a public matter and therefore a special responsibility rests upon scientific workers today, amidst the Great Patriotic War against the German fascists, the physical stranglers of everything progressive that mankind has created.

The Soviet people place all their strength and knowledge at the service of our country—its defence, its liberty, its national and civil rights. With unparalleled heroism and intrepidity the Red Army is annihilating the Hitlerites, is defending the sacred Soviet soil.

Supplying the front and rear with provisions and agricultural raw materials is one of the most important conditions upon which the defeat and destruction of Hitlerism hinges. Hence it is the sacred duty of Soviet scientists to place the agronomic and biological sciences at the service of the collective and state farms.

Even in normal times "the naturalist begins to feel himself in his element," as Kliment Arkadievich wrote, at the mere mention of the word "crop";[1] and particularly these days agrobiologists ought not for a moment to let "crops" escape their attention. In one of his lectures Timiryazev said: "... Some questions always arouse keen interest and are not affected by vogues. The problem of one's daily bread is one of them."[2] Everyone sees clearly that the food problem is of particularly vital interest during the Patriotic War.

K. A. Timiryazev wrote: "Beyond all doubt the plant is the object upon which the agriculturist must concentrate his activity; hence all his knowledge should be focussed upon this object."

Our scientific activity is directed toward the study of the requirements of plants concerning conditions of life, the study of how plant organisms respond to the action of environmental conditions. A knowledge of the requirements of plant organisms and of their response to the action of surrounding conditions enables us to take practical measures in the most diverse directions for the purpose of increasing crops and augmenting harvest totals.

"Working for science and writing for the people" makes it imperative to elaborate such theoretical questions and conduct the elaboration of them in such a way as to prevent divorcement from life, from practice.

The science of agrobiology more than any other requires an all-sided approach, an all-sided comparison of the most diverse conditions in order to give each experiment a proper setting, to carry it out, as they say, in pure form and draw the right conclusions, coordinating them with all the conditions of the plant's life and with the conditions which exist or will exist in the farming practice of the particular districts in question.

The conclusions must be popularly explained and an opportunity given

[1] К. А. Тимирязев, *Сочинения*, т. III, стр. 49.
[2] *Ibid.*, p. 48.

to help practical agriculturists make use of these conclusions. This will provide the scientist with a practical test of his conclusions, and practice is, after all, the real criterion of truth.

Such is the Soviet style of scientific work. It requires great exertion of one's mental faculties and fixity of purpose, but then it is real, profoundly scientific activity. Real science cannot be merely an amusing or aimless pastime. Any scientist who does not experience in his work a tensity of mind and fixity of purpose only thinks he is a scientist but in reality is not, for he does not work scientifically. In order to know on what question we should work, and how to go about it, we agrobiological workers who specialize in some particular department of this science must under no circumstances shut ourselves up in our narrow speciality.

The agrobiological scientist must make the precepts laid down by Kliment Arkadievich the basis of his work: that the study of the cultivated plant and its requirements is the root-problem of scientific farming. All other problems need be worked out only to the extent that they are of assistance in solving the problem of satisfying the wants of plants. Only such knowledge will help us to obtain the best and biggest crops possible.

Hence "to study organs independently of their functions and organisms independently of their life is almost as impossible as to study a machine and its parts without taking into account their action."[1] Moreover, this study must be conducted, as Timiryazev points out, "... not in the passive role of an observer but in the active role of an experimenter"; a physiologist, writes Kliment Arkadievich, "must take up the struggle against nature and by the strength of his mind and the force of his logic extort from her, wring from her, answers to his questions in order to become her master, make her subject to himself, be in a position at will to induce or stop, modify or direct, vital phenomena."

Agricultural research must be so conducted that it will widen the horizon of science and provide something to write about for the people, something for agronomists, collective-farm members and state-farm workers to read with interest and profit to themselves. "... It is not enough to toss a happy thought into the world," Timiryazev wrote; "it must, in addition, be put into the form of an irrefutable fact."[2]

Of the brilliant investigator Pasteur Timiryazev wrote: "All that Pasteur enunciated compelled assent. And this was so because he not only gave utterance to ideas but created a new method and with the aid of this method converted ideas into incontrovertible facts."[3] "His [Pasteur's—*T. L.*] most conspicuous trait was neither exceptional perspicacity nor power of creative thought, divining what to others remained concealed, but undoubtedly his marvellous ability to 'materialize' his thought, to pour it

---

[1] К. А. Тимирязев. *Сочинения*, т. IV, стр. 33.
[2] *Ibid.*, Vol. V. p. 209.
[3] *Ibid.*, p. 204.

out, if one may say so, in the tangible form of an experiment—an experiment from which nature, as if caught in a vise, could not escape without disclosing her secret."[1]

While developing Darwin's theory Timiryazev was himself an outstanding scientific experimenter, which enabled him to characterize in apt and telling words the best men of his profession. He thus showed how scientific workers ought to approach their investigations, how one should value and acquire theoretical knowledge, what to consider theory and what pseudo theory, pseudo science.

Timiryazev taught that men of science are obliged "from time to time to account to it [society—*T. L.*] as agents do to their principals. Here's what we've done, they ought to tell society; here's what we are doing and here's what we still have to do. Judge for yourselves how useful our work is at present and how promising it is for the future."[2]

No occasion could be more suitable than the present to recall the questions raised and the scientific assistance rendered to the land organs, the collective and state farms by agrobiological science during the sowing season in the spring of 1943. I shall be brief and confine myself to an enumeration of the measures in the scientific preparation and the popularization of which I took part in a leading capacity.

This year the collective and state farms, and hundreds of thousands of manual and mental workers in their individual vegetable gardens, used the tops of big food tubers to a greater or smaller extent for the planting of potatoes. We have repeatedly pointed out the practical importance of and the bright prospects held out by using the tops of big food tubers as a solution of the potato-planting problem. We are paying much attention to this measure not only because it effects a great saving in potatoes for planting (no less than a ton per hectare) but also because it makes possible at the same time the extensive use for planting purposes of the biggest tubers and the best as regards other racial properties—tubers that are always used as food. The cutting off, storing and planting of the tops of big food tubers is one of the measures that improve potato breeds and increase their yield. It is therefore extremely necessary at the present time that both scientific workers and agronomists pass on their practical experience in the planting of potato tuber tops, spreading this knowledge as widely as possible so as to ensure the practical application of this measure on a still larger scale in 1944.

This year the collective and state farms of Siberia, the trans-Ural area, North Kazakhstan and the North European part of the Soviet Union did considerable work in the sphere of increasing the germinating capacity of viable but dormant seeds of spring cereals, for dormant seeds have a low germinating capacity.

---

[1] *Ibid.*, pp. 205-06.
[2] *Ibid.*, Vol. IV, pp. 40-41.

This very feasible measure does not require much work, and is very important for increasing spring grain yields in the districts indicated. The practical solution of this problem is based on our theory of the rest period, the dormancy of seeds, tubers and the like.

The good practical results obtained in the work of raising the germinating capacity of viable seeds which was low in the control tests of sample lots made it possible to pose a new question, that of augmenting in the fields of the eastern and northern districts the germinating capacity of conditioned grain seeds, i.e., of seeds that exhibited good sprouting qualities under laboratory conditions. It was assumed that in the districts in question the relatively low germinating capacity of grain seeds in the fields was due in many cases to the failure of the seeds to complete their rest period. If the germinating capacity is ascertained at a laboratory, under conditions ensuring the cultivated seeds better aeration than when sown in the field, a high germinating capacity is obtained, but field sowing of the same seeds results in lower capacities and uneven shoots. Warming even such germination conditioned seeds in the spring air during the pre-sowing period should produce good practical results. We therefore recommended to the collective and state farms and scientific research institutes to experiment extensively along this line in the spring of the present year. The scientific elaboration of the problem of the rest period is of theoretical interest and has a direct bearing on a multitude of highly important practical plant-breeding problems.

We believe it to be of the utmost importance, both to science and to practice, to work out a regional system of agrotechnical measures that will ensure that during the period preceding the spring sowing the soil moisture is preserved and seed weeds destroyed. It is a most gratifying theme for agronomists. Unfortunately a number of scientific research institutions, as well as agronomists and heads of land bodies, are far from appreciating its full significance.

I believe it to be indispensable for a sound solution of the weed control problem that a thorough theoretical investigation be made into the phenomenon of dormancy in seeds; in the case before us, in weed seeds.

Even in peacetime millet was of no little importance among food and fodder crops. Today, in wartime, the importance of this crop is bound to increase and actually has increased greatly. Ensuring high millet crops is an exceedingly important task, especially for the arid regions of our country. On the basis of the extensive practical tests made in past years agrobiologists were in a position to submit to the land organs, collective and state farms a number of agrotechnical measures the adoption of which ensures good millet yields.

The importance that attaches to the extensive development of vegetable gardening among workers and employees is generally known. The science of agronomy must render them a maximum of assistance so that they will reap abundant crops from their plots.

I have enumerated only part of the problems that are being tackled by science, problems whose solutions are already benefiting practice.

We are conducting experiments and publishing popular articles on all these problems. They should be of assistance in the extensive practical tests to be made.

Timiryazev wrote: "...an exposition of ideas that can be grasped by all but that conceals from the reader the author's mental efforts, or a popular article in which there are incorporated independent views rarely met with even in specialized publications is usually thankless work for the scientific specialist. But the thanklessness of such work may, it seems to me, be more than compensated by appreciation of the fact that a wide dissemination of serious knowledge promotes the development in society of a veracious conception of the true tasks of science and facilitates the adoption of a conscious attitude toward it."[1]

Indeed, certain scientific workers are of the opinion that while the problems I have enumerated may be important for practical work, particularly now, in wartime, it is not, however, scientific work. Moreover, some colleagues even consider that the proposals I have submitted and which they claim have not been tested or only inadequately tested under laboratory conditions, should not be recommended to hundreds and thousands of farms for experimentation. In their opinion such methods of work contribute literally nothing to science and in practice frequently result in losses. One must, sure enough, safeguard practical farming from theoretically unsound proposals. But I do not agree that proposals must necessarily first be tested in the laboratory and only thereafter offered for practical application. The point is that whenever suggestions supposedly given their final laboratory tests are applied in production processes mistakes will invariably occur. One must therefore under no circumstances separate the working out of practical problems from extensive practical work. After all, practice, even in plant breeding, is so complicated and diversified that to anticipate or foresee everything in time is, firstly, difficult; secondly, it is literally impossible to test everything under laboratory or field conditions in research institutions. To quote K. A. Timiryazev:

"Perhaps nowhere else, in no other field of activity, is it requisite to weigh such a diversity of factors of success or to possess such many-sided information; nowhere else may the overzealous pursuit of a one-sided point of view lead to such stupendous failure as in farming."[2]

What is needed as a guarantee against an overzealous pursuit of a one-sided point of view in the practical and theoretical working out of a particular problem is a close bond between the investigator and the mass experimentation of collective and state farms.

---

[1] К. А. Тимирязев, *Сочинения*, т. III, стр. 125.

[2] *Ibid.*, p. 71.

It is the possibility of precisely such a method of work that Kliment Arkadievich dreamed of when he wrote: "If we had not only one experimental field (or what passed for one) per uyezd but scores, hundreds of cheap experimental fields our peasant would know what is required in each individual case; the plants themselves would suggest it to him."[1]

In tsarist Russia many gubernias did not have a single experimental field and at the same time K. A. Timiryazev tells us that each uyezd should have scores and hundreds of such fields, and cheap ones at that. But that is exactly what we are having now. Every collective farm, every state farm, can now have its own experimental field.

It must be firmly borne in mind, however, that if we advance science by establishing close ties with production we must not for one moment neglect the task of delving more and more deeply into the theoretical essence of questions that arise. I know from my experience in scientific work that when some profoundly theoretical problem is being solved experimentally numerous collateral laboratory experiments have to be quickly performed simultaneously. The scientist needs them to be able correctly to advise practical workers in the settlement of questions that arise in the process of performing mass experiments for which the scientifically worked-out problems serve as the theoretical basis. This is the only way to work in our science of agriculture. Without theoretical knowledge and, what is the main thing, without an advancement of theory it is not only difficult but impossible, I should say, to work with success in agrobiology, to work on the application of scientific achievements to production, as we say. Close contact between science and practice, precisely when problems are being worked out, is also a means of keeping scientific workers from performing unnecessary, purposeless experiments. It reveals vitally essential subjects for extensive laboratory experimentation. Close contact with practice apprises the investigator of the problems of theory for which solutions must be found first of all, in order that he may be of service to and meet the requirements of practice, of life.

What is the principal and determining factor in choosing problems for theoretical research? "...The cultivated plant and its requirements constitute the basic scientific task of agriculture...."[2] This, according to Timiryazev, is the pith and marrow of agrobiology. "...Everything else is important only insofar as it has a bearing upon it. This must be kept uppermost in one's mind when evaluating the importance of a particular branch of natural science to agriculture."[3]

As has already been said, all our efforts in scientific research work are directed toward the cognition of the essence of the nature of plants, their demands upon the conditions of life, and toward the elucidation of the question of how plants respond to the various environmental influences.

---

[1] К. А. Тимирязев, *Сочинения*, т. III, стр. 18.

[2] *Ibid.*, p. 52.

[3] *Ibid.*

A plant's demand for suitable conditions of life is not a whim or caprice on its part. The conditions a plant requires for its life and development are inborn, hereditary properties, historically formed in the process of development of that particular living body. The nature, the heredity of plant organisms necessitates the presence of particular environmental conditions for particular plants to develop normally and produce the biggest and best crops.

Before agrotechnically creating conditions for a plant under cultivation it is necessary to know what conditions are required by the heredity of the given plant, and when they are required for its development in general and for the development of the crop-producing organs and parts in particular.

This makes it clear that a knowledge of the natural, hereditary requirements of plants is very important to agrotechnique, one of the most important departments of agricultural science. If the science of agrotechnique fails to base its research work upon knowledge of the hereditary requirements of plants it will not be a science endowed with great prevision but merely blind empiricism. Knowledge of the hereditary requirements of plant organisms is of fundamental importance to scientific farming.

If a science that engages in the introduction, study and testing of new varieties and breeds does not base its work on a knowledge of their requirements and responses it is not a science that can foresee but one that is blindly empirical. Without a knowledge of the requirements and responses of the various strains of plants it is impossible to predict even approximately the behaviour of particular plants under any particular conditions. Such knowledge suggests the answers to such questions as in what regions certain varieties had best be tried out or what agronomic measures must be worked out for the successful cultivation of the plants whose requirements do not fit in with the usual conditions in the fields.

If the science of seed growing and plant breeding is not grounded on a knowledge of plants' requirements and does not engage in the experimental study of the origin and development of these requirements, then it also is blind empiricism. Without a profound knowledge of the origin and development of plants' requirements, breeders and seed growers will be unable to improve the old varieties and produce new good varieties according to plan, nor will it be possible for them to maintain the existing good varieties at a proper standard.

Furthermore, if scientists who study soils and their fertility do not make a knowledge of the requirements of plants the basis of their work they will inevitably reach the conclusion that what is needed is to fertilize the soil and not to feed the plant, a fallacy which V. R. Williams repeatedly pointed out. And this too leads to blind empiricism instead of the prescience without which science is not science.

In general, a knowledge of the requirements of plant organisms and of their responses to the actions of environmental conditions is essential for all departments of agrobiology. This explains why K. A. Timiryazev said that acquisition of a knowledge of the *requirements* of plants was the basic task of scientific farming.

It would indeed be difficult to point to any department of the science of agrobiology where a knowledge of the requirements of plants, as well as of the origin and development of these requirements, was not essential.

There is however a "science" which has no need of and even shuns all knowledge of the biological, natural and hereditary requirements of plants as the devil shuns contact with all things holy. I have in mind the Mendelist "science" of genetics, a pseudo science. On this score K. A. Timiryazev wrote:

"The cause of this unscientific phenomenon should obviously be looked for among circumstances of an unscientific order. The source of this pest, which the future historian will contemplate with perplexity, must be sought in another phenomenon, moving not only parallel but undoubtedly also in unison with it. This phenomenon is the intensification of clerical reaction against Darwinism."[1] And further on: "The future historian of science will probably view with regret this incursion of the clerical and nationalistic element into the brightest field of human endeavour whose sole object is the discovery of truth and its protection from all worthless extraneous matter."[2]

Mendelism lays claim to studying the problems of heredity and variation. Yet the science of heredity and variation implies a knowledge of the nature of organisms, of the conditions of life and development they require. But Mendelism in none of its variations (genetics) necessitates a study of the nature of organisms and their requirements, i. e., their heredity. Therefore to Mendelist genetics in all its variations the essence of heredity is a book sealed with seven seals.

Writing on the situation in the science of heredity Timiryazev noted: "...not one of the so-called theories of heredity advanced hitherto satisfies the primary demand made upon it, that it serve as a general working hypothesis, i.e., as an instrument for directing research toward the discovery of new facts and new generalizations."[3]

Timiryazev also assigns the logical reasons for this: "At bottom they are all only variations of one and the same theme: offspring are 'flesh of the flesh and bone of the bone' of their ancestors; but as observation progressed, the more subtle structural features 'cell from cell,' 'plasm from plasm,' 'nucleus from nucleus,' 'chromosome from chromosome,' etc.—were substituted."[4]

Timiryazev pointed out that in order to understand the phenomena of heredity one must first of all "become imbued with the idea that causes may be potential and not necessarily morphological and, in general, of a character other than the effects they produce."[5]

---

[1] К. А. Тимирязев, *Сочинения*, т. VI, стр. 264.
[2] *Ibid.*, p. 265.
[3] *Ibid.*, p. 191.
[4] *Ibid.*
[5] *Ibid.*, p. 193.

A science cannot possibly become the basis of the work of controlling the nature of organisms, if it does not take as its premise the theory of development, of dialectical materialism. Let us take, for example, so highly important a question as the inheritance of so-called acquired characters, i. e., characters which arise in organisms in the process of their development. This problem was hopelessly tangled by formal Mendelist geneticists. But in the light of the theory of development it was posed and solved quite differently in our country.

K. A. Timiryazev and I. V. Michurin repeatedly pointed out in their further elaboration of Darwin's teaching that the way to regulate the conditions of life of organisms is at the same time the way to regulate their heredity.

During K. A. Timiryazev's lifetime science was not yet in possession of facts which may be cited as unimpeachable proof that alterations in heredity can be brought about by changing the conditions of life, and that even if the changes are different in different organisms they are all adequate to meet the response of the organisms to the conditions. True, I. V. Michurin had been elaborating that problem already at that time but in tsarist Russia genuine science was smothered to such an extent that even K. A. Timiryazev never heard of his works.

What is needed for an understanding of the laws of heredity and its variability, for an understanding of how requirements for suitable conditions of life, conditions of development, arise and develop in plants is not a bare, formal, meaningless scheme to the effect that "everything comes from a chromosome and the chromosome itself from a similar chromosome," but a general biological theory that embraces all the multifarious forms of heredity. For the construction of such a theory a study of vegetative hybridization is of particular importance.

It was K. A. Timiryazev who, following Darwin, pointed out how vastly significant it was for science that such things as vegetative hybrids could exist. Experimental methods of obtaining such hybrids were first elaborated by I. V. Michurin. Michurin's mentor method is vegetative hybridization.

Comprehension of the essence of vegetative hybridization is of decisive importance, on the one hand, for a correct formulation and solution of the problem of the inheritance of so-called acquired characters and, on the other, for a more thorough understanding of heredity and the nature of organisms in general.

K. A. Timiryazev worked out a classification of the various forms of heredity, one that comprises both asexual and sexual reproduction. His development of the Darwinist idea that there is analogy and intergrading between heredity due to sexual propagation and heredity due to vegetative propagation carries immeasurably greater conviction today in the light of data now possessed by Soviet science than it did in his lifetime.

Timiryazev teaches that the basic task of scientific agriculture is the study of the *requirements of plants*. Our theoretical and practical work therefore aims to satisfy these requirements as well as to alter the nature of plants, to change their requirements in the direction needed by practical farming. Only a knowledge of the natural requirements and relation of an organism to environmental conditions makes it possible to govern the life and development of this organism. More. Such knowledge may serve as the basis for changing the heredity of organisms in a definite direction.

To ascertain the environmental conditions required by a living body (organism) for the development of particular characters or properties is tantamount to studying the nature, i.e., the heredity, of these particular characters or properties. To take heredity to mean—as has been done hitherto in genetics—only the reproduction of likes, without going into a study of the ways and the material (conditions) from which the body reproduces itself, is tantamount to barring one's own way to mastering this important and interesting phenomenon of living nature.

In the recently published booklet *Heredity and Its Variability*[1] a concise exposition will be found of our conception of these phenomena in the plant world. It differs fundamentally from the Mendelian interpretation, which is barren for the purposes of both science and practice.

We regard the heredity of the plant organism's demand for environmental conditions as a property of the living body, a property which arose and arises in the process of the organism's development. *The cause of alterations in the nature of living bodies is change in the type of assimilation, change in the type of metabolism.*

All properties, including heredity, are, naturally, obtained de novo, i. e., *the nature of the organism is obtained de novo in the new generation to the same extent* that the body of this organism builds itself up de novo in the new generation (of plants, say).

Assimilation of the conditions of its environment causes the living body to change, differentiate. A change in the conditions of life which compels plant organisms to change their development is the cause of change in their heredity, is the cause of change in the natural requirements of plants and their response to the action of environmental conditions.

The extent to which changes are transmitted to succeeding generations will depend on the extent to which the substances of the changed section of the body are included in the general chain of the process which leads to the formation of reproductive sex or vegetative cells. In free nature this depends upon the conditions that accidentally come into existence and affect the particular plant; in experiments and practical farming, it depends upon the knowledge and desires of man.

The relative constancy or conservatism of plant and animal forms as generations succeed each other is explained as follows:

firstly: by the active election from the environment of conditions for

---

[1] See present volume, p. 390.—*Ed.*

the building up of particular organs or characters, definite particles of the living organism;

secondly: by the active noninclusion in the process of conditions unsuitable to the nature of the organism. However, forcibly changed sections of the body do not include in full measure the specific substances they produce in the general chain of the process that leads to the formation of reproductive cells, and often do not include them altogether;

thirdly and lastly: in the organism, as in an integral whole, there is no "equalitarian tendency" in the supplying of the different processes with the requisite elements of food. The more important processes are supplied with greater approximation to the norm; they are protected against insufficiency as well as against a surplus of nutrition in general or of particular elements of it. Less important processes, on the other hand, are supplied less than the norm, the norm or more than the norm, depending upon the nutriment and its elements available.

Ability to analyze profoundly and with delicate precision the complicated processes of the development of organisms and the interconnections between these processes is the most important, most essential objective of theoretical agrobiology, primarily of the branch that deals with heredity and variation. To those possessing such knowledge, much both in practice and in theory becomes "possible and foreseeable."

It will not be possible for me to set forth here what knowledge our scientists have already gained in this sphere. As has already been indicated, I have systematized it in concise form in a pamphlet entitled *Heredity and Its Variability.* I should only like to say that in our country one of the theoretical parts of the science of agrobiology, the science of heredity and its variability, is becoming precisely what K. A. Timiryazev wanted it to become.

The Soviet, Michurin section of the science of heredity and its variability may already serve and does serve as a working hypothesis, as an instrument directing research toward the discovery of new facts and new generalizations in aid of collective- and state-farm practice.

I shall cite only one example. Scientific thought and research have been endeavouring for years to find a solution of the problem of cultivating winter wheat in the steppes and forest-steppes of Siberia. To physiologists it was clear that not a single one of the existing varieties of winter wheat could survive the winter in the Siberian steppe districts where the frosts reach lows of 40-45° C.

The facts we have established and which have now been experimentally confirmed may seem paradoxical. They tell us that in the severe, snow-poor open steppes and forest-steppes of the Siberian districts certain winter grain varieties can withstand these severe frosts better (with much less damage) than the considerably milder frosts in the European part of the Soviet Union where winter cereals have been extensively cultivated for ages.

It may be stated as a fact that severe frosts can be successfully withstood not only in the forest-steppe belt but also in the open steppe by various winter ryes and winter wheats as well, including even such varieties as the Crimean (or, for instance, the Kooperatorka), whose weak frost resistance is generally known in practical farming.

Moreover, sporadic late autumnal windfall shoots of spring wheat, as we assumed, frequently survive the winter despite the severe Siberian frosts. Early in spring this year these assumptions were fully borne out in the Omsk district. At that time wheat shoots that had stood the winter well and had undoubtedly appeared in autumn could be found on any unploughed field on which last year a spring wheat crop had been harvested.

At present over twenty varieties of southern winter wheats possessing low frost hardiness have been experimentally sown under our direction by A. A. Baskova in the Chelyabinsk State Plant-Breeding Station. They passed the winter very well, in fact so well that in spring one could observe no difference between such little frost-resistant varieties as Kooperatorka, Novokrymka 0204 and others, on the one hand, and such a renowned frost-resistant variety as Lutescens 0329. Even the shoots on a small plot sown to the spring variety Milturum 0321, which had fully developed in autumn, on the whole survived the winter, but were damaged. Not only the tillering nodes of winter wheats came through the winter in viable shape and without having suffered any damage but even the autumnal leaves were completely viable in spring, not in the least damaged or very little so. Yet it is a known fact that in districts where winter wheats are cultivated the autumnal leaves usually die by spring but tillering nodes remain alive and new leaves develop from them in spring.

True, in the winter of 1942-43 the snow covering in the area of the Chelyabinsk Plant-Breeding Experimental Station was above normal for that district. However, a similar group of wheat varieties which had over-wintered well can be found on our second experimental plot (worked by N. A. Belozerova and I. A. Kostyuchenko), in Omsk, in the fields of the Siberian Scientific Research Institute of Grain Husbandry. Here the 1942-43 snow covering was considerably below normal.

According to the data of the Omsk Meteorological Station precipitation was only 15 mm. here from January 1 to April 1. This information indicates the thickness of the natural snow covering in the winter of 1942-43 in the fields of the Omsk district.

Everybody knows that in all districts where winter wheats are sown on a large scale soil temperatures at the depth of the tillering nodes do not drop lower than —13°-14° C. even in years when wheat is destroyed extensively by frost. In Omsk, however, the soil temperature at the depth of the wheat tillering nodes was below —17°-19° C. on many days during the winter of 1942-43. Yet even under these conditions, such poorly frost-resistant varieties as the Kooperatorka and Krymka passed the winter well.

I shall also make mention of the fact that at the Chelyabinsk State

Plant-Breeding Station A. A. Baskova sowed in the autumn of 1942, at the same time as the above wheats, the offspring of an ear of winter barley that had been converted from the spring variety Pallidum 032; the plants of this barley also overwintered and did not perish. Yet we know that winter barley can winter only in southern districts where the climate is mild.

The factual material I have referred to argues in favour of the conclusion that under Siberian conditions, both in the forest-steppes and in open steppes, not only rye but also winter wheat can easily stand such severe frosts as agronomists, so far as I know, consider no existing wheat varieties can possibly survive. And in practical farming, too, it is common knowledge that in the steppe and forest-steppe districts of Siberia winter wheat is unable to withstand the cold winter weather. It is only in rare years, when there is good artificial snow retention (provision of warmth), that winter wheat winters well in the steppe districts; ordinarily winter wheats perish in these districts even when snow-retention measures are taken.

Another thing known to practical agriculturists is that in certain areas almost every winter Siberian varieties of rye of even high frost resistance suffer severe damage or even perish outright. Nevertheless one may safely state that in these parts the severe frosts can be rather easily resisted even by winter-wheat varieties having little frost resistance, such as the Kooperatorka. Today anyone can readily convince himself of these facts (the severe damage done to frost-hardy rye varieties and the good wintering of winter-wheat varieties with low frost resistance) by inspecting and comparing the wheats I have mentioned, sown at the Chelyabinsk Plant-Breeding Station and in Omsk, at the Institute of Grain Husbandry, as well as the rye sown after September ploughing in the collective farms, near Omsk, for instance.

The chief cause of the frequent loss of winter crops in Siberia is, as we have established, the friability of the soil in which they are planted. The more friable the soil sown to winter crops, the thinner the stand of grass of the winter crops in autumn and, consequently, the less compact the soil in which the root systems of the plants are embedded, the greater the portion of the winter crop that will perish.

We hold that, as a rule, it is not the direct action of frost that kills or damages winter plants in Siberia. Both above-ground and below-ground parts of winter plants perish in Siberia during autumn, winter and spring from mechanical injuries. In winter the part of the plant above ground is desiccated, injured and broken by the strong, dry, frosty winds. The wind sweeps aloft countless grains of sand from the soil which harm the leaves by their impact. The strong wind breaks the frozen leaves which have lost their resilience. The parts underground—the root system and tillering nodes—are damaged because the roots and tillering nodes break open.

In autumn the interstices, or empty spaces, between lumps of friable soil are filled with rain water; besides, as the frosts set in, the water which is rising in a vaporous state from the lower strata of the soil to the top layer and contained in the capillaries of the clods of earth freezes over into the hollows. The water in the hollows of the unsettled soil and also in the hollows around the tillering nodes of the winter plants forms big crystals of ice. When severe and continued frost penetrates the top layer of the soil (at temperatures of minus 15° C. to minus 20° C. and lower) cracks form, often big enough to put your hand in. The whole field is thickly covered with small crevices. They are not always noticeable as they are buried beneath dispersed soil carried along by the wind. The frozen water in the hollows between the clods presses the soil apart, bursts it asunder, at the same time tearing or injuring the parts of the plants underground.

These are the principal causes responsible for the perishing of winter plants in the Siberian districts.

In these districts conditions for hardening plants, i. e., conditions under which winter plants acquire resistance against frost, are very good. This explains why such wheat varieties as Kooperatorka and Krymka can easily withstand the impact of the severe winter frosts when field-sown for experimental purposes in Siberian districts. Hence the cultivation for consumption of Siberian winter wheats is fully possible if the existing varieties, which are considerably more frost-resistant than the Kooperatorka, are used. All that need be done is to find a way of protecting the parts of the winter-wheat growths above and below ground from mechanical injury.

Very good wintering conditions are created for winter plants in the steppe and forest-steppe regions of Siberia when the seeds of these plants are sown by tractor-drawn disk drills in unploughed stubble fields. When sown in the stubble, in unploughed fields, winter plants acquire a superior frost resistance in Siberia. Under these conditions even wheat varieties with low frost resistance become hardy enough to resist the severe frosts of the Siberian winters.

Stubbles 25-30 cm. high protect the parts of the plant above ground from the lethal mechanical action of the wind. The stubbles retain the snow, which also serves to protect the plant, against both the frosts and the action of the wind.

Unploughed and unloosened soil has practically no big vacant spaces. Therefore when sowing in stubble fields the soil has no big ice crystals that destroy or damage the roots and tillering nodes of winter plants.

This is our explanation of the fact that in the steppe districts of Siberia, on the one hand, certain autumn shoots of even windfall spring wheat quite frequently survive the winter on unploughed and unloosened plots while, on the other hand, even very frost-resistant rye varieties sown in recently ploughed friable fields, the soil of which does not become very compact by the time the frosts set in, often perish or are severely damaged.

In concluding my address permit me to state that, according to K. A. Timiryazev, the study of and attention to the *requirements of plant organisms* constitute the principal task of scientific farming, the foundation of the development of all departments of agricultural science. The ascertainment of requirements and the study of the causes of the origin and development of these requirements and of the responses of plants to the action of their environments form the basis of the theoretical work of *our Soviet science of heredity and its variability.*

This theoretical department of agrobiology is a powerful factor responsible for the emergence of science upon the scene of collective- and state-farm production to deal with various questions of practical importance. Large sections of agronomists and collective- and state-farm experimenters are thus enlisted in the scientific and practical elaboration of problems. Practice obtains from science the knowledge required to advance the development of the art of agriculture while science in its turn obtains from practice a knowledge of facts which renders it more scientific, more prescient, if one may put it so.

It is a matter of general knowledge that in the Land of Soviets science has been provided with every material and spiritual requisite for its swift development. We have every right to be proud of our classics, our biologists and Darwinian agrobiologists: K. A. Timiryazev, I. P. Pavlov, I. V. Michurin and V. R. Williams. They all spent the greater part of their conscious lives under the tsar, but it is only under Soviet rule that the country, the Party, the Government and the public at large have shown proper appreciation and esteem for the works and the activities of these men of science. Their teachings are becoming more and more an acquisition of wide sections of the working people, a guide for those who work in the field of biology and, more specifically, agrobiology.

What all of us hold dearest and cherish most is the Land of Soviets, the land of genuine liberty, the land of scientific progress. With a valour and heroism that have never been equalled the Red Army is defending the sacred Soviet soil from the brutalized German invaders and enslavers. The working people in the rear—industrial and office workers and members of collective farms—are devoting every ounce of their strength to increasing the defensive capacity of the country. To us who are engaged in the science of farming it is a matter of honour and duty to extract the utmost from Soviet agrobiological science in order to help the collective and state farms to raise and gather in the food crops and agricultural raw material needed by front and rear.

Our entire people, the Party and the Government, and Comrade Stalin personally, display each day a fatherly concern for the welfare of science and the workers engaged in it. In this hour of danger to our Homeland we must harness every bit of our strength and knowledge and place them completely at the service of our country's hallowed cause—the speediest possible discomfiture of hated fascism.

HEREDITY AND ITS VARIABILITY

THE ESSENCE OF HEREDITY

IN ALL textbooks and manuals on genetics heredity is usually taken to mean only the reproduction by living organisms of beings similar to themselves. Such a definition, in my opinion, contributes little to an understanding of the phenomenon of heredity. From time immemorial people have known that from wheat seeds you get wheat, from millet, millet, etc. This makes it possible for practical farming to propagate a particular species or variety of plant, or animal breed. No deeper conception of the phenomenon of heredity can be derived from the above definition.

Exponents of modern genetics (the science which studies the phenomena of heredity), who take as their premise the definition that heredity is only the reproduction of likes by organisms, have been studying heredity by ways and means which do not permit the investigator to find out anything about the essence of the heredity of any particular living body. What they study is not the phenomenon of heredity but the ultimate differences between organisms that differ in heredity.

The method employed in genetics to study heredity is to take two breeds, two organisms known to differ in heredity and to blend them by crossbreeding. There are those who want to find out things about the heredity of organisms they are investigating or the heredity of their characters from the diversity of the progeny obtained. All one can find out in this way is how many offspring resemble the one parent or the other. But one cannot determine, from the data furnished by such experiments, what the essence of the heredity of either parent consists in.

Our definition of the phenomenon of heredity differs from that which has hitherto been accepted in genetics. By heredity we mean *the property of a living body to require definite conditions for its life and development and to respond in a definite way to various conditions*. By heredity we mean the nature of the living body. We are therefore of the opinion that "the nature of the living body" and "the heredity of the living body" are about the same thing. For example, why do wheat plants differ from rice plants?

Because these plants differ in their natures. Similarly it may be said that wheat differs from rice because wheat has a different heredity from that of rice. To study the heredity of an organism means to study its nature.

The nature of a living body differs in principle from the nature of a nonliving body. The more a nonliving body is isolated from the action of or interaction with environmental conditions, the longer it will remain what it is. A living body, on the other hand, absolutely requires definite environmental conditions in order to be alive. If a living body is isolated from the external conditions it requires, it ceases to be alive, ceases to be what it is. Precisely herein lies the difference in principle between a living and nonliving body.

Different living bodies require different environmental conditions. We therefore know that they differ in nature, in heredity. Knowledge of the conditions required by a living body and of its responses to the operation of various conditions means knowledge of heredity properties of that body. Consequently, *to ascertain the environmental conditions required by a living body (organism) for the development of particular characters or properties is tantamount to studying the nature, i.e., the heredity, of these particular characters or properties.*

A study of the heredity (nature) of a given living body does not require the crossing of that plant or animal with one possessing another heredity. The real purpose of studying heredity is to determine the relation of an organism of a given nature to its environmental conditions. But after crossing, the offspring obtained do not possess the nature that it is desired to study. When studying heredity various crosses are needed only when one wants to determine the potency, or stability, of one heredity in comparison with another or others.

A knowledge of the natural requirements and relation of an organism to environmental conditions makes it possible to govern the life and development of this organism. More. Such knowledge may serve as the basis for changing the heredity of organisms in a definite direction.

To take heredity to mean—as has been done hitherto in genetics—only the reproduction of likes, without going into a study of the ways and the material (conditions) from which the body reproduces itself, is tantamount to barring one's own way to mastering this important and interesting phenomenon of living nature.

It has been pointed out above that according to the line formerly followed in genetics, in order to study the heredity of a given character one must take a plant possessing this character and another plant absolutely different in nature, in heredity, from the given character. After crossing them the offspring of these two parents are examined to determine how many descendant plants have the character peculiar to the one parent and how many have the character observed in the other. Such a study, however, will not show in what the heredity of either of the parents taken for the investigation consists.

The difference between our approach to the study of heredity and the methods of the Mendelist-Morganist geneticists can be illustrated by the following example. The property of winter or spring habit is doubtlessly inheritable. In their repeated studies of the heredity of these properties the geneticists took plants of the winter variety and crossed them with plants of the spring variety. They then determined in the offspring how many winter plants, i.e., plants similar, as regards this character, to the one parent, were obtained, and how many spring plants, i.e., plants similar to the other parent. In some experiments they arrived at the conclusion that the hereditary properties of the winter habit differ from the hereditary properties of the spring habit by 1, 2, 3, etc., genes, granules of some unknown substance of the living body presumably contained in the chromosomes of cells of the winter or spring plants. However, what the essence itself, i.e., the nature of the winter or spring property of cereals, consists in, how to control the development of these properties, does not appear at all from the above study. But if the heredity of an organism or of separate properties or characters of it is characterized by the environmental conditions required for the development of these properties and characters, there is revealed to us the essential nature of the given properties or characters.

Thus, on studying the causes of the failure of winter cereals to ear when sown in spring we ascertained that one of the processes of development of winter plants now called the stage of vernalization requires, in addition to the food, moisture and air existing in the fields in spring, a relatively lengthy period of time of low temperature, 0°-10°C. above zero. The absence of a lengthy period of low temperature in the fields in spring is the very reason why the process of vernalization fails to take place, and hence why all further development is retarded, why there is no earing and fruiting.

With the discovery of the nature of the vernalization stage it has become possible to compel any winter grain sown in spring to ear and bear fruit. For that purpose the seeds, properly moistened, are kept for a definite time under relatively low temperatures (vernalization) before they are sown in the field. Thus the inherited requirements for transition (development) through the indicated process are satisfied. After its completion at the growing point of the young plant or in the embryo of the seed all further inherited requirements are satisfied by the existing field conditions when such seeds are sown in the field in spring and development continues normally until it is completed, i. e., until the plants ripen. This is the kind of study we engage in to determine the essence of the heredity of the winter habit.

Upon studying a considerable assortment it appeared that some varieties of bread grains possess greater winter habit, i. e., require a longer period of low temperatures, while others possess less winter habit, require a shorter period of low temperatures. Varieties that, according to their nature, can undergo the process of vernalization under the usual spring and summer conditions are called spring varieties in practical farming.

We, on studying heredity, ascertain the conditions of life, the conditions of development, required by the organism or by separate processes, and also the relation of the organism or separate processes of it to various environmental conditions. We thus arrive at a comprehension of the essence of heredity. The geneticists, on the other hand, do not study the essence of heredity. All they find out is how many offspring resemble the one parent with regard to a particular character and how many the other.

It is well known that a living body builds itself from the conditions of its environment, from its food in the broad sense of the word. It is likewise well known that the embryos of different breeds, for instance, of particular varieties of plants living in the same environment, build their bodies differently; hence different organisms are obtained.

Each organism develops, builds its body, according to its nature, its heredity. For example, you can feed a calf and a lamb the same hay. But while they assimilate the same hay, the lamb, following its nature will develop and grow into a sheep and the calf into a cow. Everybody knows that not only do sheep and cows differ sharply as organisms but also that the quality and properties of mutton and beef differ in many respects, though both types of flesh are derived from the same fodder, in the case stated, from the same hay.

Such examples go to prove that every living body builds itself from the environmental conditions in its own fashion and in accordance with its own nature, its own heredity.

One can also readily notice—and people knew this long ago—that as a rule each generation of plants or animals develops in many respects like its ancestors, especially its nearest ones. This accounts for the definition accepted by genetics that heredity is the property of reproducing beings similar to oneself. *But reproduction of likes is a characteristic trait common to all living bodies.* Hence, the mere statement of the above-mentioned common property of living bodies, a property long known to all, cannot characterize to any extent the concrete heredity of a given living body. A study of concrete heredity requires that one follow the course of development of the organism possessing the given heredity, and determine the conditions necessary for its development as well as the reaction of the organism to the influence of its environment.

It is not only the organism as a whole that can reproduce bodies similar to itself. Every cell of the organism, every granule of a living body can reproduce its likes. For example, a cell of a young stem will reproduce stem cells, a cell of a leaf will reproduce leaf cells, a cell of a rootlet will reproduce rootlet cells. Every organism grows by its various cells reproducing cells similar to themselves.

THE ESSENCE OF VARIABILITY. GROWTH AND DEVELOPMENT

While it is known that an organism and also its individual cells and the various particles of them reproduce bodies similar to themselves, one must not forget another aspect of this property of the living body, namely, that the organism as a whole as well as the separate parts of its body reproduce, in some measure or other, bodies dissimilar to themselves. For instance, an egg or a zygote, after a definite interval of time and under suitable conditions, reproduces many thousands and even millions of cells wholly dissimilar to the first, original cell, i. e., the zygote from which they sprang. The case of a bit of begonia leaf developing into a full-grown plant may also be instanced. Here begonia leaf cells reproduce root and stem cells, i.e., cells unlike those from which they originate.

Consequently, although it is characteristic of the nature of a living body to reproduce bodies similar to itself, yet simultaneously cells and specific particles that enter into their contents are capable, in some measure or degree, of reproducing also bodies dissimilar to themselves.

The ability of the separate cells of an organism to reproduce not only likes but also unlikes has never been questioned in science. What has been disputed, and for centuries at that, was the fact that an organism as such can reproduce organisms not only similar to itself but also differing from itself. The point involved is the variability or invariability of the nature of living beings.

When Darwinism made its appearance short shrift was given to the unchangeability of living nature. Today no scientist of reputation anywhere on earth will assert that living nature does not change. The variability of living nature and the possibility that it may change are admitted. But up to now the causes of changes in the nature of organisms and the concrete ways in which these changes take place are not sufficiently known to science to make it possible to alter the heredity of an organism at will in any definite direction. Therefore modern genetics, while abstractly recognizing that living nature is variable, in practice conducts its investigations, makes its deductions and draws its conclusions on the assumption that the heredity of an organism cannot be changed by the conditions of its life. Such a science therefore maintains that it is impossible for the conditions of life to influence the variability of the nature of plants and animals in a desired direction.

Our Soviet science, the Michurin trend in science, gives a clear understanding of the way to change the nature of an organism.

Our conception of the phenomena of heredity, the changes in these phenomena and the regulation of heredity is based on the following premise:

Every living body builds itself out of nonliving material, in other words, out of food, out of the environmental conditions. The organism picks from the environment the conditions it needs; but this choosing of conditions is dependent upon the heredity of the given organism. Whenever an organism finds in its environment the conditions which it needs and which are suitable to its nature, its development proceeds in the same way as it proceeded in previous generations of the same breed (of the same heredity). When, however, organisms do not find the conditions they require and are forced to assimilate environmental conditions which, in one degree or another, do not accord with their nature, organisms or parts of their bodies result which are more or less different from the preceding generation.

If the altered part of the body is the starting point for a new generation, the latter will differ from the preceding generations in its requirements and nature. From the biological point of view we can find out the difference between these generations. It will consist of a difference in the requirement of environmental conditions. The particular conditions were unsuitable for the preceding generation and the body assimilated them of necessity, perforce, as it were. But if it imbibed them, assimilated them, a body with new properties, with a new nature, was obtained. These conditions will now be requisite for it. Thus, *the cause of change in the nature of a living body is a change in the type of assimilation, in the type of metabolism.*

Being included in, assimilated by, the living body the external conditions cease to be external and become internal conditions, i.e., they become particles of the living body, and their growth and development now require the food, the environmental conditions that they were themselves in the past. The living body is composed, in a way, of separate environmental elements which have become converted into elements of the living body. The growth of the separate parts and granules of the living body requires the same environmental conditions by means of the assimilation of which the organism originally built these parts and granules of its body. Thus, by regulating the conditions of life, new environmental conditions may be included in, or this or that element expelled from, a living body.

Whether separate elements are included in or have been excluded from a living body may be judged by the environmental conditions it requires for its growth and development. For instance, the process of vernalization of spring cereals does not require low thermal conditions. The vernalization of spring grains proceeds easily at the usual spring and summer temperatures in the fields. If spring cereals are vernalized for a long period under low thermal conditions one may not infrequently observe that one or two generations later the spring habit of the wheat will change to winter habit. And, as everyone knows, winter grains cannot undergo vernalization without low temperatures. We show by this example in what way new external conditions were included in the nature of the living body, and that thereby the progeny of the given plants acquired a new requirement, the requirement of low thermal conditions for vernalization.

Changes in requirements, i.e., in the heredity of a living body, are always adequate to the action of the environmental conditions, if these conditions have been assimilated by the living body.

As was remarked above, the different elements of an organism, its organs, cells and separate parts in cells, possess the property of reproducing themselves. For example, we know that if in leaf cells the plastids from which chlorophyll grains develop disintegrate for some reason or other, then all cells which come from the cells that lost their plastids will be albinos, i.e., white, not green. In the case stated the chlorophyll grains will not be reproduced, there will be nothing to do the reproducing.

Each molecule and atom of a living body, if one may put it that way, reproduces itself at certain moments. *But all these different molecules and cells in the organism are obtained from zygotes by reproduction of bodies that are not similar but dissimilar to themselves, by differentiation, i.e., development.*

From the initial cell (the zygote) we obtain a group of cells which are unlike the cell we started with. There are no plastids (and other things besides plastids) in the initial cell (the zygote) of the plant; they appear in the cells obtained from the zygote. In cell multiplication the plastids and all other separate parts reproduce themselves, so to say. Consequently, the *reproduction of likes is only one of the properties of the living body. Another of its properties is the reproduction of unlikes.*

The direct reproduction of likes by each cell and each molecule of the living body we call growth of the body. For instance, leaf cells reproduce likes, and as a result the leaf gets bigger, or grows, as they say. By the growth of a body we mean increment of weight and volume.

The reproduction of like by like may, however, come about not only by means of growth but also by means of development.

The reproduction of likes, not directly but through a long chain of conversions of unlikes, until beings similar to the initial one are obtained we call development. There is a qualitative distinction between these two means of reproducing likes.

To exemplify the first means of reproducing likes let us point out the following. The cell of a leaf grows and develops, then divides in two; instead of the one we have two but both remain leaf cells. The leaf increases in size, i.e., grows. This process is what we call growth. Another example might be given, that of a leaf and, of course, of its cells also reproducing bodies similar to themselves, but in a different way—through a chain of conversions. By means of grafting, A. A. Avakian substituted for the ordinary, dissected leaves of tomatoes of the Albino variety leaves of another tomato variety which resembled potato leaves, i.e., pinnate in shape. Seeds were taken from a fruit that had developed on a twig of the Albino variety. This variety, as has been stated, has dissected leaves, in accordance with its nature. After these seeds were sown in the summer of 1941 at the Experiment Base of the Lenin Academy of Agricultural Sciences of the U.S.S.R. at Gorki Leninskiye, quite a few plants were obtained whose leaves were not dissected, but were like those of the potato plant. The question may be asked why, despite the fact that it is in the nature of the Albino variety to have dissected leaves, some of its offspring happened to have nondissected leaves, leaves resembling those of the potato plant. The reason is that the plant from which the seeds were taken had leaves like those of the potato plant which had been substituted for the dissected ones by means of grafting. It was they that had reproduced themselves in the progeny.

Substances elaborated in the leaves united with substances of the neighbouring cells, were modified, were converted, developed. From these cells altered substances united with substances of other cells and altered still more. In this manner the conversion went further and further away from the leaf cells until it became a component element of the embryo. In this way, we believe, *the hereditary basis of each organ, of each character, of each property of the organism reproduces itself in the succeeding generations.*

INDIVIDUAL DEVELOPMENT OF THE ORGANISM

The development of an organism, like its growth, proceeds by means of conversion, of metabolism. The sex cells, buds or eyes from which usually entire organisms develop are, as a rule, the product of the development of the whole organism which gave rise to the particular initial bases for the new organisms. They arise and build themselves out of molecules, granules of substances—transformed many times (but in accordance with natural law)—of various organs and body parts of the organism. Therefore, all the former properties of the plant that produced the sex cells or, for example, the eyes of potato tubers are accumulated, as it were, in these cells or eyes. Hence, the initial cells also express to a greater or smaller extent the tendency of the future properties of the organism.

In development from a fertilized sex cell, i.e., from a zygote, the changes and conversions are, as it were, a repetition of the path traversed by the ancestors, particularly the nearest ones. The process which in the preceding organism went on at the very beginning will be the initial process also in the new generation; the process following the first one will be the next also in the offspring, etc. Figuratively speaking, the development of an organism may be depicted as the unwinding internally of a spiral that was wound up in the preceding generation. This unscrewing is at the same time a screwing up for the future generation. After all, the formation of a given organism proceeds on the basis of the development of the preceding one. And in the process of development of that particular organism the basis of the future generation is formed. I consider it correct to say: *to the extent that the body of a particular organism (say, of a plant) is built de novo in the new generation, to that same extent, naturally, all its properties, including its heredity, are obtained de novo, i.e., to that same extent also the nature of the organism is obtained de novo in the new generation.*

Each organ and each character reproduces itself in the course of generations both by means of growth and by means of development.

The sex cells and any other cells by which organisms are reproduced are, as a rule, created, obtained as a result of the development of the entire organism, by means of conversion, by means of the metabolism of the various organs. As a result the course of development gone through is accumulated, as it were, in the cells from which the new generation takes its start.

The primary initial cells, from which the organism develops, are the biologically most complex and possess the greatest possibilities of development. All other cells resulting from the development of the zygote when the tissues are being differentiated are biologically less complex and possess fewer potentialities of development. For instance, a whole organism may develop, be obtained, from a sex cell or a bud (eye) of a potato tuber. But from the leaf cells of many plants it is impossible to obtain whole plant organisms.

The assertion of the Mendelist-Morganist geneticists that all cells of an organism possess one and the same nature, one and the same heredity, will not hold water. Different cells of the same organism undoubtedly possess different natures, different heredities, different potentialities of development. All you need do is take from a potato tuber as initial eyes not the usual ones but such as developed from tuber cells from which they normally do not develop, and you can observe not infrequently that plants of a different nature, of a different variety, are obtained. We know of quite a number of cases when new organisms, different in their nature, may be obtained from cells of one and the same plant organism. It has already been pointed out that by far not all cells of even plant organisms can give rise to or regenerate entire organisms. This also goes to show that not all cells of an organism are of like nature. Different cells of an organism possess different natures, i.e., different heredities.

The development of an organism from a zygote is, as it were, the differentiation, the disintegration of a biologically more complex cell into simpler, more differentiated cells. The egg is biologically more complex than any cell of the organism that originated from it.

It must not be forgotten that cells differing in quality can be and always are obtained from one and the same quality of initial material, for instance, from one cell or a group of like cells, in the process of development, in the process of metabolism. These different qualities of cells are determined by the environmental conditions. *The environmental conditions are the differentiating material of the developing organism. These conditions are assimilated by the living body and thus the body changes itself, differentiates itself.*

For instance, the shoots of plants which spring up from the soil have white leaves. The cells of these leaves already contain plastids but the latter may turn into chlorophyll grains, in consequence of which the leaves become green only when subjected to the influence of light. In the case at hand light, together, of course, with the other environmental conditions, is the differentiating material of the plastids; as a result the plastids are transformed into chlorophyll grains.

The presence in plants of particular characters or properties is usually due to the fact that the latter existed in the parent organisms and, by means of conversion, of development (metabolism) were incorporated and accumulated in the sex cells, in the initial cells of the new generation. But numerous cases could be cited in which a particular character of the organism in question did not exist in the parental forms. It had been present in the older preceding generations and appeared anew only after several generations. This particular character or property had been, as they say, in a latent or recessive state. To explain this fact let us return to the example of the character of green colour in wheat leaves. When the young leaves appeared on the ground they had no green colour. There was no chlorophyll in them. But they had the substances, plastids, which are converted in these leaflets, in the light, at a proper temperature, into chlorophyll grains. You can grow part of the plant, a separate stem of it, in the dark, shut out the light from the leaves, and they will always be etiolated, yellow. In the given case no green colour will form. But if seeds are obtained on such a stem and plants grown from these seeds in the light, the leaves will be green-coloured, the chlorophyll grains will develop. In the preceding generation the green-colour character, the chlorophyll, had been missing but in the succeeding generation it appeared. There is no difficulty in understanding the cause of its appearance. The internal matter, in the present instance the plastids, which is transformed into chlorophyll grains existed in the leaves of the preceding generation. These plastids reproduced themselves, entered into metabolic relations with other substances of the living body and in the final analysis participated in the creation and development of the sex cells, the germs of the future generation. But the plants of the succeeding generations continued the normal development of the plastids into chlorophyll grains when their leaves were exposed to the action of light. The plastids possessed this property also in the preceding generation but did not develop chlorophyll grains because the necessary conditions, i.e., light, were absent. By such reasoning one can readily understand the cases where particular characters or properties of organisms do not develop for many generations and then suddenly appear and develop. The latent internal possibilities find the conditions they need for development, find a suitable environment, which explains the appearance of this or that character or property not possessed by the previous generation.

All properties and characters of an adult organism may, in a certain sense, be called latent, recessive, i.e., not manifest while the organism exists in the form of an embryo or germ. In the zygote all the characters and properties of the organism exist in latent form, as it were.

It has already been said above that a living body reproduces itself, that the different cells, granules and molecules of the body possess different natures—different heredities, different properties.

The molecules of the protoplasm and the molecules of the chromosome likewise possess, if one may put it so, different heredities, different natures. But all these living granules reproduce themselves both by means of growth and by means of development.

Proceeding from these premises we assume, and in particular instances can prove experimentally, that if you take separate groups of cells, separate parts of an organism as the fundamental, initial element you obtain a new generation having different properties and different characters, i.e., a different heredity from the one there was, or, generally speaking, from the old initial breed or variety. This can be observed, for example, in potatoes when they grow adventitious buds, i.e., eyes from the tuber pulp. After rearing plants from such eyes a new breed, i.e., a variety with different properties, is frequently obtained.

Such facts go to show that different cells of the same organism may possess different natures, different heredities. It goes without saying that organisms cannot be grown from all cells. Many cells do not possess the property of restoring the whole organism.

The same line of reasoning may be applied to difference in quality in the sense of the heredity of separate parts and separate granules of a cell. Changes in separate parts of a cell, such as, for instance, separate chromosomes, should (and this is frequently proved experimentally) bring about a change in the various organs, characters and properties of the organism obtained from this cell with altered chromosomes or sections of chromosomes, or with changed granules of protoplasm in the initial cell.

A change in this or that section or granule of the initial cell in various degrees affects the changing of the various characters and properties of the organism obtained from this cell.

Not all granules of the original cell or group of cells are to the same extent starting points for the development of the various characters and properties of the organism. At the same time it is necessary to know that individual granules of the original initial cell cannot turn, cannot develop into organisms. This requires a totality, a complex of all granules, i.e., a whole initial cell or, as for instance in vegetative multiplication, a group of cells.

ORGANISM AND ENVIRONMENT

The relative purposiveness and adaptability of the plant and animal world to external conditions and to their surroundings, as well as the harmony, the adjustment of the various organs of the organism to the discharge of particular functions, are excellently explained by Darwin's theory of natural and artificial selection. Changes beneficial to development and survival under given conditions make for the numerical increase, the propagation of such individuals, while changes deleterious to survival make for a decrease in the number of such organisms. This explains the progress, the constant process of perfection of plant and animal forms in nature. In practical farming, plant varieties and animal breeds are improved by means of artificial selection.

Three interconnected factors enter into our conception of natural and artificial selection: heredity, variation and survival. The diversity of plant and animal forms both in nature and in practical farming was created and is still created by natural and artificial selection. But the source, the material out of which organisms create or build themselves, is the environmental conditions—food in the broad sense of the word. Different living bodies elect different conditions from the environment, in accordance with their natures, assimilate them and build their bodies in conformity with the laws of their individual development, i.e., in accordance with their heredity.

Different species and genera of plants and animals require different environmental conditions for their life and development. Likewise, the same organisms require different environmental conditions in different periods of their lives. For instance, winter plants require low thermal conditions for one of their periods, now called the phase of vernalization. During the other periods of their life winter plants do not require low thermal conditions. Finally, the same plant organism requires different environmental conditions at one and the same time, but for the life and development of different organs and for undergoing different processes. Thus, for instance, the conditions required for the development of the leaves and roots of a plant differ. In general, the development of the different cells, different parts of cells and separate processes of an organism requires different conditions of the external environment. Besides, these conditions are differently assimilated.

It must be stressed *that by external we mean all that is assimilated and by internal that which assimilates.* The life of an organism is complicated and proceeds through innumerable regular processes or conversions. As a result, food imbibed or entering an organism from the external environment is assimilated by the living body through a chain of various conversions, and from being external becomes internal. This internal matter, being living matter and entering into metabolical relations with substances from other cells and particles of the body, feeds them, as it were, and thus as regards them becomes external. Organisms, beginning with zygotes (fertilized sex cells), develop by means of manifold regular changes and transformations of the body, and they become adult, capable of forming the same kind of sex cells as those from which they themselves originated. This is what constitutes the course of the individual development of plant organisms.

If a plant organism does not find in its environment the particular conditions required by the nature, i.e., the heredity of a particular process or character, the process or character in question does not develop. In these cases the internal potentialities, i.e., the heredity, for the development of the given character exist. But the character fails to develop on account of the lack of the necessary environmental conditions, i.e., of the material necessary to enable the character to build itself. In cases where the failure of a particular process or character to develop does not disturb the general life and further development of the organism, the latter may continue to live and develop normally without developing the character or property in question. However, the undeveloped characters or properties of such organisms will, as they say, be latent or recessive. These characters or properties may develop in the succeeding generations if the environment contains the requisite conditions. For instance, winter wheat plants of the Ukrainka variety on ripening produce in some years spikes with black awns and in other years with white awns. Seeds obtained from black-awned and white-awned ears of the named variety, sowing conditions being equal, produce plants with awns of the same colour. The awns are white or black depending upon the year, i.e., upon the conditions under which they are grown. This indicates that in cases where ripe plants of the Ukrainka variety with white awns are obtained, the external environment did not contain the conditions without which black pigments cannot develop. But the internal conditions, the heredity, the potentiality, the requirement for the realization of this character exist. The awn cells contain the substance which on further development might have developed into black pigment, but owing to the absence of certain external conditions this substance did not develop further and the awns remained white. Thus, in the case in question the white-awned plants possess those elements of the body which were not converted into black pigment only because their development ceased. But these elements, like all other granules and particles of the living body, may reproduce themselves in their progeny by means of metabolism, as a result of which they are incorporated, accumulated, in the sex cells.

The category in question also relates to cases of reversion, i.e., the appearance in a particular generation of characters and properties which were lacking in its immediate parents but which existed in its earlier ancestors. There are many such examples and they are generally known.

In the same way we explain the so-called fluctuating variations of plant organisms of the same nature, i.e., the same heredity. Many of the properties or characters possible in a particular variety of plant remain recessive in each concrete case, i.e., do not develop without causing substantial damage to the organism as a whole. Hence, under various environmental conditions one may observe a diversity of plants (phenotypes) belonging to the same variety, i.e., with identical heredity. The internal hereditary potentialities of development of particular characters are not realized and the characters do not develop for lack of the various environmental conditions. As a result different plants are obtained but of relatively the same nature, i.e., heredity.

No organism ever wholly realizes all its hereditary potentialities. Many properties and characters do not develop fully, remain to some extent undeveloped, recessive, without substantially affecting the development of the organism as a whole. But a plant has characters and properties, the lack of development or even underdevelopment and incompleteness of which hamper all further development and in some cases even prevent the organism from continuing to live. Such properties and characters of the organism clearly cannot be recessive, i.e., latent, since the whole organism will cease to develop if they do not develop. For instance, if you sow winter-grain seeds in spring they will produce shoots and will for a long time be in a state of tillering, and roots and leaves will develop. But such plants are unable to initiate the formation of spikes and straws or reproductive organs. In winter plants sown in spring the process called vernalization cannot take place for lack of low thermal conditions. But spikes and straws cannot develop in cereals unless the plants have undergone the process of vernalization, i.e., without a corresponding qualitative change in the content of the cells in the cone of growth, although the environmental conditions for the development of these organs exist during both the spring and summer periods. In these cases the vernalization process, which does not develop but remains recessive, as it were, is the internal cause of the failure of the plants to develop further, of their failure to advance toward the formation of new seeds. Clearly no offspring can be obtained from wheat plants which formed no seeds. In the case in question the lack of seeds was caused by the failure to go through the process of vernalization. These are the premises from which we proceed when we say that processes, characters and organs which play a substantial part in the general development of the organism (as, for instance, vernalization) cannot be recessive, or latent, in an adult organism, because without them there can be no adult organism. Such characters and properties of adult organisms can remain recessive, latent, only when these organisms possess a dual heredity with regard to such character or property. For instance, as regards the property of vernalization, the winter habit may be recessive in crosses between winter and spring plants.

In the life of an organism different degrees of importance attach to the different processes it passes through, to the different characters and organs it develops. As has already been said, the development of an organism as a whole depends upon the development of some properties or characters to a small extent, upon the development of other properties or characters to a greater extent and, finally, upon the development of still other characters to such an extent that the organism cannot develop and frequently even cannot exist without them.

The characters and properties of the first kind, when they develop or remain recessive, in the main produce the diversity generally observed in crops, particularly under varying environmental conditions. The diversity of plants obtained as a result of the various degrees of development of the characters or properties which do not substantially affect the life of the organism as a rule change the heredity of organisms but little. The various particles or granules of the body which remained recessive, undeveloped, take part in the general biological metabolism of the organism and as a result accumulate, become fixed in the sex cells. Characters which in the preceding generation were underdeveloped because of the absence of particular environmental conditions will develop in the succeeding generations if these conditions are then present. Thus, the heredity of the underdeveloped, recessive characters is reproduced in each new generation in the same way as the heredity of all the other characters and properties of the organism that were not recessive.

Development is always connected with a qualitative change of that which develops. In the development of plant organisms two kinds of such qualitative changes may be observed.

1. Changes connected with the realization of individual development, when natural requirements, i.e., heredity, are normally met by *corresponding external conditions*. The result is a body of the same breed and heredity as the preceding generations.

2. Changes in breed, i.e., changes in heredity. Such changes are also the result of the realization of individual development, but deviating from the normal, usual course. Changes in heredity are, as a rule, the result of the organism's development *under external conditions which, to one extent or other, do not correspond* to the natural requirements, i.e., its heredity.

The individual development of an organism, as has already been said, is a chain of regular transformations. If these transformations of the living body do not transcend the norm, i.e., if they are the same as they were in the preceding generation when the particular character or process was developing, there will be no change in heredity. In the generation in question the heredity is the same as in the preceding generation. However, in the development of the individual, deviations of the transformations from the norm, i.e., from the quality of analogous transformations occurring in the preceding generations, cause change in breed, in heredity.

The more the environmental conditions suit the requirements, i.e., the heredity of the organism, the more will the development of the given organism resemble that of the preceding generations and, consequently, the less will its heredity change, deviate from type, from the norm. When the organism does not find in its environment the conditions necessary for the development of particular organs or characters, these organs or characters may fail to develop altogether if they can remain recessive without detriment to the general development of the organism. If, however, the organism as a whole cannot continue to live and develop without their development, it either ceases to develop or the usual course of the process, the usual development of the organs and characters, must change and take a direction which corresponds to the new, unusual conditions. Thus, *changes in the conditions of life which necessitate changes in the development of plant organisms are the cause of change in heredity*. All organisms which cannot change in accordance with the changed conditions of life do not survive, leave no progeny.

Organisms, and hence also their nature, are created only in the process of development. A living body may change also without development but these changes will not be characteristic of living bodies. Changes occurring in living bodies without a development of these bodies will, as a rule, involve a diminution of their vitality. For instance, seeds—the embryos of organisms of different plants—when stored do not develop like organisms, but when stored too long or stored under conditions that are not normal, changes take place in the cells of the embryos. Therefore, the heredity of such seeds may also change. But such changes, as a rule, will lead to a decrease in vitality. The seeds may, due to long storage, be ruined, become less germinative, less vital.

In the development of plant organisms the heredity usually least subject to change is that of recessive characters, the undeveloped or underdeveloped state of which does not cause substantial harm to the general development of the organism. On the other hand, the heredity of those characters and properties of an organism whose development plays a substantial part in the life of the individual is in our opinion more frequently subject to change. If the external conditions do not correspond to the normal course of development of these characters or properties, either the course of their development must undergo adaptive change or the organism as a whole will cease to develop, to live.

In each new generation plants strive to manifest properties and characters which existed in the preceding generations. Some of the characters and properties which existed in a preceding generation may in the particular plant, owing to the lack of suitable external conditions, remain and as a rule always do remain in underdeveloped form; as they say, the character remained in a latent, recessive form. And, conversely, some of the characters and properties which did not manifest themselves in preceding generations may do so in the given generation. In other words, where the heredity is relatively the same, the external appearance of plants of different generations or of different plants of the same generation may (and always does) differ more or less.

The diversity of plants with relatively the same heredity (i.e., of one variety) is due to the various degrees of development of many of its properties and characters, one of which taken separately does not play an essential part in the general course of development of the organism as a whole. The heredity of such characters and organs, which easily vary in the development of the individual, is usually least conservative and lends itself most to change. Of this one can easily convince oneself by raising new plants from cuttings of tissue taken from such easily variable organs.

It is different with organs, characters and properties whose development plays an essential part in the life of the organism. Everything in the organism is directed towards providing the development of such organs or characters with conditions which do not transcend the norm. Therefore, the development of such characters varies considerably less. Their heredity is usually more conservative and lends itself less to change, as it is guarded and protected to a great extent by the entire system of the organism as a whole.

Variation in the heredity of plant varieties propagated by seeds takes place, as a rule, through a change in the heredity of conservative characters, which vary with difficulty. It depends to a considerably smaller extent on the variation of the easily variable characters and properties of the plants.

Yet it has been repeatedly stated that a change in the body is always linked up with a change in its heredity. We seem to have arrived at a contradiction. A variety changes to a greater extent in those characters which are more conservative, less capable of change in the development of the individual, and on the contrary, a variety changes to a considerably smaller extent in those characters which in the development of the individual are less conservative and lend themselves more to change (variation). On this ground geneticists arrived at a wrong theoretical conclusion. They properly attribute the fluctuating variation of the characters and properties of plants and animals to the varying conditions of the environment. But inasmuch as the alteration of a variety goes on, as a rule, in the characters and properties which vary to a considerably smaller extent in the development of the individual, they conclude that a change in variety, and hence a change in breed, i.e., heredity, does not depend on the conditions of life at all but on certain unknown causes. In their opinion the causes of mutations have not been discovered to this day. An organism's characters and properties vary on account of the variation in the conditions of life, but its heredity, its breed, does not change. Consequently, the living conditions are not the cause of changes in varieties.

As a matter of fact the heredity of a living body changes normally only when this body develops. Whatever does not develop in the living body does not change in the sense of developing. It can change only in the sense of being destroyed or fading away.

If the external environment does not contain the conditions required for the development of the various characters and properties which do not play a substantial part in the life of the organism, these characters and properties do not develop and consequently do not change. Recessive characters are, as a rule, the most stable, i.e., the least changeable. This or that character may not manifest itself for many generations in the plants of a given variety, may be latent for lack of the necessary environmental conditions. When the necessary conditions exist, recessive characters and properties can develop and assume the same form as in generations of the distant past. The heredity of such characters does not change for the reason that the latter did not develop. On the other hand, fluctuating variations in the individual development of many properties and characters of the progeny do not alter the characters or alter them but slightly for the following reason. As their properties are beyond the norm, the substances of the changed (variable) characters are not included in the processes, passage through which results in the appearance of organs or parts of plants which form the basis of future generations, as for instance seeds. Thus, a variety propagated by seed usually changes little on account of a change in the variable characters, as they are called. This is so not because changes in characters are not determined by the action of environmental conditions, of conditions of life, but because the changed nature of the particular body parts of the organism is not included, or is included to a small extent only, in the chain of processes which leads to the formation of seeds. But if the changed characters or organs are taken as the starting point, as the basis for future organisms, the progeny too will be changed, i.e., the variety will be altered.

If the environment does not contain the conditions suitable for the development of the characters and properties without which the further existence of the organism is impossible, such characters and properties cannot easily remain recessive. They develop perforce, as they say, else the organism must cease to exist. Although the action (especially the prolonged action) of unusual, unaccustomed conditions of environment does bring about the development of the particular characters or properties they will, nevertheless, develop differently from the way they developed in the preceding generations, when environmental conditions were normal. As a result, a more or less different living body is obtained possessing in consequence different properties and, of course, a different heredity, i.e., different environmental requirements.

Changes in the nature of organisms and in their various properties and characters are always in some measure enforced. For lack of necessary conditions suitable to the nature of the given living body it is compelled to assimilate conditions differing to some extent or other from those required. As a result a different body, and hence also a different nature, or heredity, of that body, are obtained. If one adopts this point of view one readily arrives at the conclusion that the heredity of the different sections of a plant from which a whole organism can be regenerated is frequently different. This can easily be corroborated experimentally in many cases. We have already pointed to the experiment of obtaining various breeds of potato tubers from one initial tuber by inducing eyes in the particular tuber from various sections of its pulp. Reference may also be made to such well-known facts as the appearance on fruit trees of separate buds or twigs, with hereditary properties and characters differing sharply from those which are characteristic of the tree as a whole. The changed characters which play an essential part in the development of an organism as a whole are usually transmitted more frequently to the seed progeny than characters of less importance. This is so for various reasons, one of which, in plants, is the multitude of characters of the same type. The more characters of one type (for instance, leaves) there are, the less a change that exceeds the norm for each character is transmitted in the progeny.

DIRECTED CHANGE IN THE BREED OF ORGANISMS

There arises the question: why then do we have relatively permanent breeds of animals and varieties of plants, i.e., heredity, in animate nature and practical farming? Everybody knows of cases where animal breeds and plant varieties, as well as species and strains, have been preserved for decades on farms and for centuries in the state of nature. In such a space of time scores and hundreds of generations succeed each other, but as concerns their nature, i.e., their heredity, they are indistinguishable or almost indistinguishable from each other. This generally observed phenomenon likewise seems to contradict the view we have expressed above that the nature, i.e., the heredity, of an organism is compelled to change when the body of the organism changes under the influence of the conditions of life. After all, many characters and organs in each generation develop differently whenever they encounter conditions which are relatively different from those obtaining in previous generations. Relatively different characters, organs and properties with consequently different heredities are obtained. Logically it would seem that these organs and characters ought to reproduce their exact copies in the following generation. In actual fact, however, the innumerable experiments made have not confirmed this. Experiments of this kind are rather easy; anybody can repeat them. For instance, take seeds of some particular variety of, say, wheat and grow plants from them: some under good and others under bad conditions of nutrition and care. The plants obtained will differ sharply in external appearance. The plants grown under good conditions may be dozens of times the weight and size of those grown under bad conditions. The difference will be not only quantitative but also qualitative. It would seem that the heredity of such different living bodies (plants) should be different. But if seeds from these different plants are sown under equal conditions the plants that grow from them will, as a rule, differ little from each other.

Hence, one may arrive at the conclusion that change in the living body apparently does not entail change in its heredity, i.e., its nature, and that consequently it is useless to look in the conditions of life of plants and animals for ways of directing change in the nature of organisms. This is precisely the erroneous conclusion arrived at by representatives of the science of genetics. Due to this cardinal error science (genetics) found itself at loggerheads with practical farming, with seed growing and pedigree stockbreeding.

I emphasize the fact that a rather large number of experiments were performed which seemed to prove beyond dispute that heredity is constant while the quality of the body is relatively variable. Moreover, they can easily be repeated if so desired. Let us refer to an experiment performed at the Belaya Tserkov Sugar-Beet Breeding Station. On a field sown throughout to one variety of sugar beet, 10,000 of the biggest roots were picked in autumn. Their average weight was 750 gr. On the same field were also picked 10,000 of the smallest roots, whose average weight was 150 gr. The two groups of roots, the biggest and the smallest, were planted apart from each other to avoid cross-pollination, a blending of the heredities of these two groups of plants possessing different weights. The seeds obtained from them were sown under identical conditions. It appeared that the average weight of the roots obtained from the big-root group was 317 gr. and of those from the small-root group 312 gr. The average weight of the roots was thus almost the same regardless of whether the seeds were taken from the biggest or the smallest sugar-beet roots. The logical conclusion may be drawn (and is drawn again and again by geneticists) that the conditions of life, of agrotechnique, undoubtedly affect crop yields, i.e., the development of the quantity and quality of the living body, but they do not exert any influence upon the quality of its nature, do not affect any alteration in its heredity.

This explains why some geneticists came to the conclusion that in sowing plants for seed as well as in breeding pedigree livestock the application of good agrotechnique or zootechnique, i.e., good feeding of pedigree animals and good tending of them, is not only unnecessary but frequently even wasteful. Such scientists are of the opinion that good farming technique or good feeding will only increase the number of seeds or the quantity of livestock products obtained, but that the quality of the heredity of the seeds or of the young stock will remain the same as with bad and cheaper agrotechnique or zootechnique. Yet it is a known fact that in *practical life good varieties of plants or animals are always produced only by the application of proper methods of cultivation or breeding. Under poor cultivation no good varieties can ever be produced out of poor ones, and in many cases even good cultivated varieties will deteriorate after a few generations.* It is a basic rule in seed growing that plants grown for seed must be tended with the utmost care. They must be provided with conditions meeting the heredity requirements of the given plants. Of well-cultivated plants the very best are selected for seed. That is the way varieties of plants are improved in practice. Under poor cultivation (i.e., when agrotechnique is bad) no selection of the best plants for seed will produce the required results—all the seeds will be poor, and even the best among them will still be poor.

It must be firmly borne in mind that while good agrotechnique, the creation of good conditions for the growing of plants, does not always improve their nature, i.e., their heredity, it never makes it worse.

If a detailed analysis of the question of change in heredity under the influence of the conditions of life of plants and animals is made it appears that there is no contradiction between the factual experimental material of the geneticists, on the one hand, and the supposedly opposite facts of practical farming, on the other. The facts set forth seem contradictory only to those geneticists who do not know life, who do not know farming practice. Therefore, the conclusions which the geneticists draw from their above-mentioned experiments radically contradict good seed growing and pedigree stockbreeding practice. They further contradict the Darwinian theory of development of plant and animal forms.

The facts show that changes in the various sections of the body of a plant or animal organism do not become fixed, are not assimilated, with equal frequency in the sex cells, i.e., in the reproductive products. The geneticists claim instead that no changes in the properties, characters or organs which arise from the conditions of life induce changes in the heredity of these properties, characters or organs. The fact that the progeny obtained in their experiments with plants and animals under various conditions remain unchanged with regard to these properties serves the geneticists as the grounds on which to base their assertion. Actually a living body that has been qualitatively changed by the conditions of life always has a changed heredity. But not always by far can qualitatively changed parts of the body of an organism enter into normal metabolic relations with a number of other parts of the body, owing to which these changes cannot always be fixed in the sex cells. The offspring will therefore frequently not possess the changed heredity of a particular changed part of the body of the parent organism or this change will be expressed in weakened form or to a minor degree.

This is explained by the fact that the process of development of each organ, of each granule of a living body, requires relatively definite environmental conditions. These conditions are elected by each process, by the development of each organ and property from the surrounding environment. Hence, if a particular part of the body of a plant organism is forced to assimilate conditions to which it is not accustomed (qualitatively or quantitatively) and if owing to this the given part of the body becomes changed, different from analogous body parts of the preceding generation, then substances passing from this part of the body to the neighbouring cells may not be elected by them, may not be included in the further chain of corresponding processes. Of course, there will be some connection between the changed part of the body of the plant organism and the other parts of the body, otherwise it could not exist; but this union may be incomplete or not mutual. The changed part of the body will receive some nourishment or other from the neighbouring parts; but it will not give up its specific substances as the neighbouring parts will not elect them. In their nature these substances are not native to the processes going on in these parts of the body. Into these processes will enter the conditions, the food suitable to them; and they can obtain this food from other, qualitatively unchanged parts of the body.

This explains the frequently observed phenomenon that certain changed organs, characters or properties of an organism fail to manifest themselves in the heredity of the progeny. At the same time we stress the point that these altered parts of the body of the parent organism possessed an altered heredity. These facts have long been known to practical horticulture and floriculture. A changed twig or bud of a fruit tree or an eye (bud) of a potato tuber cannot, as a rule, induce a change in the heredity of such offspring of the particular tree or tuber as do not derive directly from the changed parts of the parent organism. If we cut off this changed part and grow a separate, independent plant from it the latter will, as a rule, possess the changed heredity in full, i.e., the heredity of the changed part of the parent's body.

Should separate links in the general chain of development of a plant organism be unable to find the conditions their nature requires, the substances of the changed part of the body will perforce, as it were, be included wholly or partly in the chain of these processes and thereby participate in the development of the reproductive products. Therefore, a change in the nature of separate sections of the body of a plant organism may leave the heredity of its offspring entirely unaffected, may affect it partly or, lastly, may be transmitted in its entirety. *The extent of the transmission of alterations depends on the extent to which the substances of the altered section of the body join in the general process which leads to the formation of reproductive sex or vegetative cells.* In nature this depends on the accidental conditions that the plant in question encounters, while in experiments and practical farming it depends on man's knowledge and skill.

It is a known fact that environmental conditions do not depend upon the different plant organisms. Organisms only possess definite requirements for various conditions. Whether the environment will contain these conditions, and whether they will be of the requisite quality, in the requisite quantity and at the requisite time, does not depend upon the plant. At the same time the life of plant organisms and the quantity and quality of their bodies depend upon the environmental conditions. As they say in practical farming: agrotechnique determines the quantity and quality of crops. As we have already indicated, with good agrotechnique and good conditions of cultivation, plants may be obtained ten and even more times the weight and size of plants of the same variety (of the same breed) produced under exceedingly bad conditions. Here is a case in point. A millet plant which accidentally grew on a bare fallow weighed, with roots, stalks and panicles, 953 gr. Another millet plant of the same variety, which grew on a road near the field, weighed at the same stage of ripeness, together with roots, stalks and a panicle, 0.9 gr., i.e., one plant was more than a thousand times as heavy as the other. Thus, although plant organisms possess the capacity to elect environmental conditions, nevertheless, since the latter are independent of the organism, and organisms build their bodies out of the environmental conditions, the result is that the body of the organism depends largely, in both quality and quantity, upon the conditions of life. Different conditions give rise to different plants and these differences are frequently very great.

How is it then that, in spite of the marked variability of the parent organisms, the development of particular organs and characters (both in the quantitative and qualitative sense) and the nature of the offspring, i.e., the heredity of these plants, remain rather stable, relatively unaltered? Does this not argue in favour of the proposition that a change in the body of an organism does not entail a change in the nature, i.e., in the heredity, of this body? This is explained in part, as was pointed out above, by the fact that the changed parts of the body are frequently altogether excluded, or are included to a small extent only, in the metabolic relations with those links in the process as a result of which reproductive cells are obtained.

It must also be noted that not all processes in an organism, the development of not all organs and characters, are supplied with nutriment of the required quality and quantity in equal measure and with equal timeliness. Not all processes in an organism are of equal importance for the maintenance and propagation of the given species, strain or variety of plant.

It has already been stated that the characters and properties whose development does not substantially influence the life of the organism as a whole remain, as a rule, undeveloped, recessive, when there is a deficiency of the required environmental conditions. Let us add that when there is an excess of the required conditions these same characters develop, as a rule, likewise excessively, considerably above normal. In other words, the development of such characters is most variable, most fluctuating. But the characters or processes upon whose development the life of the organism as a whole largely depends vary less, fluctuate less.

If particular nutritive elements for the normal development of the entire plant are lacking, the first to starve, i.e., receive a subnormal quantity of food, will be the least essential organs, the least essential parts of the body. The processes that are more important to the organism will suffer to a smaller extent from an insufficiency of particular nutritive elements, and to a still smaller extent will those upon which the continued existence of the plant's race mostly depends. For instance, it is a well-known fact that if any farm animal is fed to excess it develops a thick layer of fatty tissue. If fed insufficiently the fatty tissue will not only cease to receive food but will itself be spent on feeding the other tissues of this organism. After the fatty tissue has been spent on feeding the organism, comes the turn of the muscular tissue, etc. In general, when animals starve the nervous and certain other tissues starve least. This is our explanation of why plants such as the two tufts of millet we took as examples, which were produced under sharply different conditions in point of nutrition and of which one exceeded the other more than a thousand times in size and weight, are far from completely transmitting these differences to their offspring. The plants were nourished quite differently, but the nutrition of the separate parts, of the separate processes of those plants, departed from normal in various degrees. The main processes of the plant that had plenty of food were protected from an excess of it, the part above the norm being absorbed by other, less important processes. On the other hand, in the case of the plant which received insufficient nutrition, the main processes starved least of all.

Therefore, although the plants sharply differed in their development and departed in opposite directions from the norm, the processes upon which the continuation of the race depends most were fed quantitatively and qualitatively approximately according to the norm. After all, the size of the seeds taken from these two plants—plants which exhibited a thousand-fold difference in weight—was almost exactly the same. Moreover, the embryos in these seeds, being the most important part, differed still less from each other. And, finally, the most essential parts of the embryos must have differed least of all.

Thus, we give the following explanation for the absence of change in the heredity of the offspring when various characters or properties of the parent plants change or when these changes are not completely transmitted (as is most frequently the case).

Firstly: the active election of suitable environmental conditions by the various processes for the development of the various organs and characters, of the various particles of the living body.

Secondly: the active noninclusion of unsuitable conditions in the process. Sections of the body changed perforce do not fully include the specific substances they produce, and frequently do not include them at all, in the general chain of the process leading to the formation of reproductive cells.

Thirdly and lastly: in the organism, as in an integral whole, there is no "equalitarian tendency" in the supplying of the different processes with the requisite elements of food. The more important processes are supplied with greater approximation to the norm; they are protected against insufficiency as well as against a surplus of nutrition in general or of particular elements of it. Less important processes, on the other hand, are supplied less than the norm, the norm or more than the norm, depending upon the amount available.

The question may arise: wherein, then, does our understanding of the interconnection of the nature, the heredity, of the organisms with the conditions of life differ from the point of view of the Morganist geneticists? The geneticists say that the conditions of life tend to cause only qualitative and quantitative changes of the body (soma) of an organism. Qualitative changes in the nature, i.e., the heredity, of the organism do not depend, however, upon the quality of the organism's conditions of life. True, we have also pointed out that an observed change in plant organisms resulting from the conditions of life is, as a rule, little reflected in the heredity of the offspring of these plants. But we maintain that a change in the body is bound to bring with it a change in the nature of that body. The offspring, however, of the organism in question, particular parts of whose body might change, will not always be altered. It would seem that such a divergence of opinion on the nature of organisms between ourselves and the Mendelist-Morganist geneticists should be immaterial in practice. For in practice you deal with seeds whose heredity, the geneticists assert, does not depend, but which we claim as a rule does depend, in one degree or another, upon the alteration of separate properties and characters of the parent plants. It would seem that this divergence of opinion is only theoretical in character. But such theoretical divergences are exceedingly important in practice.

After the victory of Darwinism the variability of the nature of plant and animal forms became a generally recognized phenomenon. But no concrete ways of changing the nature, i.e., the heredity, of plant and animal organisms were known to science, as has already been stated. Soviet agrobiology, the Michurin teaching, points out this way. A qualitative alteration of the living body is the sole way of altering the heredity of this body. However, the source of the maintenance of life and development, and this means also of change in a living body, lies in the conditions of the environment. Therefore, skilful regulation, skilful action at the requisite moments upon particular organs or parts by the requisite environmental conditions is the sole method, the sole lever for regulating not only the organism as such but also its nature, i.e., its heredity. In nature this is all done randomly. Under the influence of environmental conditions particular processes, the development of particular organs, change, adapting themselves to these conditions. If the substances of the changed organ or process are included to any extent in the chain of processes leading to the formation of reproductive cells, the changes are transmitted to the progeny. Changes which favour the survival of offspring impart to the organisms an advantage in life and development. But if the changes are harmful, the organisms which have undergone them will have less chance of survival and these changes will not be perpetuated.

The variability of the processes of development of organs and characters is always adapted to the environmental conditions, but it must be remembered that the property of adaptability will not always be analogous to purposiveness. The relative purposiveness and harmonious nature of plants and animals in nature have come into existence only by natural selection, i.e., by heredity, its variation and survival.

When we know the way of building up the heredity of an organism we can change it without waiting for suitable occasions by creating definite conditions, definite action at a particular moment of the organism's development. The better we know the concrete laws of development of the various plant organisms the sooner and the more easily we shall obtain, create the requisite forms and varieties of these plants. Up to now good practical seed growers knew only that while good agrotechnique, the proper growing of seed plants, does not always improve their nature, it at any rate does not make them worse. On the other hand, bad growing conditions quite frequently, if not always, worsen the breed of varieties and never improve it. A knowledge, therefore, of the concrete laws of development of the nature of the plants in question offers the possibility of directing, of changing the nature of organisms in the required direction at any time, without waiting for chance.

VEGETATIVE HYBRIDS

Morganist geneticists conceive of an organism as consisting of the ordinary body, known to all, and a "hereditary substance," i.e., a body which they claim is known only to them (though not one of them has so far actually seen or felt this body). The first body (soma), the ordinary body, discharges various functions of the organism. It depends upon the conditions of life and changes when these conditions change. The second body is the "hereditary substance," which, in the opinion of these geneticists, discharges only the function of reproducing the properties and characters that are similar to those of the preceding generations. Hence their definition of heredity as being only the property of an organism to reproduce beings similar to itself.

Our conception, on the contrary, is that the entire organism consists of only the ordinary body, known to all. The organism contains no special substance separate and apart from the ordinary body. But any particle, or, figuratively speaking, any granule, any droplet of the living body, once it is alive, must unfailingly possess the property of heredity, i.e., of requiring conditions suitable to its life, growth and development.

As we know, hybrids are organisms which possess the properties of two breeds, one maternal and the other paternal. In different cases particular properties of either parent predominate in the offspring in different degrees.

Hitherto sexual reproduction, the sexual union of the organisms of two breeds, was the sole method generally recognized by science of producing hybrids. Darwin and a number of other eminent biologists considered it possible to obtain also vegetative hybrids, by blending two breeds and obtaining a third, not only by means of crossbreeding but also by means of vegetative union. I. V. Michurin not only admitted the possibility of the existence of vegetative hybrids but also elaborated the mentor method. This method consists in grafting cuttings (twigs) of different varieties of fruit trees on the crown of a young variety whereby properties lacking in the latter are acquired by it, transmitted to it from the grafted twigs. This is why I. V. Michurin called this the mentor method, using mentor in the sense of trainer, improver. By this method Michurin produced many good new varieties and improved many existing ones. The Morganist geneticists of course do not deny but recognize the good varieties produced by Michurin. Yet they have refused to recognize the method of producing these varieties, and, in particular, the mentor method, i.e., vegetative hybridization, asserting, contrary to Michurin, that these varieties are obtained independently of the influence of the cuttings grafted on the crowns of the young varieties of trees possessing their own root systems.

Vegetative hybrids provide cogent proof of the correctness of our conception of heredity. At the same time they represent an insurmountable obstacle to the theory of the Mendelist-Morganists. This furthermore serves to explain why the Michurinists, who proceed from the facts and laws of objective living nature, recognize the possibility of existence of vegetative hybrids. The Mendelist-Morganist geneticists, however, deny this possibility.

The Michurinists, beginning with I. V. Michurin himself, have found methods of obtaining vegetative hybrids in large quantities. The Mendelist-Morganist geneticists, on the other hand, long denied the various cases of vegetative hybrids that had been known from time immemorial. Examples of vegetative hybrids, such as the *Cytisus Adami*, a cross between the hawthorn and the medlar, and others, were cited even by Darwin. But the geneticists considered all these cases to be not hybrids but so-called chimeras, by which they meant organisms in which the tissues of various breeds are vegetatively coalesced but not biologically blended. The geneticists used to assert that such organisms cannot sexually reproduce offspring with hybrid properties. But when the Michurinists during the last few years found a method of obtaining vegetative hybrids in large quantities and when these hybrids even in the seed offspring behaved like ordinary sexual hybrids, the geneticists could no longer find grounds for objection. They merely ignored these facts, at times calling them experimental errors. But they will not undertake to repeat these investigations for fear of obtaining vegetative hybrids.

Reference is frequently made to the generally known phenomenon that the grafting, on the most diverse stocks, of various breeds of fruit trees which in practice are propagated only in this way does not change the inherited properties of the grafted varieties. But in this instance it is forgotten that these varieties of fruit trees have already been fully formed, have already gone through their phasic development. Therefore, they cannot suffer any change in those properties and qualities which long before, prior to the grafting, had completed their cycle of development. A different result is obtained when young varieties of fruit trees, which have not yet been fully formed, are grafted on. They, in such event, change, as a rule, the entire course of their further formation.

It should be known *that the whole process of development* of plant organisms, for instance, of annual cereals, *consists of separate successively connected processes, stages, phases of development successively passing from one to the other.* It is easy enough to prove by experiment that, for instance, winter plants that have not completed the process called vernalization phase cannot pass through all the processes subsequent to that phase. Besides, after going through the processes of the vernalization phase or, for instance, the next phase, namely, the photo phase, these plants will not pass a second time through the vernalization phase or the photo phase no matter how much they are propagated vegetatively by cuttings, i.e., tissues which developed from tissues that had already gone through the vernalization or the photo phase.

All this makes it clear that in practice *old, fully formed varieties of fruit trees can and should be propagated by grafting without any risk of losing or changing their good hereditary properties*. On the other hand, *organ-*

isms which phasically have not been fully formed, which have not yet completed their cycle of development, *will always,* on being grafted, *change their development* in comparison with own-rooted, i.e., ungrafted, plants. Vegetative hybridization is not only of great importance to practical farming but also of considerable theoretical value as an aid to a correct understanding of one of the most important phenomena of nature—heredity. When we unite plants by means of grafting we obtain an organism of a different breed, namely, the breed of the scion and the stock. By sowing seed taken from either of the latter we can obtain offspring some of which will possess the properties not only of the breed from whose fruits the seeds were taken but also of the other breed with which the first was united by grafting.

Everybody knows that only plastic substances, saps, are interchanged between scion and stock. The scion and stock can interchange neither chromosomes of their cell nuclei nor protoplasm. Nevertheless, inheritable properties can be transmitted from stock to scion and reversely. Consequently, the *plastic substances elaborated by the scion and the stock also possess breed properties*, i.e., heredity. They possess the properties of the breed by which they are developed.

The numerous cases in which vegetative hybrids were obtained in recent years are clear proof of the incorrectness of the very basis of the Mendelist-Morganist theory, according to which heredity is possessed by only a certain special substance, separate and apart from the ordinary body and concentrated in the chromosomes of the cell nuclei. Any assertion that the property of heredity is bound up with some special, separate substance is wrong, in whatever part of the organism or cell it be located. *Every living particle or even droplet of a body (if the latter is liquid) possesses the property of heredity, i.e., the property of requiring relatively definite conditions for its life, growth and development.*

Annual herbaceous plants provide very suitable material for the experimental production of vegetative hybrids with which to demonstrate that this is really a change of breed (a blend of two breeds) which can be transmitted from generation to generation also by sexual reproduction, i.e., through seeds, the same as with sexual hybrids. Tomato strains would be just right for the purpose. Two strains should be chosen, with sharp, clearly visible differences in, say, the colour of the fruits: red ripe fruits of one strain and yellow or white of another. The sharp difference may be expressed in the shape of the fruit thus: round for one strain and pronouncedly elongated for the other; or in the structure of the leaves, thus: nondissected, like those of potato plants, and dissected, like the usual tomato leaf. Two strains may be taken, differing in the number of fruit chambers—two-chambered and many-chambered, etc. Indicate the character whose alteration it is desired to follow up. For instance, the task may be assigned of converting the white colouring of ripe tomato fruits of the Albino variety into a red colouring and transmitting to the latter the character of the red-fruited variety; however, this is to be done not sexually (by crossbreeding) but vegetatively, by grafting a cutting of a young Albino organism on the stem of a more mature plant of the red-fruited breed. The younger the plant whose characters it is desired to change the more successful the experiment will be. On the other hand, the plants from which it is desired to obtain a particular property or character should be older. Those of middle age will do best. Not less than 10-20 graftings should be made. They are quite easy to do and take little time. After the grafts have coalesced the best thing is to remove the twig leaves of the breed to be changed as often as possible. But the breed from which a particular character is to be taken, or transmitted, should be left as many leaves and twigs as possible. In order to make the experiment more exact, the flower buds on the grafted twig should be isolated during the flowering period by a gauze bag to protect them against insect-borne foreign pollen (though tomatoes are self-pollinators). In a number of cases in such experiments ripe fruits showing various degrees of colouring may already be obtained from the grafted twigs which, according to their breed, are characterized by white-coloured ripe fruits. After the fruits ripen the seeds should be taken from them, especially from the red ones, if there be such, and sown the following year. A number of plants from such a crop will, as a rule, bear fruits which, when ripe, will be tinted red. This colouring was transmitted by the graft component of the preceding generation through the plastic substances. The same thing may be observed with regard to any other character. For instance, seed offspring of the two-chambered variety of tomatoes obtained after it had been grafted on to a multichambered variety were multichambered without repeating the graftings. Nonerect, trailing forms, after being grafted on to erect forms, transmit the erect habit through the seeds to a considerable number of offspring. Shape of leaf, length of vegetative period (early maturity or late maturity), size of the fruits (big or small) and a number of other characters and properties were transmitted by heredity to seed offspring in the experiments performed by the Michurinists, scientific workers and experimenters.

It may now be asked: why do not all plants obtained from seeds of fruits of grafted twigs clearly manifest hybrid properties? Why is it that in a number of cases, though in the examples we have referred to this will rarely happen, one fails to discover even a single plant of the hybrid type? The following may serve as an answer. Plants of a hybrid type are not obtained in all cases because breeds and the various processes in one and the same breed, as has already been stated, possess elective capacity, show preferences as regards their conditions of life, their food. It goes without saying that plastic substances elaborated by one breed are to some extent or other unsuitable, unfit to nourish the grafted component of the other breed. The grafted component may not take or assimilate them at all, or it may elect from all substances only those which suit it most and try to obtain all the rest from the leaves or from other parts of its own breed.

This explains why as few leaves as possible should be left of the component whose breed it is desired to change.

The percentage of vegetative hybrids obtained will depend on the ability of the experimenter to compel, to coerce the grafted twig (cutting) to assimilate a maximum of the nutritive substances elaborated by the breed whose properties it is desired to transmit to the grafted breed. It is essential that the experimenter overcome the "unwillingness" (electivity) of the processes of the grafted twig to include these substances in the process of building its body.

The experiments we recommend will, as a rule, prove successful in a greater or smaller percentage of the plants. After performing them it will become clear to any geneticist still believing in the fundamentals of Mendelism-Morganism that this theory is not only fallacious but even pernicious if applied to practical pedigree stockbreeding and seed growing.

It should be emphasized that abroad in practical seed growing (including plant breeding) as well as in pedigree stockbreeding no use whatever is being made of the genetic theory. Good practical seed and pedigree stockbreeders have themselves worked out, through experiments and observations covering hundreds of years, ways and means of improving the old varieties of plants and breeds of animals, and also of rearing new ones. In foreign countries the science of genetics is divorced from practical farming, for which reason theory may develop there for many years in a wrong direction.

The vast factual material on the vegetative transmission of various characters of potatoes, tomatoes and a number of other plants with which the scientific staff under our direction had to operate has brought us to the conclusion that *vegetative hybrids do not differ in principle from sexually-propagated hybrids. Any character may be transmitted from one breed to another by grafting as well as sexual propagation.* The behaviour of vegetative hybrids in succeeding generations is likewise analogous to the behaviour of sexually-propagated hybrids. When sowing seeds of vegetative hybrids, for instance, of tomatoes (without further grafting), the hybrid properties of the plants of the preceding generation appear also in the plants of the subsequent generation. The phenomenon of segregation, as it is called, frequently met with in the offspring of sexual crosses, occurs also in the seed generations of vegetative hybrids. But what is observed much more frequently, and in a much higher degree, among the latter is so-called vegetative segregation, when the body of an organism is mosaic with regard to various characters.

An example of interest for purposes of demonstration is the grafting of white-fruited tomato cuttings upon red-fruited tomato plants. When the seeds were taken from the fruits of the white-fruited tomato twig the majority of the plants obtained in the first seed progeny bore red-coloured fruits. In a minority of the plants the fruits were white or slightly reddish. In the second seed generation the vast majority of the offspring of white-fruited plants was white-fruited. Only a few plants here and there yielded fruits which were more or less reddish. As a rule most of the offspring of red-fruited plants were red. But about 20-30% of them were white-fruited. In general the same diversity is observed as in experiments with analogous sexually-crossed tomato varieties.

Of special interest is the behaviour of the third seed generation sown in 1942 in Frunze (Kirghiz S.S.R.) by I. E. Glushchenko, a researcher of the Institute of Genetics of the Academy of Sciences of the U.S.S.R. The seeds of the second seed generation were taken from the Moscow sector of the Institute. *On some of the twigs of part of these plants the fruits obtained were red (pink), on others white.* There were several dozens of such plants. It is supposed that this property can be perpetuated and that a form of tomato is possible in which the same shrub will yield white, red and pink ripe fruits.

Vegetative hybrids deserve particular attention in studying the so-called destabilization of heredity. *They are exceedingly plastic material for the further building up of new breeds through the influence exerted by the conditions of cultivation.* Thus, for instance, the medium-maturing variety of tomatoes named Best of All when grafted on the nightshade (a weed) brought about changes in a number of characters. A vegetative hybrid was obtained. Not one of the properties inherent in the Best of All variety of tomatoes remained unchanged. A. A. Avakian selected plants which even when propagated by seed produce fruits with vastly improved flavour. The shape of the tomato fruit variety taken for grafting also changed. The vegetative hybrids of these tomatoes produced forms which first acquired early maturity from the nightshade and subsequently, under the influence of the conditions of cultivation, became still earlier-maturing plants. We obtained the earliest cultivated tomatoes known to us. When these forms were seed-planted out-of-doors (and not pricked out) in the beginning of May at the Experiment Base of the Lenin Academy of Agricultural Sciences of the U.S.S.R. in Gorki Leninskiye (near Moscow) both in 1941 and 1942, they yielded good ripe fruits before the autumn frosts set in.

In many cases vegetative hybridization is of great practical value for improving the varieties of plants bred, as well as for transmitting particular properties to the old, already existing strains of annual plants.

Vegetative hybridization may serve as a graphic demonstration of one of the most important phenomena of biology and will thus make it easier to understand this phenomenon, namely, how *the conditions of life,* the external environmental conditions, *on being assimilated, incorporated as component parts of the living body, become internal conditions.* For instance, various elements of soil solution, on being for the first time assimilated perforce by the living body of the plant, biochemically included in the composition of its body, become conditions essential for the growth and development of this altered body.

To explain this thesis let us analyze the facts relating to the transformation of the white-fruited breed of the tomato variety into the red-fruited breed by means of grafting. In accordance with its heredity the grafted twig of the white-fruited breed requires for its growth, its passage through the various processes of development, including the formation of fruits and seeds, suitable elements of food and that the food be in a suitable state. If these conditions, if food of this composition is found, the grafted cutting will develop in accordance with its nature, its heredity. If there is a shortage of the required food, the grafted cutting will build less important organs and characters out of the less suitable plastic substances. The food that suits the breed in question best will be spent on the most important organs and characters, for instance, on going through all the processes which directly lead to the formation of the sex cells. This explains why it is necessary to remove the leaves from the grafted twigs of the white-fruited breed, namely, to compel them to build their bodies to a greater extent out of the food, the plastic substances, elaborated by the roots, stems and leaves of the red-fruited component. It is self-evident that if the various substances are wholly alien, unacceptable to the white-fruited breed and there are no others, the grafted cutting must die of hunger. But if these substances prove capable of being assimilated though they do not meet the requirements of the grafted plant, a body will be built possessing properties different from those of the ordinary body of the white-fruited breed of tomatoes. Moreover, this body will resemble in one degree or another the properties of the breed which elaborated the particular plastic substances. However, the new body in question will differ to a great extent from that breed. After all, the plastic substances of the red-fruited breed were assimilated by the white-fruited breed differently from the way the red-fruited breed usually assimilates them. Each breed builds its body in its own way. The given example shows how *the living body, by assimilating this or that food, changes itself biologically. These changes consist in the acquisition of requirements for the conditions assimilated by the body.*

In vegetative hybridization experiments the scion receives its food from the twigs and roots of the stock. A scion has no roots of its own and most of its leaves are frequently gone (artificially removed). Usually, however, a plant obtains its food from its external, nonliving environment. The nutritive elements are extracted by the organism from the environment electively. Only what is suitable to the nature, the heredity, of the particular organism is taken. In the absence of suitable conditions the organism is frequently compelled, as in vegetative hybridization, to assimilate more or less unsuitable conditions. Hence, a body of a different structure is obtained. The latter requires for its growth and development conditions which have been assimilated for the first time, and perforce at that.

When the changed seeds of the vegetative hybrid obtained on a grafted twig are sown in beds, they elect from the external environment the conditions which in the final analysis are necessary for building the particular organism's body. This body, however, is similar to the body first obtained as the result of the grafting, i.e., owing to the enforced assimilation of unsuitable conditions.

Thus, if the scion was compelled to assimilate the plastic substances on account of which, as a result of a number of biochemical transformations, the ripe tomatoes obtained are red, the sown seeds from these fruits will have a tendency to elect from the external environment all the conditions which, in sum, after numerous regular conversions, will yield ripe fruits of a red colouring.

Thus, *being an external element—food—with regard to the scion, the plastic substances of the stock, after becoming a component part of the scion's body by means of assimilation, change the scion's hereditary properties.*

By analogy, as we see things, *the elements of inanimate nature, too, pass by means of an assimilation which frequently is enforced from the plant's external environment to the component parts of the living body, become living elements and acquire the property of heredity.* In future generations these external conditions are required by the living, developing body for the reproduction of bodies similar to itself.

These new elements of food are now required by the living body as a result of the processes that took place in the preceding generations, in consequence of the incorporation of the new element of the external environment. *As the inanimate elements of nature are assimilated by the living body they cease to be what they were not only externally but also in the strictly chemical sense.* At the same time they acquire a *pronounced biochemical affinity for, a gravitation toward,* that form of the elements of the external which was inherent in them before the living body had assimilated them, before they had been transformed into that particular living form.

By now much experimental material has accumulated which demonstrates the possibilty of directed change in the heredity of plant organisms by bringing the conditions of life, the environmental conditions, to bear upon it accordingly. *Vegetative hybrids are*, in science, a transition stage, as it were, *an intermediate link between the changing of the heredity of plant organisms by crossbreeding and the changing of heredity by bringing the conditions of life to bear on the organism.*

The theoretical significance of mastering the process of obtaining vegetative hybrids is obvious. These hybrids clearly show that *the heredity of plant organisms can be changed by changing their nourishment.* More. *The changes obtained correspond, are adequate, to the action of the external environmental conditions.* Thus, the action of the plastic substances of the red-fruited breed of tomatoes changes the white-fruited breed to the red-fruited breed.

The action of the plastic substances of a tomato breed with leaves resembling those of the potato plant changes the breed with dissected leaves into a breed with potato leaves, etc.

ABOLITION OF THE CONSERVATISM OF THE NATURE OF ORGANISMS

Our conception of the phenomenon of heredity enables us, through the action of environmental conditions upon plants, to elaborate methods of directively changing the nature of plant organisms, of enhancing their adaptation to conditions of cultivation in the fields. Thus, owing to their heredity, winter cereal plants cannot become vernalized, go through one of the phases of its development, if sown in spring when there is no lengthy period of low temperatures. Hence, they cannot bear fruit. But there are two ways by which they can be compelled to do so. The first way is to supply the winter plants with suitable low temperature conditions (approximately 0°-10° C. above zero) for 30-50 days, depending upon the variety. After this the winter plants will be able to continue and complete their development under the usual spring and summer field conditions. The second way is to alter their nature, after which they will cease to be winter plants with regard to their heredity. In both cases the change in the development of winter plants in spring sowings must be brought about by the action of the corresponding thermal conditions. They will differ only in the following. In the first case when vernalizing the winter variety, plants or seeds that have just begun to germinate are afforded the low temperature required by the nature of these organisms. Therefore the process of vernalization goes on normally for the development of the winter plants. The changes are the usual ontogenetic ones (growth changes). The seeds of a crop of such plants have the same heredity; they will possess the same winter habit as the seeds of the preceding generation. In the second case, at a certain moment during the passage through the phase of vernalization, the plants are afforded not the low temperature (near 0° C.) which they require for that particular process but the usual spring temperature. Two alternative possibilities arise: either the process of vernalization will not take place altogether, the plants will not complete, will not go through the process of vernalization for lack of the requisite thermal conditions and will therefore be unable to develop further; or the process of vernalization will take place, but under rather unsuitable thermal conditions. These changed thermal conditions will cause the process of vernalization to be completed differently from the way it would under normal conditions, i.e., when the temperature is low. It goes without saying that with the change in the process there will also be a change in the body, which is the result of this process. The entire further development of this body, though in external appearance indistinguishable from the development of the usual, normal, unchanged plants, will be different, as can readily be observed in the plants of the following generation. In order to pass through the phase of vernalization the next generation's plants will tend to elect the conditions which were forced on the previous generation. Instead of winter plants, plants with a tendency toward the spring habit will be obtained.

In the experiments relating to this question made in laboratories under our supervision by A. A. Avakian and other scientific workers of the All-Union Institute of Selection and Genetics, many hereditary spring forms have been obtained from winter forms. Hereditary spring forms have been obtained from all standard varieties of winter wheat experimented on. On the other hand, quite a number of spring forms of wheat and barley have been converted into hereditary winter forms.

From the point of view of mastering the process of directing changes in the nature of organisms, experiments in the conversion of winter forms into spring forms are of greater interest to the experimenter than the conversion of spring forms into winter forms. Experiments of the first kind are more convenient to perform and the results are more easily ascertained. One need only sow in spring the seeds taken from the experimental plants and the results are at once apparent. All plants which ear normally clearly indicate that their hereditary winter habit has already been altered to the spring habit. However, in experiments made to convert spring forms into winter forms it is difficult to discover changes even where the sowing material had been deliberately changed. When such material is sown in spring the experimental plants will be practically no different from the usual, unchanged spring forms because the acquired tendency toward the winter habit was not fixed. They will ear. If sown in autumn, a change in their nature will likewise be difficult to perceive even if the experimental plants pass the winter. After all, in many cases even ordinary spring plants can endure winter in the absence of severe frosts. But when there are such frosts the slight conversion of spring forms into winter forms seldom saves these plants from the devastating effect of winter. The changes must be more powerful, but this can be accomplished only over several generations.

Experiments in the conversion of spring forms of grains into winter forms are, however, of great practical interest for obtaining winter-hardy varieties. There already are a number of winter forms of wheat and barley which were obtained from spring forms by means of training, by bringing the environment to bear on them. These forms are not inferior, and some are even superior, in point of frost resistance, to the most frost-resistant varieties used in practical farming.

The science of agrobiology is confronted with the task of elaborating more and more concrete methods of altering the heredity of plant organisms in the direction we desire.

Let us briefly set forth the technique of transforming, through the action of environmental conditions, hereditary winter forms of cereals into spring forms and spring forms into winter forms. Winter forms, as is known, require a lengthy period of low thermal conditions to pass through the phase of vernalization. Spring forms do not require such conditions.

In order to transform winter forms into hereditary spring forms the process of vernalization of winter forms must be acted upon not by low (near 0° C.) temperatures but by the higher temperatures that occur in the fields in spring. According to our thesis, if a change in the processes takes place it will be adequate to the influence exerted.

In the next generation all processes of development must be gone through anew, as it were, in the same form in which they went on in the preceding generation. In the preceding generation, when not a low but a high temperature acted on the process of vernalization of winter forms, the process changed in accordance with the action of the temperature. Consequently, in the next succeeding generation the process of vernalization, which in the previous generation was traversed under the action of a higher temperature, will require the same conditions (a high temperature). This general proposition has been shown to be true in many experiments performed by us and many other scientific workers. However, despite the correctness of this general proposition, the desired result is not certain to be achieved in every particular case. The concrete possibilities and methods of altering the nature of organisms must still be worked out separately in each individual case.

In order to change winter plants to spring plants it is necessary to bring a high temperature to bear upon the process of the vernalization phase. But we know that in winter plants the process of vernalization does not take place when the temperature is high, or at least takes place very slowly. Winter wheat plants and the winter plants of other cultures can grow for months under a high temperature without going through vernalization and consequently without changing this process.

In practice many varieties of winter plants have been sown for many years on large areas in the beginning or middle of August, i.e., long before the cold winter weather sets in. The lower autumn temperatures usually set in a month or two after the sowing season. Yet in crops thus sown winter plants never turn into spring plants. Experimentally, too, winter plants can be kept for many months in a warm place (a greenhouse) but all the time they will look like grass. They will not be able to become vernalized, will not come into ears. Consequently, the process of vernalization did not change when the temperature was raised. Winter plants do not ear in the absence of the low temperatures they require for the phase of vernalization.

One may arrive at the erroneous conclusion (and the geneticists not infrequently do so) that it is impossible to direct changes in the nature of organisms by bringing conditions of life to bear upon it. As a matter of fact, however, as our numerous experiments have shown, winter plants can be converted into hereditary spring plants. Moreover, such *conversion takes place only as a result of the action of high temperatures upon the process of vernalization*, i.e., such temperatures as usually prevail in the fields in spring. The cases in which winter plants are kept for a long time under high thermal conditions and no change in heredity is obtained merely go to show that the plants, or rather their process of vernalization, did not respond to these conditions.

In the example we are analyzing, the plant organisms failed to respond to the influence brought to bear upon them because of the conservatism of heredity. The experimenter is therefore confronted with the task of finding better methods that will make it possible to induce the required action. A method already exists whereby a certain percentage of a hereditary spring form may be obtained from hereditary winter forms of any variety of cereals.

Experimental data and also a number of general biological observations have led us to conclude that *relatively high thermal conditions should be brought to bear on plants of winter habit with the aim of converting their heredity into that of spring habit not at the beginning of the process of vernalization* (and in general at no time during the process) *but only at the end, when it is being completed.* This is what success in bringing influence to bear hinges upon.

The usual duration of the process of vernalization in most winter grains, given low (0°-2° C.) thermal conditions, is 30-50 days, depending upon the variety.

The winter plants must be enabled to go through the process of vernalization at low temperatures, i.e., at temperatures suitable to their heredity. But just before the process of vernalization ends, higher thermal conditions must be created, the plants must be provided with the usual spring conditions. Usually the process of vernalization does not take place in winter plants when the temperature is high. But if high thermal conditions are created just before the completion of the vernalization process the plants will finish this process, though slowly and painfully, if one may say so. All further development will proceed normally since in spring and summer the environmental conditions in the fields are suitable for this development.

Experiments in the conversion of winter plants into spring plants were carried on as follows. Seeds of a winter variety were taken and, before spring field sowings became possible, different portions of these seeds were vernalized for different numbers of days at temperatures usual for winter plants. One sample of seeds was vernalized for 5 days before it was sown in the fields, another for 10 days, a third for 15, etc., up to 40-50 days. All these seeds vernalized to different degrees were separately sown in the fields in beds at one and the same time early in spring. The plants grown from the samples of seeds that had been completely vernalized before being sown in the fields developed normally, without delay on account of the vernalization stage (since that had already been gone through); they produced straws and spikes. However, the plants obtained from samples that did not quite complete the process of vernalization before they were sown finish it quickly if low temperatures prevail for a relatively long period under field conditions after the sowing in spring. If that is not the case the plants grown from seeds that were not completely vernalized before the sowing will be slow to finish the vernalization process. Such plants likewise ear, but with more or less delay. These plants are the most interesting for the purposes of the above-mentioned experiment. It is from them that one most frequently succeeds in obtaining hereditary spring forms. Further work along the line of obtaining spring forms from winter forms therefore requires that the seeds be taken from samples of plants not completely vernalized before the sowing but which finish the process of vernalization after the sowing, in spring, under field conditions. From the seeds of such winter plants one can obtain a certain percentage of hereditary spring forms. This was the way many spring forms were obtained from all the winter varieties of wheat experimented with at the Institute of Selection and Genetics of the Lenin Academy of Agricultural Sciences of the U.S.S.R.

It is thus clear that the hereditary winter habit can be changed to the spring habit. This change can be brought about by the action of high temperatures that suit the heredity of the vernalization stage of cereal forms called spring plants. This corroborates the correctness of the general proposition that *a change in the heredity of any property corresponds to the action of the external environmental conditions.*

As has already been stated, not all seeds obtained from grafted plants produce hybrid plants. The percentage of hybrid plants secured depends on the ability of the experimenter to overcome the resistance of the grafted breed and compel it to assimilate plastic substances to which it is unaccustomed. Analogously, spring plants will not be obtained from all seeds of winter wheat plants taken from parent samples which were known to have assimilated, i.e., finished, vernalization under high, spring temperature conditions.

In the majority of such cases what happens is something very much like the behaviour, during the spring sowing, of plants obtained from the usual, unchanged winter seeds. This takes place because when sowing seeds even from plants known to have been changed with regard to the vernalization stage one frequently obtains plants which do not yield spikes if sown in spring.

Thus, in the spring of 1936, three winter varieties of wheat were drill-sown in the fields of the Odessa Institute of Selection and Genetics; the seeds were of the usual, unvernalized kind. It was an early spring, long and cool. Usually when winter plants are sown in spring they either do not ear at all that summer or ear very sparsely and late in the season. But the plants of the above crop of all three varieties (Novokrymka 0204, Kooperatorka and Stepnyachka) eared uniformly though late, and yielded a fairly good harvest. The seeds of all three varieties of this crop were drill-sown again in the spring of 1937 in the open field without prior vernalization. As a variant of this experiment, seeds of the varieties in question taken from ordinary winter crops were all sown at the same time in one locality. One would expect that the plants of the winter variety obtained from seeds of the crop planted in the spring of the preceding year (without presowing vernalization) should in the new generation (the 1937 spring sowing) ear more uniformly, produce a greater percentage of eared plants in comparison with the second variant. In actual fact, however, the opposite proved to be the case. With regard to all three varieties the plants obtained from seeds sown in spring for the first time yielded a percentage of eared plants which, on the whole, was small. Besides, the earing was much delayed. Yet the percentage was considerably greater than that obtained from the seeds resown in spring. However, the plants from the seeds sown the second time in spring eared much earlier.

The result of this experiment patently demonstrates that the unusual termination of the process of vernalization in the winter plants of the 1936 spring sowing by the use of unvernalized seeds definitely changed the nature of the winter plants. At first sight it may seem that the change which occurred was not in the direction of the spring habit, as it should have been, but in the direction of a still greater winter habit. After all, a smaller percentage of earing plants was obtained on the plots sown in 1937 with these seeds than on the plots sown with seeds of the same varieties, but sown for the first time. As a matter of fact the change in the vernalization stage of plants of the 1936 sowing under analysis proceeded in the direction of lessening the winter habit (the requirement that the process of vernalization be traversed at low temperatures). But many experiments show that when the old, long-established property of heredity, in the case under examination the property of winter habit, is abolished, no established new heredity (in our case, the spring habit) has as yet been obtained. *In the vast majority of such cases one obtains plants possessing a so-called destabilized heredity.*

Plant organisms are said to have a *destabilized heredity* when *their conservatism has been abolished, when their capacity to elect external environmental conditions has been weakened.* Such plants, instead of retaining their conservative heredity, retain, or newly exhibit, only a tendency to give preference to some conditions over others.

Heredity can be *destabilized* in the following way:

1) *by grafting*, i.e., by uniting the tissues of plants of different breeds;

2) *by subjecting plants to the influence* of the environment at definite moments when they are undergoing *developmental processes of one kind or another*;

3) *by crossbreeding*, particularly of forms sharply differing in habitat or origin.

Much attention was paid by some of the best biologists—Burbank, Vilmorin, and particularly Michurin—to the practical value of plant organisms with destabilized heredity. Plastic plant forms with unestablished heredity, obtained in this or that way, must be sown from generation to generation under conditions the requirement or stabilization of which we want to induce in the particular organisms.

Usually when a plant with a nondestabilized heredity lacks the conditions necessary for some process to take place, as, for instance, when the low thermal conditions required for the vernalization stage of winter plants are lacking, the process does not take place. The plant waits, as it were, for the requisite conditions to arrive. If the temperature drops at night the autumn-sown winter plants go through the vernalization stage. If the temperature rises in daytime the vernalization process stops until a lower temperature sets in, even if the interval lasts many days.

But organisms whose heredity is destabilized, as for instance the progeny of the winter plants whose vernalization stage was completed not under low but under high, spring temperature conditions, do not possess an established heredity (requirement) but only a tendency to the conditions under which the vernalization process of the plants of the preceding generation was completed. If no such temperature sets in, the process will not wait but will take place under whatever temperature there is. Ordinarily, in the field, thermal and many other conditions, as a rule, vary, fluctuate. *Owing to the conservatism of their heredities plant organisms stubbornly and unswervingly elect from their varying, fluctuating environment only what is needed for particular processes to take place.* But if the heredity is destabilized, not fixed, the process fluctuates, moves in various directions, as they say. When the temperature is low it goes in one direction; when a higher temperature sets in, it takes another direction. As a result the process is not coordinated. This explains the cases of failure to ear when winter wheat whose vernalization stage has been deliberately changed is sown in spring. They remain in the tillering phase not on account of their winter habit but because the different directions which the vernalization process takes make it impossible in these cases for the process to be completed.

For plants with changed, destabilized heredity the conditions of cultivation must be selected with skill. It must be remembered that these plants are frequently most susceptible to environmental conditions. Therefore, it is necessary to provide, as far as possible, the conditions toward which it is desired to direct, to fix the heredity.

In nature the evolution of plants and animals proceeds through random changes in the old heredity, through the fortuitous building and fixation of a new heredity. In experimental as well as in practical farming the heredity of particular processes of plant and animal organisms can be made to undergo a directed change, and the building and perpetuation of the new heredity can be directed.

To obtain hereditary spring forms from the seeds of destabilized winter plants, i.e., such as completed the vernalization stage at high temperatures, they must be sown in the field in spring at different dates, the first as early as possible. This will make it possible for the vernalization process of the plants of this or that sowing date to coincide with the environmental conditions toward which they tend. Such plants ear quickly. The seeds taken from them will, as a rule, in their vast majority, produce offspring closely akin in behaviour to the spring forms. But the heredity of such forms will still be fixed to only a small extent. If the sowing conditions in spring are unusual (for instance, if the spring is too protracted and cold or too hot and short) these plants may deviate from the more or less established spring form of life. In general, after changing the heredity of the winter plants by bringing the spring thermal conditions to bear on the vernalization process at the time of its completion, the heredity of the spring habit must be fixed over a period of two or three generations. Only thereafter will the form be really established.

For practical purposes great importance attaches in a number of districts of the U.S.S.R. to the conversion of spring cereal forms into winter-hardy winter forms and winter forms into more hardy, more frost-resistant forms. These experiments in nowise differ in principle from the experimental work performed in the cases already discussed relating to the conversion of winter plants into spring plants. *Hereditary spring varieties are changed into winter varieties by sowing them late in the autumn.* The spring forms of cereals are provided with low thermal conditions at the time they go through the vernalization process, a long period (autumn, winter and early spring). A repeated sowing of the seeds obtained from these plants in late autumn fortifies the new property of winter habit. Their requirement of low thermal conditions for the vernalization process is enhanced.

As the succeeding generations are sown from year to year under increasingly severe wintering conditions, cereal plants with still unestablished (destabilized) heredity of the vernalization stage will require low temperatures to an ever-increasing degree. They will acquire the property of increasing resistance to severe frosts. At present we have a number of good forms of winter wheat obtained by various experimenters from spring wheats. These new forms are no less frost-resistant than the winter variety Lutescens 0329 of the Saratov Plant-Breeding Station, which is so far considered a most frost-resistant wheat.

Research workers A. F. Kotov and N. K. Shimansky obtained after several generations a winter form of wheat from the spring form Erythrospermum 1160 of the Institute of Selection and Genetics by means of late, prewinter sowing. When sown at the Experiment Base of the Lenin Academy of Agricultural Sciences of the U.S.S.R. at Gorki Leninskiye, Moscow Region, and also on the experimental plots of the Krasnoufimsk, Barnaul and Semipalatinsk plant-breeding stations and elsewhere, this form proved itself a promising variety for these districts.

A point of interest here is the fact that the seeds of this wheat distributed in the autumn of 1940 among the places indicated all came from one bag. But since this wheat is not yet established, is still highly plastic, it deviated at each place where it was grown in the direction of the conditions of life, of the conditions of cultivation. The conditions prevailing in each locality left their imprint upon this plastic, pliant form of plant. Under the severe wintering conditions in the districts of Siberia this wheat is becoming each year more and more frost-resistant, winter-hardy.

A. A. Avakian converted spring wheat Lutescens 1163 bred at the Institute of Selection and Genetics into a winter wheat by sowing it late in autumn. Today this wheat comes close to the most frost-resistant winter varieties in regard to its resistance to the inclemencies of winter. A number of wheats which in this respect excel the most frost-resistant Lutescens 0329 were obtained by converting natural Siberian windfalls of spring wheats into winter wheats. Thus, the wheat harvested by the collective farmer Sekisov (Michurin Collective Farm, Barnaul District, Altai Territory) already far surpasses the Saratov Lutescens 0329 in frost resistance. A number of other forms of winter wheat obtained from spring windfalls in Siberian plant-breeding stations also represent exceedingly promising material for growing highly winter-hardy varieties.

Solovey, a scientific worker, obtained a winter form by the late autumn sowing of spring barley Pallidum 032 of the Odessa station. Owing to the plasticity of this form it proved to be readily adaptable to rather severe wintering conditions. In our opinion this barley is now one of the most winter-hardy varieties among all the winter barleys known to us. It has already withstood quite well two winterings on a section of the Experiment Base of the Academy of Agricultural Sciences at Gorki Leninskiye, near Moscow, and also at the Kazan State Plant-Breeding Station. Ordinary winter barleys do not survive the winter in these regions.

The most interesting thing for practical farming in these experiments is the fact that it is quite easy to increase from year to year the resistance of the above wheat and barley forms to frost and other tribulations of winter. Unestablished forms, forms not yet consolidated after their heredity was destabilized, can easily be altered to acquire increased resistance by bringing more and more severe wintering conditions to bear on them with each passing generation. The properties acquired from generation to generation will become more and more fixed. But the acquired properties can easily be lost if the material still unestablished during the first few generations is not handled skilfully. Let us cite the following instance. The winter barley which, as already stated, Solovey obtained from spring variety Pallidum 032 when sowing it on experimental plots located in the central zone of the Soviet Union has proved to be the most winter-hardy of all winter barleys we know of. In the spring of 1940 some plots of the All-Union Agricultural Exhibition were sown with samples of this barley. For some time it behaved like a winter form. The plants trailed on the ground; no straw (flowering shoots) developed. It was assumed that, being a winter form, the plants of this barley could not pass through the vernalization stage under spring thermal conditions. However, as afterwards appeared, all plants on this 100-metre plot rapidly formed flowering shoots, eared well and yielded a good crop. This indicates that the heredity of the winter habit had not yet become consolidated in this form of barley. Having been sown in spring and having waited a while for cool thermal conditions to set in, which naturally did not happen, the plants went through a new type of vernalization, i.e., became vernalized as spring plants. The seeds harvested from these plants were sown in the autumn of the same year, 1940, by Avakian on plots located at the

Experiment Base of the Lenin Academy of Agricultural Sciences of the U.S.S.R. near Moscow. At the same time seeds of this variety taken from the Exhibition sector of the 1939 autumn sowing were planted. It appeared that the plants from the spring-sown seeds of the preceding year withstood the winter of 1940-41 incomparably worse than the variant obtained from the seeds of the crop sown in the autumn of 1939. The cultivation of only one generation of plants of the indicated barley variety under the conditions of spring sowing considerably weakened the property of winter-hardiness of the offspring of these plants. On the basis of this example we showed that *plastic, unestablished plant forms obtained in one way or another should be sown from generation to generation only under conditions the requirement or stability of which, the hardiness with regard to which, must be induced in the particular plants.*

Plant organisms that are not yet established in regard to their heredity, which are still destabilized, are in many cases very valuable material for the creation, by appropriate training, of forms and varieties we need. At the present time progress is being made in the creation of winter-hardy varieties of winter wheat for the districts of Siberia with their severe winters. Already noteworthy results have been achieved. Spring wheats that possess no winter-hardiness whatever are being converted, by means of destabilization, of altering the vernalization stage, into frost-resistant wheats. Winter wheats are being converted in this way at Siberian plant-breeding stations into wheats of increased winter-hardiness, of greater winter habit.

THE SEXUAL PROCESS

The sexual process is one of the most important in wild plant and animal organisms. All other processes are in actual fact subordinate to it. Animals and the vast majority of plants reproduce sexually.

When plant organisms are propagated asexually, vegetatively—by means of tubers, cuttings, buds, etc.—these plants do not begin to develop de novo. They continue to develop from the stage reached by the tissue taken as the basis of the new organism. *The sex cells, on the other hand, provide a new basis of development, which in many cases completely repeats all alterations and conversions undergone in the preceding generations.* This property distinguishes the sex cells in principle from all others that can serve as the basis of an organism. When plant organisms develop from seeds one can easily observe how the tissue of the developing organism changes qualitatively, beginning with the fertilized sex cell, how through a number of regular alterations and conversions ever new cells are created, tissues with their specific properties are differentiated and various organs develop. In general an ever new quality of the cells of the organism is obtained. This quality is capable of being converted into a further quality, a new quality predetermined, as it were, by its ancestors. But it is incapable of being converted into the old quality, the quality of the preceding cells that gave rise to the present ones. The sex cells, however, while new with regard to the asexual cells from which they in the final analysis were formed, are at the same time largely and frequently exactly like the initial sex cells, i.e., the old cells from which the entire development of the organism in question started. *The sex cells represent the completion of the organism's cycle of development.* At the same time *they are the beginning of the development of new organisms.*

In this light the great biological significance of the sexual process in the evolution of plant and animal forms becomes intelligible.

Natural *hereditary changes*, departures from the norm in plants and animals *are*, as a rule, *enforced changes. They take place because the conditions of life are not suited to the developmental requirements of the various organs, characters and processes* in general of plant and animal organisms. It has already been pointed out that in sexual propagation development begins de novo. Therefore, *conditions which were inappropriate, unsuitable to a particular process of the preceding forms, become normal and requisite for the new generation.*

Changes in the conditions of life, in environmental conditions, are, as a rule, independent of specific animal and plant forms. If plants and animals possessed the properties of infinite individual life they would, in popular language, be having a hard time of it all their lives. The external conditions, which are always changing at one time or other, would never suit, correspond to the requirements of the organisms. In other words, owing to changes in climate and in the conditions of life in general, organisms with very prolonged individual lives are inconceivable in free nature. Evolution, the increasing complexity of plant and animal forms, is possible only because all living forms have a succession of generations. It is a very noticeable fact that the shorter the normal individual life of plants and animals the greater adaptability to the changing environmental conditions which the species of these organisms possess. Microorganisms with brief spans of individual life adapt themselves hereditarily to the changing conditions of life most easily.

Another very important biological property of sex cells may be reduced to the following: *The sex cell is biologically (but not chemically) the most complex of cells.* In it the potential hereditary properties inherent in the whole organism are expressed in a higher degree than in any other cell of the organism.

The entire course of development traversed by the organisms of the preceding generations is accumulated, as it were, in the sex cells. From these cells development starts de novo. What takes place may be described as the unwinding from within of a chain of numerous changes and conversions which were wound up in preceding generations. We have already pointed out that this unwinding of past processes goes on only by winding up the processes for the future generation. The development takes place solely through metabolism, through assimilation and dissimilation, and this is the creation of the foundations of the future generation.

In the vast majority of plants and animals new organisms arise only after fertilization—the fusion of male and female sex cells. The biological significance of the processes of fertilization consists in the fact that organisms with a dual heredity are obtained, maternal and paternal. *The dual heredity calls forth greater vitality* (in the direct sense of the word) *of the organisms and greater adaptability to the varying conditions of life.*

The internal forces, the properties of the body itself to live, to be altered and transformed, constitute the impulse of development. Much practical and experimental material can be offered as cogent proof that fertilization, the crossbreeding of even slightly differing forms of plants or animals, produces more viable offspring. Conversely, prolonged self-fertilization, self-pollination in plants and the pairing of closely-related animals leads to an attenuation of life. Normal vital internal contradictions, the vital impulse, is created, and likewise is renovated from time to time in the plant and animal world in the vast majority of cases by means of crossing, of fertilization, by means of a sexual union of forms of plants and animals which differ at least slightly from each other.

All ordinary (asexual) cells divide in two upon completing their development; this is the way cells multiply and the body grows. *However, sex cells* not only *do not divide* in two upon completing their development but, on the contrary, normally *form two sex cells*—a female and a male—*one cell is obtained, usually a more viable one* than any of the other cells.

Both the female and the male sex cells fully possess the properties of their breeds. Breeds differ to a greater or smaller extent. After the zygote is obtained, i.e., after the female sex cell is fertilized, a single cell is formed, the basis of the organism in which all the breed properties of both initial forms are represented. It is on the basis of the contradiction arising between the two united but relatively different sex cells that the internal vital energy, the property tending to alteration and conversion, arises and gains in intensity. It is this that determines the biological necessity of crossing forms that are at least slightly different from each other. Darwin repeatedly emphasized in his works that the usefulness of crossbreeding and the biological harm of self-fertilization are a law of nature.

Renovation, intensification of vitality of plant forms, can also proceed in a vegetative, asexual way. It is attained by the living body assimilating new environmental conditions that are unusual for it. Such cases are usually rarer in nature. Nevertheless, a number of plant forms might be cited which for a lengthy period, actually for the duration of all time known to history, have been reproducing vegetatively only and all the same are not losing their vitality, their internal vital impulse. In experiments in vegetative hybridization, either to obtain spring forms from winter ones or winter forms from spring ones, and in a number of other cases of the destabilization of heredity, one may observe a renovation, an intensification of the life of organisms by the inclusion in their bodies of new conditions that are unusual for them.

The conception of the process of fertilization generally accepted in the science of genetics seems wrong to us in many respects. Cytogeneticists draw their picture of the processes of fertilization from looking into the microscope at the slide where cells in various states of development have been fixed. Everything visible is jotted down and what is not visible is drawn from imagination, is conjectured in the light of the conception, the theory of heredity propounded by the Mendelist-Morganists. The cytogeneticists proceed from the idea that heredity is a special substance different from the ordinary body and contained in the chromosomes of the cell nuclei. According to their conception the heredity confined in the chromosomes of the male sex cell nucleus and the heredity concentrated in the chromosomes of the female sex cell nucleus unite mechanically to form a single cell. The chromosome substances do not blend either in the biological or even the chemical sense. The chromosomes in the male sex cell introduced into the nucleus of the female sex cell remain there in the form in which they were and still are in the cells of the paternal organism. This postulation of the cytogeneticists is based on the fact that some time after fertilization double the number of chromosomes—the sum of the chromosomes of the female and the male sex cells—are observed under the microscope in the zygote (the fertilized sex cell). The entire conception of the process of fertilization has been built up by the cytogeneticists on this notion. Such a conception is absolutely unacceptable, especially to biologists. It corresponds neither to the sexual process nor in general to any biological process whatsoever that goes on in the living body.

Already Darwin pointed out that when vegetative hybrids become possible, physiologists will have to change their views on the sexual process radically. In actual fact, in the light of the mass of factual material on vegetative hybridization, the question of the essence of the fertilization process must be stated anew. In the first place fertilization—the union of two cells in one—is not a simple fusion of two cells physically not even soluble in each other. There is not a single normal process in the living body which does not represent an alteration or conversion, i.e., which is not an assimilation-dissimilation reaction.

The Mendelist-Morganists have actually taken away from the physiologists, and the latter have turned over to them, the examination of the problem of fertilization. All processes in the organism are transformations—metabolism. The process of fertilization is the only exception, in the opinion of formal science, and therefore the examination of it is not within the purview of physiologists. The geneticists deny that the sexual process is metabolism, is a process of assimilation and dissimilation. According to the conception entertained by the geneticists a special body, the hereditary substance, is concentrated in the chromosomes of the cells. The laws governing the life of this body are different from those governing the ordinary body.

The hereditary substance is not subject to ordinary metabolism; nothing can be included in or excluded from it. The hereditary substance is transmitted unchanged from generation to generation. In rare instances it may be lost, may perish; in rare instances, for unknown reasons, it appears de novo (mutations). Heredity is concentrated in the nuclei of the sex cells. That explains why the study of the development of sex cell nuclei has passed, during the last few decades, into the hands of the formal science of heredity, into the hands of the Mendelist-Morganists.

The numerous experiments performed during the last few years in the mass production of vegetative hybrids, and the transmission of their properties to their progeny sexually, fully entitle us to regard fertilization as an ordinary physiological process. Like any biological process, fertilization, the union of two sex cells, may be reduced to assimilation and dissimilation.

The fundamental distinction between fertilization and all other biological processes is as follows. In any physiological process one of its aspects assimilates, the other is assimilated. The assimilating body builds itself from food, starting with the elements drawn by plants from the external environment and ending with ready plastic substances. The assimilated substances serve as building material for the assimilating component. In the sexual process, however, when two cells of equal standing, so to say, unite, they mutually assimilate each other. Each one builds itself in its own way from the substance of the other. *In the final analysis neither of the two cells remains and a third, a new one, is obtained, one in place of the two.*

The Mendelist geneticists hold up the allegedly always existing multiple relations in the variation of hybrid offspring in the second and later generations as one of the principal proofs that heredity is particulate (corpuscular). They ascribe to each character and property of a living body a certain number of particles (genes) of the hereditary substance located in the chromosomes.

In fertilization, when two sex cells unite for every property, a double set of particles is obtained in the fertilized sex cell: one paternal and the other maternal.

To make this clear let us cite the classical example of Mendelist geneticism, the crossing of two forms of peas which differ, for instance, in the colour of their blossoms. Chromosomes containing particles (genes) of the hereditary substance which determine the red colour unite on fertilization in one nucleus with chromosomes containing genes of white-coloured blossoms. When the fertilized cell divides, each of the chromosomes, the maternal and the paternal, likewise divides, lengthwise, into two parts of equal value. The chromosomes move to the poles of the dividing cell, one of each pair to each pole. According to this conception all the cells of a hybrid organism possess in their pure form an equal quantity of both the paternal and the maternal hereditary substance. The result is different in reduction division, which occurs in animal organisms at the time sex cells are formed and in plants before they are formed. The chromosomes do not split lengthwise but form pairs of analogous paternal and maternal chromosomes, then separate and go to the different poles. The cells obtained contain only either the paternal or the maternal chromosome of each pair.

The geneticists are of the opinion that the chromosomes of each parental form in the hybrid cell do not lose their properties, their individuality. They are there in their pure paternal or maternal form. At the reduction division, when from each homologous pair the paternal chromosome goes to one pole of the dividing cell and the maternal chromosome to the other, a pure sex cell (gamete), nonhybrid as regards the properties whose genes are located in the particular chromosome, is obtained.

Thus, in the example we have taken of crossing white-coloured with red-coloured peas half of all sex cells will have a chromosome with a red-colour gene or genes while the other half will possess a chromosome with hereditary particles of white-coloured blossoms. When such hybrid plants are selfed the male sex cells, according to the calculus of probability, may unite with the female sex cells, i.e., the egg cells, in three combinations.

First combination: A male sex cell possessing a red-blossom gene may unite with an egg cell also containing a chromosome with a red-blossom particle (gene). A zygote is obtained whose hereditary substance has only the red-blossom character.

Second combination: A male cell having the hereditary substance of the white-blossom property unites with an egg cell which likewise possesses the white-colour property. A zygote is obtained possessing the hereditary property of white blossoms only.

Third combination: A male sex cell containing the substance which conditions red blossoms unites with egg cells possessing the white-blossom property. A zygote is obtained that has a dual heredity, both red- and white-coloured blossoms. The same happens when male white-blossom cells unite with female red-blossom cells.

Thus, self-pollination of the indicated hybrid pea plants produces the following zygotes: 25% with pure red-colour heredity, 25% with pure white-colour heredity, and 50% with dual heredity. With regard to the hereditary blossom-colour characters the following ratios are obtained: 1 red : 2 hybrid : 1 white.

It has long been known that in crossing different forms of peas and many other plants having red and white blossoms the vast majority of hybrids obtained have red blossoms. The same was observed by Gregor Mendel in his pea-crossing experiments. This phenomenon came to be called the dominance of one hereditary property over another, a contrasting property.

On the basis of the reasoning we have set forth the Mendelists arrived at the conclusion that in crossing red-blossom with white-blossom peas there should always be 75% (25% pure+50% hybrid) red plants and 25% pure white plants in the second hybrid generation. The relation of red to white plants should always be 3:1.

This "pea law," as Michurin aptly calls it, the Mendelists force upon all living nature. The fact of the matter is, however, that it is untrue even for the pea hybrids, including the factual material obtained by Mendel himself in his experiments. Even in Mendel's experiments different offspring of particular hybrid plants varied far beyond the ratio of 3:1. Thus, from the offspring of one plant 19 green seeds were obtained for 20 yellow ones and from another plant only 1 green for 32 yellows.

CATEGORIES, GROUPS AND FORMS OF HEREDITY

The various behaviours of hybrids were correctly classified by K. A. Timiryazev. He started out by dividing the phenomena of heredity into two groups: simple and complex heredity.

It is a known fact that self-pollinating plants, as, for instance, wheat or plants propagated by tubers, cuttings, layers, etc., as a rule, possess throughout their development the heredity of the maternal form, i.e., the form from which the seeds, cuttings, etc., are taken. This form of inheritance K.A. Timiryazev called *simple*.

On crossing, the heredities of two organisms usually unite. Such heredity is called *complex*, i.e., dual. It in turn may be subdivided into several groups according to the forms in which it manifests itself.

In some animals, for instance, one patch in the pelage resembles the paternal form in colour, another the maternal; or some cells of the epidermis of a leaf resemble the paternal while others the maternal, etc. This heredity is called *mixed* because one part of the organism manifests characters of the one parent and another of the other parent. These parts, or sections, may be of different sizes, from big to microscopically small.

Most frequent are the cases when the hereditary properties of both parents are blended in the offspring (do not manifest themselves in their pure form), when new properties are obtained in the progeny. Such heredity Timiryazev called *blended*, and considered it the most important.

Cases occur when parental characters expressed in opposite ways are not blended in the hybrid offspring. For instance, when crossing a pea variety having green seeds with a yellow-seed variety these characters do not blend in the offspring. Nor is a new or mean property obtained. The property obtained is that of only one of the parents. The property of the other is excluded, so to say. This form of heredity is called *mutually exclusive*.

Two categories may be observed in mutually exclusive heredity.

One of these categories comprises the cases of hybrid organisms that are homogeneous in the first and all subsequent generations. In other words, the hybrid progeny does not become heterogeneous, segregated, in the course of generations; not infrequently the properties of the one parent are completely absorbed by the other. These cases are called *Millardetism*, after the French scientist Millardet who studied this category of hybrids quite fully.

The other category of mutually exclusive heredity comprises cases of so-called *Mendelism*. Timiryazev himself pointed out that those are isolated instances, which occur only under definite conditions and actually were not discovered by Mendel at all. In those cases, beginning usually with the second generation, segregation, diversity, occurs in hybrids, some forms having paternal and others maternal characters.

Today it is clear that all the diverse forms of heredity are possible also in vegetative hybridization.

Vegetative hybrids exhibit mixed heredity when one part of the organism is represented by the properties of one breed, of the one component, and the other part by those of the other breed. Blended and mutually exclusive heredity are also encountered.

In vegetative hybrids we also have either increase in vigour of development or, on the contrary, decrease in viability, i.e., the same as we have in sexual hybridization.

All this does not mean, of course, that there is no difference whatever between vegetative and sexual hybridization. The important thing to emphasize is, however, the fact that vegetative and sexual hybrids manifest common forms of heredity. Neither of these categories of phenomena are separated from each other by an impenetrable wall but represent phenomena of one and the same order.

It has already been pointed out that from the stand they take the Mendelist-Morganists cannot admit the existence of vegetative hybrids. What defied all refutation they classed as incomprehensible, inexplicable phenomena called *chimeras*.

In actual fact the so-called chimeras may be looked upon as manifestations of mixed heredity, one part of the organism exhibiting the properties of the one component and the other part those of the other component. This phenomenon is analogous, for instance, to the case of a piebald or spotted cow one spot of whose hide is coloured like that of the maternal organism and another spot like that of the paternal one. Who would ever think of calling a spotted cow a chimera?

The facts in possession of Soviet agrobiologists are a sufficient basis for evolving a single effective theory of heredity which will fully meet the requirements set forth by K. A. Timiryazev: to serve as "a general working hypothesis, i.e., an instrument for directing the research necessary for the discovery of new facts, new generalizations."

In vegetative hybridization one component is nourished by the other; they enter into metabolic relations with each other. As a result of such interaction between the plants of two breeds a new organism is obtained combining in some measure (depending upon the conditions) the heredity of both components.

The same stand may, in our opinion, be taken with regard to *sexual hybridization*, which is also *a process of metabolic interchange between the fusing components (cells) of the cross.*

If *vegetative and sexual hybridization are phenomena of the same order*, it follows that they must have a common basis. This consists in the fact that *both vegetative and sexual hybridization are processes of reciprocal assimilative activity of the components* as a result of which the hybrid product is elaborated.

When looked at in this light Michurin's teaching acquires particular interest. By suitably preparing the organisms and giving them the necessary nutrients Michurin compelled forms which without this were biologically incompatible to crossbreed. He worked out a method of overcoming inability to cross by which each of the components of the cross is fed the products elaborated by the other. This method is a preliminary vegetative approximation. By selecting the conditions of life and the regime of feeding, the sexual process may be changed, directed, thus creating the prerequisites for the absorption of the hereditary properties of the one component by the heredity of the other. Michurin also proved that the hereditary properties of hybrid trees continue to form during the course of their individual lives until the first few years of bearing. The deviation of particular properties of the hybrid toward either component of the cross depends on how the hybrid is nourished.

From this follow the interconnection and mutual transitions that exist between vegetative and sexual hybridization, on the one hand, and between vegetative hybridization and the influence of environmental conditions, on the other.

In this connection the following fact, interesting from the point of view of theory and general biology, should be adduced. It was repeatedly observed several years ago in experiments conducted by A. A. Avakian at the Odessa Institute of Selection and Genetics and subsequently in the greenhouses of the Experiment Base of the Lenin Academy of Agricultural Sciences of the U.S.S.R. (Gorki Leninskiye).

The reference is to the following phenomenon. On crossing Hostianum 0237, a winter wheat, with spring wheat 1160 and 1163 (the latter two wheats being full sisters), the seeds obtained were normal. At first sprouts normal in their external appearance developed from them. But when the third leaf appeared the first shrivelled up. As soon as the fourth appeared the second shrivelled up. All the time only the last two leaves remained alive on the plant. In the end the plant perished. Thousands of such plants were under experiment at one time or another and not one of them lived long enough to ear; all died. The Mendelist-Morganists would attribute such a phenomenon to the action of lethal genes. But they would have nothing to offer wherewith to combat this action. They would declare it fatal, irresistible, and would endeavour to show that in such cases there is only one solution: don't take for crossbreeding plant or animal organisms that have lethal genes. Yet a cross of the same combination, Hostianum 0237 and 1160, produced hybrids which vegetated splendidly in these very same greenhouses and yielded viable, nonperishable plants. The point is that one of the components (paternal form 1160) is a spring variety, but for two generations before crossing it was sown in Odessa, not in spring but in the autumn. Then a cross was effected. This proved sufficient to obtain viable offspring. A different rearing of wheat 1160 altered the plant's sex cells; hence the difference in the result of hybridization.

In other experiments made by A. A. Avakian castrated Hostianum 0237 plants were fertilized with a mixture of pollen of the Erythrospermum 1160 variety and of Hostianum 0237, the maternal form. The plants bred from the seeds obtained were of hybrid origin. They were spring plants while the maternal form was a winter plant. But these plants too proved viable, nonperishing. Thus the presence of pollen of the maternal form, in the case in question Hostianum 0237, influenced the result of fertilization with Erythrospermum 1160 pollen. The offspring obtained were not lethal, as is usual in such a cross, but viable.

Michurin, too, pointed out the expedience in certain cases of mixing pollens. By this means he succeeded in crossing species and genera which otherwise could not cross.

All these facts show once more that fertilization, like vegetative hybridization, is a peculiar process of assimilation, of metabolism.

The categories of phenomena connected with cross-pollination also argue in favour of such a conception of the sexual process. As Darwin proved irrefutably, *cross-pollination* is, as a rule, *good for the organism*. Offspring from seeds obtained by means of cross-pollination possess *greater vitality*. According to the explanation given by Darwin, different organisms developing under relatively different conditions build themselves differently from the environing nutrients. Relatively different organisms are obtained and hence also different sex cells. A union of such sex cells differing somewhat in their heredities produces more viable organisms. This is the proposition on which the measure we propose, intravarietal crossing of self-pollinating field crops, rests.

The basis for intravarietal crossbreeding is elective fertilization. It has been pointed out above that each organism, depending upon its breed, its heredity, requires relatively definite conditions for its life and development. An organism usually does not imbibe nutrient elements that are bad for it, if better elements are available. Herein lies the historically evolved adaptability of organisms. Every process in an organism is endowed with relative elective capacity with regard to conditions. In spite of the assertion of the Mendelists to the contrary the sexual process is no exception in this regard.

The study of elective, free fertilization in plants is of great practical and theoretical importance also for an understanding of the laws governing the behaviour of hybrid offspring.

At the Odessa Institute of Selection and Genetics D. A. Dolgushin performed the following experiment: On the plots assigned for the testing of varieties of winter wheat several dozen spikes taken from each of the twenty varieties experimented with in this crop were castrated in 1938. The castrated spikes were given the opportunity of being pollinated by the pollen of any variety. For each castrated flower there was many times more pollen of other varieties than there was of the uncastrated plants of its own type.

In the first generation the seeds obtained from the castrated spikes produced plants whose only point of distinction was somewhat greater vitality, greater vigour than that of the maternal forms sown alongside. All these plants (except a small percentage) showed no morphological differences from the maternal forms, though some of the latter had recessive characters (for instance, awns, white spikes, etc.). In all the twenty varieties the plants of the second generation obtained from the free, elective intervarietal crosses withstood the inclement winter of 1939-40 better than the maternal forms.

The wheat assortment taken included the most frost-resistant variety, Lutescens 0329. According to Morganist conceptions this wheat could not acquire greater hardiness from anywhere on being pollinated by the other varieties inasmuch as all the other varieties undergoing the test were considerably inferior to it with regard to this particular property. It is also of interest to note that in free, elective intervarietal crossbreeding not a single variety of interior winter-hardiness, like the Kooperatorka, increased its hardiness to any great extent. But in artificial (enforced) crossing of Kooperatorka with more frost-resistant varieties the hybrids, as a rule, considerably surpass the Kooperatorka in hardiness.

This and a number of similar experiments show that frequently in elective, unrestricted fertilization of plants seeds are obtained which produce plants differing little from the maternal type but unfailingly (though not to a great extent) more vital, more resistant to climatic inclemencies.

In our view free, unrestricted, elective fertilization in plants leads, as a rule, to the almost complete absorption of one heredity by another. Much more often than not, the maternal heredity absorbs the paternal. We repeatedly observed this phenomenon in experiments with self-pollinating plants, as for instance in the free wind-pollination of castrated wheat with the pollen of various other strains. The same result was obtained in experiments with cross-pollinated plants—rye. I shall refer to an experiment which A. A. Avakian made at the Institute of Selection and Genetics of the Lenin Academy of Agricultural Sciences of the U.S.S.R. Strips 0.5 metres wide and 25 metres long were sown to spring rye, alternating with various winter varieties. At a distance of 3-4 metres from these strips a strip 5 metres wide was sown to Pullman, a variety of winter rye. The plants of all the varieties that were included in the experiment (both winter and spring varieties) flowered simultaneously. Hence, the air contained simultaneously a mixture of pollen from all these varieties. On testing the offspring it was found that all winter varieties produced more than 90% winter plants. For example, the Pullman variety produced no more than 1.5% spring plants; all the rest were winter plants, as is usual for this variety. On testing three generations the offspring of the spring variety similarly proved to be almost exclusively spring plants. Only a few solitary plants were of the winter habit. The preservation in the offspring of the maternal plant forms in such experiments can under no circumstances be explained solely by the plants electing the pollen of their own variety. Here undoubtedly there were also such phenomena as the almost complete absorption, assimilation, of one heredity by another, i.e., the maternal heredity absorbed, assimilated the paternal heredity.

From this point of view it is easy to understand the facts observed: prolonged stability of varieties (for instance, annual varieties) of cross-pollinated plants in free nature. They can be freely pollinated by the wind or insects with the pollen of other varieties closely related to them and growing together with them. Nevertheless, year after year, the plants, within the limits of the variety in question, are, as a rule, to outward appearance, of the same form, relatively speaking. At the same time they differ from the other varieties growing together with them. You need only collect seeds of a wild-growing isolated plant of a particular variety (for instance, white poppy), which was surrounded by plants of another variety (red poppy), and sow these seeds, when, as a rule, you obtain, in a considerable majority, plants of the maternal form and only a minority can exhibit the properties of mixed breeds (hybrids). Such experiments with the sowing of seeds of various wild plants were performed by E. M. Temirazova at the Experiment Base of the Lenin Academy of Agricultural Sciences, near Moscow.

It is known that in the vast majority of the numerous crosses performed by geneticists and at plant-breeding stations plants with hybrid properties are obtained. In the course of generations these plants show variation (segregate) to a greater or smaller extent. Proceeding from the principles underlying the theory that the hereditary substance is corpuscular (particulate), the Mendelist geneticists assert that the products of any cross between forms that differ from each other are bound, in the course of generations, to result in segregation, i.e., are bound to separate into paternal and maternal characters, and, moreover, in the ratio of $(3:1)^n$. But in actual fact such a separation is not at all necessary, neither in elective nor artificial crossbreeding.

We are in possession of numerous facts which go to show that when castrated flowers are intentionally pollinated with alien pollen, seeds are obtained from which seemingly pure maternal plants grow. The latter, in their turn, in the succeeding generations also produce purely maternal forms. The same results were obtained by P. N. Yakovlev on a section of the Michurin Central Genetic Pomological Laboratory. The castrated flowers of *Cerasus Besseyi Bail* were pollinated with the pollen of the peach. The stones obtained after sowing produced plants differing in nowise from *Cerasus Besseyi.* It was conceivable that in the case in question the plants were not hybrid because of bad castration. Although *Cerasus Besseyi* cannot be fertilized with its own pollen, the flowers on separate branches of these plants were castrated anew and pollinated a second time with peach pollen. The offspring obtained the second time also could not be distinguished in any way from the maternal form.

Six generations of hybrids were successively castrated and pollinated with peach pollen. Only in the fifth generation two samples with characters of the paternal form—peach—were found among the many plants obtained as a result of planting the stones of such crosses.

Many other cases may be mentioned, for instance, the crossing of currants with gooseberries, apples with pears, etc., where the influence of one of the parents (usually the male) is frequently almost entirely absent in the offspring. To "explain" such cases as instances of parthenogenesis, i.e., the obtaining of seeds without a process of fertilization, is beneath all criticism. The plants referred to will not yield seeds parthenogenetically.

The ineptness of "explaining" cases as parthenogenesis where the heredity type of one of the parents predominates is most patent when the organism obtained in the cross deviates entirely in the direction of the paternal and not the maternal form.

In an experiment conducted by Kh. K. Yenikeyev at the Michurin Central Genetic Pomological Laboratory, a 16-chromosome American plum, Cheresota, was crossed with a 48-chromosome Michurin plum variety, Reine Claude Reforma. The 16-chromosome plum was taken as the maternal form and the 48-chromosome as the paternal. The plant obtained from this cross had the paternal habitus, including the 48 chromosomes, i.e., as many as the paternal plant had.

All these examples clearly attest to the diversity of the biological process of fertilization, which cannot be fitted into the cytogenetic pattern invented by the Morganists.

We have already noted above that fertilization, like every other process in the living organism, is subject to the laws of assimilation and dissimilation. The fusion of two sex cells is a process of assimilation, a process of mutual absorption, as a result of which in place of two sex cells (a male and a female) we obtain a third, a new cell, which we call a zygote. The hybrid embryo obtained inclines more or less in the direction of the nature of the sex cell which in its own way assimilates its partner more fully, so to speak, than the latter assimilates it. If, for instance, both sex cells possess equal power of mutual assimilation (absorption) a zygote (fertilized cell) is obtained which produces an organism with an approximately equal distribution of maternal and paternal properties and characters. If the power of assimilation of one of the sexual components predominates, the hybrid obtained exhibits greater inclinations in the direction of this parent, to the point of the complete absorption of the hereditary properties of the other.

On this basis it becomes possible, when crossing plant forms, to mould the nature of hybrid embryos, causing them to deviate more or less in the direction of the maternal or paternal form. This should be borne in mind when it is necessary to transmit to the hybrid only a certain few properties (such as resistance to adverse climatic conditions). Michurin suggests in his works that in such cases one had better take the pollen from young plants, plants flowering for the first time, whose nature has not yet become settled. Conversely, the flowers of the other component, to which it is desired to add only a few properties of the first parent, must be picked from a strong tree that has already fruited several times, and they must be so placed on the branch that they are assured of the best influx of nutrients. This will create the conditions for the predominance in the offspring of the properties of the one (the desired) variety and the considerable absorption of the properties of the other.

In a number of cases Michurin earnestly advises the choice of forms for crossing that are remote in regard to place (conditions) of origin not only from each other but also from the place (conditions) where the new variety will be formed. This is necessary when a cultivated variety of southern origin yielding good fruits but unable to endure severe winter conditions is taken as one of the parents, and it is desired to obtain a variety that bears good fruits and can resist severe conditions. If such a southern variety is crossed with a frost-resistant local breed, which, however, yields inferior fruits, the conditions (climate, food, etc.) will promote, enhance the absorbing and assimilating capacity of the local variety's sex cells and a rather undesirable hybrid will be obtained. In this case it is advisable not to take parents (neither the hardy nor the nonhardy) of local origin in order that the external conditions may in equal measure be relatively unsuitable—be averse to the development of the properties of both components in the formation of the embryo. From such hybrid seeds it is easier to grow varieties that yield good-quality fruits and can withstand unfavourable conditions, provided the plants obtained from these seeds are skilfully trained.

The sexual process of plants can be regulated. It is possible to obtain hybrids which obviously deviate in some measure or other in the direction of either parent. A hybrid generation can be bred that varies in only a minor degree. Not infrequently hybrids are produced which from the first generation are practically stable and transmit this property through their seeds from generation to generation.

It now becomes clear in what cases spatial or other isolation of crops of cross-pollinating plants from pollination by other varieties is absolutely necessary and in what cases not. In all cases where the biological usefulness of a plant property conflicts with its economic usefulness, isolation of the seed plants of the variety in question during flowering from foreign pollen is essential. This is of particular importance, for instance, in the seed growing of garden and technical crops. Spatial isolation during flowering is an absolute requisite for plants like cabbage, carrots, red beets, sugar beets, hemp and many others. On the other hand, in cases where the biological usefulness of a character or property coincides with its economic usefulness spatial isolation is not only not useful but frequently even harmful. For example, when the resistibility of a particular variety of rye to severe wintering conditions has to be increased, growing such rye near crops of other varieties will certainly be useful. On this account it will be advisable also in the case of self-pollinators of field cultures, whose hardiness and resistance to climatic inclemencies frequently have to be increased, to castrate part of the plants of these varieties and give them the opportunity of being fertilized electively by the pollen of other varieties sown alongside.

By selecting the conditions which most effectively "humour" the plant (elective fertilization, superior agrotechnique, etc.) the breed properties of plants can slowly, gradually but continually be improved, perfected.

By selecting the conditions which force a plant to abandon the fixed trend of its adaptability and thus destabilizing, abolishing the conservatism of its heredity (either by sharply changing the conditions of cultivation or by enforced fertilization, especially in distant crosses) it is possible in subsequent generations, by a proper choice of the conditions of training, rapidly to create new requirements of the plant, to create new breeds and varieties differing radically from the initial ones.

Regulation of environmental conditions, the conditions under which plant organisms live, makes possible directed change, the creation of varieties possessing the heredity we desire. *Heredity is the concentrate, as it were, of the environmental conditions assimilated by plant organisms in a series of preceding generations.*

By means of skilful hybridization, by the method of sexual conjugation of breeds, it is possible at once to unite in the organism that which has been concentrated, assimilated and fixed in its passage from nonliving to living material by many generations. But no hybridization will produce the desired results unless the conditions are created which will promote the development of the characters which we want the newly-bred or improved variety to inherit.

It must be remembered that nonliving nature is the prime source of living nature. The living body builds itself from the environmental conditions and thereby changes itself.

NATURAL SELECTION AND INTRASPECIFIC COMPETITION[1]

DARWIN created the theory of development of the organic world, the theory of development of plants and animals. It consists of the doctrine of evolution, the doctrine of natural and artificial selection. According to Darwin all forms of plants and animals have been and are being evolved by means of natural and artificial selection. In free nature, where there is no interference by man, wild plants and animals assume form only by means of natural selection. But forms of plants and animals connected with human activity originate through both natural and artificial selection. It would be wrong to consider the forms of agricultural plants as the product of artificial selection only. This applies particularly to field crops where natural factors that man has subjected but slightly to his control play a considerable part in the formation of types and varieties.

We fully share the view that natural and artificial selection are the sole factors entering into the building of plant and animal forms.

But it ought to be known that in the science of biology there exist two opposite views on the role of natural and artificial selection in the creation of organic forms.

According to one view, subscribed to by our Soviet biologists, natural and artificial selection are the creators of new forms.

According to the other view, disseminated among biologists who entertain the conceptions of the formal geneticists, the Morganist-Mendelists, nothing creative can be expected from natural and artificial selection. These scientists assign to selection merely the role of sorting out but not of creating new forms or new properties and characters.

To anyone who approaches biological problems from the standpoint of the metaphysician it seems that the views held by these scientists on the role of selection are nearer the truth than our views. Indeed, can anything new be created by sorting (or selection, if you want to call it so)? One can sort only what already is in existence. Granted that you can select a pair of shoes in a store, and not only select them but make them fit your feet, as

[1] Lecture delivered on November 5, 1945, at the Improvement of Qualifications Courses for the Staffs of State Plant-Breeding Stations of the People's Commissariat of Agriculture of the U.S.S.R.—*Ed.*

they say, yet no one is going to say in such a case that the shoes came into existence through selection. They were made in some factory before they ever got to the store.

And if red spikes are selected out of a wheat population containing, let us say, red and white spikes and are then sown pure, there are those who will say that this is the same as selecting shoes. Here too nothing new is created by selection, because the red spikes were mixed with the white spikes before any selection took place. Can it be said, then, that red-spiked wheat or wheat varieties with any other characteristics are created by selection?

Yes, if you approach biological phenomena formally and not with the logic of dialectics you will be sure not to discover any creative role in either natural or artificial selection.

However, inasmuch as Darwinian selection is not a selection of dead, congealed objects but a selection of plants or animals, i.e., living organisms, different conclusions follow. Living organisms or their offspring continue to live and consequently continue to change and, as a rule, to change in the direction already taken.

After all, there was no need to wait until clearly visible red spikes made their appearance in the population by natural selection. One could have begun by artificially selecting only slightly reddish spikes and sowing their seeds under conditions conducive to the development of red spikes. In this way one would have been able in a few years (wheat generations) to get red-spiked wheat by the repeated selection of increasingly red spikes. Such wheat, in the first place, was not contained in the original population and, in the second place, it might not have appeared if a repeated artificial selection had not taken place and the population had been abandoned to the action of natural selection only.

Without selection and proper tending the animal breeds and plant varieties established by practical farming would never have come into existence.

Darwin pointed out again and again that, as a rule, variation takes the direction of selection. He built up his theory to a large extent on the regular augmentation of the characters that are subjected to selection. The experienced eye of an animal or plant breeder detects in individual specimens insignificant changes in the direction in which he wants to steer a particular breed. Such specimens are selected for breeding purposes. After the passage of several generations these at first barely perceptible deviations gain such proportions in skilled hands that anyone can now perceive them. This signifies that *as a result of selection, augmentation, i.e., the creation of a character or property, takes place.* Consequently, natural and artificial selection create breeds and varieties.

The metaphysicians in biological science either do not understand or pay no attention to the fact that with natural and skilful artificial selection the variation of organisms takes the same direction as the selection. That explains why they conclude that selection creates nothing new because only that which already exists in the population can be selected. This is also the source of the well-known false theory that pure lines are immutable and that therefore selection is useless within the limits of the so-called pure-line varieties. Yet it is a well-established fact that if seeds of rye or some other cultivated plant are selected year after year for size and sown, there will, as a rule, be an increase in the size of the grains of that variety.

I emphasize "*as a rule*," because this does not happen in every case or with every grower. Why not in every case or with every grower? Because not everyone who knows how to sort knows how to raise plants in such a way that the character in question will vary in the direction of its augmentation. The ability to use selection to advantage requires the possession of a correct theory, the ability to comprehend the laws of development of plant forms. However, if one does not know these laws or is unable to discover them, success will attend only those who, as they say, have luck, i.e., everything except the sorting of the plants or seeds comes about without the deliberate action of the selector.

Selection without a knowledge of the laws of plant development can produce only accidental results and *only when plants are grown for the direct practical purpose of yielding good crops*. This explains why practical farming, unaided by any theory, has created good varieties of plants and breeds of animals by means of selection. And these results were achieved in spite of the numerous attempts of the Mendelist scientists to prove by experiments conducted for this special purpose that it is impossible to create anything new by the method of selection.

The mistake made by the geneticists, who refuse to recognize the creative role played by selection, consists in that they understand Darwinian selection literally. In their opinion biological selection is not a whit different from a selection of inanimate objects, whereas the concept *natural and artificial selection* is considerably wider and deeper. *Darwin used the term natural selection metaphorically, figuratively, as including three factors operating constantly and in unison: variation, heredity and overpopulation.*

Let us take, for example, *The Historical Method in Biology,*[1] an admirable book written by that fine Darwinist Kliment Arkadievich Timiryazev. In this work K. A. Timiryazev points out repeatedly that selecton must be taken as a metaphorical expression.[2] Kliment Arkadievich used to state emphatically: "Darwin himself again and again overruled this apparent objection, asserting quite definitely that he himself used the expression 'natural selection' in a figurative, metaphorical sense."[3]

In describing the three factors which in their totality and by their continuous operation constitute the concept of natural selection, K. A. Timiryazev pointed out that the variability inherent in living organisms "furnishes material for the formation of new peculiarities in the structure and functioning of

[1] К. А. Тимирязев. *Сочинения*, Сельхозгиз, 1939 г., т. VI.
[2] *Ibid.*, pp. 123, 153, 154, 199, 200, 201, 204, 206, 207, 216, 217.
[3] *Ibid.*, p. 123.

organisms," heredity "consolidates and accumulates them," and overpopulation "removes all beings unsuited or ill-suited to the conditions of existence."[1]

I want to stress once more that *by selection breeders ought to understand something else besides sorting.* Of course sorting, too, in the literal sense, i.e., by means of sieves, according to length and size of ears, according to colour, hardiness, etc., should not be forgotten, if for no other reason than that in practical farming new forms needed by man can be created (though this may largely be purely accidental) by sorting alone.

For ages practice that knew no biological theories created relatively good forms of plants and animals adapted to the satisfaction of man's requirements. By way of contrast to this instances can be cited of animal- and plant-management "theoreticians," followers of false theories, devotees and disseminators of the science of formal genetics, who for decades have been working to create better varieties but have brought forth nothing so far save barren promises.

Plant breeders who work for the practical purpose of creating better varieties, of growing good, élite seeds, must have a clear conception of not only the benefits to be derived from correct theory but also the harm of incorrect theory. It is necessary for them to know that incorrect theory retards the practical work of creating varieties. Contrariwise, theory which correctly reveals the laws of plant development assures to the breeder, when applied in practice, appreciable success in his work.

Let us return to Darwinian selection, to our understanding of the three factors of which it consists. Let us analyze the problem of the last factor—*overpopulation.* It is with this factor, as being apparently the most easy to comprehend, that an exposition of Darwinism ordinarily begins. I cannot agree to the generally accepted treatment of it. Nor do I agree to the way the overpopulation factor is treated because, as I see it, *there never has been, nor can there be*, as a rule, *any such thing in nature as overpopulation.*

I therefore understand by Darwinian natural selection the following jointly operating factors: variation, heredity and survival (instead of overpopulation). I consider that such an understanding comes closer to reality and the general correct Darwinian theory of development, to creative Darwinism.

As is well known, Darwin and Darwinists point to the universally observable great discrepancy between the number of embryos of organic forms that appear and the number of organisms that attain mature or old age.

For example, in plants, insects and fishes the number of organisms of mature age is hundreds or thousands of times less than that of the embryos engendered. But I consider the interpretation of this phenomenon as given by Darwin and subsequently repeated by many (if not all) Darwinists, and which is based upon intraspecific competition, to be a wrong one. In my opinion this interpretation runs counter to the principal laws of biology, counter to the fundamentals of Darwinism.

---

[1] К. А. Тимирязев, *Сочинения*, Сельхозгиз, 1939 г. т. VI. стр. 154.

This interpretation has its origin in the so-called Malthusian "theory"; it contradicts the very essence of Darwin's theory of evolution.

"Darwin," wrote Karl Marx, "failed in his excellent work to see the fact that by discovering 'geometrical' progression in the animal and plant world, he was refuting the theory of Malthus. The Malthusian theory is based precisely on the point that he counterposes Wallace's geometrical progression of man to the chimerical 'arithmetical' progression of animals and plants. Darwin's book for example, where the causes of species' extinction is discussed, contains, if one leaves aside his main principle, a detailed refutation, based on natural history, of the Malthusian theory."[1]

Malthus put forward his "theory" to justify the phenomenon generally observed in bourgeois society of the vast majority of people receiving insufficient material values for the normal satisfaction of their requirements in spite of an overproduction of these values.

Bourgeois science of course sought and still seeks an explanation of this not in the laws of development of a society built on the principle of exploitation. It is looking for such an explanation in the "law of nature," according to which, it is claimed, every species of plant or animal on earth, as well as man, produces offspring considerably in excess of the food available for them (meaning by food conditions of life in general).

Once there is a multitude of consumers (say, plants of a given species) and only a limited quantity of conditions of life (say, light), competition, struggle, is inevitable. Moreover, the more the organisms approximate each other in their requirements, in their interests, the fiercer the struggle among them. Hence the assertion that the struggle for conditions of life among individuals, if not direct then indirect, is more intense within a species or variety than between species, and that this can readily be observed in nature.

As a matter of fact, however, this phenomenon cannot be observed in nature because it does not exist. For it would be wrong to consider that rabbits, for instance, suffer more, even indirectly, from each other because of their kindred requirements than from animals of other species, as for instance wolves or foxes, not to mention all kinds of infectious diseases inflicted upon them by organisms quite distant from them in respect to species or genus.

In general, as we attentively observe the life and development of plants and animals and through practice gain increasing knowledge of the laws governing natural and artificial selection as the creators of organic forms, we arrive at the following conclusion: The discrepancy between the great number of embryos produced and the number of adult individuals obtained from them is in no degree a condition that gives rise, as is alleged, to a struggle or competition among the individuals most closely related in their biological requirements. The question at issue concerns the laws of struggle

[1] K. Marx, *Theorien über den Mehrwert.* 2. Band, 1. Teil, Stuttgart 1921, S. 315.

and competition which are motive forces in the process of evolution, the process of changing the nature of plant and animal forms.

Even if fortuitously it is possible to observe overpopulation among the most closely related plant or animal forms on small areas for short periods of time, such overpopulation will still not be the motive force of evolution. On the contrary, all organisms evince, though in various degrees, a diminution of vitality. All closely related organisms which have experienced the action of accidental overpopulation will always be less adapted to survive than before they were subjected to the influence of overpopulation. Overpopulation of individuals within a single species or variety which suffer from each other directly or indirectly is not the rule in nature but an extremely rare exception. But even when it does occur, overpopulation is in no degree a factor of natural selection, a factor of progress. For this very reason I do not include it among the factors of Darwinian natural selection. In all my works I substitute survival for the factor of overpopulation.

Let us return to a fact universally observed in nature—the colossal birth rate of plants and animals as compared with the number of organisms that attain maturity. If closely related forms do not (with rare exceptions) interfere with one another, do not compete with each other for the conditions of life, why then is there so great a discrepancy between the number of embryos and the number of adult organisms? This question deserves to be carefully studied by biologists, and particularly agrobiologists.

We are told that living beings tend toward overpopulation but that since the conditions of life are limited a most intense struggle or competition goes on between closely related organisms, within a species or a variety. But what we actually encounter quite frequently in practical agronomy is a very different phenomenon, namely, difficulty in obtaining the requisite seed crop in the case of a number of plants. There is a fairly large number of agricultural plants whose whole crop of seeds is used for sowing. They include alfalfa and clover. The areas sown to these plants depend to a considerable extent on the amount of seed obtained from the preceding crops. What one observes in these cases is not a surplus of seeds, of embryos (or germinating plants) and not a lack of space to live in but the very opposite. The lack of seeds of these plants makes it necessary to sow the designated areas to other plants.

The factors which keep down, limit, the number of individuals in a species or variety are of course different in each particular case. But in our opinion competition within a species or variety is practically never a limiting factor.

Besides climatic and soil conditions, the number of individuals that make up any particular species of plant depends primarily upon the presence and number of individuals of other, as a rule remotely related, species whose life activity in one or another aspect is not a matter of indifference to the individuals composing the species we are discussing. For instance, the timely presence of bumblebees, common bees and other insects which facilitate the pollination (fertilization) of clover and alfalfa increases the seed production of these plants. On the other hand, the presence of insects which feed on flower buds or the ovaries of clover or alfalfa lowers and frequently almost completely destroys the seed crop. Hence, clover and alfalfa seed crops depend in the main on the extent to which the seed plots are provided with pollinator insects and protected against attacks by pests. If practical agriculturists do not interfere with this process of ordinary biological life, do not promote the development of useful insects, do not prevent the development of pests or destroy them it will, as a rule, be impossible to get a sufficiency of seeds, not to speak of a surplus of them.

It was Darwin's greatest merit that he furnished the theory of the evolution of plants and animals with a true foundation, gave a materialist, historical explanation of the so-called purposiveness that we meet at every step and see in the structure of the forms of plants and animals and in their behaviour. The discovery of the causes of these and other concrete phenomena, the regulation of which is required by practical agriculture, constitutes the principal task of agrobiology and at the same time develops and deepens the theory of Darwinism.

The biologists whose work is not connected with the solution of practical farming problems find it difficult to agree that the number of individuals in a species does not depend on intraspecific competition, on competition between closely related individuals. But practical agriculture clearly shows that always the number of individuals in a species or variety depends solely upon success in the struggle for existence which the individuals of a particular species or variety carry on with environing inanimate and animate nature. The struggle for life carried on by species and varieties includes both symbiosis, harmony in the broad sense, and antagonism. All this goes on because the members of any species of plant, animal or microbe live at the expense of the vital activity of the members of some other species. In nature the interconnection between the species of plants and animals represents a very complicated chain. Herein included are both harmony and collision in the life and in the number of individuals of any species of plant or animal in nature. But competition, particularly of the severest kind, between individuals within a species or variety, has no bearing whatever on biological phenomena.

In order to secure the quantity of particular plants and animals that practical breeders need, it is extremely necessary that the science of agrobiology comprehend the complicated biological interconnections, the laws governing the life and development of plants and animals. This is essential in order to be able to provide needed plants with the conditions necessary for them, to do so in the best possible way and with the greatest benefit to man, and to protect these plants against all normal biological and climatic adversities.

We must study the harmony of life existing between the species we need and other species for the purpose of creating in the fields the conditions necessary for growing good crops.

We must study the collisions, the tribulations, which the species we need suffers in the general chain of biological regularities so as to be able to protect our crops in the fields with the least expenditure of human effort from both pests and disease.

Practical agriculture will never require a study of the biological laws of the overpopulation of this or that species. But if one is well acquainted with living nature it will not be difficult to discover that here too there is no such overpopulation of individuals of a species as calls forth intraspecific competition among them. The extent to which the inherent possibility (the possibility peculiar to living bodies) of multiplying infinitely is made a reality is always strictly conditioned and controlled by nonliving and living environment. Hence there neither is, nor can there be normally, overpopulation that leads to competition within a species. If overpopulation can be observed in exceedingly rare cases, it occurs in a purely accidental way and not on the basis of biological necessity (law), and does not constitute part of the chain of laws governing evolution.

Theory requires close connection with practice if it is to develop with success. Darwin built his theory of evolution on a generalization of agricultural practice. The best way to develop this splendid theory is in close association with practice.

It is well known that various weeds are antagonists of cultivated plants. Many are the agrotechnical measures that aim to protect cultivated plants from being overwhelmed by weeds.

It is likewise well known that many weeds do not grow wild, in a state of nature, or if they do then very sparsely. This signifies that the nature of these plants, like that of cultivated plants, is closely connected with practical agriculture.

In studying the life of weeds one will find many an interesting example showing that they behave as if in the struggle for existence they deliberately set out to vanquish cultivated vegetation. The following phenomenon will serve to illustrate the point: in the struggle for life these plants seem to refrain altogether from struggling with their competitors; they seem to relinquish all vital conditions to the cultivated plants. But all this, as it turns out, is only for a certain time. If shoots of grain, for instance of spring wheat, appear simultaneously with weed shoots—wild oats (*Avena fatua*) or others—the wild oats or other weeds, as a rule, emerge the victor in the competition (provided there is no weeding). But if by chance or intentionally, owing to the application of some agrotechnical measure, the grain shoots appear considerably earlier than, say, the wild oat shoots and the grain stand has time to develop well, the young shoots of many weeds will find it difficult to compete with the cultivated plants. In these cases the cultivated plants emerge the victors. But here many weeds will fail to germinate. The seeds, though alive, fail to sprout although there is moisture, warmth and access to air. However, after the grain plants (their competitors) are mown a mass of weed shoots come up at once if there is moisture. They did not germinate, did not compete, until the grain was ripe, but then took complete hold of the field. In such cases new-sown cultivated plants cannot hold out in the struggle without agrotechnical interference for the purpose of weed control. Therefore, after the grain is harvested stubble-ploughing or a fine mellowing of the soil is recommended to cause still more shoots to spring up so as to be able some time later to destroy them by deep ploughing.

Numerous analogous examples of the behaviour of weeds could be cited. They all go to show that in the struggle for life a plant, as if foreseeing the situation, strictly computes (being the product of natural selection) its forces and potentialities.

The path that leads to the discovery of the natural causes of these and all other biological phenomena is shown us by the theory of natural and artificial selection. Taking this teaching as our basis we perceive that a plant foresees nothing and expects nothing. As a result of selection it has acquired the property, for instance, of not growing in soil in which there are many living roots of competing plants. In the struggle for existence this property has proved useful to its possessor, therefore individuals endowed with this property survived better and in greater numbers, and left offspring. It is of course possible to discover the specific causes of the failure of the seeds to germinate. The causes may possibly be certain substances secreted by the living roots and inhibiting further growth. If we know the specific causes we can "outsmart" the smartest. If we know what substances inhibit the germinating capacity of the seeds of the weeds in question one can neutralize them by introducing other substances into the soil and thus compelling the seeds to sprout without waiting for their competitors, i.e., the cultivated plants, to quit the field. Where there is a powerful development of cultivated plants young weeds perish.

Another example of plants seemingly possessing natural foresight may be found in alfalfa, a perennial fodder grass. But the "foresight" displayed by these plants is a misnomer for miscalculation.

The varieties of this plant are the product of natural selection; artificial selection was little concerned with fodder grasses. Therefore the behaviour of these plants follows in the main the laws that took shape under natural, and not under artificial, selection, although these valuable plants have long been used in practical farming.

In the central zone and in some of the other districts of the Soviet Union alfalfa, of which there usually is a lack of seeds as has already been stated, will, if sown pure, frequently fail to form seeds, although the grass stand may be good, thick and strong. This happens even when the plants are normal, strong and during growth do not oppress each other.

Under these conditions the seeds will be useless, biologically, for the propagation of the species. Alfalfa seeds are not scattered by wind or animals. They fall round about the maternal plants. But when sown thick (normally) these places are occupied by plants of the same perennial alfalfa.

A plant expends much of its most valuable energetic substances on the production of seeds and if under the particular conditions the seeds are useless for the propagation of the species, it is more useful for the survival of the species if the substances that ought to be expended on seeds are deposited in the root neck as reserves for the sprouts of the following year. In the districts indicated alfalfa plants actually behave in this way.

But as soon as alfalfa is sown decidedly more thinly than normal, i.e., made sparse, it will develop seeds in abundance even where the spaces free from alfalfa are occupied by other plants, by fodder grasses, for example. On the whole, in the districts mentioned alfalfa produces seeds when there is free space, not occupied by other alfalfa plants, around its tufts, even if these spaces are occupied by plants of other species. It prepares seeds precisely for the purpose of populating this space. No one will say that the above-described property of alfalfa to yield seeds only when there is free space is not useful for the survival of the species.

But it would be still more useful for alfalfa as well as for the species, if it produced seeds even when there is no free space. For people would gladly sow these seeds in other fields where there is no alfalfa. This would be more useful for the species, as it would be more widely disseminated. But the nature of these plants was formed by natural, not artificial, selection and the property of not producing seeds in pure alfalfa fields was and still is biologically useful to the alfalfa as a species in respect to its survival. The above refers of course to natural conditions and not to practical farming.

Taking the observed phenomenon as their starting point the men of practice and science must discover its immediate causes so that they may be removed, leaving plots of good, pure alfalfa for seed growing. In the meantime, while we know only this phenomenon, it is absolutely necessary to follow the urgent advice of V. R. Williams, namely, not to sow alfalfa pure but mixed with fodder grasses. Such a mixture, as Williams indicates, creates better soil fertility conditions than sowing alfalfa pure. As has been shown by observations and experiments, for instance, those of F. I. Filatov,[1] sowing alfalfa mixed also yields more seeds. Of course, in order to obtain good seed crops, seeds of alfalfa and perennial grasses must be sown in different proportions in the different districts.

The advantage of deciding purely theoretical questions of biology in close connection with practice can be demonstrated with particular clarity by taking kok-saghyz, a species of dandelion, as an example.

We know that in Darwinian literature the dandelion is often taken as a striking demonstration of the general tendency of living beings toward overpopulation. As a result of this tendency "readily observable" competition for the conditions of life inevitably arises everywhere, it is claimed, among the members of the species or variety. A minority of the individuals, the most fit, those which possess the personal qualities that give advantage

---

[1] Research worker of the Institute of Grain Husbandry of the Southeast of the U.S.S.R.

in the struggle, survives; the majority, the less fit, which do not possess the particular fitness that gives such advantage, perishes. The dandelion, a widely-known plant, is taken to clearly illustrate the point. A theoretical, abstract calculation of the number of seeds obtained from one dandelion plant and the further calculation of the number of seeds of all plants obtained from the first seeds, etc., are used to show the alleged inevitability of intraspecific competition. Such a calculation really indicates that the progeny of a single plant can occupy the surface of the whole earth in less than ten years' time.

"Thus," writes K. A. Timiryazev, "the tenth generation of one dandelion family would require an area fifteen times as great as that of all the land on earth."... "Let us return to our dandelion; let our thoughts carry us to the period (between the 9th and the 10th year) when its progeny will have spread to the entire land surface of the earth. What will happen after that? Each plant on completing its life cycle will perish,[1] leaving a hundred descendants and a bit of land enough for one.

"Who will get this inheritance? Whose portion will be life and whose death at the very threshold of life? This is decided in fierce combat from which only one will emerge the victor".... "What determines who shall be the chosen one? His own merits."[2]

It would seem that overpopulation of a species and *intraspecific* competition as its consequence might really be observed in nature.

But no sooner was this same dandelion—kok-saghyz—sown for the practical purpose of getting as many of these plants as possible in special fields than the erroneousness of abstract calculation was at once discovered. Many things proved to be the opposite of what had been expected.

Plants and animals really do possess the inherent faculty of multiplying without end. This property is useful to each species but the point is that the environmental conditions essential for a realization of *unlimited* multiplication never exist. Hence species and varieties never reach the point of overpopulation. On the contrary, everywhere, as a rule, underpopulation is to be observed.

According to the doctrine of natural selection it can never be useful but, on the contrary, only harmful for a species to perpetuate an adaptation for intraspecific competition. Perpetuation of properties harmful to a species and leading to a diminution in the number of its members (and this would, after all, be the inevitable result if there were intraspecific competition) contradicts the entire spirit of the Darwin-Timiryazev teaching on natural selection.

Let us return to the example of the dandelion and analyze it, starting with the results of agricultural practice.

---

[1] For the sake of simplicity we consider the dandelion an annual plant. (Note by K. A. Timiryazev.)

[2] К. А. Тимирязев, *Сочинения*, Сельхозгиз, 1939 г., т. VII, стр. 131, 132, 133.

I happen to be participating in the scientific elaboration of a practical farming method of raising kok-saghyz, a dandelion species. It may, indeed, be easily observed that a kok-saghyz plant can produce many hundreds of seeds each year, and it is not an annual plant. It may bear during the first, second and third years of its life. It is not very exacting with regard to climate and soil, in the agronomic sense, and can germinate in a very wide zone.

It would seem that all that had to be done was to remove the over-population factor, i.e., not to sow the seeds too thickly at any one place (as it is claimed is the case in free nature), but provide the plants with as much room as possible (to prevent competition among them), and the needed kok-saghyz area could quickly be brought into existence. At any rate it seemed that the quantity of seeds available would not be a factor limiting the area under kok-saghyz needed for practical purposes.

In actual fact it proved not so simple to obtain the requisite quantities of kok-saghyz seeds. Kok-saghyz plants, whose nature formed in the lower layer of the vegetal cover composed of other grass species, fare considerably worse when sown singly, separately, on the plantations than when sown in groups. Therefore kok-saghyz seeds should not be sown evenly in the field prepared for it but at the rate of 100-200 seeds at a time (a pinch of seeds for one hill)

Practice has shown that if kok-saghyz is sown in rows, each seed lying by itself in a line, and not in groups, in hills of 100-200 seeds, it is exceedingly difficult to obtain even germination of the plant. As a result, under field conditions, growth is frequently so sparse that even the quantity of seeds sown is not reproduced.

On these grounds we proposed that kok-saghyz be sown in hills instead of in rows, placing 100-200 seeds in a hill (occupying an area of 5-10 sq. cm.). Then good sprouts are obtained in bunches or small groups. This method of sowing also fully takes care of the kok-saghyz plant's inconsiderable natural requirement for shade against bright sunlight. The 50-100 young plants growing in a group shade each other and in this way provide the microconditions which the nature of kok-saghyz requires. As the hills in the rows are 40 cm. apart and the rows 60 cm. apart, and as the plantation is kept well weeded and its soil friable by cultivation, there is always sufficient moisture and nutriment in the soil for the kok-saghyz plants formed into groups as required by their nature. Under such conditions of cultivation kok-saghyz grows better than in free nature where it finds itself under the cover of plants of other species. The natural vegetal cover not only affords the light shade which kok-saghyz needs but is, at the same time, oppressive to it both because it gives too much shade and because it intercepts the food and moisture in the soil.

Hill planting of kok-saghyz yields without much trouble fine crops of the marketable product, i.e., the roots. The seed crop per unit of area is also considerably in excess of naturally grown crops or on farms where kok-saghyz is grown singly and not bunched in rows.

But even in such seed spotting, seed multiplication is on the average much less than 100-fold, the figure taken in the abstract example of the dandelion to prove the existence of violent intraspecific competition as an inevitable result of the unlimited capacity to multiply supposedly inherent in species.

In actual fact kok-saghyz is excellent proof not of the existence, but of the nonexistence of intraspecific competition in free nature. The example of kok-saghyz not only proves convincingly the ability of organic forms to multiply unlimitedly under proper conditions but also argues in favour of the proposition that no species or variety can actually ever reach a stage of overpopulation that will call forth intraspecific competition impeding the further numerical increase of individuals in the particular species.

To prove that it is inherently possible for kok-saghyz to multiply without limit let us point out the following. In view of the fact that under the given conditions of kok-saghyz cultivation the rate of seed multiplication is insufficient for practical needs, which is frequently due not to biological causes but to the difficulty of gathering in the seeds which ripen at different times, we have worked out a method of propagating kok-saghyz vegetatively by using bits (cuttings) of its roots. It had been found that kok-saghyz can be propagated by planting root cuttings (weighing 0.2 to 0.3 grams) and that from these tiny slits plants are obtained more vigorous than those grown from seeds. Thus, kok-saghyz plants can be grown not only from seeds but also from small cuttings of vegetative parts, i.e., the possibilities of propagating this plant are colossal indeed.

Yet in spite of this we observe no intraspecific competition.

Tables 1 and 2 give the weights of the roots of experimental kok-saghyz plants, the number of which per hill (spot) at harvest time in autumn ranged from 1 to 37. (See Table 1.) Consequently, the plants in the various hills

Table 1

KOK-SAGHYZ GROWN FROM SEEDS[1]

No. of plants in hill	No. of hills	Average wt. of roots of all plants in hill[2]	Average wt. of roots of 1 plant	Average wt. of roots of 5 biggest plants in each hill	Average wt. of roots of 1 of 5 biggest plants of each hill
1	2	3	4	5	6
1—5	183	35.2	12.7	52.3[3]	10.5[3]
6—10	106	59.6	7.9	49.2	9.8
11—15	55	76.4	5.8	48.5	9.7
16—20	40	82.0	4.6	44.5	8.9
21—25	25	90.0	4.0	42.8	8.6
26—37	19	103.2	3.4	42.9	8.6

[1] Sown at the Experiment Base of the Lenin Academy of Agricultural Sciences of the U.S.S.R. Roots analyzed by I. E. Glushchenko and R. A. Absalyamova.

[2] Weight everywhere in grams.

[3] This figure applies only to the group of hills having 5 roots.

Table 2

KOK-SAGHYZ GROWN FROM CUTTINGS[1]

No. of plants in hill	No. of hills	Average wt. of roots of all plants in hill	Average Weight of Roots of 1 Plant[2]												
			The heaviest in all hills	2nd heaviest in all hills	3rd heaviest in all hills	4th heaviest in all hills	5th heaviest in all hills	6th heaviest in all hills	7th heaviest in all hills	8th heaviest in all hills	9th heaviest in all hills	10th heaviest in all hills	11th heaviest in all hills	12th heaviest in all hills	13th heaviest in all hills
1	2	3	4	5	6	7	8	9	10	11	12	13	14	15	16
1	78	65.8	65.8	—	—	—	—	—	—	—	—	—	—	—	—
2	190	79.8	51.4	28.4	—	—	—	—	—	—	—	—	—	—	—
3	297	92.4	48.4	27.0	17.0	—	—	—	—	—	—	—	—	—	—
4	401	98.8	43.7	26.1	17.7	11.2	—	—	—	—	—	—	—	—	—
5	402	114.2	42.6	27.7	20.0	14.6	9.2	—	—	—	—	—	—	—	—
6	316	117.7	38.3	25.1	19.1	14.9	11.8	8.2	—	—	—	—	—	—	—
7	273	128.8	38.5	25.7	19.8	15.5	12.2	9.9	6.8	—	—	—	—	—	—
8	115	130.0	34.4	24.7	19.7	14.9	12.2	10.5	8.0	5.3	—	—	—	—	—
9	74	135.7	34.6	23.6	17.4	14.9	13.2	10.4	9.0	7.1	5.3	—	—	—	—
10	24	143.2	33.9	21.4	17.4	16.1	13.1	11.1	9.6	7.6	6.9	5.6	—	—	—
11	18	123.5	29.7	21.1	15.2	11.1	9.3	9.7	7.2	6.1	5.7	5.4	3.1	—	—
12	5	143.4	35.6	19.8	17.2	13.6	12.0	10.0	7.0	7.0	6.6	5.8	5.2	3.6	—
13	4	171.5	34.2	23.7	19.2	18.0	13.2	10.7	10.5	9.5	8.0	7.5	7,5	5.0	4.2

[1] The experiment was performed in 1945 at the Institute of Genetics of the Academy of Sciences of the U.S.S.R. by Senior Research Worker T. L. Ivanovskaya.

[2] Weight everywhere in grams.

experienced various group influences. On two hill-sown kok-saghyz fields (grown from seeds—Table 1, and grown from cuttings—Table 2; see pp. 459-60) in autumn 1945 the plants were harvested separately by hills. The number of plants in each hill and the crop obtained from their roots were recorded. The crop data relating to all hills with the same number of plants were summarized and the average weight of the roots per hill (spot) calculated.

On sowing, about the same number of cuttings or seeds were placed in each hill. But at harvest time the number of plants in each hill was naturally different. The causes of this numerical inequality of the plants in the various hills were numerous. Due to accidental causes the number of shoots that sprang up and the loss of plants during vegetation undoubtedly differed in the different hills. The nutrimental area for all plants in each hill, regardless of their number, was, as has already been stated, an equal 40 cm. × 60 cm. With such intervals between rows and between the hills in the rows there can be no question of the plants of one hill having influenced those of another.

But in all cases, whether the hills contained a few or many plants, the latter were very close together. After all, when planting seeds, 100-200 of them are put in the ground at one spot, and the same applies to propagation by cuttings, the only difference being that each hill gets not 100-200 seeds but 8-15 cuttings. If sown in this way the plants in a hill are so close to each other that in ordinary harvesting for production all the plants of a hill are treated as one. This greatly facilitates the harvesting of root crops as they are closely intertwined. No time need therefore be spent on picking out the roots of each plant separately.

In hill planting, as already mentioned, 100-200 seeds are planted at one spot in the ground, i.e., as many as or even twice as many as the number of dandelion seeds taken to demonstrate theoretically that competition within the species is inevitable. Consequently, if intraspecific competition absolutely exists it should manifest itself in the highest degree in the hills we have sown. As a result of such competition there should be a sharp drop in hill-planted kok-saghyz crops in comparison with crops obtained from the same number of seeds sown evenly (singly) in a plantation of the same size. But in practice opposite results have been obtained. In hill-sowing, when 100-200 seeds are sown in one spot, the resultant crop is several times as big as in row sowing, when the same quantity of seeds is evenly distributed singly throughout the rows.

From the tables it appears that the more plants there were in a hill at harvest time the bigger was the average root crop from one spot. Thus, where there was 1 plant in a hill grown from cuttings, the average crop weighed 65.8 gr., where there were 5 plants—114.2 gr., where 10 plants—143.2 gr. (See Table 2, column 3.) In seed-sown crops the picture is the same. When the plants remaining in the hill at harvest time numbered 1-5 the average crop per hill weighed 35.2 gr., with 11-15 plants it weighed 76.4 gr. and with 26-37 plants—103.2 gr.

In our experiment no count was taken of the seeds. But the figures obtained from large collective-farm areas sown in Kiev Region clearly show that the more plants there are to a hill—in certain cases there were as many as 150—the greater the total seed crop from all the plants of the hill.

It would seem that one might well end here the analysis of the question of whether there is intraspecific competition among kok-saghyz plants, i.e., one of the dandelion species.

For the purposes of practical agriculture the question is clear. So long as it possesses the nature which took form in the wild thickets kok-saghyz must be sown in groups, or hills, of 100-200 seeds (or even 200-300 if there are enough seeds), discarding all thought of intraspecific competition.

This method of sowing, which involves a considerably smaller expenditure of labour, yields a crop of both roots and seeds considerably greater than crops obtained from the same number of seeds planted singly at even distances in rows. If the root and seed crops are bigger when the plants grow thick, in hills, this means that *in that particular case there is no question of intraspecific competition so far as practical farming is concerned.*

The fact however that dandelion (kok-saghyz) possesses the unlimited inherent faculty of multiplying not only sexually but also asexually indicates not that intraspecific competition must exist in nature but that if this faculty is made proper use of in practice any requisite area can actually be sown.

But the average kok-saghyz root crop figures given in the above tables and their dependence on the number of plants in the hill may at first glance convey the impression of being proof not of the nonexistence but, on the contrary, of the existence of intraspecific competition.

The fourth column of Table 1 gives the average weight of the roots of one plant for every hill category. The figures given in this column make it perfectly clear that the more plants a hill contains the less the average weight of the roots per plant. Thus, when the hill consists of 1-5 plants the average weight of the roots per plant is 12.7 gr., and when the hill has 26-37 plants this average drops to 3.4 gr.

However, the figures of the fourth column alone are not enough to determine whether there was competition or not. For the decline in the average weight can also be explained as follows: The more the plants were crowded together, sown in one hill, the greater their mutual oppression (not competition) or rather the less nutriment was available per plant and hence the smaller the average root crop per plant.

Such an explanation would be acceptable if the weight of the roots of each plant were at least relatively close to the average root weight per plant in the hill. But actually the sizes of the roots of the different plants in one hill differ sharply in all cases. Thus, the average root weight per plant for each hill category is only very slightly indicative of the root weight of the biggest and smallest plants. Hence the figures in the fourth column do not indicate whether what really occurred here was a general mutual oppression of the plants or competition among them—a suppression of some plants by others; or perhaps there was neither general oppression nor competition but a different development of each plant in the numerically different groups. The fifth and sixth columns of the same table give the average aggregate weight of the roots of the 5 biggest plants in each category of hills (column 5) and the average weight of the roots of one plant (column 6). When there were 5 plants to a hill the average aggregate weight of their roots was 52.3 gr. and the average weight of the roots of one plant—10.5 gr.; when the hill contained 26-37 plants the average root weight of the 5 biggest plants in each hill was 42.9 gr. and correspondingly the average weight for one plant was 8.6 gr.

If it was still possible to interpret the average root weights for one plant in each hill category as the consequence of possible mutual oppression among the plants, the average root weights for one of the 5 biggest plants in each hill category leaves almost no grounds for such a supposition. The decrease in the average root weight of one of the 5 biggest plants due to the increase in the total number of plants in the hill is not great in the given case—from 10.5 gr. when there are 5 plants in the hill to 8.6 gr. when there are 26-37 plants in the hill. Consequently, in the case in question, mutual oppression and insufficient food or moisture as influencing the development of plants must be discarded, making room, it would seem, for a different phenomenon—competition among the plants that grow so close to each other in the hills. The roots of the 5 biggest plants in the 21-25-plant hill category weighed 42.8 gr.—almost as much as the roots of its remaining 16-20 plants (47.2 gr.). In the hill category containing 16-20 plants the roots of the 5 biggest plants weighed 44.5 gr. while the remaining roots weighed on the average only 37.5 gr., although there were 2.5 times as many plants (11-15). It would seem safe to arrive at the following conclusion: the stronger, fitter kok-saghyz plants oppressed the less strong, the less fit. Obvious competition among the plants of one species!

This is "corroborated" still further by the figures contained in Table 2 where the crop is grown not from seeds but from cuttings, which produce bigger roots.

In this table the average root weights of the plants in each hill category are given in the order of decreasing weights (sizes). From these figures one seems to be able to note the degree of competition. For this one need only examine the horizontal columns of Table 2, which as was said above, give the average weights of the roots in descending order, for each hill category. When the hill contained 1 plant, the average weight of the roots of that plant in all (78) hills was 65.8 gr.; when there were 2 plants in a hill the roots of the stronger one weighed on the average 51.4 gr. and of the weaker one only 28.4 gr.; when 5 plants, the strongest had roots weighing 42.6 gr., the second strongest 27.7 gr., the third strongest 20 gr., the fourth strongest

14.6 gr. and the fifth strongest 9.2 gr. All the other variants (hill categories) present analogous pictures in the above table. On the whole these figures seem to confirm in toto the existence here of intraspecific competition, of the oppression of the weaker plants by the stronger. It may be thought that this oppression continues until some of the plants have completely ousted, extruded the rest. As one observes kok-saghyz plants from sprouting to maturity one can easily see that there is a constant diminution in the number of individuals, a process of self-decimation.

It would seem that for the science of biology these data alone would suffice to answer the question of whether intraspecific competition exists or not.

But such a conclusion as to the existence of intraspecific competition, drawn from the figures contained in the tables giving the weights of kok-saghyz roots, would be absolutely wrong.

Practice, as we already said, has shown that in seed spotting (several hundred seeds sown in one spot) the root and seed crops are bigger than if the same quantity of seeds is sown over the same planting area singly and close to each other in rows. Hence, practice tells us that in this particular case not only is there no oppression of some kok-saghyz plants by others but that they grow better in compact groups (hills or spots). How then explain the bigger crops of roots and seeds in hill planting as compared with row planting? We can find no other explanation for this phenomenon. It is, of course, impossible to reconcile oneself to obviously opposite conclusions, one drawn by biological science and the other by practical farming. The Marxist Lenin-Stalin methodology, the only correct one, clearly tells us that to suppose one basis exists for the realities of life and another for science is completely false. There should not be two bases; there must be unity between theory and practice. In the example under analysis the conclusion that follows from the practical cultivation of kok-saghyz is more scientific and is more in congruence with the biological laws of development of kok-saghyz than the conclusion drawn by the biologists who in the figures we cited above (Tables 1 and 2) see proof of the existence of intraspecific competition.

Indeed, even if a cursory analysis of the figures in Table 2 read horizontally seems to warrant the conclusion that competition does exist among kok-saghyz plants, a reading of these figures vertically indicates something else. Thus, the average weight per root in hills where at harvest time there was only one plant was 65.8 gr., i.e., considerably more than the weight of the biggest root in those hills which had many plants. However, in the same column, for instance in the 8-plant hill category, the average root weight of the heaviest plant (34.4 gr.) can already be compared with the root weight of the heaviest plant in the hill category containing 13 plants. The average root weight of the second heaviest plants in the hills which at harvest time had only 2 plants each was 28.4 gr., i.e., only a little more than in those categories where there were many more plants. In the last category, which contained 13 plants to each hill, the second heaviest plant root weighed 23.7 gr.

However, beginning with the weights of the roots of the third biggest plants it may regularly be observed that each succeeding root weight (taking the order of decreasing size) was relatively bigger in the hills which at harvest time had more plants. Thus in those hills which at harvest time had only 4 plants each the average weight of the roots of the fourth heaviest plant was 11.2 gr. and in the hills which had 10 plants each the weight of the roots of the fourth heaviest plant was 16.1 gr. The root weight of the fifth heaviest plant in the 5-plant hill category was 9.2 gr. and in the 10-plant hill category—13.1 gr., etc.

But after all, the roots, say, of the fourth heaviest plant in the hills having only 4 plants each formed in the congested condition caused by the 87.6 gr. which the remaining 3 plant roots weighed, while the roots of the fourth heaviest plant of the hills with 13 plants each experienced the oppressive action not of 87.6 gr., but of 153.5 gr., the aggregate weight of the remaining plant roots. And nevertheless the root weight of the fourth heaviest plant in the second case was 18 gr. as against 11.2 gr. in the first case.

Where, then, does oppression come in here, not to speak of competition? It is not in evidence. On the contrary, it is quite apparent that if the weights of the plant roots are arranged in descending order the root of the last plant in any hill will weigh less than the ordinally corresponding plant in those hills that have plants left.

In our particular experiment the more plants there were in the hill the greater was the weight of the roots of the plants bearing the same ordinal number.

I would like, in addition, to point out that in the wild-growing kok-saghyz thickets the average root weight per plant is usually 3-4 gr. This weight equals the average root weight of one plant taken from the variant with the largest number of plants per hill. (See Table 1, column 4, hill containing 26-37 plants.) Even of these most congested hills of our experiment it cannot be said that the plants obtained were oppressed. Quite the opposite. As practice teaches us, those individual roots of particularly large size which occur with increasing frequency lately in experimental plots are biologically such serious departures from the norm that often they produce no seeds at all or those produced are few and not viable. From the economic point of view these roots are of the greatest interest to us as we are now changing to the vegetative propagation of kok-saghyz.

All these data, taken in their entirety, indicate, firstly, that there is no intraspecific competition and, secondly, that the laws of life of species have really not yet been studied at all, a fact which Engels pointed out in his day.

"For that matter, the organisms of Nature also have their laws of population, which have been left almost entirely uninvestigated, although their formulation would be of decisive importance for the theory of the evolution of species. But who was it that gave the most definite impulse to work in this direction? No other than Darwin."[1]

The laws of the numerical increase of the organisms of a species are today almost as little investigated as in Engels' time.

But to investigate them from the point of view of the existence of competition is wrong, is not in accordance with reality and runs counter to agricultural practice.

To prove the nonexistence of intraspecific competition we shall adduce one more example taken from practical agriculture.

Field cultures, if thinly sown, ordinarily produce a small crop; and if the soil is weedy they produce no crop at all.

This, of course, does not mean that in practice this or that crop should be sown more thickly than is required under the particular conditions. It only means that the plants of the particular species, if sown thinly, are almost always defeated, in the fierce competitive struggle, by individuals of other species, in the case cited by weed plants. One must proceed from this premise and the concrete practical conditions in deciding the question of the rate of sowing, upon which the crop so largely depends. For example: in nonarid districts not less than 150 kg. of spring wheat seeds are sown per hectare while in arid districts only 50-60 kg. If the sowing rates are smaller the stand will be too sparse in both cases and will be overgrown with weeds, i.e., it will succumb in the competitive struggle with other plant species. There is therefore good reason for sowing weedy soils more thickly than the generally accepted standard for weed-free soil. In practice one always has in mind *not intraspecific but interspecific competition among the plants*, one always thinks of the competitive struggle for the conditions of life not among individuals of the same species but among individuals of different species which require the same environmental conditions.

If grain is sown too thickly in an arid district one will observe here too not intraspecific competition among the plants, but a general affliction: all plants will suffer from lack of moisture.

As a result not a single plant will produce a normal seed crop. Hence planting crops too thickly is also harmful in practice, not at all, however, because of the alleged existence in nature of intraspecific competition but simply because of the mistake made by man in sowing the particular plant species too thickly. Such incorrectness is corrected by subsequent practice.

It happens in free nature too, though rarely, in our opinion, that there are many more germs at one spot than there should be, but these cases have the same sequel as those that occur in practice. The multitude of individuals that accidentally happen to be at one spot either will not produce, in their entirety, any new seeds (germs) at all, or if they will produce any they will

[1] F. Engels, *Anti-Dühring*, Moscow 1947, pp. 105-06.

be comparatively small, and, besides, feeble and of low viability. The species in question will of course continue to live, not however on account of these rare cases that produce offspring possessing little viability but on account of all the remaining cases, where sowing was normal. Hence, the "mistake" of sowing too thickly is corrected also by nature, i.e., by natural selection.

Many biologists think that intraspecific competition in nature is inevitable, and think so for the sole reason that on seeing the numerous germs produced by every individual they are at a complete loss to know (and it is not easy to know) how many of these germs are really necessary for a species to at least maintain its numbers under the given conditions. The example of the dandelion has already shown us that the one hundred seeds, which, it is claimed, would be sufficient, if there were no intraspecific competition, rapidly to create overpopulation, actually are barely enough to reproduce the original number.

Is it not the same with all other species and varieties established in free nature? We believe it is.

The absence of intraspecific competition among plants and animals also explains, in our opinion, the fact that all examples cited as proof of its existence are abstract, purely theoretical, though some of them are claimed to have been verified in practice. Thus, there is the example, often referred to, of sowing an artificially composed mixture of equal amounts of different wheat varieties. As can easily be observed, in a few generations the ratios between the individuals of the different varieties will, as a rule, be greatly changed as compared with the original ratio. The explanation for this must be sought, true enough, in the fact that a struggle or competition was going on, but not among the individuals of the same species, or in the present case between the varieties of the same species.

Every species wages a struggle for existence through its individual members. This struggle is waged both against the numerous calamities brought on by the nonliving environment and still more against the living environment surrounding it and inhabited by members of other species, some of which directly devour individuals of the species in question while others compete with them for the conditions of life.

Let us analyze from this point of view the wheat example to which Darwin and Timiryazev refer in seeming confirmation of the existence of even severe intraspecific struggle or competition.

In *The Origin of Species* we read: "But the struggle will almost invariably be most severe between the individuals of the same species, for they frequent the same districts, require the same food, and are exposed to the same dangers. In the case of varieties of the same species, the struggle will generally be almost equally severe, and we sometimes see the contest soon decided: for instance, if several varieties of wheat be sown together, and the mixed seed be resown, some of the varieties which best suit the soil or climate, or are naturally the most fertile, will beat the others and so yield more seed, and will consequently in a few years supplant the other varieties."[1]

K. A. Timiryazev states in his article, "Natural Selection": "In the dandelion we have seen the simplest example of struggle—the struggle among the individuals of a species. But the reader may object that after all this was a purely theoretical example, that it was only a deduction, a conclusion drawn from the law of the rapid multiplication of organic beings. Are there no facts which directly confirm the truth of our conclusion, that a struggle is really going on in nature, that some organisms are vanquished and ousted by others? A very simple experiment may supply the factual proof desired. If a few varieties of any plant, for instance, wheat, are sown together, some of them, most likely those more adapted to the soil or climate or more fertile, soon gain the upper hand over the rest and finally extrude them completely."[2]

Any seed grower interested in the production of wheat is certain to confirm the fact that the original ratio in which the individuals of two or more varieties are taken will undergo a sharp change after even only a few generations. But the seed grower should disagree with the statement that some varieties disappear altogether from the mixture. Practical seed growing goes to show the very opposite. Let us suppose, for instance, that a seed-growing farm has two varieties of wheat, say, awnless Lutescens and awned Erythrospermum; that the former, the awnless variety, is more and better adapted to the field conditions of the particular farm and that the second, the awned variety, is worse adapted and hence less fertile than the former. But if there is an admixture of even only 0.1% of the second, the less productive awned variety, in the fields sown to the first, the more productive awnless variety, it is absolutely necessary that this admixture be eliminated by weeding. This is a rather laborious task, since even a small area of, let us assume, 50 ha. requires much time to clear it of this trifling (0.1%) admixture. But in seed growing this work is essential, for if the numerically small admixture of an unproductive variety is not eliminated at the outset, the admixture will increase in 2 or 3 years about tenfold or even more. The 0.1% will increase to 1-1.5%.

It is of course impossible to understand and explain, from the standpoint of intraspecific competition, why a less fertile variety on a given farm should show a relative numerical increase, especially a manifold increase, when grown in the midst of a more fertile variety. Always and everywhere in seed growing one may observe that admixtures of less fertile varieties increase relatively more than the basic more fertile varieties. This phenomenon, well known to practical seed growers, is unimpeachable evidence that not only is there no intraspecific competition but, on the contrary, there is something else, something of an opposite nature in the phenomenon under

[1] Charles Darwin, *The Origin of Species*, London 1929, p. 56.
[2] К. А. Тимирязев, *Сочинения*, Сельхозгиз, 1939 г., т. VII, стр. 137.

analysis. Agronomists and agrobiologists can understand this "something else" only by taking as their starting point the theory of development, the theory of creative Darwinism.

Practical cereal seed growers know definitely that if a farm is to produce pure-strain seed, the slightest admixture, even if of a notoriously less fertile variety, must immediately be eliminated. Of itself this admixture will not only not be extruded by the more productive variety but in two or three years will become relatively more numerous and the big-crop-producing variety demanded of the seed-growing farm will be considerably adulterated by the less fertile one. Such seeds are culled.

If the existence of intraspecific competition is recognized the problem is solved in a simpler and cheaper way. Once the particular variety of wheat under the particular conditions is more fertile and the admixture from another variety is less fertile and quantitatively so inconsiderable (0.1%, say) that the untrained eye of the layman does not notice it, it will be completely forced out in 2 or 3 years. Consequently, no special expenditures on weeding need be incurred. However, no seed grower reasons that way and none of them act accordingly. They all do varietal weeding. But those seed growers of little experience who after reading about it in books, believe that a law of intraspecific competition does exist in nature and consider it possible to dispense with varietal weeding end by either never becoming real seed growers or by forgetting all about this fictitious proposition and always carrying out the varietal weeding of small admixtures of other varieties even if their yielding capacity is lower.

One might also point out other frequent mistakes committed by agrobiologists who believe in the law of intraspecific competition. Seeing that in practical seed growing a numerically small admixture of another variety of wheat is annually weeded out of the basic variety of wheat (as otherwise it rapidly increases in proportion to the basic variety) they arrive at the conclusion that the culled admixture is considerably more fertile than the strain that is being protected from it. They therefore propose that what should be culled is not the admixture but the basic strain. In other words, the variety that constitutes the admixture should be selected, not thrown away but sown, since it is capable of producing the bigger crop. But a practical seed grower will not do so, for he knows quite well that the culled variety is less productive under the conditions of his farming establishment and the conditions of the district he caters for.

Nobody has yet explained scientifically or practically why a variety with a low yielding capacity should, when admixed in a small quantity to another, more productive variety of the same species, rapidly increase in numerical proportion. It is of paramount importance for both science and practice to know this. However, hitherto biologists not only did not try to explain this phenomenon but did not even know that it always existed in practical seed growing. In their opinion, which is based on the "law" of intraspecific competition, it is imposible for this phenomenon to exist.

The wheat example cited by Darwin and Timiryazev as being allegedly easily proved in practice as well as the explanation of the ousting of some varieties by others on the basis of competition is as wrong and abstract as the dandelion example.

The wheat used in exemplification can also be formally, but more correctly, explained without calling in intraspecific competition. As a matter of fact, a wheat variety possessing lower yielding capacity and less adaptiveness to the climate will diminish relatively if raised in a mixture. This diminution is, however, not due to an extrusion by varieties which have a higher yielding capacity and are more fit, as the law of intraspecific competition would have it. It diminishes solely because it is less productive.

To illustrate this point let us make the following theoretical calculation. Let the variety bearing the greater yield produce 15 c. of seed per ha. and the variety bearing the smaller yield only 5 c. Suppose the sowing norm to be 100 kg. of seed per ha., to consist of a mixture of 50 kg. of the one and 50 kg. of the other variety. Theoretically the crop per ha. should be neither 15 c. nor 5 c. but 10 c., if each variety will yield according to its nature, as if sown purely.[1] Consequently, leaving all intraspecific struggle or competition out of account it clearly follows that the mixture of varieties obtained from the crop for the next sowing will be in the ratio of 3:1 instead of the original 1:1; after another year (with the same theoretical calculations) the ratio should no longer be 3:1 but 9:1; in the fourth sowing year the ratio will be 27:1. Apparently, in another 2 or 3 years the less productive variety will really completely vanish, even though we eliminated the extruder (intraspecific competition) from our calculations. But the point is that in practice this low-producing variety, which has decreased to a fluctuating fraction, will persist for decades without further diminution.

In the fourth year of sowing the less productive variation was, according to our theoretical calculations, to drop from 50% to 3-4% without any pressure on the part of intraspecific competition. But in actual fact practice in analogous cases has shown that the percentage of the less productive variety when mixed with a more productive variety will be small but considerably above 3-4%, and no further sowings can reduce this proportion. Therefore seed growers can remove the admixture by weeding only.

A correct understanding of the phenomenon we have analyzed is very important for agrobiology. Indeed, we have here a numerically small admixture of a variety known to be less productive to a variety of superior yielding capacity—an admixture which constantly increases in numerical proportion to some extent (varying in each particular case) with each new sowing instead of, even without the operation of the law of intraspecific competition, being completely extruded by the fitter variety.

[1] The 50 kg. of seed of the first variety should yield a 7.5 c. crop and the 50 kg. of seed of the second variety should yield a 2.5 c. crop. In actual practice the crop in this case will, as a rule, be over 10 c., i.e., more than we had figured out theoretically, though it will be less than 15 c., i.e., less than the crop of the better variety.

One of the serious obstacles to an analysis and correct understanding by biologists of the question indicated is the absolutely incorrect extension of one of the laws of development of class, capitalist society—namely, of the factor of struggle and competition among antagonistic classes—to the field of biology, to the life of species of plants or animals.

Not a single species of plant or animal constitutes, nor can it constitute, a class society. Therefore, there neither is nor can there be any class struggle here, even if it has been called intraspecific competition in biology.

"In Darwin," Karl Marx writes, "whom I have looked up again, I am amused by his statement that he is applying the 'Malthusian' theory *also* to plants and animals, as if in Mr. Malthus the whole point did not consist in the fact that it is *not* applied to plants and animals but only to people—in geometrical progression—as opposed to plants and animals. It is remarkable that Darwin recognizes in the beasts and plants his English society with its division of labour, competition, opening up of new markets, 'inventions' and the Malthusian 'struggle for existence.' It is Hobbes' bellum omnium contra omnes[1] and is reminiscent of Hegel in his *Phänomenologie*, in which civil society is depicted as a 'spiritual animal kingdom' while in Darwin the animal kingdom is depicted as civil society."[2]

A species is not an abstraction but a really existing unit. The life and development of a species and the emergence of other species and varieties go on through the individuals of that species. The numerical composition of a particular species, as the theory of Darwinism correctly states, depends in the main not so much on a high birth rate as on the conditions which make it possible for the greatest number of individuals to survive.

Proceeding from this premise it is possible to explain why a less fit, less productive variety of wheat, a small quantity of which (an admixture) is compactly surrounded by a more highly productive, a fitter variety, increases in relative proportion. If sown pure the less fit variety may have difficulty in withstanding the attacks of pests and disease and competition with plants of other species (weeds), etc. As a result of it all this variety yields a relatively small harvest. If a small number of representatives of this variety (in the form of an admixture) get in between representatives of a more productive, better adapted variety of the same species, they greatly benefit by the improvement in the environmental conditions due to the absence of intraspecific competition. The individuals of the admixture thus sharply increase their yielding capacity in comparison with their capacity when surrounded by individuals of their own, little adapted variety. Very frequently and even almost always, as has already been said, intraspecific admixtures are more productive in practical seed growing also in comparison with plants of the variety (strain) with which they mixed. For each species and variation the environmental conditions are

---

[1] A war of all against all.—*Tr.*

[2] Marx-Engels, *Briefwechsel*, Band III, M. 1937, S. 94. Marx and Engels, 18 Juni, 1862

the determining and limiting factors of multiplication or, what amounts to the same thing in this particular example, of its yielding capacity. Suppose the shoots of a wheat variety or strain to be free or almost free from attack by the harmful Hessian fly. Spring wheat Odesskaya 13 (a variety of Erythrospermum) is such a strain. Another spring strain, Lutescens 062, suffers from this pest. In districts where the said fly is widespread it may happen in years of its mass occurrence, and in the fields of the Odessa Institute of Selection and Genetics it actually was repeatedly observed, that of two strains sown side by side one is not affected and yields a crop several times as big as the second, the plants of which were severely weakened by the larvae of the Hessian fly. Now let us imagine that both of these strains are sown separately in two big fields. This pernicious fly will clearly concentrate on the field containing the plants of the easily vulnerable strain. Conversely, it will not concentrate on the field sown to the strain that is little affected or almost not at all affected by the fly.

In this event a relatively small admixture of plants of the susceptible strain to the nonsusceptible strain will also not be affected in view of the total or almost total absence of this pest. Therefore the admixture will produce a crop several times the size of that which it would have produced had it been sown pure.

The same thing applies to the liability of wheat to be infected by rust. For example, when the plants of a very susceptible strain are slightly mixed with a stand of a resistant or slightly susceptible strain they remain, in view of the absence of a large concentration of rust spores, as unaffected as the rust-resistant strain to which they have been admixed.

We have taken only two examples: first—susceptibility to an internal parasite, and, second, susceptibility to plant disease, and of course each plant's life is affected by more than one harmful insect or disease. Proceeding from this alone it now becomes intelligible, if viewed generally, why admixtures of less productive strains become more productive not only in comparison with plants of their own strains sown pure in the field but also in comparison with the plants of the strain with which they were admixed. Admixtures, in the first place, have the advantage, pointed out by us in the examples we have given, of ridding themselves of the action of some of the diseases and pests peculiar to them, as the latter do not concentrate on the fields where the strain with which they have been mixed is grown. Secondly, they prove unfailingly to be more resistant, at least to some of the numerous diseases and pests from which the basic strain suffers to a greater or smaller extent.

The same applies to the conditions of the inanimate environment. It may be observed with great frequency that when winter-wheat plants of low frost resistance are mixed in small quantities with plants of a more frost-resistant strain they stand the frosts much better than when they are among plants of their own strain.

That is why, in our opinion, plants of numerically inconsiderable admixtures are at times more productive than the plants of this variety sown pure and than the plants of the basic variety to which they have been admixed.[1]

As has been pointed out, every member of a species of field culture has not only one but many parasites and diseases as well as many competitors for the same conditions of life as he himself enjoys. *But all these pests, diseases and competitors are always individuals that belong to other, relatively remote, species, but not to the species in question.*

In the struggle for existence in living nature species have formed and are forming which consist of different subspecies and variations and, in the final analysis, of relatively different individuals. Nature, they say, does not like uniformity. All individuals differ from each other. In every species there is variety—in some more, in others less—but always variety. *But all this diversity of varieties within a species and of individuals within its varieties makes up one relative whole—the species.*

In plants any change in heredity takes place perforce, and only by the action of environmental conditions, when relatively new conditions of the external environment, in excess of the requisite norm, are assimilated. A change in the course of assimilation and dissimilation, i.e., in metabolism, leads to a change in the heredity of that part of the living body where the change in metabolism has occurred. Alterations in the living body are the result of alterations in their metabolism. These bodily changes always correspond to the action of the environmental conditions. If the changed part

[1] For the purpose of developing theory in the direction of a more profound understanding of the life of species of plants and simultaneously for the purpose of solving the practically important problem that always confronts science but has not yet been solved by it—the problem of when and what admixtures (populations) of strains possess greater yielding capacity than pure strains—it may be of great benefit to perform the following special and quite simple experiment at a number of places, both at experimental seed-growing stations and by collective-farm experimenters. I believe this experiment may supply valuable material for the above purpose. Take in pairs wheat strains (varieties) that are easily distinguished by their spikes—spring plants for spring sowing and winter plants for autumn sowing. Then, keeping strict account of the germination percentages and the weights per thousand grains of each strain, mix sowing seed so, according to weight, that in each of the following variants each of the two varieties (strains) be represented by the following percentages of germinating grains. First mixture: first variety, 99%, second variety, 1%; second mixture: first variety, 98%, second variety, 2%; then 97% and 3%; 96% and 4%; 95% and 5%, etc. The last two mixtures: first variety, 2% and 98% of germinating grains; 1% and 99%. Hence, there is a total of 99 different mixtures of the two varieties (strains). Each one of the two varieties taken will be represented, with regard to the number of germinating grains, by combinations ranging from 1%-99%. The weight of each mixture depends upon the area of the plot expected to be sown. It is best to take plots not less than 10 sq. m., and if conditions allow even 100 sq. m. each. In the first case, with two replicates of the sowing, the experiment will take 2,000 sq. m. (1,000 sq. m. for each replicate); in the second case, 2 ha. The sowing norm should be the one accepted in the district. As for the method of sowing: on the 10 sq. m. plots, on which it is difficult to use a seed-drill, ordinary (but good) hand broadcasting will do. The 100 sq. m. plots must be drill-sown. When harvesting the crop, keep account of the spike and seed percentages of the two varieties in each mixture. The performance of such an experiment, especially by different persons and at different places, will provide a wealth of material for both biological science and seed-growing practice.

of the body is the basis, the genesis of a new plant organism, the latter will likewise possess the changed heredity. But the new character acquired as a result of a change in heredity (though the latter always changes in accordance with the influence of conditions) will by far not always promote the further survival of both the particular changed individual and his progeny. The usefulness, the relative purposiveness, of each alteration is assessed and decided only by subsequent survival. If the particular alteration is a hindrance in the struggle for life, individuals possessing it do not survive or leave no offspring, i.e., disappear from the scene of life. If the particular change is of assistance in the struggle for life carried on by the individual and his offspring, it is useful and is perpetuated by being repeated in the individuals of the progenies. Therefore alterations deleterious to the survival of the individual and his offspring cannot become fixed. Consequently, according to Darwin's doctrine of evolution, only changes that promote the survival of individuals and their progeny perpetuate themselves, are transmitted to the descendants. Harmful changes pass, are not perpetuated. Thus, the organic world of plants and animals, caught in an intricate biological chain of interconnections, is polished off more and more by natural selection and inevitably advances in the sense of achieving an ever greater, or rather ever new, adaptability of the individuals of each species to the animate and inanimate surrounding environment.

Species and varieties inevitably progress and develop in consequence. The species and varieties whose development reaches a standstill sooner or later depart from the scene of life.

Beneficial changes accumulated in the progenies of any original number of individuals, create new varieties and the latter may gradually become, change into new species.

As has already been pointed out, what really exists among species (of course by far not among all of them) is interspecific competition, not infrequently severe, for the conditions of life. New species come into existence and are individuated through varieties of the old species. Hence in all cases where the old and the new species are competitors, and this is far less often the case than not, the birth of the new species within the old is accompanied by a simultaneous birth of competition for the conditions of life. Insofar as the nascent new species still forms part, or, to be more exact, is still one might say a variation, of the old species, this competition may be termed intraspecific, yet at the same time it is already interspecific, being waged between the old species and the new, individuated species. In other words, intraspecific competition is nonexistent to the same extent as species of plants in nature are constant. The impermanence of species, the formation of new species from old, in a number of cases (but of course not in all) gives rise to apparent intraspecific competition. But in reality these are germs of an interspecific competition for the conditions of life.

GENETICS[1]

GENETICS is that department of biology which deals with the development of organisms. It may also be called the department of science which studies heredity and its variability. Today two kinds of genetics exist: the old and the new. They are sharply opposed to each other in their fundamental principles. The first of these two kinds, named Mendelist-Morganist genetics, recognizes the existence in the organism of a special germ-plasm which differs in principle from the body of the organism and which, unlike the ordinary body, alone possesses heredity. According to T. H. Morgan, "heredity is the name given to express ... [the] relation of continuity of the germ-plasm material and its consequences in the successive generations that arise from the germ-plasm." Alterations of the germ-plasm (mutations) are, it is asserted, absolutely independent of the body (soma) of the organism. Hence it is self-understood that alterations (mutations) of the germ-plasm, or of the hereditary substance, are independent of the conditions of life that act upon the organism's body. Hence, no new properties or characters acquired by the organism as a result of the action of the conditions of life are ever inherited.

The reproduction of characters in successive generations is determined not by the body of the parents, but by the germ-plasm, changes in which are allegedly independent of the organism's body. Hence the theory of Mendelism-Morganism categorically rejects the possibility of directed change in the nature of plant and animal organisms by regulating their conditions of life and development. Changes in the heredity (mutation) of organisms are considered by this science not to depend on the conditions of life. Only the development of the organism's body depends on the conditions of life; the development of its heredity does not. Heredity may change (mutation) but the quality of these alterations does not depend on the specific character of the action of the conditions of life in which the organisms that gave rise to the alterations existed. According to this science man can benefit only from mutations, inheritable alterations, that appear randomly and are not

[1] This article was written for the third edition of the *Selsko-Khozyaistvennaya Entsiklopedia* (Vol. 1, under "Genetika").—*Ed.*

controlled by him. This bars the way to finding any means or methods of directing changes in the nature (heredity) of organisms. The theory of Mendelism-Morganism has therefore always been essentially contradictory to the demands and requirements of plant-breeding and seed-growing practice as well as of pedigree stockbreeding.

A counterblast to Mendelism-Morganism is contained in I. V. Michurin's motto: "We cannot wait for favours from Nature; we must wrest them from her."

The new genetics, genetics of the Michurin trend, rejects the basic tenet of the old Mendelist-Morganist genetics—the complete independence of the properties of heredity from the conditions of life of plants and animals. Michurinian genetics does not recognize the existence in the organism of any hereditary substance of any description whatsoever separate and distinct from the organism's body. This science means by heredity the principal characteristic property of the living body which expresses itself in the ability of this body to live, feed itself, grow and multiply in accordance with its own nature. An alteration of the heredity of an organism or of the heredity of a particular part of its body is always the result of an alteration of the living body itself. But an alteration of the living body originates from a changed type (a type deviating from the norm) of assimilation and dissimilation, from a changed metabolism (a metabolism deviating from the norm). Although a change in an organism or in a particular organ or property of it is not always or not completely transmitted to the offspring, changed germs of new organisms are obtained only as a result of a change in the body of the parent organism, as a result of the direct or indirect action of the conditions of life upon the development of the organism or its separate parts. A change in heredity, the acquisition of new properties and their accentuation during several succeeding generations, is always determined by the conditions of life of the organism. Heredity changes and becomes more complex by reason of the characters and properties acquired by the organism over several generations.

Only by controlling the conditions of life and the development of plants and animals can their natures be comprehended in an ever higher degree and methods be found of changing it in the direction required by farm practice.

Thus the principal premises of old and new genetics are opposites of each other.

MENDELISM-MORGANISM

(THE CHROMOSOME THEORY OF HEREDITY)

For an exposition of the essence of Mendelist-Morganist genetics, let us quote from Morgan's article, entitled *Heredity*, in the *Encyclopedia Americana*, 1945: "Beginning in 1883 August Weismann in a series of essays that were in part speculative, yet backed by a constant appeal to observation and experiment, attacked the prevailing idea that characteristics acquired by the individual are transmitted to the germ cells and may reappear in the offspring. The germ cells had been shown in many cases to separate themselves at any early stage in the development of the embryo from the rest of the cells, remaining in an unspecialized condition while the remaining cells from which the body of the individual is to be formed were differentiating. The germ cells become later the essential parts of the ovary and testis respectively. In origin, therefore, they are independent of the rest of the body and have never been a constituent part of it. They are protected and nourished by the body, but not otherwise influenced [i.e., changed—*T.L.*] by it. The germ tract is the imperishable stream which throws off in each generation the body cells whose destiny is to maintain the germ cells. All new modifications arise primarily in the germ cells, and first show themselves as characters in the individuals that develop from these germ cells. Evolution is germinal in origin and not somatic [i.e., bodily—*T.L.*] as had been earlier taught. This idea of the origin of new characters is held almost universally today by biologists. Heredity is due, therefore, to the conservation in the germ plasm of those elements that have appeared in it from time to time—the old as well as the new. The germ plasm is the capital of the race of which the interest only is spent in each generation in producing new individuals . . . the real hereditary mechanism was discovered by Mendel. . . . It has been found that Mendel's laws apply not only to characteristics of cultivated plants and animals, not only to superficial things such as color, but to the characters of wild animals, to differences that distinguish species, and to the most fundamental attributes of living things. Mendel's law of segregation states that the elements that the two parents contribute to the offspring constitute pairs, and that the members of each pair separate in the formation of the germ cells in the offspring so that each germ cell contains one member only of each pair. For example, Mendel crossed a race of edible peas having green seeds to one having yellow seeds. All of the offspring seeds were yellow. Yellow dominates green. If plants from these hybrid peas are self-fertilized (or crossed to each other) they produce both yellow and green peas in the ratio of three yellows to one green. These green peas are pure and never produce anything but green peas. The yellow peas, however, were found to be of two kinds, some were pure yellow always producing all yellow descendants, others were hybrids, producing both yellow and green peas in the ratio of three to one. Taken altogether these second generation peas appeared in the proportion of one pure yellow, two hybrid yellows, one pure green. Mendel pointed out that if the original green grandparent contributed an element for green and the yellow grandparent an element for yellow, these contrasted elements form a pair in the hybrid—a pair whose members separate (segregate) from each other when the germ cells (gametes) are produced. In consequence, half of the ovules will contain the element for yellow and half the element for green; likewise half of the pollen grains will contain the element for yellow and half the element for green. Chance meetings of the ovules and pollen, thus will give: 1 green green; 2 green yellow; 1 yellow yellow.

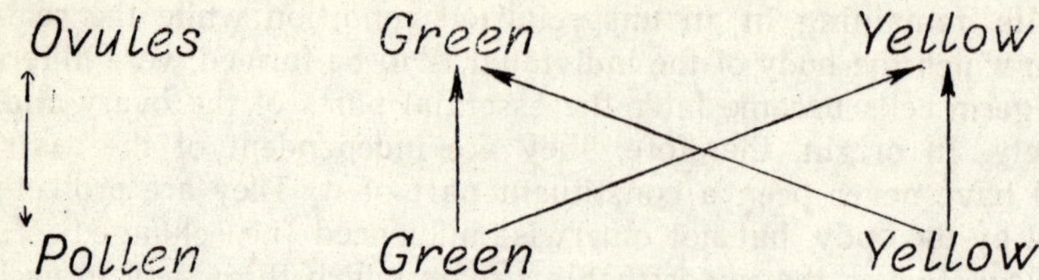

"Mendel's second law deals with cases when more than one pair of characters are involved. It has been found that tall and short races of peas are contrasted characters that segregate in the same way as do yellow and green. If a tall race with yellow peas is crossed to a short race with green peas, the segregation of each pair is independent of that of the other, so that one quarter of the egg cells of such a hybrid contains the elements for tall and yellow; another quarter the elements for tall and green; another quarter the elements for short and yellow, and another quarter for short and green. Similarly in the formation of the pollen the same four kinds of gametes are produced. Chance meetings of ovules and pollen give 16 combinations....

"Since yellow dominates green and tall dominates short, there are in this second filial (F_2) generation nine tall yellow; three short yellow; three tall green; one short green. Thus, while members of each pair of factors of the hybrid segregate at the time of ripening of the germ cells, the separation is independent for each pair. This is Mendel's second discovery that may be called the law of independent assortment.

"Mendel showed that three pairs of characters behave in the same way, i. e., their genes assort independently, and there is reason to suppose that this law holds whenever the genes for the two or more pairs of characters are carried by separate pairs of chromosomes, but, as will be shown below, whenever the genes are carried by the same pair of chromosomes the distribution is controlled by a third law of heredity, viz., the law of linkage.

"The elements that are supposed to represent, in a sense, the hereditary characters are commonly called genes and the word Genetics, or the study of behaviour of the genes, has come to replace, in the modern work on inheritance, the older term Heredity with its numerous connotations. The Mendelian characters have been spoken of as unit characters, and it is sometimes implied that the gene directly produces each such character. The clearest evidence, however, indicates that the so-called unit character is only one of the many effects which the gene may produce in conjunction always with many, perchance with all of the other genes. Thus, the germ plasm is looked upon as the sum total of the genes whose combined effect is responsible for each character of the body. While the body is built up through the interaction of the materials that the gene produces, yet in the formation of the germ cells the genes act as independent units that collect in pairs, then segregate, those pairs in separate pairs of chromosomes assorting independently, those pairs in the same chromosome pair being linked.

"Modern work on the cell has pointed unmistakably to the mechanism by means of which both segregation of the genes and the assortment of the chromosomes takes place. Every cell in the body or immature germ cell contains a double set of chromosomes (except in the male of certain groups, which lacks one of the sex chromosomes); one member of each pair has come from the father and one from the mother. During the maturation process the maternal and the paternal chromosomes mate with each other—like with like. Subsequently at the so-called reduction division one member of each pair goes to one daughter-cell and the other member to the other daughter-cell. If the chromosomes carry the Mendelian genes the maternal and the paternal genes will be segregated when the chromosomes are reduced to produce the gametes; but at the reduction division there is not a separation of all the maternal from all of the paternal chromosomes as a group, but each pair of chromosomes segregates independently of the others, so that the daughter-cells may get any possible assortment of the chromosomes of paternal and maternal origin, but always one or the other member of each pair. This condition fulfills all the requirements of Mendel's second law of free assortment.

"But obviously if, as assumed, the chromosome threads are the bearers of the genes and if, as generally held today, the thread is a structural element that remains intact even in the resting stages of the cell, then the genes must be inherited in groups corresponding in number to the number of chromosomes. In a word, all the genes in a given chromosome will be linked together. The most recent evidence shows that this is the case, and that there are as many groups of linked genes as there are kinds of chromosomes. Since 1906 the number of demonstrated cases of linked genes has steadily increased so that it can no longer be questioned that this relation is a characteristic feature of Mendelian inheritance. In one instance, that of the fruit fly, Drosophila ampelophila, it has been shown that the 200 known hereditary differences are inherited in four groups corresponding to the four pairs of chromosomes. Thus, Mendel's law of segregation has found its justification in the cytological mechanism of reduction in the germ cells; while his law of free assortment has been confirmed in the method of assortment of the chromosome. Later the discovery of the meaning of linkage phenomena has brought all the fundamental properties of heredity into complete harmony with the chromosomal mechanism. The individuality of the chromosomes that is responsible for linkage has been found, however, not to be absolute, for the members of a pair have been shown to interchange equivalent parts at times, but the interchange has been found to follow a predictable course, and while it complicates the results, it in no sense undermines the general principle involved. In some species the interchange (crossing over) takes place only in the female sex (Drosophila), in others in the male sex (silkworm moth), while in still others it takes place in both sexes, as in some hermaphroditic plants.

"The inheritance of sex has been one of the great biological discoveries of the present century. The factor or factors of sex have been shown to be carried by special chromosomes called the sex chromosomes. In certain great groups (mammals, most insects, etc.) the occurrence of two of these chromosomes, called the X chromosomes, produces a female, the presence of one of them produces a male. Thus the female is XX, the male X. At the reduction division in the female one X is eliminated from the egg so that each egg contains but one X chromosome. In the male there is only one X, this X chromosome goes in the reduction division to only one sperm cell of the two formed, so that two classes of spermatozoa result. At the time of fertilization random meeting of any egg with any spermatozoon will give two classes of individuals, those with two X's (females) and those with one X (males). This mechanism preserves the numerical equality of the sexes. In other groups (birds, moths) the relation is reversed, the male containing two X's, the female one; hence all of the spermatozoa contain one X; half of the eggs contain but one X, the other half none."

Such are the principal theses of the chromosome theory of heredity as expounded by T. H. Morgan, its founder.

CRITICISM OF THE CHROMOSOME THEORY OF HEREDITY

Weismannism—which came into existence at the turn of the century—and subsequently Mendelism-Morganism were spearheaded against the materialist elements in Darwin's theory of evolution.

The chromosome theory is based on Weismann's absurd proposition regarding the continuity of the germ-plasm and its independence of the soma, a proposition which K. A. Timiryazev already condemned. In line with Weismann, the Morganist-Mendelists take it for granted that parents are genetically not the progenitors of their offspring. Parents and children, according to their teaching, are brothers and sisters. Furthermore, neither parents nor children are really themselves. They are only by-products of the inexhaustible germ-plasm. Variations in the latter are absolutely independent of its by-product, i.e., of the body of the organism.

All this can be found in the principal postulates contained in T. H. Morgan's article from which we have quoted. One need only peruse the first part of the excerpt, the passage which explains Weismannism as the basis of the chromosome theory of heredity. Let us consider the following, for instance:

"The germ cells become later the essential parts of the ovary and testis, respectively. In origin, therefore, *they are independent of the rest of the body and have never been a constituent part of it.... Evolution is germinal in origin and not somatic* [bodily—*T.L.*] *as had been earlier taught.* [*Italics mine.—T.L.*] This idea of the origin of new characters is held almost universally today by biologists."

The same thing, only in greater detail, is stated in an article, *Genetics,* by W. E. Castle, also contributed to the *Encyclopedia Americana,* like T. H. Morgan's article, *Heredity.* After stating that usually the organism develops from a fertilized egg, Castle goes on to set forth the "scientific" foundations of genetics as follows:

"In reality the parent does not produce the child nor even the reproductive cell which functions in its origin. The parent is himself merely a by-product of the fertilized egg (or zygote) out of which he arose. The direct product of the zygote is other reproductive cells, similar to those from which it arose.... Hence heredity (that is, the resemblance between parent and child) depends upon the close connection between the reproductive cells which formed the parent and those which formed the child, one being the immediate and direct product of the other. This principle of the 'continuity of the germinal substance' (reproductive cell material) is one of the foundation principles of genetics. It shows why body changes produced in a parent by environmental influences are not inherited by the offspring. It is because offspring are not the product of the parent's body but only of the germinal substance which that body harbours.... To August Weismann belongs the credit for first making this clear. He may thus be regarded as one of the founders of genetics."

This mystical Weismannist scheme of things was swallowed whole and even exacerbated, one might say, by Mendelism-Morganism. The fundamental theses which we have cited and which are the point of departure of Mendelism-Morganism (the chromosome theory of heredity) are utterly false. They are at variance with reality. Therefore, although the Mendelist-Morganists in the Soviet Union fully subscribe to these principles they, as a rule, keep quiet about them. In their articles and lectures on Mendelism-Morganism they do not expound the basis of this science as they are afraid of being ridiculed by their readers and hearers who know for a certainty that the germs of organisms or sex cells are one of the results of the vital functions of parent organisms. Only if silence is maintained on the basic theorems of Mendelism-Morganism can the chromosome theory of heredity appear to constitute a harmonious and, at least to some extent, true system to people who do not have a detailed knowledge of the life and development of plants and animals. But as soon as we embrace the absolutely true and generally known proposition that sex cells, or the germs of new organisms, are procreated by the organism, by its body, and not by the sex cell from which that particular organism, now mature, originated, the entire "harmonious" chromosome theory of heredity is at once thrown out of gear. The role and significance of chromosomes in the development of cells and organisms is of course not in the least impaired thereby.

Michurinian genetics recognizes the existence of chromosomes, but it rejects the chromosome *theory* of heredity, rejects Mendelism-Morganism.

While claiming to have discovered the laws of development of living bodies (the laws of heredity) Mendelism-Morganism completely denies development itself. This science would have us believe that each chicken comes (develops) from an egg but that not a single egg develops from a chicken. Eggs originate directly from eggs and only from eggs. The body of the chicken forms by means of development but this development cannot exert any influence on the offspring inasmuch as the organism, they allege, cannot produce any offspring whatever. The offspring derives directly from the same egg as its parent. In other words, that which develops does not enter into the offspring whereas the fictitious immutable "continuous germ-plasm" produces offspring. This is the scholastic foundation on which the chromosome theory of heredity is built. For continuous life which is brought about through the development of living matter (egg—organism—egg) the Mendelist-Morganists substitute continuous "germ-plasm" (egg—egg). That is the reason why the development of the living body disappears from their field of vision.

The chief characteristic of the Mendelist-Morganist theory is the divorcement of the organism from the environmental conditions. But since living matter not only cannot grow and develop divorced from the conditions of life but cannot even live and consequently possess the property of heredity, Mendelist-Morganist genetics, in building up its theory of heredity, was compelled to recognize the connection between the body of an organism and the conditions of life. But in taking this step in the right direction this science at the same time made a big blunder. It severed the principal property of living matter, heredity, from the body of the organism. According to the data of this science only the chromosomes of the cells are "hereditary substance." Hence the name "chromosome theory of heredity."

Underlying this theory is the fictitious proposition that part of the chromosome substance cannot be considered as belonging to the ordinary body, that heredity inheres only in this special substance, and that the rest of the organism's body does not possess this property. Hence, the conclusion was drawn that the organism and everyone of its cells consist of ordinary body (soma) and hereditary substance, contained in the chromosomes.

But according to Michurin's teaching the organism consists only of the ordinary body. There is no hereditary substance in the organism or its cells separate and apart from the ordinary body.

Heredity is possessed not only by the chromosomes but by the living body as a whole and by every particle of it. It is therefore wrong to consider the chromosomes a special hereditary substance or organ of heredity in the organism and cell just because they possess the property of heredity. Various organs, including organs of propagation, can and do exist in the organism but there is no organ of heredity and there can be none. To look for a special organ of heredity in the organism is like looking there for an organ of life.

All facts relating to heredity changes and connected with chromosome changes militate not for but against the chromosome theory of heredity, which maintains that changes in the living body do not entail changes in the properties of heredity.

Indeed, how many facts have been accumulated (and not by just anybody but by the Morganists themselves) which go to prove that any morphological change in one of the organs or organella of the body, namely, the chromosomes, is transmitted hereditarily with rather great accuracy by the action of environmental conditions. A chromosome change acquired in the process of individual development of a cell or organism is, as a rule, always transmitted by inheritance to the daughter cells. Does this not argue in favour of the transmission of acquired characters? Are the mutability of the chromosomes and the inheritance of these changes not refutations of the chromosome theory of heredity?

The facts cited by the cytologists in support of the chromosome theory only formally confirm, as it were, the theory underlying Mendelist-Morganist genetics. Actually these facts are utterly at variance with this theory.

It is true that chromosomes exist. In the gametes their number is half that of what it is ordinarily. If the gametes bear chromosomal aberrations of any kind they give rise to altered organisms. It is true that the various visible morphological changes of a given chromosome often, and even always, bring about changes in various characters of an organism. It has been proved that the presence of two X-chromosomes in the fertilized egg of *Drosophila* usually determines the development of a female from that particular egg and not of a male.

These facts as well as other data are correct. Beyond doubt, changes in the egg or in its chromosomes cause a change in the development either of the organism as a whole, or of its separate parts; but *it is likewise an indisputable fact that altered environmental conditions may change the process of the building up of the body, including the building up of the chromosomes and of germinal cells in general for the future generation.* In the first case the rudiments (egg) that had been altered by environmental conditions produce altered organisms; in the second—the organism that was modified by external factors may produce altered gametes. But the latter proposition is denied in toto by the Mendelist-Morganists. Herein lies their fundamental error.

It is impossible correctly to understand the development of an organism and to discover the laws that govern it unless the organism is investigated in its dialectical unity with the conditions of life. If a living body is isolated from the conditions it needs to sustain life it ceases to live. This fact alone is enough to show that the organism and its living conditions constitute an inseparable dialectical whole.

Living matter not only depends on environmental conditions, on the conditions of life, but originally, under certain conditions, came into existence from nonliving matter. Plant and animal forms have been and are being built by the conditions of life, by the conditions of the surrounding environment. An organism is an integral whole, a system, only in unity with the conditions of life it needs. For it is from the conditions of life that the organism, having originated from the egg, builds its entire body with all its properties, including its principal property, heredity.

Heredity is the property of a living body to live, grow, develop and multiply in a definite way. Therefore the properties of heredity can be apprehended only by studying the conditions which the organism requires to build a living body, i.e., a body possessing the property of heredity. New methods of increasing the productivity of plants by means of agrotechnique and the regulation of the conditions of life are constantly found in practical farming. This is the only possible way of controlling and changing the nature (heredity) of organisms in a definite direction. These are the general principles from which Michurinian genetics proceeds in working out concrete ways and means of governing the nature of plant organisms.

MICHURINIAN GENETICS

Upon the foundation of his own highly fruitful labours in the field of breeding varieties of fruits and berries, I. V. Michurin created a new agrobiological theory which has been widely accepted and creatively followed in the Soviet Union.

At the session of the Lenin Academy of Agricultural Sciences of the U.S.S.R. held in July 1948 the question of "The Situation in Biological Science" was discussed. At that session the materialist Michurinian teaching triumphed completely in our country over the idealist teaching of Weismannism-Morganism.

In the discussion of biological questions our Party made its directing influence felt. Comrade Stalin's guiding ideas were of decisive weight here, opening up, as they did, vast new perspectives of scientific and practical work.

According to the principles of Michurin's theory the organism and the conditions necessary to ensure its life must be considered as a unity. Different living bodies require different environmental conditions for their development. Hence our knowledge that they have different natures, different heredities. Heredity is the *property of a living body to require definite conditions for its life, its development, and to respond definitely to the influence of particular conditions.*

A knowledge of the natural requirements of the organism and of its relation to environmental conditions makes it possible to control the life and development of this organism. Such knowledge may serve as the basis for altering the heredity of organisms in a definite direction.

The following example may be taken to illustrate the distinction between the above approach to the study of heredity and the approach of the Mendelist-Morganist geneticists. The Mendelist geneticists repeatedly made a study of the heritability of the winter habit and spring habit of cereals. For that purpose they crossed winter varieties with spring varieties. In the offspring they determined how many plants of the winter habit, i.e., plants similar in regard to this character to one of the parents, were obtained, and how many plants of spring habit, i.e., plants similar to the other parent. They arrived at the conclusion that the hereditary properties of the winter habit differed from those of the spring habit by one or more genes, i.e., granules of some unknown substance allegedly present in the chromosomes of the winter and spring plant cells. *But such a study will not tell anyone wherein the essence of the winter and spring habits of plants lies, or how to control the development of these properties.*

Michurinian genetics demands a wholly different approach to the study of heredity.

Thus, on investigating the causes of the failure of winter grains to ear when sown in spring it was ascertained that one of the processes of development of winter plants, now called the vernalization phase, requires, in addition to such available spring field conditions as food, moisture and air, also a relatively lengthy period of low temperature (0°-10° C.). The absence of a prolonged low temperature period in the fields in spring is the very reason why the process of vernalization does not take place, and consequently why the entire further development is retarded and earing and fruiting prevented.

With the discovery of the nature of the phase of vernalization it has become possible to compel any spring-sown winter cereal to ear and fruit. For this purpose properly moistened seeds are kept (vernalized) for a definite period under relatively low thermal conditions in the field before sowing. This satisfies the hereditary requirements for the development of the said process. Upon its completion all further hereditary requirements of winter plants can be satisfied even when spring-sown by the field conditions available and therefore development proceeds normally until it is completed, i.e., until the seeds are ripe. This sort of study has revealed the *essence of the heredity of the winter habit and spring habit.*

Each living body builds itself out of the conditions of its environment after its own fashion, according to its heredity. That is why different organisms live and develop in the same environment. As a rule, each given generation of a plant or animal develops largely in the same way as its predecessors, particularly its close predecessors. *Reproduction of beings similar to itself is a general characteristic of every living body.*

When an organism finds in its environment the conditions suitable to its heredity, its development proceeds in the same way as it proceeded in previous generations. When, however, organisms do not find the conditions they require and are forced to assimilate environmental conditions which, to some degree or other, do not accord with their nature, then the organisms or sections of their bodies become more or less different from the preceding generation. If an altered section of the body is the starting point for the new generation, the latter will, to some extent or other, differ from the preceding generations in its requirements and nature.

The cause of changes in the nature of a living body is change in the type of assimilation, in the type of metabolism. For example, the process of vernalization of spring cereals does not require lowered temperatures. Normally it proceeds in temperatures such as obtain in the spring and summer in the fields. But if spring cereals are vernalized for a long period, say, two or three generations, under low temperatures, they not infrequently turn into winter plants. For it is a well-known fact that winter cereals cannot pass through the process of vernalization without lowered temperatures. Here is a concrete example showing how a new requirement is induced in the offspring of these particular plants—the requirement for lowered temperatures as a condition for vernalization; at the same time changes in requirements, i.e., in the heredity of a living body, always correspond to the action of the environmental conditions.

Sex cells and any other cells through which organisms propagate are produced as the result of the development of the whole organism, by means of conversion, by means of metabolism. The phases in the development of an organism are accumulated, as it were, in the cells from which the new generation originates.

We may therefore say that to the extent to which in a new generation the body of an organism (a plant, say) is built anew to that same extent also all its properties, including heredity, develop.

In one and the same organism the development of the different cells and of different parts of cells, the development of individual processes, requires different external conditions. Besides, these conditions are assimilated in different ways. It should be stressed that in this case we mean by *external that which is assimilated, and by internal, that which assimilates.*

The life of an organism proceeds through innumerable law-governed processes and conversions. The food that enters the organism from the external environment undergoes a series of conversions whereby it is assimilated by the living body, changing from external to internal. This internal, since it is living matter, enters into metabolic relations with the substances of other cells and particles of the body, feeding them and thus becoming external with regard to them.

No organism ever realizes all its hereditary possibilities in full. Many properties and characters develop incompletely, remain to some extent underdeveloped, recessive, without substantially affecting the development of the organism as a whole. In the following generations these characters or properties may develop if the environment contains the necessary conditions.

Different significance attaches to the different processes undergone, to the development of the different characters and organs in the life of the organism. Upon the development of some properties or characters the development of the organism as a whole depends to a smaller extent, upon the development of other properties and characters, to a larger extent; and finally upon the development of still other characters the organism is so greatly dependent that without them it cannot develop, and frequently cannot exist.

Two kinds of qualitative changes are observed in the development of plant organisms:

1. Changes connected with the realization of individual development, when natural requirements, i.e., heredity, are normally met by the *corresponding external conditions*. The result is a body of the same breed and heredity as the preceding generations.

2. Changes in breed, i.e., the changes in heredity. Such changes are also the result of the realization of individual development, but deviating from the normal, usual course. Changes in heredity are, as a rule, the result of the organism's development *under external conditions which, to one extent or other, do not correspond* to the natural requirements, i.e., its heredity.

Changes in the conditions of life compel changes in the development of plant organisms. They are the root cause of changes in heredity. Organisms which cannot change in accordance with the changed conditions of life do not survive, leave no progeny.

Organisms, and hence also their nature, are created only in the process of development. Outside of development a living body may also undergo an alteration but such alterations will not be characteristic of the living body.

Numerous facts go to show that changes in various sections of the body of a plant or animal organism are not fixed, assimilated, with equal frequency in the sex cells, i.e., the reproductive material.

This is explained by the fact that the process of development of each organ, of each granule of the living body, requires relatively definite external conditions. These conditions are elected from their environment by the development of each organ and property. Therefore, if this or that section of the body of a plant organism is forced to assimilate conditions relatively unusual for it, and as a result undergoes alteration and becomes different from the analogous section of the body in the preceding generation, the substances which it sends forth to neighbouring cells may not be elected by the latter, may not be joined into the further chain of corresponding processes. Of course, there will still be a connection between the altered section of the plant organism and the other sections of the body, for otherwise the organism could not exist at all; but this connection may not be fully reciprocal. The altered section of the body will be receiving this or that food from the neighbouring sections; but it will not be able to give away its own specific substances, because the neighbouring sections will refuse to elect them.

This explains the frequently observed phenomenon when various altered organs, characters or properties of an organism do not appear in the progeny. But the altered sections of the body of the parent organism always possess an altered heredity. Fruit growers and horticulturists have long known these facts. An altered twig or bud of a fruit tree or the eye (bud) of a potato tuber cannot, as a rule, influence the alteration of heredity of the offspring of the given tree or tuber which are not directly generated from the altered sections of the parent organism. If, however, the altered part is cut away and grown separately as an independent plant, the latter, as a rule, will possess a changed heredity in its entirety, the heredity which the altered part of the parent body contained.

The extent of the transmission of alterations depends on the extent to which the substances of the altered section of the body join in the general process which leads to the formation of reproductive sex or vegetative cells.

Variation of the processes of development of organs and characters is always adaptive to the environmental conditions under the influence of which it occurs, but it must be remembered *that an adaptive variation is by far not always advantageous for the organism as a whole.* Relative fitness, harmony of plants and animals in free nature, is produced only by natural selection, which includes heredity, variation and survival.

Once we know how the heredity of an organism is built up, we can change it in a definite direction by creating definite conditions at a definite moment in the development of the organism.

Good varieties of plants or animals are always produced only by the application of proper methods of cultivation or breeding. Under poor cultivation no good varieties can ever be produced out of poor ones, and in many cases even good cultivated varieties will deteriorate after a few generations. It is a basic rule in seed growing that plants grown for seed must be tended with the utmost care. They must be provided with conditions meeting the hereditary requirements of the given plants. Of well-cultivated plants the very best should be and are selected for seed. That is the way varieties of plants are improved in practice. Under poor cultivation (i.e., when agrotechnique is bad) no selection of the best plants for seed will produce the required results—all the seeds will be poor, and even the best among them will still be poor.

According to the chromosome theory of heredity, hybrids can only be produced by sexual reproduction, although Darwin and a number of other eminent biologists have acknowledged the possibility of obtaining also vegetative hybrids. They have admitted that it is possible to mix two breeds not only by sexual crossing but also by vegetative coalescence. I. V. Michurin not only recognized the possibility of producing vegetative hybrids, but elaborated the mentor method. This method consists in the following: by grafting cuttings (twigs) of different varieties of fruit trees on the branches of a young variety, the latter acquires properties which it lacks, these properties being transmitted to it through the grafted twigs. That is why I. V. Michurin called this the mentor method. By this method he produced new and improved existing varieties.

I. V. Michurin and the Michurinists have found methods of obtaining vegetative hybrids in large quantities.

The vegetative hybrids are cogent proof that Michurin's conception of heredity is correct. At the same time they represent an insuperable obstacle to the theory of the Mendelist-Morganists.

Organisms grafted before they have been completely formed phasically, before they have completed their cycle of development, will always undergo changes of development as compared with own-rooted plants, i.e., ungrafted plants. In the union of plants by means of grafting, the product is a single organism of mixed breed, that of the scion and that of the stock. By planting the seeds from the scion or the stock it is possible to obtain offspring, some representatives of which will possess the characteristics not only of the breed from which the seed has been taken, but also of the other with which it has been united by grafting.

The stock and the scion could not have exchanged chromosomes of the cell nuclei; yet hereditary characters may be transmitted from stock to scion and vice versa. Consequently, *the plastic substances produced by the scion and the stock also possess the characters of the breed,* i.e., of the heredity.

The wealth of factual material concerning vegetative transmission of various properties of potatoes, tomatoes and a number of other plants leads us to the conclusion that vegetative hybrids do not differ in principle from hybrids obtained sexually. Any character may be transmitted from the one breed to another by means of grafting just as well as by the sexual method. The behaviour of vegetative hybrids in succeeding generations is likewise analogous to that of sexual hybrids. If the seeds of vegetative hybrids are sown without further grafting, for instance, the seeds of tomatoes, the hybrid properties of the plants of the preceding generation are present also in the plants of the succeeding generation. The so-called segregation phenomenon, frequently met with in the offspring of sexual crossings, also takes place in the seed generations of vegetative hybrids. But in the latter, vegetative segregation is observed much more frequently and to a considerably greater extent. Here the body of the organism obtained is a mosaic of various characters.

The representatives of Mendel-Morgan genetics are not only unable to obtain alterations of heredity in a definite direction but categorically deny that it is possible to change heredity so that it will adequately correspond to the action of the environmental conditions. But if one proceeds from the principles of Michurin's teaching it will be found possible to change heredity in correspondence with the influence exerted by the conditions of life. For instance, A. A. Avakian and other scientific workers by means of proper rearing produced numerous hereditary spring forms from winter forms. Hereditary spring forms were obtained from all standard winter-wheat varieties experimented with. On the other hand, a great number of spring wheats and barleys were converted into hereditary winter forms.

The experiments performed in converting spring forms of cereal grains into winter forms and winter forms into forms of still greater winter habit in regions of Siberia, for example, where winters are severe are of great practical importance for the production of winter-hardy varieties. There already exist a number of winter wheats derived from spring wheats which in frost resistance are not inferior and in some cases even superior to the most frost-resistant varieties known to practical farming.

Many experiments show that when an old, established property of heredity, for example winter habit, is being eliminated, no established new heredity will be obtained at once. In the vast majority of such cases the plants obtained possess a so-called destabilized heredity.

Plant organisms are said to have a destabilized heredity when their conservatism has been eliminated and their electivity with regard to external conditions is weakened. Instead of conservative heredity such plants preserve, or there appears in them, only a tendency to show some preference for certain conditions.

Heredity can be destabilized in the following way:

1) By grafting, i.e., by uniting the tissues of plants of different breeds;

2) By subjecting plants to the influence of the environment at definite moments when they are undergoing developmental processes of one kind or another;

3) By crossbreeding, particularly of forms sharply differing in habitat or origin.

Much attention was paid by some of the best biologists—Burbank, Vilmorin and particularly Michurin—to the practical value of plant organisms with destabilized heredity. Plastic plant forms with unestablished heredity, obtained in this or that way, must be sown from generation to generation under conditions the requirement or stabilization of which we want to induce in the particular organisms.

In most plant and animal forms new organisms develop only after fertilization—the fusion of female and male sex cells. The biological significance of the process of fertilization is that thereby organisms are produced with a dual heredity—maternal and paternal. Dual heredity lends greater vitality to organisms and widens the range of their adaptability to varying conditions of life.

All ordinary (nonsex) cells divide in two at the end of their development; this is the way in which cells multiply and the body grows. Sex cells however not only do not divide in two at the end of their development but on the contrary from two sex cells, a male and a female, one cell is normally obtained, and this one cell is usually more viable than either of the other two.

Both the female and male sex cells fully possess the properties of their breeds. The breeds differ in one degree or another. After the zygote is obtained, i.e., after the female sex cell is fertilized, the two cells form one cell—the inception of an organism in which the breed properties of both forms are represented. The contradiction between the two united but relatively different sex cells gives rise to or intensifies viability, the capacity of undergoing change or transformation. It is precisely this that determines the biological necessity of crossing forms that differ, even if slightly, from each other.

Darwin repeatedly emphasized in his works that the usefulness of crossbreeding and the biological harmfulness of self-fertilization are a law of nature.

The vitality of plant forms may be renovated and strengthened also vegetatively, asexually. This is brought about by the living body assimilating new external conditions, conditions unusual for it. In experiments in vegetative hybridization, in experiments with the aim of producing spring forms from winter forms or vice versa, and in a number of other cases of the nature of organisms being destabilized, we may observe the renovation and strengthening of the vitality of organisms.

K. A. Timiryazev correctly classified the various behaviours of sexual hybrids. He divided the phenomena of heredity into two groups: simple and complex heredity.

It is well known that self-pollinating plants, such as wheat, or plants propagated by tubers, cuttings, layers, etc., possess, as a rule, during their development a heredity that is largely maternal, i.e., the heredity of the form from which the seeds, cuttings, etc., were taken. This form K. A. Timiryazev called *simple* heredity.

Crossbreeding usually unites the heredity of two organisms. Such a heredity is called *complex*. It in turn may be subdivided into several groups according to the different forms in which it manifests itself.

Among certain animals a spot in their pelage may, for instance, resemble in colour the paternal form; another, the maternal. Or among plants some cells of the leaf epidermis may be like the paternal cells, others like the maternal, etc. Such heredity is called *mixed* because one part of the organism exhibits characters of the one, and another part characters of the other parent. These parts or sections may be of different sizes—from big to microscopic.

Most frequent are those cases in which the hereditary properties of both parents blend in the offspring and do not appear in pure form, cases where new properties result in the offspring. Such heredity Timiryazev called *blended* and considered most important of all.

Cases occur in which contrasting parental characters do not mix in hybrid offspring. For instance, on crossing a pea variety having green seeds with one having yellow seeds these characters do not blend in the offspring. No new or mean property is obtained. What results is the property of only one parent. The property of the other is excluded, as it were. This form of heredity is called *mutually exclusive.*

Two categories of facts may be observed in mutually exclusive heredity.

One category embraces cases of hybrid organisms which are uniform in the first and all succeeding generations. In other words, there is no diversity, no segregation in the generations of hybrid offspring. Not infrequently the properties of the one parent are completely absorbed by the other. This is called *Millardetism* after the French scientist Millardet, who made a quite thorough study of this category of hybrids.

The other category of mutually exclusive heredity embraces cases of so-called *Mendelism*. Beginning with the second generation these cases of hybrids exhibit segregation, diversification, some forms having paternal and others maternal characters.

Today all the different forms of heredity may likewise be found in vegetative hybridization.

In vegetative hybrids a mixed form of heredity occurs when one part of the organism exhibits the properties of the one strain, the one component, and the other part exhibits those of the other component. Cases of blended and of mutually exclusive heredity also occur.

Such hybrids also exhibit unusually vigorous development or on the other hand decreased viability, i.e., the same as in sexual hybridization.

All this does not mean of course that there is no difference between vegetative and sexual hybridization, but it is important to emphasize that vegetative and sexual hybrids exhibit common forms of heredity. No impassable barrier separates these two categories of phenomena; they are phenomena of one and the same order.

Regulation of environmental conditions, the conditions under which plant organisms live, makes possible directed change, the creation of varieties possessing the requisite heredity, inasmuch as the inheritability of properties acquired by plants and animals in the process of their development is a law of living nature.

Heredity is the concentrate, as it were, of the environmental conditions assimilated by plant organisms in a series of preceding generations.

By means of skilful hybridization, by the method of sexual conjugation of breeds, it is possible at once to unite in the organism that which has been assimilated and fixed in its passage from nonliving to living material by many generations of the breeds taken for the cross. But, according to Michurin's teaching, no hybridization will produce the desired results unless the conditions are created which will promote the development of the characters which we want the newly-bred or improved variety to inherit.

It must be remembered that nonliving nature is the prime source of living nature. The living body builds itself from the environmental conditions and thereby changes itself.

THE TASKS OF THE LENIN ACADEMY OF AGRICULTURAL SCIENCES OF THE U.S.S.R.[1]

THE DECISION adopted by the February Plenum of the Central Committee of the Communist Party of the Soviet Union (Bolsheviks) on the report of A. A. Andreyev arms all agriculturists with a program of action for the next several years. This decision confronts agricultural research institutions, including the Lenin Academy of Agricultural Sciences of the U.S.S.R., with serious tasks. The Plenum decision also makes it the duty of the Ministry of Agriculture of the U.S.S.R. and the Ministry of State Farms of the U.S.S.R., as well as the local Party and Soviet organizations, "to take measures to ensure the speediest practical application of the achievements of agricultural science, considering this work a paramount condition for the improvement of agriculture."

Successful fulfilment of this demand will strengthen the bond between science and production, enhance its role in the further development of socialist agriculture and serve as a stimulus to work with still greater perseverance and industry.

Men of science should not stand aloof when it comes to applying agronomic measures. All measures for the improvement of agriculture should be drawn up with an eye to the conditions prevailing in the locality, zone or district. This requires the participation of scientific workers so that the theory underlying the problems that are being worked out may be applied correctly, concretely and in accordance with the particular zonal conditions.

MILLET YIELDS MUST BE GREATLY INCREASED

The decision of the February Plenum of the Central Committee of the Communist Party of the Soviet Union (Bolsheviks) makes it the duty of the Ministry of Agriculture of the U.S.S.R. and the Ministry of State Farms of the U.S.S.R. to organize in 1947, with the assistance of the Lenin Academy of Agricultural Sciences of the U.S.S.R., the production on collective and state farms of an average millet crop of 15 centners per hectare on an area of

---

[1] Revised stenographic report of an address delivered at an open Party meeting of the Academy on the tasks in the light of the decision of the Plenum of the Central Committee of the Communist Party of the Soviet Union (Bolsheviks) entitled "On the Measures to be Taken to Improve Agriculture During the Postwar Period."—*Ed.*

one million hectares. This of course implies the task of considerably raising yields also on all other fields that will be sown this year to millet. The discharge of this task makes it necessary to utilize primarily the experience gained by the collective and state farms and agricultural science in their struggle for big millet crops in 1939 and 1940.

Since the war and the devastation brought on by the Nazi occupation and since last year's drought a number of districts have been experiencing difficulty with regard to seed in general and have not always fulfilled their assigned agricultural work on time. Under these conditions some districts have fallen into the bad habit of compensating for their "sowing shortages" by extending their millet areas. I am therefore of the opinion that while aiming at higher millet yields we should prevent the indiscriminate increase of areas under this crop. We must do our utmost to enlarge the area sown to wheat and other basic grains.

It should be firmly borne in mind that good millet crops require well-weeded fields sown, as a rule, in far-between rows; also timely cultivation and weeding.

If millet is provided with the conditions it needs, a very high yield can be obtained.

The late Chaganak Bersiyev, a Kazakh collective-farm member and one of the foremost millet growers, achieved most admirable results. He obtained yields for millet that have not been matched anywhere in the world for any grain whatever, namely, 1,200 to 1,300 poods per hectare. This record even surpassed the yields that theoretical calculations had forecast as the highest possible.

But when, as is often the practice in various districts, millet is looked down upon as not being a worthwhile grain, its yield is the lowest of all cereals. That explains why it is so extremely necessary for agronomists and scientific workers to give this crop the benefit of their particular care and attention, to exercise a sort of guardianship over it, by their timely, correct and concrete advice to the collective and state farms.

Last year the Kiev area achieved major successes in the struggle for high millet yields. Thus, despite the severe drought, Shpola District produced 27 c. of millet per ha. over an area of more than 2,000 ha., the equivalent of 50 c. per ha. for a normal, nondrought year.

In 1945 meteorological conditions in the Kiev Region were considerably better than in 1946. Nevertheless, the 1945 millet yield for that region averaged only half of the 1946 figure.

This sharp difference is explained by the fact that in 1945 no attention was paid to millet in the Kiev Region and a number of other regions. Last year, however, on the advice of I. D. Kolesnik, a research worker of the Lenin Academy of Agricultural Sciences of the U.S.S.R., the Party and Soviet organizations of Kiev Region took an interest in this crop and were able to enlist the broad mass of collective-farm members in the struggle for high yields. The millet was planted in wide rows, good, vernalized seeds were used and the requisite time limits were observed. Provision was made for the proper care of the crops. As a result the millet yield increased sharply even though there was a fearful drought.

I shall mention only the most important of the basic measures indispensable for a good millet crop.

First of all, millet-cultivating teams must be organized in time and sowing plots assigned to them in advance, as otherwise the millet fields may not be properly attended to and produce low yields in consequence. The law allowing additional pay for overfulfilling the millet-yield plan must be explained to the teams and strict control instituted to ensure the prompt payment of any additional items that may be due to collective farmers.

It must be fully borne in mind that millet responds much more vigorously than other plants to proper presowing tillage, to sowing good, vernalized seed at the right time and proper tending of the crops. And, vice versa, millet crops often shrink, even to paltry proportions, if the requisite agrotechnical methods are not employed to create the conditions they require. Comparatively little effort and means are needed to establish these conditions if effort and means are applied properly and at the right time. A crop of 25 to 30 c. per ha. is easy to obtain. If you add manuring to good tending, even 40 to 50 c. per ha. is possible, and some expert millet growers raise even 80 to 100 c. and more per ha.

Millet is raised not only as a food but also as a fodder crop. A sharp increase in millet yield—which, as has been said above, requires no great expenditure of energy or means but, on the other hand, imperatively demands great care and attention—may also be greatly instrumental in solving the meat problem. The 40-50 c. of millet obtained per ha. can readily be converted into 10-12 c. of pork and fat.

Weeds are millet's worst enemy. Millet sprouts are small; at first they grow slowly and if the necessary measures are not taken the weeds will choke the millet plants and the crop will be greatly reduced. Therefore, in picking plots the weed factor must be given first consideration. Millet requires a field that is as clear of weeds as possible. Besides, as soon as spring sets in, the moisture must be retained in the autumn-ploughed soil by double harrowing.

For millet an autumn-ploughed field must be picked. If it was not ploughed in autumn, plough as early as possible in spring but *under no circumstances wait until the early grains have been sown.*

Millet seeds are small; they should not be sown too deep. It therefore is a rule of prime and fundamental importance that steps be taken to preserve the moisture in the topmost layer of the soil until the millet is sown.

If on a plot set aside for millet the moisture is not retained at the very beginning of spring the upper layer of the soil may dry up and then it will depend upon the caprices of nature whether the sown millet will sprout or not. If there is rain after the sowing shoots will appear; if not, not. This explains why land sown to millet must be cultivated—harrowed after autumn ploughing or spring-ploughed—early in spring though millet is sown somewhat later than the other early crops.

The next fundamental agrotechnical rule for successful millet growing is careful weeding before sowing. It is then that as many weeds as possible must be induced to germinate and must be destroyed by cultivating twice: the first time about 10 days after the moisture is retained, the second time immediately before sowing.

For weed killing use duck-foot cultivators which cut the weeds without turning and drying the soil. Autumn-ploughed plots should first be cultivated to a depth of 6-8 cm., and at the same time be gone over once with a harrow; the second cultivation should be 5 cm. deep.

After cultivating the soil prior to the sowing it will be necessary, if the weather is dry, to go over the surface with a wooden roller so as to draw the moisture toward its top stratum where the seeds lie after sowing.

In the various districts of the European part of the Soviet Union millet should not be sown too early, when the soil has not yet become warm; nor, on the other hand, should the sowing be delayed. Millet must be sown as soon as the temperature of the top layer of the soil reaches 12°-15° C. and generally no longer drops below 10°.

When millet is sown too late there is danger that the top layer of the soil may become too dry, as a result of which it will be difficult to obtain sprouts. Besides, in belated millet sowing the emergence of the panicles and the flowering may occur at a time when the harmful gall midget Stenodiplosis panici swarms about, which in many districts may destroy such late-sown crops completely.

In the trans-Ural area, Siberia and North and Central Kazakhstan the summers are short and millet often will not ripen. Here millet must be sown not later than May 20-25, i.e., not later but earlier than oats.

To make the weeding of millet easier it should be planted in far-between rows. Millet should be sown close only if the field is absolutely weedless or in virgin soil where no weeding is required.

Tractor- and horse-drawn cultivators and also hand cultivators (planettes) should be extensively used to loosen the soil between the rows and kill the weeds. Even horse-drawn, not to speak of tractor-drawn cultivators, speed up the work tenfold in comparison with the time it takes to do it by hand, using a chopper.

In case of a shortage of horse-drawn cultivators and factory-made hand cultivators they should be manufactured locally, in workshops and smithies.

Millet seeders must be very carefully examined to see if they are in good condition, without openings through which seeds might spill. In order to ensure even sowing it is best to put equalizers into the seeders. The best seed variety for the district must be used: smut-free seeds with good sprouting capacity and cleaned of weeds.

From 7-10 days before sowing the seeds must be vernalized.

In 1947 a sharp increase in millet yield must be achieved.

The experience of 1939 and still more of 1940, when collective and state farms reaped an average of 15 c. per ha. on 500,000 ha. and of 20 c. on 200,000 ha., convincingly showed that sharp increases of millet yields over considerable areas are quite feasible, are a task that can be shouldered by our agricultural personnel. It is the duty of our research workers to render all possible agronomic assistance to the collective and state farms in the execution of the task set by the Plenum of the Central Committee of the C.P.S.U. (B.)—that of securing an average of 15 c. of millet per ha. on an area of 1,000,000 ha.

THE SUMMER PLANTING AND VERNALIZATION OF POTATOES

By decision of the Plenum it is proposed to restore in the course of two years the prewar area under summer-planted potatoes in the southern and southeastern districts. This is a very important measure and, of course, the Academy cannot afford to ignore it. I consider the summer planting of potatoes a splendid achievement of Soviet agrobiology. Collective-farm practice over a period of seven years before the war showed that if seed potatoes are obtained each year by means of summer planting the breed qualities of these tubers constantly improve. Planting potatoes in summer eliminates the need for renewing the planting material in the South by fresh shipments from northern regions.

Abroad no really effective practical means of combating the degeneration of seed potatoes has yet been found for southern regions. Hence, seed potatoes have to be shipped there annually from northern or mountainous regions. As a result southern districts have, as a rule, much less potatoes than they require. The sole cause is degeneration of the planting material, i.e., the lack of some method of growing good seed potatoes under the conditions prevailing in southern districts.

As has already been said, in our country such a method has been found. Planting potatoes in the South in summer constitutes one of the chief measures for the development of potato growing in these districts.

How badly this measure is needed to improve potato growing in the South is well illustrated by the following crop averages for any southern region of the European part of the U.S.S.R. where the summer planting of potatoes was widely practiced before the war. Let us take Crimea Region, for example. According to data supplied by A. M. Favorov (Odessa All-Union Institute of Selection and Genetics), in 1936 there were altogether 7,100 ha. under potatoes on the collective and state farms of Crimea Region, of which 150 ha. were summer-sown (for the first time in the Crimea). In 1936 the spring-sown potato crop averaged 34 c. per ha. This low yield was due to degeneration of seed potatoes caused by the hot weather conditions in the South. Each succeeding year, as summer plantings for the production of seed potatoes were extended, the breed qualities of the potatoes improved and yield averages on Crimean collective and state farms regularly rose from year to year. In 1938 the yield amounted to 70 c. per ha. and in 1940 to 88 c. over an area of 12,900 ha., of which 7,300 ha. were now summer-planted.

It must be noted that as reproduction by means of summer planting increases there is a rise not only in the average yield of spring-planted crops but also of summer-planted crops. Thus in 1936, as mentioned above, 150 ha. were summer-planted, exhibiting an average yield of 65.5 c. per ha. (that year the spring-planted crop yielded 34 c. per ha.). In 1940, 7,300 ha. were planted in summer, the average yield rising to 92 c. per ha. As the result of four successive reproductions by means of summer planting the yield increased from 65.2 c. to 92 c. per ha. For an arid, unirrigated district an average yield of 92 c. is not bad. Moreover, there was every reason to suppose that as summer reproductions increased in number and the breed qualities of the planting material improved, average yields would continue to augment. The temporary occupation by the enemy completely destroyed potato seed growing in the South which was well organized before the war. Now this work has to be done all over again.

Let us take another example to illustrate the immense improvement in seed potatoes derived from summer planting. The following facts are generally known. If an early potato variety, such as the Early Rose raised in Moscow Region, is grown in the South for a year or two, it being planted as usual in spring, and part of this material is shipped back to Moscow Region and grown under comparable conditions together with the Moscow reproduction of the same potato strain, the yield from the planting material brought from the South will be 3-4 times less than the yield from tubers grown under the conditions obtaining in Moscow Region. The opposite results if potatoes of, for example, the same Early Rose variety are summer-planted in the South and the planting material is then shipped to Moscow Region for the purpose of comparing yields. Such an experiment was performed in 1940 in the fields of the Institute of Genetics of the Academy of Sciences of the U.S.S.R. It was found that the seed potatoes of the Early Rose variety shipped from Odessa after four years of reproduction produced a yield of 480 c. per ha. The same variety, planted under similar conditions but grown all the time in Moscow Region yielded a crop of only 220 c. per ha.

The decided improvement in the planting qualities of the potatoes grown by the summer-planting method is due, in the main, to the unusually good nutrition the potato plants receive in the South when planted by this method. The fact of the matter is that in the arid southern regions of the Ukrainian S.S.R., Crimea Region and North Caucasus it is necessary to set aside for summer sowing plots of autumn-ploughed land which must be cultivated until the middle of summer like the best bare fallow in arid districts in order to conserve and accumulate moisture. By midsummer, i.e., the time when potatoes are planted, a large quantity of easily assimilated nutritive matter accumulates in the soil thus cultivated. When planted in a properly prepared field the potato, from the moment it is planted until the time it is picked, finds itself under exceptionally favourable conditions, having an abundance of readily assimilable food, so that in autumn, when temperatures are lower, it quickly forms very big and healthy tubers. These good conditions which summer planting creates for the potato explain, by and large, the pronounced improvement in its seed qualities.

The urgent necessity of developing potato growing in the southern industrial districts and in the environs of the southern cities and health-resort centres is a matter of general knowledge. Hence the practical importance of planting potatoes in summer in these districts, so that they may serve as bases for breeding early-variety seed potatoes for spring-sown food crops.

This year the vernalization of potato tubers acquires great importance. Tuber vernalization must be extensively practiced in all districts where there is a shortage of seed potatoes. This measure will not only increase the potato crop per unit of area but will make it possible to increase the planting area.

Properly vernalized tubers should before planting be cut into pieces containing 2-3 eyes. Each such piece will yield a crop no smaller than an ordinary whole but unvernalized tuber would produce.

On the basis of the vast prewar practical experiments of collective-farm members in the South of the Ukrainian S.S.R. one may boldly assert that only half as many potato tubers are needed per hectare for planting when they are well vernalized than when the usual unvernalized tubers are used. This explains the practical importance of vernalizing potatoes, especially this year. But care must be taken to ensure that the vernalization is done properly, that the tubers do not rot during the process. This requires that vernalization premises be well aired.

Scientific workers must come to the aid of the collective-farm members in this matter. In every collective farm somebody must be specially assigned to this job and provided with literature and instructions on vernalization. If nobody is put in charge of this work or the person in charge does not know his business everything may be spoiled.

In prewar years all potatoes in the Southern Ukraine were, as a rule, vernalized before planting. Today it has become absolutely necessary to train collective-farm vernalizers, to teach new people, to make full use of people with previous experience in the field of vernalization.

THE ALFALFA SEED CROP MUST BE INCREASED

The Plenum decision points out that alfalfa must be extensively planted in summer in well-cultivated bare fallow in the various districts of the Ukraine, North Caucasus, the Crimea and the Moldavian Soviet Socialist Republic. This measure will double and triple yields of alfalfa seed in the districts mentioned. In summer alfalfa is sown close, not in far-between rows. Fields sown in this way need no weeding and no ploughing. Sowing alfalfa in summer requires well-cultivated fallow fields no less than 100 m.

away from the alfalfa fields of the preceding year to prevent the seeds and flowers from being attacked by pests.

Alfalfa sown at the right time in properly cultivated plots guarantees seed crops of 2-5 c. per ha. Summer-sown alfalfa can therefore produce tens of thousands of centners of seed each year in the southern districts. Even in so dry a year as 1946 summer-sown alfalfa stood high in the fields of the Odessa All-Union Institute of Selection and Genetics, showed vigorous development and yielded 1.5 c. of seed per ha.

The fight to apply the proper method of summer-sowing alfalfa in the southern regions is a fight for more alfalfa seeds, i.e., a struggle for efficient farming, for the introduction of correct crop rotation, for extending the area under spring wheat, which, to increase its yield, requires soil on which perennial grasses were grown. Herein lies the practical importance of sowing alfalfa in summer in the southern districts of the European part of the U.S.S.R. And this makes it incumbent upon the staffs of research institutes to take all the necessary measures to help the collective and state farms introduce summer sowing of alfalfa in their fields as quickly as possible.

SKIM COULTERS

Practical and scientific workers must appreciate the utter importance of the instructions of the February Plenum concerning skim coulters. We are directed to make full use in 1947 of all available skim coulters, in 1948 to use skim-coulter-equipped ploughs in no less than half and in 1949 in all tractor ploughing. Spreading the use of skim coulters is a most important measure: it will ensure the receipt of many additional tens of millions of centners of grain. Skim-coulter ploughing must become an index of the efficiency of our agriculture and our agronomic science. Here is a problem that still requires much elaboration.

It would be absurd to suppose that in many instances skim coulters were not used because tractor drivers did not want to use them, did not want to increase the crop. In some cases skim-coulter ploughs did not work, as for instance where the stubble was more than knee-high and elsewhere for other reasons, such as the imperfect construction of the coulters themselves. Our agricultural mechanization and agrotechnical institutes must make a real effort to work out a solution for this highly important problem.

RUBBER-BEARING PLANTS

The decision of the Plenum of the Central Committee of the Communist Party of the Soviet Union (Bolsheviks) charged the Lenin Academy of Agricultural Sciences of the U.S.S.R. with the task of accelerating the elaboration of a method of vegetative propagation of our most important rubber-bearing plant, kok-saghyz, and the simultaneous application of this method in practice. The words "and the simultaneous application of this method in practice" require our attention. I am convinced that by far not every problem connected with the vegetative propagation of kok-saghyz has already been worked out in our country, and this only because we have worked exclusively in experimental institutions, out of touch with the thousands of collective farms. The vegetative propagation of kok-saghyz is a measure with such excellent prospects that it is worth fighting for. If cuttings are used for multiplication, the crop of kok-saghyz roots will be double that obtained by the usual method of seed sowing. Besides, much less labour is needed to plant cuttings.

Scientists have also been commissioned to work out a method of growing another native rubber-bearing plant, tau-saghyz, the prospects of which, in my opinion, are very bright in the steppe districts of the Ukrainian Soviet Socialist Republic. The Odessa Institute of Selection and Genetics experimented with tau-saghyz on a small scale even before the war. Although prewar tau-saghyz plantations were absolutely neglected and not once cultivated during the occupation period they withstood these adverse conditions well. This testifies to the great biological hardiness and adaptability of this plant to the aridity of the Ukrainian steppe.

The rubber-growing state farms of South Kazakhstan with their old tau-saghyz fields have shown that after the roots are grubbed tau-saghyz plantations are restored and again ready for exploitation in two or three years.

After grubbing and removing the tau-saghyz roots the plantations are restored by regeneration of the plants from the root parts left in the soil.

There is thus every reason to assume that once tau-saghyz plantations are laid out it will be possible, under the conditions prevailing in the south of the Ukraine, to exploit them many times, i.e., the roots will be dug up for the production of rubber while the plants will restore themselves after each grubbing. This is the direction in which our scientists must develop their work, and do so with the greatest possible speed and efficiency.

HYBRID SEEDS

Soviet agrobiologists know full well that plants grown from hybrid seeds are, as a rule, more resistant to adverse conditions than the initial forms taken for hybridization and are superior to them in yield.

According to Darwin's teaching, prolonged self-pollination weakens the vitality of plant forms and, vice versa, a crossing of relatively differentiated forms produces a progeny of greater vitality.

The more closely related, the less dissimilar the forms entering into the crossing are, the less the effect obtained by hybridization.

Hybrid maize seeds exhibiting great increases in yield may be derived from a cross of two local (sown in the district) but very dissimilar varieties or from the crossbreeding of a local variety (maternal form) with a specially selected inbred line (paternal form) which is not employed in practical farming because of its inferior yield or finally from crossing specially selected inbred lines.

Plant-breeding stations must greatly extend their work of producing and selecting initial material for growing hybrid maize seeds. In 1947 the Brown County and Grushevskaya varieties, tested by the Dniepropetrovsk Grain Institute for hybridization purposes, must be sown over a considerable area of the steppe districts in the Ukrainian Soviet Socialist Republic.

This will make possible the extensive use of hybrid maize seed in 1948, in a number of districts, to increase the yield of this plant.

The introduction into agriculture of hybrid seed for growing maize is a measure that will raise the yield of this crop and the productivity of collective- and state-farm labour.

In addition, scientific workers must pay great attention to the experimental solution of the problem of the utilization in practice of intravarietal and intervarietal hybrid wheat seeds.

I am certain that in capitalist countries hybrid wheat seeds are not used in practical farming, as is extensively done in the case of maize in the U.S.A., for the sole reason that there is no practical method of rapidly producing these seeds in quantity. In our country, however, a method of rapidly producing hybrid wheat, particularly the winter form, in quantity was worked out already before the war at the Odessa Institute of Selection and Genetics in conjunction with a large number of collective farms. By this method any farm can easily obtain dozens of centners of hybrid wheat seed of the third and fourth generations. Therefore, parallel with the introduction into practical farming of the sowing of hybrid maize seed, in accordance with the decision of the Plenum of the Central Committee of the Communist Party of the Soviet Union (Bolsheviks), our plant-breeding and seed-growing institutions must rapidly solve the problem of the advisability of using hybrid seeds of wheat and other plants in agricultural production.

THE BREEDING OF GRAIN

The decision of the February Plenum of the Central Committee of the Communist Party of the Soviet Union (Bolsheviks) records that the breeding and varietal testing of cereals still do not meet the demands of agriculture. It is likewise pointed out that even at the present time the collective farms of Siberia have no winter-hardy varieties of winter wheat.

The growing of different varieties of agricultural plants, their reproduction and distribution (districting) throughout the territory of our country are links of a single chain—the work of plant breeding and seed growing.

The proper choice of varieties of agricultural crops for the various districts of the Soviet Union with their sharply differing conditions for plant growth is a most important and complicated matter.

In my opinion the main reason for the shortcomings in growing varieties and assigning them to the different districts is the fact that this work is done separately and apart from the practical farm work of the collective and state farms. Varieties, as a rule, are distributed among the districts merely according to the data of the varietal test network, without preliminary verification at the collective and state farms. I consider this the main reason why it takes so long to test varieties at the variety test plots and why, despite the long test period, there is legitimate doubt as to the suitableness of varieties for the districts to which they are assigned. Moreover, in a very great number of cases varieties, even after such districting, are for a long time—10 years and even longer—actually not sown in the districts to which they have been assigned, the reason being lack of seed. As a result a very great interval elapses from the time that the station breeds the variety in question (or the time the seeds of that variety are brought from another locality to the given district for varietal testing) to the time this variety is made use of on areas of practical importance, such as at least 100,000-200,000 hectares.

Regional plant-breeding stations and regional agricultural boards ought to take a leading part, I believe, in promoting the cultivations of new good varieties in their districts—locally-bred varieties and those brought from elsewhere—and in drawing up proposals for distribution in the zones served by the respective stations.

Parallel with the plant-breeding stations and regional agricultural boards, the management of the State Cereal Variety Trial Commission should submit its own proposals, based on its own data, for distributing varieties in the districts and thus act as a governmental body controlling the allocation of varieties to the heterogeneous districts of our country.

Obviously it should not only be possible but compulsory for the staff of the State Cereal Variety Trial Commission to make use of the data of the plant-breeding stations and, inversely, the plant-breeding stations should in turn make use of the data of the State Cereal Variety Trial Commission.

There is no reason to fear the discrepancies which will undoubtedly arise between the districting proposals proceeding from the plant-breeding stations and those proceeding from the state network. Such discrepancies should be clarified and removed by the all-Union and Republican Ministries of Agriculture.

It seems to me that the process of breeding, reproducing, testing and districting varieties should be arranged in about the following way. A plant-breeding station or a plant breeder who knows the conditions of the districts and is breeding a new variety of his own must study and test this variety in his own fields. If in the process of breeding he becomes convinced that he has obtained a promising variety he, by prior agreement with several collective or state farms, should supply them with small quantities of seeds of this variety for sowing 0.5-1 ha. each while testing it at his station.

If in the course of a year the reports from the plant-breeding station's fields and from the various farms on the new variety are favourable, the plant breeder, in accordance with arrangements made with other farms, supplies the latter also seed of the new variety for sowing 0.5-1 ha. plots. Farms which sowed the new variety the year before either discontinue sowing it or enlarge the area under it to 3-5 ha., as they see fit, being guided in their decision by the results achieved. The collective farms may use the remainder of the seeds for commercial purposes or, if there are any other collective farms that want to try out this variety, they may, by way of exchange, supply them with seeds of the new variety for sowing a few hectares each.

If the second year also yields good results the breeder of the new variety should extend the area sown with it by the inclusion of additional farms, of course only with their full consent, farm areas already under this new crop being simultaneously increased.

Thus, each collective and state farm with which the plant-breeding station has concluded an agreement is able to determine the worth of the new variety.

Until the new variety has passed the state test, farms must not be advised to devote more than 30-40% of their areas to this plant even if it produced good results on the farm in question for 3-5 years in succession.

After the variety has been tested for 3-4 years at the station and on the farms it is submitted to the State Cereal Variety Trial Commission where it is tested for another 3 years and then either districted or barred.

Let no one think that our proposal delays even for one year the submission of any variety to a state test. It has been the rule hitherto that plant-breeding stations may submit their varieties for state testing only after they tested them for 3-4 years. And that is as it should be. We propose the same thing with the sole difference, and that a fundamental one, that when the variety is submitted to the State Cereal Variety Trial Commission there must already be on hand a 2-3 years' appraisal of its quality not only on 100-metre plots at the stations but also on regular farm areas, even if small.

If during the three-years' state test the variety proves to be really good the collective-farm areas sown to it will be increased.

If after the three years it is found that the new variety deserves to be districted and it is districted, then thousands of hectares will already be producing it on farms. Consequently, immediately after it is districted the State Variety Fund will be able to buy tens of thousands of centners of seed of the new variety, at once supply in full all seed-growing agencies (district seed farms) and turn the surplus seed over to collective farms for their seed plots.

It may be objected that if, as we propose, plant-breeding stations take an active part in the districting and reproduction of their varieties, good varieties of other plant-breeding stations are likely to be barred from the respective districts. But such undesirable occurrences can easily be prevented.

In the first place, besides promoting its own really good variety, a station must take an equal interest in and promote those of other stations in the zone it serves if they are actually better than its own variety. In the second place, if a variety of a particular plant-breeding station proves suitable, according to the observations of the station and the data obtained from the collective and state farms, for a zone not served by the originating station and if the local plant-breeding station of that zone ignores this good new variety, then the originating station is obliged to appeal the case to the management of the State Cereal Variety Trial Commission or to the Minister of Agriculture.

I am of the opinion that the arguments advanced against permitting plant-breeding stations to give to collective and state farms varieties that have not yet been tested for the particular district are wholly unfounded. This is not likely to damage the farms, as is asserted.

No one is really proposing that considerable areas be sown at once with varieties which are new to a district or collective farm without first testing them or verifying their economic value. Nobody is proposing this and moreover it is utterly impracticable.

Everybody knows that a plant breeder never has more than a few score kilograms or, at best, a few centners of seed of a new variety. For that reason he cannot offer a collective farm at one time a large amount of seed of the new variety for sowing a large area. Besides, a collective farm will never agree to sow at the start large areas with seeds of a variety unknown to it. Finally, what breeder would assume the risk of advising even one collective farm to seed down a large plot at once, without prior test, to a new variety?

According to our proposal a variety, during the six years that it undergoes the obligatory test (three years of testing by the station and three years by the state), will be appraised with sufficient accuracy by practical agriculturists to allow a decision to be made on whether it should be districted or barred. Moreover, if the variety proves bad the collective farms themselves will stop sowing it while it is being tested in the State Cereal Variety Trial Commission. On the other hand, if the variety is a good one but the Trial Commission decides not to district it, the collective farms, agronomists and land administration workers will insist that its wrong decision be rectified. In general, cooperation with the collective and state farms makes possible many improvements in the breeding, districting and rapid introduction of new varieties into practical farming.

On the basis of facts with which I am well acquainted I believe that our scientists in conjunction with our practical workers are capable of rapidly solving also so difficult a problem as the introduction of winter wheat on Siberian collective and state farms. Take this example: Five years in succession the Karaganda State Farm has been producing winter wheat of the Alabasskaya variety on thousands of hectares. It must be borne in mind in this connection that while the fields of this state farm are located in a zone where winter conditions are more severe than in the fields of any other Siberian state or collective farm, the Karaganda State Farm has completely mastered the cultivation of winter wheat on the above area.

Winter wheat has also been wintering well annually for the last five years in the fields of the Siberian Scientific Research Institute of Grain Husbandry at Omsk on an area covering scores of hectares.

A solution of the very difficult problem of the wintering of winter wheat under the severe conditions prevailing in Siberia has, I believe, been successfully worked out. Now scientific workers must widen the scope of the problem so as to give also Siberian collective farms ample opportunity to sow and harvest good winter-wheat crops. The solution of this problem should, in my opinion, take the course of sowing the principal Siberian grain—spring wheat—in good fallow fields and also in virgin soil and land long left unploughed, while winter wheat should be sown in the stubble of the spring wheat. This is how, upon our advice, the above-mentioned Karaganda State Farm solved the problem of planting winter wheat, and it is along these lines that the Siberian Scientific Research Institute of Grain Husbandry is working out the winter-wheat problem on an extensive scale.

Thus, the problem of providing good overwintering conditions for a number of varieties of winter wheat existing in Siberia has already been solved by sowing them in stubble. A mass of coordinated data, accumulated during the last five years, may be adduced in confirmation of this. But finding the conditions which ensure the overwintering of winter wheat in the districts of Siberia is not equivalent to solving the entire problem of cultivating this form of wheat in those sections of the country. As I see it, only one, but the most important part of this problem has been solved.

We must now strive to ensure good annual yields from well-wintered stubble-sown crops of winter wheat. I believe that the solution for this aspect of the problem is already clear to men of science.

Experimental stubble sowings of winter wheat have been overwintering well, but in a number of cases their yield has been low. These cases are explained by a deficiency of nutritive substances for wheat in the soil in spring and early summer. This seems to me to be due to the fact that during the winter the soil freezes severely and on account of the low soil temperature in spring and early summer the activity of the useful microflora of the soil revives and develops slowly in Siberia. The lack of intensive activity of the useful microflora spells starvation for the winter-wheat plants which have passed the winter well. In these cases the winter-wheat plants grow slowly and wither. When the summer warmth sets in and the activity of the soil's microflora develops apace the soil acquires a sufficiency of nutriment but the withering plants are now beyond recovery Under these conditions, when the winter wheat develops feebly, it is but natural that weeds should grow rampant. This, it seems to me, explains all the cases where experimental winter wheat sown in stubble at Siberian plant-breeding stations perishes or yields small crops in summer after successful wintering.

The explanation we give is, I believe, substantiated also by the practical experience of the Karaganda State Farm, where year after year winter-wheat crops of fair size for a very droughty district are raised. Wintering conditions on the fields of this state farm are not less but more severe than in other Siberian districts. At the same time spring in the Karaganda State Farm district is warmer and sets in at once, the soil becomes warm much sooner, and the activity of the useful soil microflora manifests itself more rapidly in the spring, for which reasons plants in the fields of that state farm are much less liable to starve than elsewhere in Siberia.

The same can be said of the experiments made in 1946 by A. Kochergin at Omsk in the fields of the Siberian Scientific Research Institute of Grain Husbandry. A small amount of superphosphate was introduced just before the winter wheat was sown in stubble. The plots thus fertilized at the rate of 1 c. of superphosphate per ha. produced, under the conditions prevailing last year, almost double the grain crop of unfertilized plots.

The experiments conducted by N. A. Belozerova (Siberian Scientific Research Institute of Grain Husbandry at Omsk) for a period of 5 years convincingly show that high winter-wheat yields may be procured in Siberia by sowing various winter wheats in stubble. These experiments produced crops of 25-30 c. per ha.

Good winter-wheat crops were obtained in the fields of the institute mentioned when spring and early summer were warm, in other words, when the winter-wheat plants did not starve during the first half of the summer. It is perfectly clear that well-wintered winter wheat sown in stubble can be made to yield big crops every year if the plants' nutrition is skilfully managed. In solving the problem of obtaining good yields from crops sown in stubble, which ensures the overwintering of winter wheat, it is imperative that scientific workers in Siberia pay particular attention to experimenting in the fertilization of the fields under these crops.

There are grounds for assuming that to apply manure or humus to fallow fields sown to spring wheat and to sow winter wheat in the stubble of this spring wheat would be especially effective. Under such conditions winter wheat will not suffer in spring from lack of nutrition as the easily assimilable reserves of it accumulated since the previous autumn still remain in the soil.

Preliminary experiments and observations also show that overwintered winter wheat sown in stubble responds favourably to a surface application in autumn of straw ash, with which winter-wheat fields can easily be supplied in Siberia.

This is approximately the method which, in my opinion, must be tried out as soon as possible, parallel with the growing of new winter-wheat varieties suitable for Siberia, if the assignments of the February Plenum of the Central Committee of the Communist Party of the Soviet Union (Bolsheviks) with respect to the question of winter wheat for Siberian collective farms are to be fulfilled.

Let me add a few words on spring wheat for the Ukrainian S.S.R.

Measures must be taken to ensure that the collective and state farms of the Ukraine, in addition to enlarging areas and raising the yields of their principal grain, winter wheat, considerably increase spring-wheat areas and yields. Soviet agrobiology can and must give the collective and state farms substantial assistance in this matter. Let me remind you that as early as 1933 the Institute of Selection and Genetics promised to produce soon, in two and a half years, a new variety of spring wheat for the southern districts of the Ukraine. As is known, the following varieties were forthcoming at the appointed time: Lutescens 1163, and a still better variety, Odesskaya 13. A good feature of the spring-wheat variety Odesskaya 13 is its immunity from the Hessian fly. Before the war the above varieties were assigned to the southern districts. Now it is necessary rapidly to restore the seed growing of these and other varieties, as well as to breed new varieties of spring wheat and other plants within fixed time limits (2-3 years) by the employment of approved methods.

QUESTIONS OF AGRICULTURAL ECONOMICS

The importance of that part of the decision of the February Plenum of the Central Committee of the Communist Party of the Soviet Union (Bolsheviks) which deals with work in the sphere of agricultural economics is self-evident. Our scientific research institutions must intensify this work.

No matter what sphere we work in, we must start and end every problem with economics, must examine it from the point of view of fulfilment of state assignments, of increasing collective- and state-farm labour productivity and of increasing output.

AGROBIOLOGICAL SCIENCE MUST BE DEVELOPED

Only in the Land of Socialism has the science of agriculture begun to serve the whole of the people, has it become a collective science, a science of the masses. In addition to the men and women of science many thousands of labouring folk in the collective-farm villages are drawn in our country into the research and experimental work carried on in agriculture.

The science of biology, which reveals the laws of development of living bodies, must form the basis of the science of agronomy. The more profoundly and veraciously biological laws are revealed the more certain is the knowledge imparted by the science of agronomy and the more effective it becomes.

Biological phenomena can be correctly understood only from the standpoint of dialectical materialism. And, contrariwise, it is particularly harmful to examine the processes going on in living organisms not in their development but in their immobile form, detached from their interconnections with the environment.

In capitalist countries, I believe, biological science holds a special place among the other natural sciences. In all departments of natural science except biology major discoveries become the basis for a rapid advance of science in general. But since the bourgeois socio-economic sciences do not recognize the true laws of development of society, bourgeois biological science cannot, in my opinion, base itself on the true laws of development of living bodies, cannot make these laws the foundation of its knowledge. Thus, Darwinism is not officially recognized by bourgeois biologists as the theory of development of the organic world. Nor do they recognize the theoretical generalizations and discoveries of the most eminent explorers of living nature, such as Michurin, Burbank, Daniel and others.

It is therefore impossible under capitalism to acquire genuine biological knowledge from books or college courses but only, as Burbank said, directly from nature's university, i.e., from practice. Nor is it fortuitous that all outstanding biologists in capitalist countries acquired an understanding of the laws of the development of nature only after they reached advanced old age, after traversing a lengthy road, as a result of their own painstaking study of living nature. Yet bourgeois science has considered and still considers such colossi of genuine biological science to be merely practical farmers and in no sense scientists.

Bourgeois science recognizes nothing that is not in the interests of the ruling class of the capitalist countries. Therefore the science of biology in these countries examines living bodies statically, i.e., in immobile, petrified form. It is but natural that such biology could not and cannot now become the basis of agronomic science. No wonder such bourgeois theories as, for instance, Mendelism-Morganism in the sphere of genetics, came into flagrant contradiction with agronomic practice. Formal genetics based its theory solely on the morphology of living bodies, i.e., only on their form, completely ignoring the content of living bodies.

In our country the collective-farm system has opened up unexampled prospects for the development of genuine agronomic theory, for a profound comprehension of the nature of plant and animal organisms. Thanks to these possibilities a new science has recently developed—agrobiology, the science of agronomy, which takes the correctly expounded laws of biology as its theoretical foundation.

We are firmly convinced that if the living body is viewed as a dialectical *unity*, the *body* must be taken as the *form* and the *conditions of life* of that body as the *content* of this unity.

By the conditions of life of living bodies we do not mean external environment in general but only those of its material factors which in unity with the body create the vital process of assimilation and dissimilation.

A living body is a body in unity with the conditions of its life. As soon as a living body is deprived of its conditions of life it becomes nonliving, dead. As soon as an organism is deprived of the conditions of life it needs it ceases to be an organism and becomes a corpse. In science a living body must always be considered only in unity with its conditions of life.

The forms of living bodies have been and are still created only by their conditions of life. It is therefore intelligible that changes in plant and animal forms can be controlled only by a skilful control of the conditions of life of the plants and animals concerned.

In contrast with this proposition the Morganist neo-Darwinists (followers of Weismann in science) both abroad and in our own country assert that changes in plant and animal forms are purely accidental, qualitatively independent of the conditions of life and are hence independent of the will of men; or, as they say, mutations are not directed.

This science would have it that marsh plants, for instance, were not created, or formed, by the conditions of the marsh. They were created, formed in consequence of certain "unknowable" causes and the conditions of the marsh only selected the forms able to survive in this medium.

If this "theory" were taken as the guiding principle practical agriculturists would have to sit idly by and wait until some needed change in the form of a plant or animal made its appearance.

As against this erroneous conception Ivan Vladimirovich Michurin enunciated the motto: "We cannot wait for favours from Nature; we must wrest them from her."

Soviet agricultural scientists are in possession of all conditions, both moral and material, necessary for their work, for the successful fulfilment of the tasks assigned by the decision of the February Plenum of the Central Committee of the Communist Party of the Soviet Union (Bolsheviks). This decision creates such conditions for our scientific workers in agriculture as will still further enhance the success of their activities. It will undoubtedly prove a powerful stimulus for elevating the plane of all scientific research work.

We must make a much better study of the works of Darwin, Michurin, Burbank, Williams and the other classics of agrobiology, and develop their principles.

We must bear in mind that what we need is uninterrupted development of agrobiological theory. In all countries it lags behind such sciences as physics, chemistry and other departments of natural science. In the Soviet Union biological and agrobiological science has been provided with every facility for genuine development.

The vast body of our scientific workers and agronomists who have been armed with the advanced theory of Marxism-Leninism will, under the leadership of the great Stalin, undoubtedly discharge the duties imposed upon them by the February Plenum of the Central Committee of the Communist Party of the Soviet Union (Bolsheviks).

WHY BOURGEOIS SCIENCE IS UP IN ARMS AGAINST THE WORKS OF SOVIET SCIENTISTS

A CORRESPONDENT of the *Literaturnaya Gazeta* submitted to Academician T. D. Lysenko a number of questions relating to one of the scientist's recent works which dealt with the nonexistence of intraspecific competition in the plant and animal world. Below we reproduce this interview.

Question: It is well known that in the Ukraine, on your recommendation, kok-saghyz is sown in hills and that the correctness of this method has been splendidly confirmed. No root crop is less than 30-40 c. per ha. and in many localities the yield is as high as 60-80 c. The collective farmers themselves speak of the marvellous revolution wrought in the agrotechnique of kok-saghyz of which they formely harvested, as a rule, 3-4 c. per ha., and only in rare instances 10-20 c.

Would you be good enough to explain the theoretical basis of the new method of sowing rubber-bearing plants and why, according to the press, bourgeois scientists, particularly American, are up in arms against your biological works.

Answer: As far as the hill sowing of kok-saghyz is concerned, I only suggested the idea. For the elaboration and practical application of this method credit must be given to Stalin Prize winner I. D. Kolesnik and to the collective-farm members of Kiev Region.

Hill sowing is based theoretically on the nonexistence of intraspecific competition. The correctness of this proposition was proved for the first time in our country. Bourgeois biological science not only did not know this but is incapable even of accepting our estabished thesis that there is no intraspecific struggle in nature.

At first glance it may seem that bourgeois science, in its attempt to prove the existence of intraspecific competition, proceeds from natural selection, a correct thesis of Darwinism. After all anybody can see that an eternal struggle between organisms is going on in nature. And organisms whose requirements coincide (for instance, carnivorous animals of various species), carry on this struggle directly or indirectly, compete among themselves for the capture of food, while organisms whose requirements do not coincide (for instance, carnivorous animals and plants) wage no struggle among themselves.

All this can easily be observed in nature.

But the bourgeois scientists ignore the fact that in both cases, whether a struggle is going on or not, the individuals concerned are not of the same species but of different species of animals and plants. They stress the point that the more alike the requirements of the organisms (we add what they omit—not of one but of various species) the fiercer the struggle inevitably is. From this they draw a conclusion which not only is not confirmed by the phenomena observed in nature but which directly contradicts the laws of development of plants and animals. They say: Inasmuch as the organisms belonging to the same species show the closest similarity in their requirements it is between them that the fiercest struggle exists. Here nothing is said about the fact that no one has ever yet succeeded or ever will succeed in seeing or showing to others an example of this supposedly greatest of competitions in nature, namely, competition among individuals of the same species. For instance, who can demonstrate that rabbits interfere with each other more than they are interfered with by wolves or that wolves harm each other more than they are harmed by rabbits which, having fine ears and long legs, run away from them and leave them hungry.

Anyone will believe that weeds, being different in species from wheat, for instance, interfere with it and kill it. But nobody is going to believe that a thin and hence weed-choked stand of wheat fares better in the field than thick-grown and hence pure wheat. I affirm once more that no one has ever yet produced, or ever will produce, any scientific proof that competition within a species exists in nature.

How explain the predilection of bourgeois biological science for the "theory" of intraspecific competition? It has to justify the state of affairs in capitalist society, in which the vast majority of people, particularly during an overproduction of material values, are in a sorry plight.

All humanity belongs to one biological species. That is the reason why bourgeois scientists found it necessary to invent the intraspecific struggle. They say that a fierce struggle for food, of which there is an insufficiency, goes on in nature, within the species, among its individual members—a struggle for the conditions of life. The stronger, fitter individuals win. The same thing, they aver, goes on among human beings: the capitalists possess millions but the workers are poor because the capitalists, you see, are brainier, are more capable by nature and heredity.

We Soviet people know full well that the oppression of the working people, the domination of the capitalist class and imperialist war have nothing in common with the laws of biology. These phenomena are all governed by the laws of decaying bourgeois, capitalist society, which has outlived its day.

Nor is there any intraspecific competition in nature itself. What does exist is competition among species: the rabbit is eaten by the wolf but does not eat other rabbits; it eats grass. Likewise wheat does not crowd wheat out of existence. Now couch grass, goose foot and sow-thistle are members of other species and when they appear in wheat or kok-saghyz fields they capture food from them, struggle with them.

In order that the weak kok-saghyz plants may be able not only to hold their own in this severe interspecific struggle but to produce greater crops we have come to their assistance. The collective-farm members have begun to sow kok-saghyz in hills: 100-200 kok-saghyz seeds are placed in one hole and around it 1/4 sq. m. of free space is left. The weed attacks the hill but on encountering a mighty wall of resistance on the part of the numerous kok-saghyz plants it cannot make its way into the hill. And the kok-saghyz, having rid itself of its worst enemy, keeps on growing in bunches (association) by using up the nutriment and moisture of the entire free space given to it.

But even sixty to eighty centners of kok-saghyz root per hectare is not the limit. More can and must be grown!

Question: What you have told me, Comrade Lysenko, is really new, but the main thing is that your scientific papers do not merely represent a "point of view." They are corroborated by a wealth of practical experience. I would very much like you to supplement your remarks by an explanation of how kok-saghyz behaves within the hill itself.

Answer: How kok-saghyz fares within the hill? Not only not badly but very well, indeed. In nature the life of every individual is wholly subordinated to the interests of its species. This is something everyone should know. In nature every plant and animal pursues but one aim, which is created by natural selection—to produce beings similar to itself. A wolf has feet, hair, ears—all of which serve the one purpose of propagation, of increasing the number of wolves. The ephemera butterfly lives one day only for the purpose of leaving offspring. To multiply at the expense of other species and to the detriment of other species is a biological law.

Now let me come back to your question: how does the kok-saghyz live in the hill? The seeds planted in the hole are of one species; they are all governed by one law—to propagate their species, in this particular case kok-saghyz, to the utmost. Since there is enough food and moisture in the free space provided for them and the collective-farm members can weed the wide spaces between the rows with horse- or tractor-drawn cultivators, the kok-saghyz seeds grow excellently.

If the kok-saghyz is not hill-sown but planted evenly over the whole field the weak single seeds are choked off by the weeds.

At present I am giving much thought to the hill planting of forests, especially for our treeless steppe districts. The hill planting of forest trees is a matter that in my opinion opens up wide vistas. Lay out 100-200 one- or two-metre patches to a hectare and plant 50-100 saplings to a patch and in a comparatively short time you will have a forest without having to do any cultivation. And this is the main thing, for it is still difficult to cultivate saplings when the collective-farm members are busy cultivating sunflower, maize and the like. This is the reason why it has been so difficult hitherto to make headway in forest planting. However, young forest trees which grow in groups, in hills, themselves prevent the encroachment of grass, their most dangerous enemy. After the passage of only 3-5 years the newly-planted trees begin to be of service: they retain the snow and protect the fields from strong winds. Such plantations must be tried out in the steppes.

Question: The other day Ivan Danilovich Kolesnik, a Stalin Prize winner, read a paper at a meeting held in the editorial offices of the newspaper *Sotsialisticheskoye zemledeliye*, scientific workers from various agricultural institutes and offices, among them Professor A. A. Sokolov, attending. The latter took the floor during the ensuing discussion declaring that hill sowing was nothing new; that ten years ago he had already heard that this method was being applied in America.

I. D. Kolesnik very properly asked Professor Sokolov where he had been these ten years; why if he knew about the hill-sowing method he had kept quiet all this time. It might have increased the yields of a number of crops in our country. Professor Sokolov made no reply. But what we are interested to know in this case is something else: is it true that the hill method was discovered as long as ten years ago in America?

Answer: Bourgeois biological science has been and still is by its very nature incapable of making any discovery based on the nonexistence of intraspecific competition, a principle which it does not recognize. Hence, American scientists could not have occupied themselves with hill sowing. What suits the book of these servants of capitalism is not a real struggle with the elemental forces of nature but an invented struggle between white-spiked and black-spiked wheat, both of which belong to the same species. They endeavour to justify both the class struggle and the oppression of the Negroes by the whites in the United States by such figments of their imagination as intraspecific competition and the "eternal laws of nature." Therefore, how, then, can they be expected to recognize the nonexistence of any struggle within any species?

Last Question: Why, then, did Professor Sokolov consider it necessary to make a statement in which he tried to detract from the importance of your scientific work by insinuating that the Americans were ahead of you?

Answer (after a pause): I was not at that conference and do not know what was said there. But I do know that even in our midst some biologists, for instance, Professor P. M. Zhukovsky, still recognize intraspecific competition. I call this a survival of bourgeois mentality. Intraspecific competition does not exist in nature and there is no reason for fabricating it in science. A keen struggle of ideas is in progress and the new always meets with resistance from the old. But here in the Soviet Union the new always wins.

THE SITUATION IN BIOLOGICAL SCIENCE[1]

1. BIOLOGY, THE BASIS OF AGRONOMY

AGRONOMY deals with living bodies—plants, animals, microorganisms. A theoretical grounding in agronomy must, therefore, include knowledge of biological laws. And the more profoundly the science of biology reveals the laws of the life and development of living bodies, the more effective is the science of agronomy.

In essence, the science of agronomy is inseparable from biology. When we speak of the theory of agronomy we mean the discovered and comprehended laws of the life and development of plants, animals, and microorganisms.

The methodological level of biological knowledge, the state of the biological science treating of the laws of the life and development of plant and animal forms, i.e., primarily of the science known for half a century now as genetics, is of essential importance for our agricultural science.

2. THE HISTORY OF BIOLOGY: A HISTORY OF IDEOLOGICAL BATTLE

The appearance of Darwin's teaching, expounded in his book, *The Origin of Species*, marked the beginning of scientific biology.

The leading idea of Darwin's theory is the teaching on natural and artificial selection. Selection of variations favourable to the organism has produced, and continues to produce, the fitness which we observe in living nature; in the structure of organisms and their adaptation to their conditions of life. Darwin's theory of selection provided a rational explanation of the fitness observable in living nature. His idea of selection is scientific and true. In substance, his teaching on selection is a summation of the age-old practical experience of plant and animal breeders who, long before Darwin, produced varieties of plants and breeds of animals by the empirical method.

Darwin investigated the numerous facts obtained by naturalists in

---

[1] A paper read at the Session of the Lenin Academy of Agricultural Sciences on July 31, 1948.—*Ed.*

living nature and analyzed them through the prism of practical experience. Agricultural practice served Darwin as the material basis for the elaboration of his theory of evolution, which explained the natural causes of the purposiveness we see in the structure of the organic world. That was a great advance in the knowledge of living nature.

In Engels' opinion, three great discoveries enabled man's knowledge of the interconnection of natural processes to advance by leaps and bounds: first, the discovery of the cell; second, the discovery of the transformation of energy; third, "the proof which Darwin first developed in connected form that the stock of organic products of nature environing us today, including mankind, is the result of a long process of evolution from a few originally unicellular germs, and that these again have arisen from protoplasm or albumen, which came into existence by chemical means."[1]

The classics of Marxism, while fully appreciating the significance of the Darwinian theory, pointed out the errors of which Darwin was guilty. Darwin's theory, though unquestionably materialist in its main features, is not free from some serious errors. A major fault, for example, is the fact that, along with the materialist principle, Darwin introduced into his theory of evolution reactionary Malthusian ideas. In our days this major fault is being aggravated by reactionary biologists.

Darwin himself recorded the fact that he accepted the Malthusian idea. In his autobiography we read:

"In October 1838, that is, fifteen months after I had begun my systematic enquiry, I happened to read for amusement Malthus' *On Population*, and, being well prepared to appreciate the struggle for existence which everywhere goes on from long-continued observation of the habits of animals and plants, it at once struck me that under these circumstances favourable variations would tend to be preserved, and unfavourable ones to be destroyed.... Here then *I had* at last *got a theory by which to work*."[2] (My emphasis—*T.L.*)

Many are still not clear about Darwin's error in transferring into his teaching Malthus' preposterous reactionary ideas on population.

The true scientist cannot and must not overlook the erroneous aspects of Darwin's teaching.

Biologists should always ponder these words of Engels: "The entire Darwinian teaching on the struggle for existence merely transfers from society to the realm of living nature Hobbes' teaching on *bellum omnium contra omnes* and the bourgeois economic teaching on competition, along with Malthus' population theory. After this trick (the absolute justification for which, as indicated in point 1, I deny, particularly in regard to Malthus' theory) has been performed, the same theories are transferred back from organic nature to history and the claim is then made that it

---

[1] F. Engels, ***Ludwig Feuerbach und der Ausgang der klassischen deutschen Philosophie,*** **Moskau 1946, S. 44.**

[2] ***The Life and Letters of Charles Darwin,*** **London 1887, Vol. 1, p. 83.**

has been proved that they have the force of eternal laws of human society. The childishness of this procedure is *obvious*, and it is not worth while wasting words on it. But if I were to dwell on this at greater length, I should have started out by showing that they are poor *economists* first, and only then that they are poor naturalists and philosophers."[1]

For the propaganda of his reactionary ideas Malthus invented an allegedly natural law. "The cause to which I allude," he wrote, "is the constant tendency in all animated life to increase beyond the nourishment prepared for it."[2]

It must be clear to any progressively thinking Darwinist that, even though Darwin accepted Malthus' reactionary theory, it basically contradicts the materialist foundation of his own teaching. Darwin himself, as may be easily noted, being as he was a great naturalist, the founder of scientific biology, whose activity marks an epoch in science, could not be satisfied with the Malthusian theory, since it is, in fact and fundamentally, at variance with the phenomena of living nature.

Under the weight of the vast amount of biological facts accumulated by him, Darwin felt constrained in a number of cases radically to alter the concept of the "struggle for existence," to stretch it to the point of declaring that it was just a figure of speech.

Darwin himself, in his day, was unable to fight free of the theoretical errors of which he was guilty. It was the classics of Marxism that revealed those errors and pointed them out. Today there is absolutely no justification for accepting the erroneous aspects of the Darwinian theory, those based on Malthus' theory of overpopulation with the inference of a struggle presumably going on within species. And it is all the more inadmissible to represent these erroneous aspects as the cornerstone of Darwinism (as I. I. Schmalhausen, B. M. Zavadovsky, and P. M. Zhukovsky do). Such an approach to Darwin's theory prejudices the creative development of its scientific core.

Even when Darwin's teaching first made its appearance, it became clear at once that its scientific, materialist core, the theory of the evolution of living nature, was antagonistic to the idealism that reigned in biology.

Progressively thinking biologists, both in our country and abroad, saw in Darwinism the only right road to the further development of scientific biology. They took it upon themselves to defend Darwinism against the attacks of the reactionaries, with the Church at their head, and of obscurantists in science, such as Bateson.

Such eminent biologists as V. O. Kovalevsky, I. I. Mechnikov, I. M. Sechenov, and particularly K. A. Timiryazev, defended and developed Darwinism with all the passion of true scientists.

---

[1] F. Engels, letter to P. L. Lavrov, 12-17 November, 1875.

[2] T. R. Malthus, *An Essay on the Principle of Population*, London, New York and Melbourne. 1890, Book 1, p. 2.

K. A. Timiryazev, that great investigator, saw distinctly that only on the basis of Darwinism could the science of the life of plants and animals develop successfully, that only by further developing Darwinism and raising it to new heights would biological science become capable of helping the tiller of the soil to obtain two ears of corn where there was formerly only one.

Darwinism as presented by Darwin contradicted idealistic philosophy, and this contradiction grew deeper with the development of the materialist teaching. Reactionary biologists have therefore done everything in their power to empty Darwinism of its materialist elements. The individual voices of progressive biologists like K. A. Timiryazev were drowned by the chorus of the anti-Darwinists, the reactionary biologists the world over.

In the post-Darwinian period the overwhelming majority of biologists—far from further developing Darwin's teaching—did all they could to debase Darwinism, to smother its scientific foundation. The most glaring manifestation of such debasement of Darwinism is to be found in the teachings of Weismann, Mendel, and Morgan, the founders of modern reactionary genetics.

3. TWO WORLDS—TWO IDEOLOGIES IN BIOLOGY

Weismannism, which made its appearance at the turn of the century, and Mendelism-Morganism, which followed, were both primarily directed against the materialist foundations of Darwin's theory of evolution.

Weismann named his conception Neo-Darwinism, but, in fact, it was a complete denial of the materialist aspects of Darwinism. It insinuated idealism and metaphysics into biology.

The materialist theory of the evolution of living nature necessarily presupposes the recognition of hereditary transmission of individual characteristics acquired by the organism under definite conditions of its life; it is unthinkable without recognition of the inheritance of acquired characters. Weismann, however, set out to refute this materialist proposition. In his *Lectures on Evolutionary Theory,* he asserts that "not only is there no proof of such a form of heredity, but it is inconceivable theoretically."[1] Referring to earlier statements of his in a similar vein, he declares that "thus war was declared against Lamarck's principle of the direct transforming effect of use and disuse, and, indeed, that marked the beginning of the struggle which is going on to this day, the struggle between the Neo-Lamarckians and the Neo-Darwinists, as the contending parties are called."[2]

---

[1] A. Weismann, *Vorträge über Deszendenztheorie*, Bd. I, Jena 1904, S. 198.
[2] *Ibid.*

Weismann, as we see, speaks of having declared war against Lamarck's principle; but it is easy enough to see that he declared war against that without which there is no materialist theory of evolution, that under the guise of "Neo-Darwinism" he declared war against the materialist foundations of Darwinism.

Weismann denied the inheritability of acquired characters and conceived the idea of a special hereditary substance "to be sought for in the nucleus."[1] "*The sought-for bearer of heredity*," he stated, "*is contained in the chromosome material.*"[2] The chromosomes, he said, contain units, each of which "determines a definite part of the organism in its appearance and final form."[3]

Weismann asserts that there are "two great categories of living material: the *hereditary substance*, or *idioplasm*, and the '*nutrient substance*,' or *trophoplasm*. . . ."[4] He declares that the bearers of the hereditary substance, "*the chromosomes, represent a separate world, as it were*,"[5] a world independent of the body of the organism and its conditions of life.

Having thus disposed of the living body as being merely a nutritive soil for the hereditary substance, Weismann proclaims that the hereditary substance is immortal and is never generated de novo.

Thus, he asserts, "the germ-plasm of a species is never generated de novo, it only grows and multiplies continually, handed down from generation to generation. . . . Looked at only from the point of view of propagation, the germ cells are the most important element in the individual specimen, for they alone preserve the species, whereas the body is reduced practically to the status of a mere breeding ground for the germ cells, the place in which they form and, under favourable conditions, feed, multiply, and ripen."[6] The living body and its cells, according to Weismann, are but the *container and nutritive medium* of the hereditary substance; they themselves can never produce the latter, they "can never bring forth germ cells."[7]

Weismann thus endows the mythical hereditary substance with the property of continued existence; it is a substance which does not itself develop and at the same time determines the development of the mortal body.

Further: ". . . the hereditary substance of the germ cell, *prior* to the reduction division, potentially contains all the elements of the body."[8] And although Weismann does state that "in the germ-plasm there is no determinant of a 'hooked nose' just as there is no determinant of the wing of a butterfly with all its parts and particles," he goes on to emphasize that, never-

[1] *Ibid.*, S. 277.

[2] *Ibid.*

[3] *Ibid.*, S. 305.

[4] *Ibid.*, S. 279.

[5] *Ibid.*, S. 239.

[6] *Ibid.*, S. 339-40.

[7] *Ibid.*, S. 339.

[8] *Ibid.*, S. 282.

theless, the germ-plasm "... contains a certain number of determinants which successively determine the development of an entire group of cells in all its stages, leading to the formation of the nose in such a mode as to result in a hooked nose, exactly in the same way as the wing of a butterfly, with all its little veins, cells, nerves, trachea, glandular cells, form of scales, and pigment deposits, comes into being by the successive action of multitudinous determinants upon the course of the proliferation of the cells."[1]

Hence, according to Weismann, there can be no new formations of the hereditary substance, it does *not* develop with the development of the individual, and is *not* subject to any dependent changes.

An immortal hereditary substance, independent of the qualitative features attending the development of the living body, directing the mortal body, but not produced by the latter—that is Weismann's frankly idealistic, essentially mystical conception, which he disguised as "Neo-Darwinism."

Weismann's conception has been fully accepted and, we might say, carried further by Mendelism-Morganism.

Morgan, Johannsen, and other pillars of Mendelism-Morganism, declared from the outset that they intended to investigate the phenomena of heredity independently of the Darwinian theory of evolution. Johannsen, for example, wrote in his principal work: "... one of the major aims of our research was to put an end to the harmful dependence of the heredity theories on speculations in the field of evolution."[2] The purpose of the Morganists in making such declarations was to wind up their investigations by assertions which in the final analysis denied evolution in living nature, or recognized it as a process of purely quantitative changes.

I have already said that the conflict between the materialist and the idealist outlook in biological science has been going on throughout its history.

In the present epoch of struggle between two worlds the two opposing and antagonistic trends, penetrating the foundations of nearly all branches of biology, are particularly sharply defined.

Socialist agriculture, the collective- and state-farm system, has given rise to a Soviet biological science, founded by Michurin—a science new in principle, developing in close union with agronomic practice, as agronomic biology.

The foundations of Soviet agrobiological science were laid by Michurin and Williams, who generalized and developed the best of what science and practice had accumulated in the past. Their work has enriched our knowledge of the nature of plants and soils, our knowledge of agriculture, with much that is new in principle.

Close contact between science and the practice of collective and state farms creates inexhaustible opportunities for the development of theoretical

---

[1] A. Weisman, Vortrage über Deszendenztheorie, Bd. 1, S. 314.

[2] W. Johannsen, *Elemente der exakten Erblichkeitslehre*, Jena 1926, S. 248.

knowledge, enabling us to learn ever more and more about the nature of living bodies and the soil.

It is no exaggeration to state that Morgan's feeble metaphysical "science" concerning the nature of living bodies can stand no comparison with our effective Michurinist agrobiological science.

The new vigorous trend in biology, or more truly the new Soviet biology, agrobiology, has met with bitter opposition on the part of representatives of reactionary biology abroad, as well as of some scientists in our country.

The representatives of reactionary biological science—Neo-Darwinists, Weismannists, or, which is the same, Mendelist-Morganists—uphold the so-called chromosome theory of heredity.

Following Weismann, the Mendelist-Morganists contend that the chromosomes contain a special "hereditary substance" which resides in the body of the organism as though in a case and is transmitted to succeeding generations irrespective of the qualitative features of the body and its conditions of life. The conclusion drawn from this conception is that new tendencies and characteristics acquired by the organism under the influence of the conditions of its life and development are not transmissible and can have no evolutionary significance.

According to this theory, characters acquired by plant and animal organisms cannot be handed down, *cannot be inherited.*

The Mendelist-Morganist theory does not include in the scientific concept "living body" the conditions of the body's life. To the Morganists, environment is only the background—indispensable, they admit—for the manifestation and operation of the various characteristics of the living body, in accordance with its heredity. They therefore hold that qualitative variations in the heredity (nature) of living bodies are entirely independent of the environment, of the conditions of life.

The representatives of Neo-Darwinism, the Mendelist-Morganists, hold that the efforts of investigators to regulate the heredity of organisms by suitably changing the conditions of life of these organisms are utterly unscientific. They therefore call the Michurin trend in agrobiology Neo-Lamarckian, which, in their opinion, is absolutely fallacious and unscientific.

Actually, it is the other way round.

First, the well-known Lamarckian propositions, which recognize the active role of external conditions in the formation of the living body and the inheritance of acquired characters, unlike the metaphysics of Neo-Darwinism (or Weismannism), are by no means fallacious. On the contrary, they are quite true and scientific.

Secondly, the Michurin trend cannot be called either Neo-Lamarckian or Neo-Darwinian. It is creative Soviet Darwinism, rejecting the errors of both and free from the defects of the Darwinian theory insofar as it included Malthus' erroneous ideas.

Furthermore, it cannot be denied that in the controversy that flared up between the Weismannists and Lamarckians in the beginning of the twentieth century, the Lamarckians were closer to the truth; for they defended the interests of science, whereas the Weismannists were at loggerheads with science and prone to indulge in mysticism.

The true ideological content of Morgan's genetics has been well revealed (to the discomfiture of our Morganists) by the physicist Erwin Schrödinger. In his book, *What Is Life? The Physical Aspect of the Living Cell,* he draws some philosophical conclusions from Weismann's chromosome theory, of which he speaks very approvingly. Here is his main conclusion: "...the personal self equals the omnipresent, all-comprehending, eternal self." Schrödinger regards this conclusion as "the closest a biologist can get to proving God and immortality at one stroke."[1]

We, the representatives of the Soviet Michurin trend, contend that inheritance of characters acquired by plants and animals in the process of their development is possible and necessary. Ivan Vladimirovich Michurin mastered these possibilities in his experiments and practical activities. The most important point is that Michurin's teaching, expounded in his works, shows every biologist the way to regulating the nature of plant and animal organisms, the way of altering it in a direction required for practical purposes by regulating the conditions of life, i.e., by physiological means.

A sharp controversy, which has divided biologists into two irreconcilable camps, has thus flared up over the old question: *can characters and properties acquired by plant and animal organisms in the course of their life be inherited?* In other words, whether qualitative variations of the nature of plant and animal organisms depend on the nature of the conditions of life which act upon the living body, upon the organism.

The Michurin teaching, which is in essence materialist and dialectical, proves by facts that such dependence does exist.

The Mendelist-Morganist teaching, which in essence is metaphysical and idealist, denies the existence of such dependence, though it can cite no evidence to prove its point.

4. THE SCHOLASTICISM OF MENDELISM-MORGANISM

The chromosome theory is based on Weismann's absurd proposition regarding the continuity of the germ-plasm and its independence of the soma, a proposition which K. A. Timiryazev already condemned. In line with Weismann, the Morganist-Mendelists take it for granted that parents are genetically not the progenitors of their offspring. Parents and children, according to their teaching, are brothers or sisters.

Furthermore, neither parents nor children are really themselves. They are only by-products of the inexhaustible and immortal germ-plasm. Variations

---

[1] E. Schrödinger, *What Is Life? The Physical Aspect of the Living Cell,* Cambridge University Press, 1945, p. 88

in the latter are absolutely independent of its by-product, i.e., of the body of the organism.

Let us turn to the Encyclopedia where we naturally may expect to find the quintessence of the question under discussion.

In the 1945 edition of *The Encyclopedia Americana*, T. H. Morgan, founder of the chromosome theory, writes in the article entitled "Heredity": "The germ cells become later the essential parts of the ovary and testis respectively. In origin, therefore, *they are independent of the rest of the body, and have never been a constituent part of it.... Evolution is germinal in origin and not somatic as had been earlier taught.* (My emphasis—*T. L.*) This idea of the origin of new characters is held almost universally to-day by biologists."

The same idea differently worded is propounded in the same *Encyclopedia Americana* by Professor Castle in the article on "Genetics." After stating that usually the organism develops from a fertilized egg, Castle goes on to set forth the "scientific" foundations of genetics as follows:

"In reality the parent does not produce the child nor even the reproductive cell which functions in its origin. The parent is himself merely a by-product of the fertilized egg (or zygote) out of which he arose. The direct product of the zygote is other reproductive cells, similar to those from which it arose.... Hence heredity (that is, the resemblance between parent and child) depends upon the close connection between the reproductive cells which formed the parent and those which formed the child, one being the immediate and direct product of the other. This principle of the 'continuity of the germinal substance' (reproductive cell material) is one of the foundation principles of genetics. It shows why body changes produced in a parent by environmental influences are not inherited by the offspring. It is because offspring are not the product of the parent's body but only of the germinal substance which that body harbours.... To August Weismann belongs the credit for first making this clear. He may thus be regarded as one of the founders of genetics."

To us it is perfectly clear that the foundation principles of Mendelism-Morganism are false. They do not reflect the reality of living nature and are an example of metaphysics and idealism.

Because this is so obvious, the Mendelist-Morganists of the Soviet Union, though actually fully sharing the principles of Mendelism-Morganism, often conceal them shamefacedly, veil them, conceal their metaphysics and idealism in a verbal shell. They do this because of their fear of being ridiculed by Soviet readers and audiences who are firm in the knowledge that the germs of organisms, or the sex cells, are a result of the vital activity of the parent organisms.

It is only when no mention is made of the fundamentals of Mendelism-Morganism that persons having no detailed knowledge of the life and development of plants and animals can be led to think of the chromosome theory of heredity as a neat system, as in some degree corresponding to the truth,

But once we accept the absolutely true and generally known proposition that the reproductive cells, or the germs, of new organisms are produced by the organism, by its body, and not by the very same reproductive cell from which the given, already mature, organism arose, nothing is left of the "neat" chromosome theory of heredity.

Naturally, what has been said above does not imply that we deny the biological role and significance of chromosomes in the development of the cells and of the organism. But it is not at all the role which the Morganists attribute to the chromosomes.

Plenty of examples can be cited to show that our home-grown Mendelist-Morganists accept in its entirety the chromosome theory of heredity, its Weismannist foundations and idealistic conclusions.

Academician N. K. Koltsov, for example, asserts: "Chemically, the genoneme with its genes remains unchanged in the course of the entire ovogenesis and is not subject to metabolism—oxidizing and reduction processes."[1] This assertion, which no literate biologist can accept, denies the existence of metabolism in a section of the living and developing cells. It must be obvious to everyone that N. K. Koltsov's conclusion is fully in line with the Weismannist and Morganist idealist metaphysics.

N. K. Koltsov's false assertion dates back to 1938. It has long since been exposed by the Michurinists, and it would, perhaps, not have been worth while going back to the past if not for the fact that the Morganists persist in holding on to their antiscientific positions to this day.

We can find further proof of this by turning once more to Schrödinger's book mentioned above. Schrödinger says in substance the same things as Koltsov. Since he shares the idealistic conception of the Morganists, he also asserts that there exists an "*hereditary substance, largely withdrawn from the disorder of heat motion. . . .*"[2] (My emphasis—*T.L.*)

The Russian translator of Schrödinger's book, A. A. Malinovsky (a scientific worker in N. P. Dubinin's laboratory), in his "Postscript" to the book, subscribes—and with good reason—to Haldane's opinion, linking Schrödinger's idea with N. K. Koltsov's views.

In that "Postscript," written in 1947, Malinovsky says: "The view accepted by Schrödinger according to which the chromosome is a gigantic molecule (Schrödinger's "aperiodic crystal"), was first put forward by the Soviet biologist, Professor N. K. Koltsov, and not by Delbrück, with whose name Schrödinger associates this conception."

There is no point, in this case, in going into the question of who is entitled to claim credit for the authorship of this scholastic view. A more important point is the high appreciation of Schrödinger's book by one of our domestic Morganists, A. A. Malinovsky.

---

[1] Н. К. Кольцов, «Структура хромосом и обмен веществ в них», *Биологический журнал*, том VII. вып. 1. 1938 г., стр. 42.

[2] E. Schrödinger, *What Is Life? The Physical Aspect of the Living Cell*, p. 85.

Here are a few samples of the praise he showers on this book:

"In a fascinating form, accessible both to the physicist and the biologist, Schrödinger reveals to the reader a new trend rapidly developing in science, a trend largely combining the methods of physics and of biology."

"Strictly speaking, Schrödinger's book represents the first coherent results of this trend.... Schrödinger makes a big contribution of his own to this new trend in the science of life, and this quite justifies the enthusiastic opinions voiced about his book in the foreign scientific press."

Since I am no physicist, I shall say nothing concerning the methods of physics which Schrödinger combines with biology. As for the biology in Schrödinger's book, it is Morganist pure and simple, and this, in fact, is what makes Malinovsky go into raptures over it.

The enthusiastic praise of Schrödinger's book in Malinovsky's "Postscript" speaks eloquently enough of our Morganists' idealistic views and positions.

M. M. Zavadovsky, Professor of Biology in the University of Moscow, writes in an article entitled "The Creative Road of Thomas Hunt Morgan": "Weismann's ideas found a wide response among biologists, and many of them have taken the road suggested by that highly gifted investigator.... Thomas Hunt Morgan was one of those who highly appreciated the main content of Weismann's ideas."[1]

Now what "main content" is meant here?

What is meant is an idea of prime importance to Weismann and all Mendelist-Morganists, including Professor M. M. Zavadovsky. The latter formulates that idea as follows: "What came first, the chicken or the egg? And," writes Professor Zavadovsky, "to this clearly put question Weismann gave an explicit, categorical reply: the egg."[2]

It is obvious to anyone that both the question and the answer which Professor Zavadovsky, following Weismann, gives are nothing but a revival, and a belated one at that, of old scholasticism.

In 1947 Professor M. M. Zavadovsky repeats and defends the ideas he set forth in 1931 in his work *Dynamics of Development of Organisms*. There M. M. Zavadovsky considered it necessary to "firmly join with *Nussbaum* who maintains that sexual products do not develop from the maternal organism, but from the same source as the latter,"[3] that "the seminal corpuscles and eggs do not originate in the parent organism, but have a common origin with the latter."[4] And in his "General Conclusion" Professor Zavadovsky wrote: "Analysis leads us to the conclusion that the cells of the germ track cannot be regarded as products of somatic tissue. The germ cells and the cells of the soma should be regarded not as daughter and parent

---

[1] *Бюллетень Московского общества испытателей природы*, том LII, вып. 3, 1947 г., стр. 86.

[2] *Ibid.*

[3] М. Завадовский, *Динамика развития организма*, 1931 г., стр. 321.

[4] *Ibid.*, p. 313.

generations, but as twin sisters, of which one (the soma) is the feeder, protector, and guardian of the other."[1]

The geneticist N. P. Dubinin, Professor of Biology, wrote in his article, "Genetics and Neo-Lamarckism": "Genetics quite rightly divides the organism into two distinct sections—the hereditary plasm and the soma. More, this division is one of its fundamental principles, one of its major generalizations."[2]

We need not continue the list of authors who, like M. M. Zavadovsky and N. P. Dubinin, frankly expound the ABC of the Morganist system of views. In college textbooks on genetics this ABC is called the "Mendelian laws" (dominance, segregation, purity of gametes, etc.). An example of how uncritically our Mendelist-Morganists accept idealistic genetics is the fact that the standard textbook on genetics in many of our colleges has until quite recently been a translated American, strictly Morganistic, textbook—by Sinnott and Dunn.

Fully in line with the main theses of this textbook, Professor N. P. Dubinin wrote in that same article of his ("Genetics and Neo-Lamarckism"): "Thus the facts of modern genetics rule out any recognition of the 'foundation of foundations' of Lamarckism—*the idea that acquired characters are inherited.*"[3] (My emphasis—*T.L.*)

The Mendelist-Morganists have thus thrown overboard one of the greatest acquisitions in the history of biological science—the principle of the inheritance of acquired characters, first put forth by Lamarck and subsequently organically incorporated in Darwin's teaching.

To the materialist teaching that it is possible for plants and animals to inherit individual variations of characters acquired under the influence of conditions of life, Mendelism-Morganism opposes an idealistic assertion, dividing the living body into two separate substances: the mortal body (or soma) and an immortal hereditary substance, germ-plasm. It is further categorically maintained that changes in the soma, i.e., in the living body, have no effect whatever upon the hereditary substance.

5. THE IDEA OF UNKNOWABILITY IN THE DOCTRINE OF "HEREDITARY SUBSTANCE"

Mendelism-Morganism endows the postulated mythical "hereditary substance" with an indefinite variation property. Mutations, i.e., changes of the "hereditary substance," are supposed to have no definite tendency. This assertion of the Morganists is logically connected with the underlying basis of Mendelism-Morganism—the principle that the hereditary substance is independent of the living body and its conditions of life.

---

[1] М. Завадовский. *Динамика развития организма, 1931, стр. 326.*

[2] Журнал *Естествознание и Марксизм*, 1929 г., № 4, стр. 83.

[3] *Ibid.*, p. 81.

The Morganist-Mendelists, who proclaim that hereditary alterations, or "mutations" as they are called, are "indefinite," presume that such alterations *cannot as a matter of principle be predicted.* We have here a peculiar conception of unknowability; its name is idealism in biology.

The assertion that variation is "indefinite" raises a barrier to scientific prediction, thereby handicapping practical agriculture.

Proceeding from the unscientific and reactionary Morganist teaching concerning "indefinite variation," the head of the Department of Darwinism at the University of Moscow, Academician I. I. Schmalhausen, asserts in his *Factors of Evolution* that hereditary variation, in its specific features, does not depend on the conditions of life and therefore has no definite tendency.

"Factors unassimilated by the organism," writes Schmalhausen, "if they reach the organism at all and influence it, can have but an indefinite effect.... Such influence can only be indefinite. Consequently, all new alterations in the organism, which as yet have no past history, will be indefinite. This category of alterations will include, however, not only mutations as new 'hereditary' changes, but any new (i.e., appearing for the first time) modifications."[1]

On a preceding page in the same book Schmalhausen writes: "In the development of any individual, environmental factors perform, in the main, only the role of agents liberating the course of certain form-building processes and the conditions which make it possible to consummate their realization."[2]

This formalistic, autonomistic theory of a "liberating cause" in which the role of external conditions is reduced to the realization of an autonomous process, has long been demolished by the advance of progressive science; it has been exposed by materialism as unscientific in essence, as idealistic.

Schmalhausen and others among our domestic followers of imported Morganism cite Darwin as their authority. In proclaiming the "indefiniteness of variation," they invoke Darwin's statements on the subject. Darwin indeed spoke of "indefinite variability." But that was due to the *limitations* of selection practice in his days. Darwin was aware of that himself and wrote that "we cannot at present explain either the causes or nature of the variability of organic beings."[3] "The subject," he said, "is an obscure one; but it may be useful to probe our ignorance."[4]

The Mendelist-Morganists cling to everything that is obsolete and wrong in Darwin's teaching, at the same time discarding its living materialist core.

In our socialist country, the teaching of the great transformer of nature, I. V. Michurin, has created a fundamentally new basis for directing the variability of living organisms.

Michurin himself and his followers have obtained and are obtaining

---

[1] И. И. Шмальгаузен, *Факторы эволюции*, Изд. АН СССР, 1946 г., стр. 12-13.

[2] *Ibid.*, p. 11.

[3] Ch. Darwin, *The Variation of Animals and Plants Under Domestication*, Vol. II, London 1885, p. 282.

[4] *Ibid.*, p. 237.

directed hereditary changes in plant organisms literally in immense quantities. Yet Schmalhausen still asserts that:

"The appearance of individual mutations is by all indications a case of chance phenomena. We can neither predict nor deliberately induce this or that mutation. So far it has been found impossible to establish any causal connection between the quality of mutation and definite changes in the factors of the environment."[1]

On the basis of the Morganist conception of mutations, Schmalhausen has formulated the theory of so-called "stabilizing selection"—a theory profoundly wrong ideologically and hamstringing practical activity. According to Schmalhausen, the formation of breeds and varieties proceeds—presumably inevitably—in a declining curve: the formation of breeds and varieties, rapid at the dawn of culture, increasingly expends its "reserve of mutations" and gradually declines. "Both the formation of breeds of domestic animals and the formation of varieties of cultivated plants," writes Schmalhausen, "proceeded with such exceptional speed mainly, apparently, because of the previously accumulated reserve of variability. Further strictly directed selection is slower. . . ."[2]

Schmalhausen's assertion and his entire conception of "stabilizing selection" follow the Morgan line.

As we know, Michurin, in the course of his lifetime, produced more than three hundred new plant varieties. Many of them were produced without sexual hybridization, and all of them were the result of strictly directed selection, including systematic training. It is an insult to progressive science to assert—in face of these facts and subsequent achievements of followers of Michurin's teaching—that strictly directed selection must progressively decline.

Schmalhausen obviously finds that Michurin's facts do not fit in with his theory of "stabilizing selection." In his book, *Factors of Evolution*, he gets out of the difficulty by making no mention of Michurin's work or of the very existence of Michurin as a scientist. Schmalhausen has written a bulky volume on factors of evolution without ever once mentioning—not even in his bibliography—either K. A. Timiryazev or I. V. Michurin. Yet K. A. Timiryazev bequeathed to Soviet science a remarkable theoretical work bearing practically the same title: *Factors of Organic Evolution*. As for Michurin and the Michurinists, they have put the factors of evolution to work for agriculture, revealed new factors and given us a deeper understanding of the old ones.

Schmalhausen has "forgotten" the Soviet advanced scientists, the founders of Soviet biological science. But at the same time he quotes profusely and repeatedly statements of big and small foreign and native representatives of Morgans metaphysics and leaders of reactionary biology.

---

[1] И. И. Шмальгаузен, *Факторы эволюции*, стр. 68.
[2] *Ibid.*, pp. 214-15.

Such is the style of Academician Schmalhausen, who calls himself a "Darwinist." Yet at a meeting of the Faculty of Biology at the University of Moscow his book was recommended as a masterpiece in the creative development of Darwinism. The book has been given a high rating by the deans of the Faculties of Biology at the Universities of Moscow and Leningrad; it has been praised by I. Polyakov, Professor of Darwinism at the University of Kharkov, by the Pro-Rector of the University of Leningrad, Y. Polyansky, by the member of our Academy, B. Zavadovsky, and by other Morganists who sometimes pose as orthodox Darwinists.

6. THE STERILITY OF MORGANISM-MENDELISM

The Morganist-Weismannists, i.e., the adherents of the chromosome theory of heredity, have repeatedly asserted—without any grounds and often in a slanderous manner—that I, as President of the Academy of Agricultural Sciences, have used my office in the interests of the Michurin trend in science, which I share, to repress the other trend, the one opposed to Michurin's.

Unfortunately, so far it has been exactly the other way round, and it is of that that I, as President of the Lenin Academy of Agricultural Sciences, may and should be accused. I have been wanting in strength and ability to make proper use of my official position to create conditions for the more extensive development of the Michurin trend in the various divisions of biological science, and to restrict, if ever so little, the scholastics and metaphysicians of the opposite trend. As a matter of fact, therefore, the trend so far repressed—repressed by the Morganists—happens to be the one which the President represents, namely, the Michurin trend.

We, the Michurinists, must squarely admit that we have hitherto proved unable to make the most of the splendid possibilities created in our country by our Party and the Government for the complete exposure of the Morganist metaphysics, which is in its entirety an importation from foreign reactionary biology hostile to us. It is now up to the Academy, to which a large number of Michurinists have just been added, to tackle this major task. This will be of considerable importance in the matter of training forces and providing more scientific aid to collective and state farms.

Morganism-Mendelism (the chromosome theory of heredity) is to this day taught, in a number of versions, in all colleges of biology and agriculture, whereas the study of Michurin genetics has in fact not been introduced at all. In the higher official scientific circles of biologists, too, the followers of Michurin and Williams have often found themselves in the minority. They were a minority in the Lenin Academy of Agricultural Sciences, too. But the situation in the Academy has now sharply changed thanks to the interest taken in it by the Party, the Government, and Comrade *Stalin* personally. A considerable number of Michurinists have been added as members and corresponding members of our Academy, and we expect that more will be added shortly, at the coming elections. This will create a new situation in the Academy and new opportunities for the further development of the Michurin teaching.

There is no truth whatever in the assertion that the chromosome theory of heredity, with its underlying metaphysics and idealism, has hitherto been repressed. The very opposite is the truth.

In our country the practical achievements of the Michurin trend in agrobiological science have been standing in the way of Morganistic cytogenetics.

Aware of the practical worthlessness of the theoretical postulates of their metaphysical "science," and reluctant to give them up and to accept the effective Michurin trend, the Morganists have bent and are bending all their efforts to check the development of the Michurin trend which is inherently opposed to their pseudo science.

It is a calumny to assert that somebody has been preventing the cytogenetic trend in biological science from associating itself with practical agriculture in our country. There is no truth whatever in the assertion that "the right to the practical application of the fruits of their labours has been a monopoly of Academician Lysenko and his followers."

The Ministry of Agriculture might tell us exactly what the cytogeneticists have offered for practical application, and, if there have been such offers, whether they were accepted or rejected.

The Ministry of Agriculture might also tell us which of its scientific research institutes (to say nothing of colleges) have not engaged in cytogenetics in general and, particularly, in the polyploidy of plants obtained by the application of colchicine.

I know that many institutes have been engaged and are engaged in this sort of—in my view—scarcely productive activity. More, the Ministry of Agriculture set up a special institution, headed by A. R. Zhebrak, to study questions of polyploidy. I think that this institution, though it has for some years done nothing besides its work on polyploidy, has produced literally nothing of practical value.

Here is one example which might be cited to show how useless is the practical and theoretical program of our domestic Morganist cytogeneticists.

Professor of Genetics, N. P. Dubinin, Corresponding Member of the Academy of Sciences of the U.S.S.R., who is regarded by our Morganists as the most eminent among them, has worked for many years to ascertain the differences in the cell nuclei of fruit flies in urban and rural localities.

For the sake of complete clarity, let us mention the following. What Dubinin is investigating is not qualitative alterations—in this case, in the nucleus of the cell—resulting from the action of qualitatively different conditions of life. What he is studying is not the inheritance of characteristics acquired by fruit flies under the influence of definite conditions of life, but changes, recognizable in the chromosomes, in the composition of the population of these flies as the result of the simple destruction of a part of them, for one thing, during the war. Dubinin and other Morganists call such destruction "selection." (*Amusement.*) It is this sort of "selection," identical with an ordinary sieve, which has nothing in common with the truly creative role of selection, that constitutes the subject of Dubinin's investigations.

His work is entitled: "Structural Variability of Chromosomes in Populations of Urban and Rural Localities."

Here are a few quotations from it:

"During the study of various populations of *D. funebris* in the work of 1937 the fact was noted that there were noticeable differences as regards concentration of inversions. Tinyakov stressed this phenomenon on the basis of extensive material. However, only the 1944-45 analysis has shown us that these substantial differences are due to the differences of conditions of habitat in town and in countryside.

"The population of Moscow has eight different orders of genes. In the second chromosome there are four orders (one standard and three different inversions). One inversion in the III chromosome and one in IV . . . Inv. II—1 has its limits from 23 C to 31 B. Inv. II—2, from 29 A to 32 B. Inv. II—3, from 32 B to 34 C. Inv. III—1, from 50 A to 56 A. Inv. IV—1, from 67 C to 73 A/B. In the course of 1943-45 the karyotype of 3,315 individuals in the population of Moscow was studied. The population contained immense concentrations of inversions, which proved to be different in various sections of Moscow."[1]

Dubinin went on with his investigations during and after the war and studied the problem of the fruit flies in the city of Voronezh and its environs. He writes:

"The destruction of industrial centres during the war upset the normal conditions of life. The *Drosophila* populations found themselves in severe conditions of existence which, possibly, surpassed the severity of wintering in rural localities. It was of profound interest to study the influence of the changes in the conditions of existence caused by the war upon the karyotypical structure of urban populations. In the spring of 1945 we studied populations from the city of Voronezh, one of the cities that suffered the worst destruction as the result of the fascist invasion. Among 225 individuals only two flies were found to be heterozygous for inversion II—2 (0.88%). Thus the concentration of inversions in this large city proved to be lower than in some rural localities. We see here the disastrous influence of natural selection upon the karyotypical structure of the population."[2]

Dubinin, as we see, writes so that outwardly his work may appear to some to be even scientific. As a matter of fact, this was one of the main works on the basis of which Dubinin was elected Corresponding Member of the Academy of Sciences of the U.S.S.R.

---

[1] *Доклады Академии Наук СССР*, 1946 г., том LI, № 2, стр. 152.
[2] *Ibid.*, p. 153.

But if we were to put it all in plainer terms, stripping it of the pseudo-scientific verbiage and replacing the Morganist jargon with ordinary Russian words, we would arrive at the following:

As the result of many years of effort Dubinin "enriched" science with the "discovery" that during the war there occurred among the fruit-fly population of the city of Voronezh and its environs an increase in the percentage of flies with certain chromosome structures and a decrease in the percentage of flies with other chromosome structures (in the Morganist jargon that is called "concentration of inversions" II—2).

Dubinin is not content with these discoveries, "highly valuable" from the theoretical and practical standpoint, which he made during the war. He sets himself further tasks for the restoration period. He writes:

"It will be very interesting to study in the course of several coming years the restoration of the karyotypical structure of the urban population in connection with the restoration of normal conditions of life."[1] (*Animation. Laughter.*)

That is typical of the Morganists' "contribution" to science and practical activity before the war and during the war, and those are the vistas of the Morganist "science" for the restoration period! (*Applause.*)

7. MICHURIN'S TEACHING, THE FOUNDATION OF SCIENTIFIC BIOLOGY

Contrary to Mendelism-Morganism, with its assertion that the causes of variation in the nature of organisms are unknowable and its denial that directed changes in the nature of plants and animals are possible, I. V. Michurin's motto was: "We cannot wait for favours from Nature; we must wrest them from her."

His studies and investigations led I. V. Michurin to the following important conclusion: "It is possible, with man's intervention, *to force* any form of animal or plant *to change more quickly and in a direction desirable to man.* There opens before man a broad field of activity of the greatest value to him."[2]

The Michurin teaching flatly rejects the fundamental principle of Mendelism-Morganism that heredity is completely independent of the plants' or animals' conditions of life. The Michurin teaching does not recognize the existence in the organism of a separate hereditary substance which is independent of the body. Changes in the heredity of an organism or in the heredity of any part of its body are the result of changes in the living body itself. And changes of the living body occur as the result of departure from the normal in the type of assimilation and dissimilation, of departure from

---

[1] *Доклады Академии Наук СССР*, 1946 г., том LI, № 2, стр. 153.
[2] И. В. Мичурин, *Сочинения*, том IV, стр. 72.

the normal in the type of metabolism. Changes in organisms or in their separate organs or characters may not always, or not fully, be transmitted to the offspring, but changed germs of newly generated organisms always occur only as the result of changes in the body of the parent organism, as the result of direct or indirect action of the conditions of life upon the development of the organism or its separate parts, among them the sexual or vegetative germs. Changes in heredity, acquisition of new characters and their augmentation and accumulation in successive generations are always determined by the organism's conditions of life. Heredity changes and its complexity increases as the result of the accumulation of new characters and properties acquired by organisms in successive generations.

The organism and the conditions required for its life constitute a unity. Different living bodies require different environmental conditions for their development. By studying the character of these requirements we come to know the qualitative features of the nature of organisms, the qualitative features of heredity. Heredity is *the property of a living body to require definite conditions for its life and development and to respond in a definite way to various conditions.*

Knowledge of the natural requirements of an organism and its response to external conditions makes it possible to govern the life and development of the organism. By regulating the conditions of life and development of plants and animals we can probe their nature ever more deeply and thus establish what are the means of changing it in the required direction. Once we know the means of regulating development we can change the heredity of organisms in a definite direction.

Each living body builds itself out of the conditions of its environment after its own fashion, according to its heredity. That is why different organisms live and develop in the same environment. As a rule, each given generation of a plant or animal develops largely in the same way as its predecessors, particularly its close predecessors. *Reproduction of beings similar to itself is a general characteristic of every living body.*

When an organism finds in its environment the conditions suitable to its heredity, its development proceeds in the same way as it proceeded in previous generations. When, however, organisms do not find the conditions they require and are forced to assimilate environmental conditions which, to some degree or other, do not accord with their nature, then the organisms or sections of their bodies become more or less different from the preceding generation. If the altered section of the body is the starting point for the new generation, the latter will, to some extent or other, differ from the preceding generations in its requirements and nature.

The cause of changes in the nature of a living body is a change in the type of assimilation, in the type of metabolism. For example, the process of vernalization of spring cereals does not require lowered temperatures. Normally it proceeds in temperatures such as obtain in the spring and summer in the fields. But by using lower temperature conditions in the vernalization of spring cereals it is possible, after two or three generations, to turn them into winter cereals. And winter cereals cannot pass through the process of vernalization without lowered temperatures. Here is a concrete example showing how a new requirement is induced in the offspring of these particular plants—the requirement for lowered temperatures as a condition for vernalization.

Sex cells and any other cells through which organisms propagate are produced as the result of the development of the whole organism, by means of conversion, by means of metabolism. The phases in the development of an organism are accumulated, as it were, in the cells from which the new generation originates.

We may therefore say that to the extent to which in the new generation the body of an organism (a plant, say) is built anew to that same extent also all its properties, including heredity, develop.

In one and the same organism the development of different cells and of different parts of cells, the development of individual processes, requires different external conditions.

Besides, these conditions are assimilated in different ways. It should be stressed that in this case we mean by *external that which is assimilated, and by internal that which assimilates.*

The life of an organism proceeds through innumerable correlated processes and conversions. The food that enters the organism from the external environment undergoes a series of conversions whereby it is assimilated by the living body, changing from external to internal. This internal, since it is living matter, enters into metabolic relations with the substances of other cells and particles of the body, feeding them and thus becoming external with regard to them.

Two kinds of qualitative changes are observed in the development of plant organisms.

1. Changes connected with the process of the realization of the individual cycle of development, when natural requirements, i.e., heredity, are normally met by the *corresponding external conditions.* The result is a body of the same breed and heredity as the preceding generations.

2. Changes in the nature of the organisms, i.e., the changes in heredity. Such changes are also the result of individual development, but deviating from the normal, usual course. Changes in heredity are, as a rule, the result of the organism's development *under external conditions which, to one extent or other, do not correspond to the natural requirements of the given organic form.*

Changes in the conditions of life make the very type of development of plant organisms change. A changed type of development is thus the primary cause of changes in heredity. Organisms which cannot change in accordance with the changed conditions of life do not survive, leave no progeny.

Organisms, and hence also their nature, are created only in the process of development. Of course, a living body may undergo an alteration also outside the process of development (a burn, a break in joints, in roots, etc.), but such alterations will not be characteristic or necessary for the vital process.

Numerous facts go to show that changes in various sections of the body of a plant or animal organism are not fixed by the reproductive cells with the same frequency or to the same extent.

This is explained by the fact that the process of development of each organ, of each particle of the living body, requires relatively definite external conditions. These conditions are elected from the environment by the development of each organ and minutest organelle Therefore, if a section of the body of a plant organism is forced to assimilate conditions relatively unusual for it and as a result undergoes alteration and becomes different from the analogous section of the body in the preceding generation, the substances which it sends forth to neighbouring cells may not be elected by the latter, may not be joined into the further chain of corresponding processes. Of course, there will still be a connection between the altered section of the plant organism and the other sections of the body, for otherwise it could not exist at all; but this connection may not be fully reciprocal. The altered section of the body will be receiving this or that food from the neighbouring sections; but it will not be able to give away its own specific substances, because the neighbouring sections will refuse to elect them.

This explains the frequently observed phenomenon when altered organs, characters, or properties of an organism do not appear in the progeny. But the altered sections of the body of the parent organism always possess an altered heredity. Fruit growers and horticulturists have long known these facts. An altered twig or bud of a fruit tree or the eye (bud) of a potato tuber cannot, as a rule, influence the alteration of heredity of the offspring of the given tree or tuber which are not directly generated from the altered sections of the parent organism. If, however, the altered part is cut away and grown separately as an independent plant, the latter, as a rule, will possess a changed heredity, the one that characterized the altered part of the parent body.

The extent of the hereditary transmission of alterations depends on the extent to which the substances of the altered section of the body join in the general process which leads to the formation of reproductive sex or vegetative cells.

Once we know how the heredity of an organism is built up, we can change it in a definite direction by creating definite conditions at a definite moment in the development of the organism.

Good varieties of plants or animals are always produced only by the application of proper methods of cultivation or breeding. Under poor cultivation no good varieties can ever be produced out of poor ones, and in many cases even good cultivated varieties will deteriorate after a few generations. It is a basic rule in seed growing that plants grown for seed must be tended will the utmost care. They must be provided with conditions meeting the optimum of the hereditary requirements of the given plants. Of well-cultivated plants the very best should be and are selected for seed. That is the way varieties of plants are improved in practice. Under poor cultivation, i.e., when agrotechnique is bad, no selection of the best plants for seed will produce the required results—all the seeds obtained will be poor, and even the best among them will still be poor.

According to the chromosome theory of heredity, hybrids can only be produced by sexual reproduction. That theory denies the possibility of obtaining vegetative hybrids, for it denies that the conditions of life have any specific influence upon the nature of plants. I. V. Michurin, on the other hand, not only recognized the possibility of producing vegetative hybrids, but elaborated the "mentor" method. This method consists in the following: by grafting cuttings (twigs) of old varieties of fruit trees on the branches of a young variety, the latter acquires properties which it lacks, these properties being transmitted to it through the grafted twigs of the old variety That is why I. V. Michurin called this method "mentor." The stock is also used as a mentor. By this method Michurin produced new and improved existing varieties.

I. V. Michurin and the Michurinists have found methods of obtaining vegetative hybrids in large quantities.

The vegetative hybrids are cogent proof that Michurin's conception of heredity is correct. At the same time they represent an insuperable obstacle to the theory of the Mendelist-Morganists.

Organisms grafted before they have been completely formed phasically, before they have completed their cycle of development, will always undergo changes of development as compared with own-rooted plants, i.e., ungrafted plants. In the union of plants by means of grafting, the product is a single organism of mixed breed, that of the scion and that of the stock. By planting the seeds from the scion or the stock it is possible to obtain offspring, some representatives of which will possess the characteristics not only of the breed from which the seed has been taken, but also of the other with which it has been united by grafting.

Obviously, the stock and the scion could not have exchanged chromosomes of the cell nuclei; yet hereditary characters have been transmitted from stock to scion and vice versa. Consequently, *the plastic substances produced by the scion and the stock possess the characters of the breed*, are endowed with definite heredity *just as the chromosomes, and just as any particle of the living body.*

Any character may be transmitted from the one breed to another by means of grafting just as well as by the sexual method.

The wealth of factual material concerning vegetative transmission of various properties of potatoes, tomatoes, and a number of other plants leads us to the conclusion that vegetative hybrids do not differ in principle from sexual hybrids.

The representatives of Mendel-Morgan genetics are not only unable to obtain alterations of heredity in a definite direction, but categorically deny that it is possible to change heredity so that it will adequately correspond to the action of the environmental conditions. The principles of Michurin's teaching, on the other hand, tell us that it is possible to obtain changes in heredity *fully corresponding to the effect of the action of conditions of life.*

A case in point is the experiments to convert spring forms of cereal grains into winter forms, and winter forms into forms of still greater winter habit in regions of Siberia, for example, where the winters are severe. These experiments are not only of theoretical interest. They are of considerable practical value for the production of winter-hardy varieties. We already have winter forms of wheat obtained from spring forms, which are not inferior, as regards frost resistance, to the most frost-resistant varieties known in practical farming. Some are even superior.

Many experiments show that when an old established property of heredity is being eliminated, we do not at once get a fully established, solidified new heredity. In the vast majority of cases, what we get is an organism with a plastic nature, which I. V. Michurin called "destabilized."

Plant organisms with a "destabilized" nature are those in which their conservatism has been eliminated, and their electivity with regard to external conditions is weakened. Instead of conservative heredity, such plants preserve, or there appears in them, only a *tendency* to show some preference for certain conditions.

The nature of a plant organism may be destabilized:

1. By grafting, i.e., by uniting the tissues of plants of different breeds;
2. By bringing external conditions to bear upon it at definite moments when the organism undergoes this or that process of its development;
3. By crossbreeding, particularly of forms sharply differing in habitat or origin.

The best biologists, first and foremost I. V. Michurin, have devoted a great deal of attention to the practical value of plant organisms with destabilized heredity. Plastic plant forms with unestablished heredity, obtained by any of the enumerated methods, should be further bred from generation to generation in those conditions, the requirement of which, or adaptability to which, we want to induce and perpetuate in the given organisms.

In most plant and animal forms new generations develop only after fertilization—the fusion of female and male reproductive cells. The biological significance of the process of fertilization is that thereby organisms are produced with a dual heredity—maternal and paternal. Dual heredity lends greater vitality to organisms and widens the range of their adaptability to varying conditions of life.

It is the usefulness of enriching heredity that determines the biological necessity for crossbreeding forms differing from each other even if ever so slightly.

The vitality of plant forms may be renovated and strengthened also vegetatively, asexually. This is brought about by the living body assimilating new external conditions, conditions unusual for it. In experiments in vegetative hybridization, in experiments with the aim of producing spring forms from winter forms or vice versa, and in a number of other cases of the nature of organisms being destabilized, we may observe the renovation and strengthening of the vitality of organisms.

By regulating external conditions, the conditions of life of plant organisms, we can change varieties in a definite direction and create varieties with desirable heredity.

Heredity is the effect of the concentration of the action of environmental conditions assimilated by the organism in a series of preceding generations.

By means of skilful hybridization, by the method of sexual conjugation of breeds, it is possible at once to unite in one organism that which has been assimilated and solidified in the crossed breeds by many generations. But, according to Michurin's teaching, no hybridization will produce the desired results, unless the conditions are created which will promote the development of the characters which we want the newly-bred or improved variety to be able to pass on.

I have here propounded Michurin's teaching in most general outline. The important point that must be stressed here is that it is absolutely necessary for all Soviet biologists to make a profound study of this teaching. The best way for scientific workers in various branches of biology to master the theoretical depths of the Michurin teaching is to study Michurin's works, to read them over again and again, and to analyze them with a view to solving problems of practical importance.

Socialist agriculture stands in need of a developed, profound biological theory which will help us quickly and properly to perfect the methods of cultivating plants and obtaining plentiful and stable crop yields. It stands in need of a profound biological theory which will help workers in agriculture to obtain in a short time the highly productive forms of plants they need, to correspond to the high fertility which the collective farmers are creating on their fields.

Unity of theory and practice—that is the highroad for Soviet science. The Michurin teaching best embodies this unity in biological science.

In my speeches and writings I have cited numerous examples of the successful application of the Michurin teaching in solving questions of practical importance in various departments of plant breeding. Here I shall take the liberty to dwell briefly on some questions of animal breeding.

As in the case of plant forms, the development of animal forms is closely linked with their conditions of life, with the conditions of their environment.

The basic factors for increasing the productivity of domestic animals, for improving existing breeds and producing new ones, are their food and the conditions in which they are kept. This is particularly important if the effectiveness of crossbreeding is to be heightened. Various breeds of domestic animals have been and are produced by men for various purposes and under various conditions. Each breed therefore requires its own conditions of life, those that contributed to its formation.

The greater the divergences between the biological properties of a breed and the conditions of life provided for the individual animals, the less will be the economic value of the given breed.

For example, the advantages—from an economic standpoint—of rich pastures and good feeding with succulent and concentrated fodders are smaller in the case of cattle which by nature cannot give much milk than in the case of cattle with high milking capacities. In the former case we obviously have a breed which, in the economic respect, does not justify the conditions provided for it. Such a breed should be improved by crossbreeding so as to adjust it to the conditions of feeding and maintenance.

On the other hand, a breed noted for its milk-yielding properties, when placed in conditions of poor feeding and maintenance, will not only fail to live up to its reputation as a milk producer, but its chances of survival will be diminished. In such cases the conditions of feeding and maintenance should be improved so as to adjust them to the breed.

Our science and practice of animal breeding, in line with the state plan for obtaining produce in the required quantities and of proper quality, must be guided by the principle: *to select and improve breeds in accordance with the conditions of feeding, maintenance and climate, and at the same time to create conditions of feeding and maintenance most suitable to the given breeds.*

The principal method of constantly improving breeds is to select pedigreed animals best suited for the required aim and at the same time to improve the conditions of feeding, maintenance and care that are most conducive to the development of the animals in the desired direction.

Crossbreeding is a radical and quick method of changing breeds, that is to say, the progeny of the given animals.

In crossbreeding we get, as it were, a union of two breeds evolved by man in the course of a long period of time by creating various conditions of life for the animals. But the nature (heredity) of crosses, particularly in the first generation, is usually unstable and easily responds to the action of the conditions of life, feeding, and maintenance.

Therefore, in crossbreeding it is of especial importance, when choosing a breed for the improvement of a local breed, to bear in mind the conditions of feeding, maintenance, and climate. At the same time, in order to develop the characters and properties which we want to induce in the local breed by crossbreeding, we must provide conditions of feeding and maintenance conducive to the development of the new improving breed properties; otherwise, we may fail to establish the desired qualities and the local breed may even lose some of its good qualities.

I have given an example of the application of the general principles of the Michurin teaching to animal husbandry to show that Soviet Michurin genetics, revealing as it does the general laws of the development of living bodies in order to cope with problems of practical importance, is also applicable to stockbreeding.

When we speak of mastering the teaching of Michurin we also mean the development and deepening of this teaching, the development of scientific biology. That is the line along which we must secure the growth of the forces of our Michurinist biologists so as to provide ever-increasing scientific assistance to the collective and state farms in coping with the tasks set by the Party and the Government. (*Applause.*)

8. YOUNG SOVIET BIOLOGISTS SHOULD STUDY THE MICHURIN TEACHING

Unfortunately, so far the Michurin science has not been taught in our universities and colleges. We Michurinists are greatly to blame for this. But it will be no mistake to say that it is also the fault of the Ministry of Agriculture and the Ministry of Higher Education.

To this day Morganism-Mendelism is taught in the majority of our universities and colleges in the departments of genetics and selection, and in many cases also in the departments of Darwinism, whereas the Michurin teaching, the Michurin trend in science, fostered by the Bolshevik Party and by Soviet reality, remains in the shade.

The same may be said of the position with regard to the training of young scientists. By way of illustration, we shall cite the following. In an article "On Doctors' Theses and the Responsibility of Opponents," printed in issue No. 4 of the *Vestnik vysshey shkoly* (*Higher Education Messenger*) for 1945, Academician P. M. Zhukovsky, who is the Chairman of the Biology Experts' Commission under the Highest Committee on Academic Degrees, wrote: "A deplorable situation has developed in the matter of theses on genetics. *Theses on genetics are very rare*; they represent, in fact, solitary instances. This is to be explained by the abnormal relations, which have assumed the character of enmity, between the adherents of the chromosome theory of heredity and its opponents. The truth of the matter is that the former somewhat fear the latter, who are very aggressive in their polemics. It would be better to put an end to this situation. Neither the Party nor the Government forbid the chromosome theory of heredity, and it is freely propounded in universities and colleges. So let the controversy go on."[1]

Let us first note that P. M. Zhukovsky confirms that the chromosome theory of heredity is freely taught in universities and colleges. That is true.

---

[1] *Вестник высшей школы*, № 4, 1945 г., стр. 30.

But he wants more: he wants Mendelism-Morganism to be still more widely propounded in our colleges. He wants us to have many more Mendelist-Morganist Masters and Doctors of Science who would still more extensively propagate Mendelism-Morganism in our universities and colleges. That, in fact, is what Academician Zhukovsky is driving at in a large section of his article, and that reflects his general line as Chairman of the Biology Commission.

No wonder therefore that the Commission set up all sorts of obstacles in the case of theses on genetics whose authors attempted, even if ever so timidly, to develop this or that principle of Michurin genetics. On the other hand, theses by Morganists, enjoying P. M. Zhukovsky's patronage, appeared and were passed on favourably not at all so rarely—in any event, oftener than the interests of true science required. True enough, theses with a Morganist tendency appeared more rarely than Academician P. M. Zhukovsky would have liked. But there are reasons for this. Under the influence of the Michurin criticism of Morganism young scientists with philosophical training have in recent years come to realize that the Morganist views are utterly alien to the world outlook of Soviet people. In this light the position of Academician P. M. Zhukovsky is rather dubious, seeing that he advises young biologists to pay no heed to the Michurinists' criticism of Morganism, but to go on developing the latter.

Soviet biologists are right when they are suspicious of the Morganist views and refuse to listen to the scholasticism of the chromosome theory. They stand to gain, always and in everything, if they will ponder more often on what Michurin said of this scholasticism.

I. V. Michurin held that Mendelism "... contradicts the truths of nature, before which no artful structure reared out of wrongly understood phenomena can stand up. What I would like," he wrote, "is that the thinking unbiassed observer should ponder over this and personally test the truth of these conclusions; they represent a basis which we bequeath to naturalists of coming centuries and millenniums."[1]

9. FOR A CREATIVE SCIENTIFIC BIOLOGY

I. V. Michurin laid the foundations for the science of regulating the nature of plants. These foundations have wrought a change in the very method of thinking when dealing with problems of biology.

A knowledge of *causal* connections is essential for the practical work of regulating the development of cultivated plants and domestic animals. For biological science to be in a position to render the collective and state farms ever greater assistance in obtaining higher crop yields, higher yields of milk, etc., it must comprehend the complex biological interrelations, the laws of the life and development of plants and animals.

---

[1] И. В. Мичурин, *Сочинения*, Сельхозгиз, 1940 г., том III, стр. 308-309.

A scientific handling of practical problems is the surest way to a deeper knowledge of the laws of development of living nature.

Biologists have paid very little attention to the study of the interrelations, the natural and historical connections that exist between individual bodies, individual phenomena, parts of individual bodies and links of individual phenomena. Yet only these connections, interrelations, and natural interactions enable us to understand the process of development, the essence of biological phenomena.

But when living nature is studied in isolation from practical activity the scientific principle of the study of biological connections is lost.

The Michurinists, in their investigations, take the Darwinian theory of evolution as their basis. But in itself Darwin's theory is absolutely insufficient for dealing with the practical problems of socialist agriculture. That is why the basis of contemporary Soviet agrobiology is Darwinism transformed in the light of the teachings of Michurin and Williams and thereby converted into Soviet creative Darwinism.

Many problems of Darwinism assume a different aspect as the result of the development of our Soviet agrobiological science, of the Michurin trend in agrobiology. Darwinism has not only been purified of its deficiencies and errors and raised to a higher level, but has undergone a considerable change in a number of its principles. From a science which primarily *explains* the past history of the organic world, it is becoming a creative, *effective* means of systematically mastering living nature, making it serve practical requirements.

Our Soviet Michurinist Darwinism is a creative Darwinism which poses and solves problems of the theory of evolution in a new way, in the light of Michurin's teaching.

I cannot in this report touch on many of the theoretical problems of great practical significance.

I shall dwell briefly on only one of them—namely, the question of intra- and interspecific relations in living nature.

The time has come to consider the question of speciation, approaching it from the angle of the transition of quantitative accumulation into qualitative specific distinctions.

We must realize that speciation is a transition—in the course of the historical process—from quantitative to qualitative variations. Such a leap is prepared by the vital activity of organic forms themselves, as the result of quantitative accumulations of responses to the action of definite conditions of life, and that is something that can definitely be studied and directed.

Such an understanding of speciation, an understanding of natural laws, places in the hands of biologists a powerful means of regulating the vital process itself and consequently speciation as well.

I think that in posing the question this way we may assume that what leads to the appearance of a new specific form, to the formation of a new species out of an old one, is not the accumulation of quantitative distinctions by which varieties within a species are usually recognized. The quantitative accumulations of variations which lead to the leap which changes an old form or species into a new form are variations of a *different order*.

Species are not an abstraction, but actually existing links in the general biological chain.

Living nature is a biological chain broken up, as it were, into individual links, or species. It is therefore wrong to say that a species does not retain the constancy of its qualitative definiteness as a species for any length of time. To insist on that would be to regard the evolution of living nature as proceeding along a plane, without any leaps.

I am confirmed in this opinion by the data of experiments for the conversion of hard wheat (durum) into soft (vulgare).

Let me note that all systematists admit that these are good, indisputable, independent species.

We know that there are no true winter forms among hard wheats, and that is why in all regions with a relatively severe winter hard wheat is cultivated only as a spring, not a winter, crop. Michurinists have mastered a good method of converting spring into winter wheat. It has already been mentioned that many spring wheats have been experimentally converted into winter wheat. But all those belonged to the species of soft wheat. When experiments were started to convert hard wheat into winter wheat it was found that after two, three or four years of autumn planting (required to turn a spring into a winter crop) durum becomes vulgare, that is to say, one species is converted into another. Durum wheat with 28 chromosomes is converted into several varieties of soft 42-chromosome wheat; nor do we, in this case, find any transitional forms between the durum and vulgare species. *The conversion of one species into another takes place by a leap.*

We thus see that the formation of a new species is prepared by an alteration of vital activity under definite new conditions in a number of generations. In our case it is necessary to bring autumn and winter conditions to bear on hard wheat in the course of two, three or four generations. Then it can change by a leap into soft wheat without any transitional forms between the two species.

I think that it may be pertinent to note that what led me to study the essentially theoretical problems of species and of intraspecific and interspecific relations among individuals, has never been mere curiosity or a fondness for abstract theorizing. I was and am led to study these questions of theory by my work in the course of which I have to find answers to purely practical problems. For a correct understanding of the relations among individuals within a species and between species it was necessary to have a clear idea of the qualitative distinctions of intraspecific and interspecific diversities of forms.

It thus became possible to find new solutions to such problems of practical importance as weed control in farming, or the choosing of ingredients for the sowing of grass mixtures, of the speedy and extensive afforestation of steppe areas, and many others.

That is what led me to make a new study of the problem of intra- and interspecific struggle and competition, and after a thorough and comprehensive investigation I have come to the conclusion that there exists no intraspecific struggle and mutual assistance among individuals within a species, and that there does exist interspecific struggle and competition and also mutual assistance between different species. I regret that I have so far done very little to elucidate the theoretical implications and practical significance of these questions in the press.

* * *

I shall now conclude. Thus, comrades, as regards the theoretical line in biology, Soviet biologists hold that the Michurin principles are the only scientific principles. The Weismannists and their followers, who deny the heritability of acquired characters, are not worth dwelling on at too great length. The future belongs to Michurin. (*Applause.*)

V. I. Lenin and *J. V. Stalin* discovered I. V. Michurin and made his teaching the possession of the Soviet people. By their great paternal attention to his work they saved for biology the remarkable Michurin teaching. The Party, the Government, and *J. V. Stalin* personally, have taken an unflagging interest in the further development of the Michurin teaching. There is no more honourable task for us Soviet biologists than creatively to develop Michurin's teaching and to follow in all our activities Michurin's style in the investigation of the nature of the development of living beings.

Our Academy must work to develop the Michurin teaching. In this it ought to follow the personal example of concern for the work of I. V. Michurin shown by our great teachers—*V. I. Lenin* and *J. V. Stalin.* (*Loud applause.*)

CONCLUDING REMARKS

Comrades, before I pass to my concluding remarks I consider it my duty to make the following statement.

The question is asked in one of the notes handed to me: What is the attitude of the Central Committee of the Party to my report? I answer: The Central Committee of the Party examined my report and approved it. (*Stormy applause. Ovation. All rise.*)

I shall now take up some of the points brought out at our session.

The adherents of the so-called chromosome theory of heredity who spoke here denied that they were Weismannists and all but proclaimed themselves antagonists of Weismann. On the other hand, it has been clearly shown in my report and in many of the speeches of representatives of the Michurin trend that Weismannism and the chromosome theory of heredity are one and the same thing. Mendelist-Morganists abroad make no secret of this. In my report

I quoted articles by Morgan and Castle published in 1945, in which it is plainly stated that the so-called teaching of Weismann is the basis of the chromosome theory of heredity. By Weismannism (which is the same as idealism in biology) is meant any conception of heredity which maintains that the living body is divided into two substances which are different in principle: the ordinary living body, presumably possessing no heredity but subject to alterations and transformations, that is to say, to development; and a specific hereditary substance, presumably independent of the living body and not subject to development under the influence of the conditions of life of the ordinary living body, or the soma. That much is beyond any doubt. No efforts of the advocates of the chromosome theory of heredity, neither those who spoke nor those who did not speak at the session, to lend their theory a materialist appearance can change the character of this theory, which is essentially idealistic. (*Applause.*)

The Michurin trend in biology is a materialist trend, because it does not separate heredity from the living body and the conditions of its life. There is no living body without heredity, and there is no heredity without a living body. The living body and its conditions of life are inseparable. Deprive an organism of its conditions of life and the living body will die. The Morganists, however, maintain that heredity is isolated, something apart from the mortal living body, from what they call the soma.

Those are the principles on which we differ with the Weismannists. And connected with them is also our difference on a question which has a long history behind it, namely, the question of inheritance of characters acquired by plants and animals. The Michurinists say that inheritance of acquired characters is possible and necessary. This principle has once more been fully confirmed by the abundant factual material demonstrated at this session. Morganists, among them those who spoke at our session, cannot comprehend this principle so long as they have not fully discarded their Weismannist notions.

It is still not clear to some that heredity is inherent not only in the chromosomes, but in any particle of the living body. They therefore want to see with their own eyes cases of hereditary properties and characters transmitted from generation to generation without the transmission of chromosomes.

These questions, so incomprehensible to the Morganists, can best be answered by demonstrating and explaining experiments in vegetative hybridization carried on extensively in our country. It was I. V. Michurin who elaborated vegetative hybridization. And experiments in vegetative hybridization show incontrovertibly that heredity is a property not only of the chromosomes, but of every living thing, every cell and every particle of the body. For heredity is determined by the specific type of metabolism. You need but change the type of metabolism in a living body to bring about a change in heredity.

Academician P. M. Zhukovsky, as becomes a Mendelist-Morganist, cannot conceive transmission of hereditary properties without transmission of chromosomes. He cannot conceive that the ordinary living body possesses heredity. According to his views, that is the property of the chromosomes only. He therefore does not think it possible to obtain plant hybrids by means of grafting, he does not think it possible for plants and animals to inherit acquired characters. I promised Academician Zhukovsky to show him vegetative hybrids, and I have now the pleasure of demonstrating them at this session.

In this case one of the participating plants was a variety of tomatoes with leaves not dissected, as usual, but like those of the potato. Its fruits are red and oblong in shape.

The other variety that participated in the grafting was one with the usual dissected tomato leaves. The fruits when ripe are not red, but yellowish, white.

The variety with the potato leaves was used as the stock, and the variety with the dissected leaves was the scion.

In the year when the graft was made no changes were observed either in the scion or in the stock.

Seeds were gathered from the fruits that had grown on the scion and from those that had grown on the stock. These seeds were then planted.

Most of the plants that grew from the seeds taken from the fruits of the stock did not differ from the initial variety, that is to say, they were potato-leafed and their fruits were red and oblong in shape. Six plants, however, had dissected leaves, and some of them had yellow fruits, that is to say, both the leaves and the fruits had changed under the influence of the other variety, the one which had been the scion.

Academician P. M. Zhukovsky has expressed doubt as to the purity of the experiments in vegetative hybridization, pointing out that cross-pollination of the varieties might have occurred—in other words, that it was a case of sexual hybridization. But how, Comrade Zhukovsky, can the results of the experiments I demonstrate be explained by cross-pollination?

All who have had anything to do with the hybridization of tomatoes know that when the plants with dissected leaves and yellow fruits are cross-pollinated with the plants with potato leaves and red fruits, the first generation will have dissected leaves, but invariably red fruits.

But see what we have got in our experiments. The leaves are indeed dissected, but the fruits are not red but yellow. How, then, can these results be explained by accidental cross-pollination?

Here are the fruits of the other of these vegetative hybrids. The leaves of this plant are also dissected, but of the ripe fruits on the cluster, one, as you see, is red and the other yellow. Variety within a single plant is a quite frequent phenomenon among vegetative hybrids. It should be borne in mind that vegetative hybridization is not the usual mode of union of breeds, not the one that has developed in the course of their evolution. That is why as the result of grafting we often get organisms that are destabilized and therefore prone to vary.

It is not in all plants by any means that we can observe easily perceptible alterations in the year of the grafting or even in the first seed generation. Nonetheless we already have every ground to assert that every graft of a phasically young plant produces changes in heredity. To prove this point we are going on with our work on vegetative hybrids of tomatoes at the Institute of Genetics of the Academy of Sciences of the U.S.S.R.

I shall now show you plants of the second seed generation obtained from the same graft; but these are from seeds taken from plants which gave no visible alterations in the first seed generation. On a number of plants from the second seed generation the leaves are changed—they are not like potato leaves in appearance, but dissected, and the fruits are not red but yellow. In this case, too, there is no reason to doubt the purity of the work or to suspect cross-pollination. In the first generation these plants had potato leaves and red fruits. If the dissected leaves in the plants of the second generation are the result of cross-pollination, why are the fruits not red but yellow?

We thus see that as the result of grafts we obtain directed, adequate alterations; we obtain plants combining the characters of the breeds joined in the grafting, that is to say, we get true hybrids. New formations are also observed. For example, among the progeny of the same graft there are plants that have borne small fruits, like those of uncultivated forms. But we all know that in the case of sexual hybridization, too, we observe, besides the transmission to the progeny of characters of the parent forms, also the appearance of new forms.

I could cite many more cases of the production of vegetative hybrids. It is no exaggeration to say that there are hundreds and thousands of them in our country. The Michurinists not only understand how vegetative hybrids are produced, but produce them in large numbers from numerous varieties.

I have dwelt at length on vegetative hybrids because they provide instructive material of great significance. For not only Mendelists, but even materialists who have not seen vegetative hybrids, may refuse to believe that everything that is alive, every particle of a living body, possesses heredity as well as the chromosomes. This can be easily demonstrated by the examples of vegetative hybridization. Chromosomes cannot pass from stock to scion and vice versa—that is a fact no one disputes. Yet hereditary properties, such as the colouring of the fruit, its shape, the shape of the leaves, and others, are transmitted from scion to stock and from stock to scion. Now show us any properties of two breeds united into one by means of sexual hybridization—in the case of tomatoes, for instance—which could not be united or have not been united by the Michurinists, by means of vegetative hybridization.

Thus experiments in vegetative hybridization provide unmistakable proof that any particle of a living body, even the plastic substances, even the sap exchanged between scion and stock, possesses hereditary qualities.

Does this detract from the role of the chromosomes? Not in the least. Is heredity transmitted through the chromosomes in the sexual process? Of course it is.

We recognize the chromosomes. We do not deny their existence. But we do not recognize the chromosome *theory* of heredity. We do not recognize Mendelism-Morganism.

Let me remind you that Academician P. M. Zhukovsky promised that if I showed him vegetative hybrids, he would believe and revise his position. I have now kept my promise and shown him vegetative hybrids. But I must remark, firstly, that dozens and hundreds of such hybrids could be seen in our country for at least a decade now; and, secondly, is it possible that Academician Zhukovsky, a botanist, does not know what is known to many, even if not all, horticulturists—namely, that in decorative horticulture a great deal has been done, and is being done, to change the heredity of plants by means of grafting?

Some of the Morganists who spoke at this session alleged that, together with the chromosome theory of heredity, Lysenko and his followers reject all the experimental facts obtained by Mendelist-Morganist science. Such allegations are false. We do not reject any experimental facts, and this holds good for the facts concerning chromosomes.

Some go so far as to assert that the Michurin trend denies the action upon plants of the so-called mutagenic factors, such as X-rays, colchicine, etc. But how is it possible to assert anything of the sort? Certainly, we Michurinists cannnot deny the *action* of such factors. We recognize the action of the conditions of life upon the living body. Why then should we refuse to recognize the action of such potent factors as X-rays or a strong poison like colchicine, etc.? We do not deny the action of the so-called mutagenic substances. But we insist that such action, which penetrates into the organism not in the course of its development, not through the process of assimilation and dissimilation, can only rarely and only *fortuitously* lead to results useful for agriculture. It is not the road of systematic selection, not the road of progressive science.

The protracted and numerous efforts made in the Soviet Union to produce polyploid plants with the aid of colchicine and similar potent factors have in no way led to the results so widely advertised by the Morganists.

A great deal has been said and written to the effect that a geranium began to give seeds after its chromosome complement had been increased. But this geranium is not being grown for the market, and I, as a scientist, venture the opinion that it never will be so grown, because it is much more practical to propagate geraniums by cuttings. Currants, for example, can be grown from seeds, but in practice they are propagated by cuttings. Potatoes can also be grown from seeds, but it is more practical to plant tubers. As a rule, plants which can be propagated both by seeds and by cuttings (i.e., by the vegetative method) are propagated for practical ends by the latter method.

This does not mean that we minimize the importance of the fact that a geranium has been obtained which is capable of producing seeds. If not for practical ends, this form can be of use in the study of plant breeding.

And what I have said of geraniums applies also to mint.

What other polyploids are often represented by the Morganists as highly important achievements? Wheat, millet, buckwheat, and a few other plants. But, according to the statements which we have heard here from the Morganists themselves (A. R. Zhebrak, for example), all these polyploids—wheat, millet, buckwheat—have so far, as a rule, been found to be of small fertility, and their authors themselves have refrained from recommending their cultivation for practical ends.

There only remains the tetraploid kok-saghyz. This is the first year it is being tested on collective farms. It goes without saying that, if it proves to be good, it ought to be introduced in practical farming. So far, however, according to the data of three years of government seed testing, it is not superior to the ordinary diploid varieties, such as Bulgakov's, for example. This is the first year tetraploid kok-saghyz is being tested on collective farms. In another two or three years we shall have practical proof of how good it is. I sincerely wish that it may prove to be the best of all kok-saghyz forms. Our agriculture can only gain thereby.

At the same time we must not forget that among the varieties of cultivated plants there are plenty of polyploids whose origin has nothing to do with colchicine and the "mutagenic" theory, nor, for that matter, with the theory of Morganism-Mendelism as a whole. For centuries people did not know that many good varieties of pears, for example, are polyploids. But we have also as many equally good varieties of pears which are not polyploids. These facts alone provide enough grounds for the conclusion that it is not the number of chromosomes that determines the quality of a variety.

There are good and bad varieties of durum wheat with 28 chromosomes, and there are good and bad varieties of soft 42-chromosome wheat.

Is it not obvious that breeding must be conducted, not with a view to the number of chromosomes, not with a view to polyploidy, but with a view to inducing good qualities and properties?

When a good variety has been produced, we can also determine the number of its chromosomes. But no one, certainly, will think of discarding a good variety only because it has turned out to be a polyploid or not a polyploid. No Michurinist, no serious-minded person generally, can approach the question from such an angle.

Our Morganists, among them some who spoke at this session, in order to adduce proof that their theory is effective, often point to some varieties of cereal grains which are widespread in practical farming, as, for example, Lutescens 062, Melanopus 069, and some other varieties of long standing which they claim have been produced on the basis of Morganism-Mendelism. But actually Mendelism has nothing to do with the production of these varieties. How, for example, have varieties like Lutescens 062, Melanopus 069, Ukrainka, and some others been produced? They were produced by the ancient method of selection from local varieties.

I shall quote here Professor S. I. Zhegalov, who wrote in his work, *An Introduction to the Selection of Agricultural Plants*: "Under ordinary farming conditions we have to deal, not with pure forms, but with 'varieties' representing more or less complex combinations of various forms.... The first, perhaps, to draw attention to this fact in the first quarter of the nineteenth century (long before the appearance of Weismannism—*T. L.*) was the Spanish botanist Mariano Lagasca, who published his observations in Spanish. There is an interesting story extant about a visit he paid to his friend, Colonel Le Couteur, at the latter's estate on Jersey Island. During an inspection of the fields he drew the attention of his host to the considerable diversity of forms among the plants and suggested that individual forms be selected for further pure breeding. The idea appealed to Le Couteur, who selected twenty-three different forms and began to test their relative merits. As a result of the tests, he found one of the forms to be the very best, and in 1830 put it on the market as a new variety named 'Talavera de Bellevue.' Since then this kind of work has been tried many times, and it has led to the production of many valuable varieties. In substance, it consists in separating the initial mixtures into their component parts. That is why this method is known as '*analytical selection.*' At present it is the principal method employed in work with self-pollinating plants and is systematically applied by all stations, particularly in the early stages of the work on plants formerly little affected by selection."[1]

A little farther Professor S. I. Zhegalov writes: "The method of analytical selection lends meaning to an aphorism credited to Jordan: 'In order to produce a new variety, we must first have it.' "[2]

Comrade Shekhurdin, was the form of wheat now called Lutescens 062 to be found among the local Poltavka variety or not? (*Voice from the audience*: "Yes, positively.") The same is true of the forms called Ukrainka and Melanopus 069.

That is why S. I. Zhegalov accepts the aphorism that in applying the method of analytical selection it is necessary, in order to produce a new variety, first to have it. The named varieties, to which our Mendelists usually point, have indeed been obtained in this manner.

We Michurinists, however, cannot agree with Professor S. I. Zhegalov and his interpretation of Darwinian selection. For it is possible to begin to select plants with scarcely perceptible, still feeble useful characters, in order to reinforce and develop these useful characters by repeated selection and proper cultivation. But, as is obvious to anyone, the described Darwinian method of selection has no bearing whatever on the Mendelist-Morganist theories.

It should be mentioned that formerly varieties were bred only on the basis of the above method. For that matter, this method is being applied today and will be applied in future. It is a useful method, and practical breeders who successfully apply it should be appreciated and encouraged.

---

[1] С. И. Жегалов, *Введение в селекцию сельскохозяйственных растений*, 1930 г., стр. 79-80.

[2] *Ibid.*, p. 83.

Far from rejecting the method of continuous improving selection, we, as is well known, have always insisted on it. The Morganists, on the other hand, have ridiculed the application of repeated improving selections in practical seed growing.

Weismannism-Morganism has never been, nor can it be, a science conducive to the systematic production of new forms of plants and animals.

It is significant that abroad, in the United States for example, which is the home of Morganism and where it is so highly extolled as a theory, this teaching, because of its inadequacy, has no room in practical farming. Morganism as a theory is being developed *per se,* while practical farmers go their own way.

Weismannism-Morganism does not reveal the real laws of living nature; on the contrary, since it is a thoroughly idealistic teaching, it creates an utterly false idea about natural laws.

For instance, the Weismannist conception that the hereditary characteristics of an organism are independent of environmental conditions has led scientists to affirm that the property of heredity (i.e., the specific nature of an organism) is subject only to chance. All the so-called laws of Mendelism-Morganism are *based entirely on the idea of chance.*

Here are a few examples.

"Gene" mutations, according to the theory of Mendelism-Morganism, appear fortuitously. Chromosome mutations are also fortuitous. Due to this, the direction of the process of mutation is also fortuitous. Proceeding from these invented fortuities, the Morganists base their experiments too on a fortuitous choice of substances that might act as mutagenic factors, believing that they are thereby acting on their postulated hereditary substance, which is just a figment of their imagination, and hoping to obtain fortuitously what may by chance prove to be of use.

According to Morganism, the separation of the so-called maternal and paternal chromosomes at reduction division is also a matter of pure chance. Fertilization, according to Morganism, does not occur selectively, but by the chance meeting of sex cells. Hence, the segregation of characters in the hybrid progeny is also a matter of chance, etc.

According to this sort of "science" the development of an organism does not proceed on the basis of the selection of conditions of life from the environment, but again on the basis of the assimilation of substances fortuitously entering from without.

In general, living nature appears to the Morganists as a medley of fortuitous, isolated phenomena, without any necessary connections and subject to no laws. Chance reigns supreme.

Unable to reveal the laws of living nature, the Morganists have to resort to the theory of probabilities, and, since they fail to grasp the concrete content of biological processes, they reduce biological science to mere statistics. It is not for nothing that statisticians like Galton, Pearson, and latterly Fisher and Wright, are also regarded as founders of Mendelism-Morganism. Probably that is also the reason why Academician Nemchinov has told us here that, as a statistician, he found that he could easily take in the chromosome theory of heredity. (*Amusement, applause.*)

Mendelism-Morganism is built entirely on chance; this "science" therefore denies the existence of necessary relationships in living nature and condemns practical workers to fruitless waiting. There is no effectiveness in such science. With such a science it is impossible to plan, to work toward a definite goal; it rules out scientific prediction.

A science which fails to give practical workers a clear perspective, the power of finding their bearings and confidence that they can achieve practical aims does not deserve to be called science. (Applause.)

Physics and chemistry have rid themselves of fortuities. That is why they have become exact sciences.

Living nature has been developing and is developing on the basis of strict laws inherent in it. Organisms and species develop in line with natural necessities inherent in them.

By ridding our science of Mendelism-Morganism-Weismannism we will expel fortuities from biological science. (Applause.)

We must firmly remember that *science is the enemy of chance.* (*Loud applause.*) That is why Michurin, who was a transformer of nature, put forward the slogan: "We cannot wait for favours (i.e., lucky chance—*T.L.*) from Nature; we must wrest them from her." (*Applause.*)

Aware of the practical sterility of their theory, the Morganists do not even believe in the possibility of the existence of an effective biological theory. Ignorant even of the ABC of the Michurinist science, they cannot to this day imagine that for the first time in the history of biology a truly effective theory has come into being—the Michurin teaching. (*Applause.*)

A great deal can be scientifically predicted on the basis of the Michurin teaching, thus freeing practical plant breeders to an ever-increasing extent from the elements of chance in their work.

Michurin himself elaborated his theory, his teaching, only in the process of solving problems of practical importance, in the process of the production of good varieties. That is why *the Michurin teaching is, by its very spirit, inseparable from practice.* (*Applause.*)

Our collective-farm system, our socialist agriculture, created all the conditions for the flowering of the Michurin teaching. Let us recall Michurin's words: "In the person of the collective farmer the history of agriculture of all times and all nations has an entirely new type of farmer, one who has taken up the struggle with the elements marvellously armed technically and acting on Nature as a man with the views of a renovator."[1]

"I see," wrote I. V. Michurin, "that the system of collective farming, by means of which the Communist Party is inaugurating the great work

---

[1] И. В. Мичурин, *Сочинения*, Сельхозгиз, 19 9 г., том I. стр. 477.

of renovating the land, will lead labouring humanity to real dominance over the forces of Nature.

"The great future of our entire natural science is in the collective and state farms."[1]

The Michurin teaching is inseparable from the practical collective- and state-farm activity. It is the best form of unity of theory and practice in agricultural science.

It is clear to us that the Michurin movement could not develop extensively, if there were no collective and state farms.

Without the Soviet system, I. V. Michurin would have been, as he himself wrote, "an obscure hermit of experimental horticulture in tsarist Russia."[2]

The strength of the Michurin teaching lies in its close association with the collective and state farms, in the fact that it *elucidates profound theoretical problems by solving important practical problems of socialist agriculture.*

Comrades, our session is drawing to its close. This session has vividly demonstrated the strength and potency of the Michurin teaching. Many hundreds of representatives of biological and agricultural science have taken part in it.

They have come here from all parts of our vast country. They have taken an active part in the discussion on the situation in biological science and, convinced in the course of many years of practical activity that the Michurin teaching is right, are ardently supporting this trend in biological science.

The present session has demonstrated *the complete triumph of the Michurin trend over Morganism-Mendelism.* (*Applause.*)

It is truly a historic landmark in the development of biological science. (*Applause.*)

I think I shall not be wrong if I say that this session has been a great occasion for all workers in the sciences of biology and agriculture. (*Applause.*)

The Party and the Government are showing paternal concern for the strengthening and development of the Michurin trend in our science, for the removal of all obstacles to its further progress. This imposes upon us the duty to work still more extensively and profoundly to arm the state and collective farms with an advanced scientific theory. That is what the Soviet people expect of us.

We must effectively place science, theory, at the service of the people, so that crop yields and the productivity of stockbreeding may increase at a still more rapid pace, that labour on state and collective farms may be more efficient.

---

[1] *Ibid.*

[2] И. В. Мичурин, *Сочинения*, Сельхозгиз, 1941 г., том IV, стр. 116.

I call upon all academicians, scientific workers, agronomists, and animal breeders to bend all their efforts and work in close unity with the foremost men and women in socialist farming to achieve these great and noble aims. (*Applause.*)

Progressive biological science owes it to the geniuses of mankind, *Lenin* and *Stalin*, that *the teaching of I. V. Michurin has been added to the treasure house of our knowledge, has become part of the gold fund of our science.* (*Applause.*)

Long live the Michurin teaching, which shows how to transform living nature for the benefit of the Soviet people! (*Applause.*)

Long live the Party of Lenin and Stalin, which discovered Michurin for the world (*applause*) and created all the conditions for the progress of advanced materialist biology in our country! (*Applause.*)

Glory to the great friend and protagonist of science, our leader and teacher, Comrade Stalin! (All rise. Prolonged applause.)

EXPERIMENTAL HILL SOWING OF FOREST BELTS[1]

THE decision of the Party and the Government "On the Plan for the Planting of Shelter Belts, Introduction of Travopolye Crop Rotations and Building of Ponds and Reservoirs in Order to Ensure High and Stable Harvests in the Steppe and Forest-Steppe Areas of the European Part of the U.S.S.R." has imposed upon the Lenin Academy of Agricultural Sciences a number of very important tasks. Point 26 of this decision reads:

"It shall be the duty of the Ministry of Agriculture of the U.S.S.R., the Ministry of Forestry of the U.S.S.R. and the Ministry of State Farms of the U.S.S.R. to ensure the experimental hill sowing of forest belts in the fields of each scientific or experimental institution, as well as in the state forest farms of the steppe and forest-steppe areas. The Lenin Academy of Agricultural Sciences of the U.S.S.R. shall, within a period of two months, draw up instructions concerning the carrying out of these sowings."

In execution of the task set by the Party and the Government, our Michurin biology must and can harness the full strength of its collective thought and assemble all the necessary theoretical and practical knowledge for the purpose of drawing up such instructions concerning the experimental planting of forest belts in the fields of scientific-research institutions and state forest farms as will in the next few years convincingly demonstrate to practical agriculturists the surest way of cultivating extensive forests in the steppe. In this manner we shall be in a position to render scientific assistance to the collective and state farms of the steppe areas in creating the best conditions for the growing of forests with the least expenditure of human labour and material.

For this purpose we must make the best possible use of the theory of Michurin agronomic biology, linking it up still more closely with concrete collective-farm practice, with life, in this particular question with the practical problems of steppe forest cultivation. Genuine science tolerates no

---

[1] A paper read at a conference of scientific workers of the Lenin Academy of Agricultural Sciences of the U.S.S.R. held on November 23, 1948.—*Ed.*

fortuities, is averse to haphazard methods. Its aim is to foresee, and this is its duty to practical farming.

A knowledge of the *causes* of phenomena and of the effects that follow from the interconnections of phenomena makes it incumbent upon Michurinist scientific workers to devise such working plans for the solution of major and minor practical problems as will yield the best results at the present time. In the case at hand the plan of our scientific work and its fulfilment should help the collective and state farms to set up good permanent forest belts with the least expenditure of labour and material.

Permit me to say a few words at the outset about a generally known phenomenon, namely, the struggle of the steppe against the forest and vice versa. Hitherto the steppe in most cases vanquished the forest. Responsible for this was not the fact that the forest, as a natural phenomenon, was always too weak to combat the steppe, but the fact that due to the anarchy of capitalist economy man's interference with nature almost always facilitated the victory of the steppe over the forest. The opposite occurred but rarely. Indeed, until recently man in the vast majority of cases engaged in felling timber, doing but little to grow it. Therefore man intentionally or not helped the steppe in its struggle against the forest. True, in clearing land of forest to make room for cultivated field crops man constantly took measures to prevent wild steppe plants from entering fields sown to cultivated crops.

It is well known, for instance, that enemies of cultivated crops are couch grass and catchweed—the vanguard of steppe plants in their struggle with the forest and annual cultivated crops.

Consequently, wild steppe vegetation is the common foe of the forest and of agricultural crops. But man always protected the latter by agrotechnical measures from weeds, including such pioneers of steppe vegetation as couch grass and catchweed.

We likewise know that forests in steppe areas create favourable conditions for cultivated agricultural plants. Forests moderate and even eliminate inclemencies of the steppe climate, such as strong winds, dust storms and dry winds.

The climatic tribulations of the arid steppes hinder the development of agricultural cultivated plants, which means that they lower the productivity of human labour. Consequently, the steppe uses its vegetation and its entire complex of climatic conditions to fight both against the forest and against cultivated, agricultural plants.

Is it therefore really impossible for us scientific workers to combine the growing of sown and planted forests with the cultivation of various agricultural field plants in combat against their common enemies, the wild steppe plants and the inclemencies of the weather, and is it impossible to derive practical advantages from such a combination?

I for one believe it to be possible. Even without turning to biological theory it may be said as a mere matter of practice that if one thing hinders two the two can always combine, at least temporarily, against their common enemy. I shall confine myself for the time being to this simple point to substantiate the measures for the hill sowing of protective forest belts in the steppe on long-cultivated land. I shall not speak in my report on the width of shelter belts or how they should be distributed in the fields. I shall only speak on how to create the best conditions for growing the kinds of woody plants we need, mainly oak, with the least expenditure of labour and material.

On a well-ploughed cultivated field prepared for the sowing of winter or spring cereal or any other agricultural crop, a strip of land should be marked off for the sowing of the trees. The marking must be in two directions. In one direction a space of 5 m. must be allowed between rows; in the other direction, at right angles to the first, a space of 3 m.

At each intersection, of which there will be 667 to a hectare, sow 35-40 germinative acorns.

Such sowing requires about one centner of acorns per hectare of forest belt. The technique and time of sowing will be given below.

Thus, there will be 667 acorn-sown patches per hectare.

The question inevitably arises why some agricultural crop should not be sown in a ready, well-tilled field, where only 667 patches of about one square metre each per hectare are sown to oak, in strictly rectangular order, the centre of each of these patches being 5 m. from the centre of the next in one direction and 3 m. in the other. It is obvious that in such cases agricultural crops not only may but should be planted.

After sowing acorns in hills, any row crop—melons, potatoes, root crops, maize, sunflower—or wide-row crops—millet and buckwheat—or nonrow cereals—wheat, barley or oats—should be sown in the field in the 4-m-wide spaces between the rows, about 1 m. out of the 5 being taken up by the acorns.

If after the acorns are sown the plot is left to be sown or planted late annual crops, as for instance, millet or buckwheat or summer-sown potatoes, the spaces between the patches sown to acorns must, after rains or after the appearance of weed sprouts, be cultivated with implements provided with duck-foot cultivators which neither turn nor dry up the soil.

If, on the other hand, after the sowing of the acorns early cereals are to be sown they must be sown on time and only in the wide interspaces.

The outside distance between the wheels of a tractor-drawn 24-row disc seeder is 4.1 m.

If row crops or solid cereal areas are thus sown in the wide interspaces the result is a field occupied by strips slightly less than 4 m. wide sown to annual crops alternating with strips slightly more than 1 m. wide on which every 2 m. of still free, unsown area alternates with 1 m. already occupied by a patch (hill) of sown acorns. We recommend that the 2 m. free patches be sown in all cases to maize or sunflower in hills

50 cm. apart. On a patch a little over 1 m. in width and 2 m. in length there will be room for 3 hills of maize or sunflower. In each hole 3-5 maize or sunflower plants should be placed. Thus 3 hills of maize or sunflower will lie between each 2 hills of oak sprouts. These sunflower lie between each 2 hills of oak sprouts. These sunflower or maize crops occupying approximately 1,500 sq. m. per hectare of sown forest must be hand-tilled during the period of vegetation. When gathering in the harvest their stalks must not be cut. They must be left standing to accumulate snow on the tree-sown area in winter.

Thus, the first year after the acorns are sown there will be only 667 patches (hills) of oak sprouts per hectare of forest belt. All the remaining area will be sown to annual agricultural crops.

After the annual crops have been harvested on the wide, 4 m. interspaces this area must at once be stubble-ploughed by means of disc implements.

In the autumn the stubble-ploughed spaces between the patches of oak seedlings, each 4 m. wide, must be sown to rye by a tractor-drawn 24-row disc seed drill. Of the drill's 24 furrow openers three, namely, the 6th, 12th and 19th, must sow not rye but brushwood seeds. In most cases yellow acacia seeds with an admixture of other bushes should be taken.

In this way the end rows of the sown bushes (yellow acacia, etc.) will be at a distance of 152.5 cm. from the centres of the oak patches. The distance between the rows of bushes will be 90 and 105 cm.

To sow yellow acacia and other brushwood seeds the three sowing apparatuses in the seed box of the drill which correspond to the discs that are to cover the brushwood seeds in the ground are partitioned off. The seed box is filled with rye seeds for all the sowing apparatuses except the three indicated. For these three (6th, 12th and 19th) the compartments partitioned off are filled with brushwood seed. To regulate the rate of sowing brushwood seed the requisite amount of rye seed is added.

Early in spring the maize or sunflower stalks left during the winter to accumulate snow must be removed from the narrow strips between the hills of oak seedlings. As has already been stated, these strips are only a little over a metre wide and the hills are 2 m. apart. In spring one hill of Norway maple seed must be sown in shovel-turned soil between each two oak hills. Hence the maple spots will be 1 m. away from the edges of the nearest hills of oak yearlings.

The second year after the oak is sown the belts will have the following appearance:

1) The centres of the patches (hills) containing the year-old oak seedlings will be 5 m. apart on two sides and 3 m. apart on the other two sides;

2) In the wide interspaces there will be 3 rows of yellow acacia seedlings mixed with other brushwood; the rows of brush will be 90 and 105

cm. apart; the end rows of brushwood will be 152.5 cm. away from the centres of the oak patches;

3) On the sides where the centres of the adjacent hills of oak seedlings are 3 m. apart there will be one Norway maple hill between each two oak hills;

4) The wide interspaces (approximately 4 m. in width) between the edges of the hills of oak seedlings, along which 3 rows of yellow acacia mixed with other brushwood are sown, will be occupied by a solid stand of rye.

When harvesting the rye it must be cut as high as possible so as to leave high stubbles for snow retention on the young forest belt.

In the autumn, after the rye has been gathered in, the 4-m-wide interspaces already sown to rye must be sown a second time to rye, but without any presowing cultivation of the soil. Sow directly into the stubble and the yellow acacia sprouts. The latter will not be damaged by the disc drill or will suffer very little.

During the second year of the yellow acacia its shoots will already be higher than the cutting line of the moving machine (self-propelled combine). In harvesting the rye the tops of the yellow acacia will therefore be cut off. This will only be to the good, for the lower parts of the yellow acacia stalks will branch better as a result of the cutting.

In the autumn of the second year of the yellow acacia (and the third year of the oak) rye must be stubble-sown once more. In harvesting the rye the acacia tops will be clipped once more, which will augment the ramification of the shoots.

The following year, after the ripening of rye, the oak seedlings will be four years old, and the maple and acacia seedlings three years old. It seems to us that thereafter the forest belt may be left to grow clean, i.e., without sowing the interspaces with grain. In its fourth year yellow acacia whose tops were clipped the two preceding years will be able to close in the entire free area and keep out steppe weeds, particularly couch grass and catchweed.

What advantage is obtained by combining the planting of forest trees, in this particular case dominant oak, with that of annual agricultural crops?

Forest trees are mortally afraid of steppe grasses, particularly couch grass and catchweed. Couch grass, catchweed and dog's-tooth grass are different plants. They usually grow under different climatic conditions, in different areas, but they perform the same role in the development of the steppe—that of pioneers, of the first detachment of steppe vegetation in the struggle between steppe and forest.

Among wild grasses couch grass and catchweed are the fittest to struggle against forest trees, particularly during the first few years of the lifetime of the latter.

The forest also has its different breeds and species which in the general combat between forest and steppe act as pioneers, as the vanguard of a forest forcing back the steppe vegetation.

By this I only mean to say that we biologists should know that not all species of steppe plants and not all species of forest plants are equally persistent in this struggle between steppe and forest.

When the proposals I have submitted are carried out the young seedlings or plantations of forest trees will be protected against their worst enemies—couch grass and catchweed—by annual agricultural crops. This protection will consist both in the cover afforded by the cultivated annual agricultural plants and in the tilling of the soil sown to annual agricultural crops. And this will constitute the advantage which the young forest tree seedlings secure by uniting with the various annual agricultural crops until the canopy of the tops of the trees and shrubs closes. After the crowns of the above-indicated combination of woody plants—oak, maple and underbrush—have become contiguous, the forest belt will resist the encroachments of the steppe plants by itself and will prevent the intrusion of couch grass and other enemies of the forest.

Practically our proposals concerning the method of carrying out the experimental sowing of forests in steppe and forest-steppe areas amount to the following: a) the recommendation of suitable woody plants: oak, Norway maple, yellow acacia and other underbrush, and the suitable distribution of these trees and underbrush over the area in question: oak and maple—in hills, yellow acacia mixed with other underbrush—in rows; b) the protection of the tree seedlings during the first few years of their lives from the inclemency of the steppe, mainly from wild steppe grasses, by sowing various annual agricultural plants.

What are the theoretical premises from which we proceed in our proposals?

We proceed from the qualitative difference that exists between interconnections or interrelations among individuals within a single species and those that exist among different species. We take as our starting point the fact that all intraspecific interconnections between individuals, like the interconnections between the organs of an organism, tend to ensure the existence and welfare of the species, which means, tend to increase the number of individuals in the species. Not a single instance can be cited from wild plant or animal life of any organ of an organism or any property of a plant or animal tending in any degree to harm its own species. This is one of the principal theses of Darwinian natural selection.

The life of a species is composed of the lives of its individual members and the lives of all of them taken together. In nature there is no struggle or mutual assistance among individuals of the same species, nor can there be. Intraspecific struggle and competition—reactionary Malthusian propositions—were introduced into biological science, into Darwinism, from without. They impeded the scientific comprehension of the laws of living nature.

Recognition by science of intraspecific struggle is particularly injurious in the practical work of afforestation, of which brief mention will be made below when the practical aspect of forest development in the steppes during the past century will be discussed.

Biology must start from the premise that species are not only units in botanical and zoological systematics. Species are qualitatively distinct states of living matter; hence species exist in nature as separate links of the multiplex chain of living nature in development. Living nature is represented not by a continuous line but by a single chain consisting of separate, qualitatively different links—species.

It is a well-known biological fact that every species of plant or animal lives at the expense and to the detriment of other species. Hence, no species is exempt from the inroads of other species which live at its expense and to its detriment. Some species, for instance, are carnivorous, beasts of prey, feed on other species of animals. Hence, a struggle assuming various forms goes on between the former and the latter. Herbivorous species of animals feed on various species of plants; between these too a struggle is in progress. For instance, some plants develop burrs, others elaborate various substances which are poisonous to animals. Different species feed on identical species; members of different species demand the same conditions of life. Hence, competition arises among various species of plants, for instance, for light, moisture or nutrition. To be successful in this competitive struggle different species evolved different organs through natural selection. Parallel with interspecific struggle and competition we biologists can and should observe different degrees of union among the individuals of different species to wage a common struggle for life, these unions being directed both against the individuals of other species, their enemies and competitors, and against the inclemency of inanimate nature. All this goes to show that what exists in nature is struggle, competition and mutual assistance among the individuals of the various species.

However, the intraspecific interrelations among individuals come neither under the concept struggle nor under the concept mutual assistance inasmuch as all these interrelations tend solely to secure the existence of the species, its well-being, to increase the number of its members.

It was on the basis of the theoretical and biological principles that in nature there is no intraspecific struggle or competition, nor any intraspecific mutual assistance, but only interspecific struggle, competition and mutual assistance that the above program of experiments in forest development was drawn up.

Indeed, no forester will deny that couch grass, catchweed and other wild steppe plants are the greatest scourge of tree and brush seedlings (particularly if thinly sown). We propose to keep away plants that have such a baneful effect on forests by sowing various annual cultivated species of plants which are not inimical to forest species and do not have special organs with which to attack the roots of trees and brushwood. Good agrotechnique, good care of these cultivated plants should richly repay the tree and brush seedlings for the soil moisture consumed by the crops. Foresters know that seedlings of woody plants require shade. The top growths of cultivated plants provide in their mass beneficial shade for the tree and brush seedlings and increase the moisture of the air round about these seedlings.

For the sowing of forest belts we recommend that such trees and brushwood be selected as will soonest be able collectively to resist the inclemency of the steppe and at the same time produce a permanent forest in the steppe and yield good wood. Hence, we recommend oak, Norway maple and rapidly growing underbrush such as yellow acacia. Yellow acacia intermixed with other brushwood should quickly supply shade for soil where there are no oaks or maples and not allow wild grasses to penetrate. In order that the oak may not perish, be choked off by the maple, yellow acacia or other underbrush, we recommend that the oak be sown thickly, in groups, in hills of 35-40 acorns. Besides, since oak trees grow slowly at first, we suggest that they be sown a year earlier than the maples and yellow acacias.

I believe that the proposal to sow oak in hills will also arouse no objection on the part of practical foresters. They know that in natural mixed forests too the different kinds (species) of trees and brushes when young always come in clusters or groups. Single young trees, oaks for example, will always be obstructed in a forest by other species of woody plants.

There is no reason to be afraid that 20-30 oak or maple seedlings will be cramped on a small, say, one sq. m. patch. After all, we need not rear all the oak and maple seedlings. What we need is an oak forest with maple as a codominant. All we want is that when the forest is 30-50 years old, one or two oaks should be left on each square-metre patch, i.e., each hill, and one or two maples out of each maple hill. This will be quite enough for the area in question, each hectare of which will contain about a thousand oaks evenly distributed, to constitute an oak forest with an undergrowth of maple and brushwood.

Wild plants, particularly various species of forest trees, possess the biologically useful property of self-thinning. The property of self-thinning consists in the ability of dense sprouts of a particular species by reason of their mass to successfully withstand other species and at the same time not to interfere, to compete, with each other. The reason why this happens is that as the young trees grow, fewer woody plants than those that exist can provide the requisite contiguity of crowns (branches); therefore a number of trees normally are eliminated, die off. When the growth of trees is dense a process of differentiation, as practical foresters call it, separating the trees into an upper, intermediate and lower layer, goes on within each species. Trees of the lower layer have outlived their usefulness and wither away while those of the intermediate layer pass on to the lower or upper layer as circumstances may determine. Wild plants, particularly forest trees, as has already been stated, possess the property of timely self-thinning in so well-expressed a form that even experimentally they cannot be sown so densely as to cause the kind of tree (species) in question to perish on that account in the given area. Quite the opposite: the more densely the species in question is sown the greater the probability of its favourable development in the particular area.

Such cultivated plants as, for instance, wheat and a number of others do not possess the biological property of self-thinning. If these plants are sown too thickly they do not differentiate into strata and not one of their great number can develop normally and produce a normal crop of seeds. Too thickly sown crops of cereals, for instance, particularly in arid regions, perish to the last plant, without producing a seed crop.

As was pointed out above, wild plant species, including forest trees, behave differently in this respect. That is the reason why in nature they worst their competitors, other species.

Dense sprouts of a wild plant species so regulate their number by means of self-thinning that its individual members cannot interfere or oppress each other and at the same time the entire area is occupied by this particular species. Other species, competitors of the given species, are not admitted to this area.

Let us now briefly review the experience gained in the cultivation of steppe forests during the last one hundred years or so from the point of view of the thesis that there is no intraspecific competition or mutual assistance but that there is interspecific struggle, competition and mutual assistance. As this experience has been repeatedly dealt with by various authors in books on silviculture and as it is well known to those who specialize in this field, I shall not dwell on its history. I only want to call attention to the fact that this vast practical material accumulated during so long a period of time will easily convince anyone of the harm done to agricultural practice by the recognition in biological theory of a ruthless intraspecific struggle which does not exist in nature while ignoring the existence of interspecific struggle, competition and mutual assistance.

It is well known that during the course of some one hundred years of steppe-forest cultivation several fine forest massifs were created, but there were also many failures: planted forests perished.

The progress achieved by our Michurin biology enables me to see clearly that the principal cause of these failures in steppe-forest cultivation was the fact that the old biological science recognized fictitious intraspecific competition and ignored interspecific competition. On the other hand, all successful cases of steppe-forest cultivation which resulted in fine forest massifs can be explained in the main by the fact that practical foresters intentionally or unintentionally disregarded the erroneous recommendations of theory.

Indeed, how else if not by the recognition of intraspecific competition and the ignoring of the interspecific struggle can one explain the type of planting called the Don method and afterwards the "normal" method of steppe-forest planting, the types which in the past were compulsory for government forestries. After all, these methods consisted essentially in the rotational planting, in the row, of single trees of different species. As we know, the rows were planted 1.5 m. apart and in each row the saplings were 60 cm. apart. In order that the young trees of the same species, say oak, might not compete with or oppress each other, the rows were planted as follows: one oak sapling, 60 cm. away a young tree of an elm species, 60 cm. further a young ash tree, then a young maple, then an elm of another species and then, finally, again a young oak.

Thus the dominant species for the steppe, the oak, was, according to the fictitious "theory" of intraspecific struggle, given apparently favourable conditions. The young oaks were planted far apart so that they might not hinder each other, so that there might be no struggle or competition among them. The young oaks were so to speak placed under the protection of other species—elms, ashes and maples. But elms and ashes in close proximity to single oak saplings do not protect the latter but on the contrary are their worst competitors and oppressors. To place single oak seedlings under the protection of ash trees, for instance, comes close to putting chickens under the protection of foxes.

What, then, was bound to become of such plantations in practice if there is no intraspecific but only interspecific competition in nature?

Though of various species the young trees could not hinder each other during the first few years of their lives since their rows were 1.5 m. apart and they were planted 0.6 m. apart in these rows. Repeated cultivation throughout the year protected these plantations from the steppe grasses. The belts continued to be tilled until the young trees locked horns, after which such tilling became impossible. The contiguity of the planted trees should by now itself keep out the steppe grasses. But when the branches of intermixed young trees of various species touch, fierce interspecific competition among them for light and moisture sets in.

Foresters have long known from observation and experience that the oak is the principal, the dominant species for forming permanent forests in the steppe. Other trees play a secondary though also important role in the cultivation of steppe forests.

But inasmuch as the oak during the first five years or so of its life grows exceedingly slowly and thus has time to develop deep roots any other kind of tree situated close to a single oak will obstruct, oppress it and in the end the oak will succumb. In all the forests planted by the so-called "normal" method described by us the oaks perished rapidly. Having been planted singly in between other species of trees the oak saplings were unable to compete with them. But after the soil ceased to be cultivated the power of resistance of the other tree species, those which had destroyed the young oaks, proved insufficient to withstand the inclemency of the steppe. That is just why all such plantations in the beginning, during the first few years, while they were being cultivated, gladdened the hearts of foresters, but afterwards brought them disappointment, for they began to wither and perish. On the other hand, planted forests thrived where the dominant species, the oak, for some reason or other was not oppressed by other species of forest trees when it was young. In all these cases a number of other woody plant varieties such as the Norway maple as well as many bushes found themselves in favourable conditions under the protection of the oak trees. These successful plantations in various steppe localities are unimpeachable proof that it is perfectly possible for good, permanent forests to grow in arid steppes. The lack of success in the cultivation of steppe forests is to be attributed to the false theory of the existence of fierce intraspecific competition and an ignoring of the existence of interspecific competition, struggle and mutual assistance.

Some forestry experts like Morozov, Vysotsky and Ogiyevsky, who were well acquainted with forest life, made correct practical recommendations but at that time it was beyond their power to change biological theory, to throw overboard the reactionary thesis of intraspecific struggle. Therefore, the practical recommendations of these scientists were shelved indefinitely, while the false theoretical propositions on forest cultivation persisted until the recent past.

In order to show the practical results achieved in steppe afforestation when the theory of intraspecific competition was recognized and interspecific competition ignored, i.e., to show the results of the so-called "normal" method of planting, according to which saplings of different varieties were planted singly and in rotation, permit me to quote the following excerpt from M. K. Tursky's textbook *Silviculture*, pp. 307-08, published in 1929:

"This was called the 'normal' method and at the end of the eighties of last century was made the obligatory pattern for all foresters of the steppe belt.

"In such plantations elms, by virtue of their rapid growth, began to choke off oaks when they, the elms, were only 3-4 years old. Aid in the shape of a special method of tending oaks—opening of the canopy—had to be speedily rendered. Opening the canopy meant doing the following: at first, when separate elm branches hung over the oaks, they were lopped off. Then, as the elms and oaks grew, such pruning was insufficient and the elms were cut down halfway in such a manner that the tops of the oaks were entirely free. Later they cut the whole tree down to the ground. The first kind of canopy opening was called *weak*, the second *medium* and the third, cutting down the whole tree, was called *strong*.

"Seven or eight years later the oak plantations mixed with elms in accordance with the above method, began perceptibly to ail and their tops to wither, particularly where elms had been introduced, and there were hardly any oaks left. They had been choked until they perished in spite of the light penetration. In 12-15 years the process of dying off of the plantations could no longer be checked.

"... Where the elm type was departed from, where tatarian maples and particularly yellow acacias replaced part of the elms, the plantations, especially the oaks, looked healthy. These observations were the occasion for a special report by G. N. Vysotsky in 1893 in which he developed his idea that underbrush should be introduced instead of elms. In his opinion the underbrush would shade the soil during the first few years, the same as the elms did, but would not choke off the oaks."

I have quoted this passage from Tursky's textbook in order to show that in practice some foresters discerned, practically sensed, the existence of interspecific struggle and mutual assistance. They also knew that different species under different conditions behave differently toward each other. Practical forest cultivation shows that combinations of secondary forest-tree species must be chosen with skill so that they may help and not hinder the dominant species, such as oaks and pines.

Some foresters recommended that oaks be sown or planted not singly but patchwise. Ogiyevsky, true enough, experimented in patchwise sowing of oak on a rather large scale, on hundreds of hectares, not for the steppe but for the forest zone (Tulskiye Zaseki). He sowed about 200 acorns on each 2 sq. m. patch. He saw and realized that in a forest zone the oaks' chief enemy is the aspen and in order to protect the oak from the aspen he sowed the former thickly in patches in the expectation that a great number of oak sprouts on a small patch of land would be able to withstand the pressure of other species. As we know, this experiment of Ogiyevsky's proved a splendid success.

That Ogiyevsky's experiment in thickly planting forests patchwise should be made use of in our practical work is not the only point here. This old-time experiment also implies that its author realized from his observation of forest life that what existed in nature was not intraspecific but interspecific competition while in science false theses continued to exist.

A scientific analysis of steppe-forest cultivation from the standpoint of the nonexistence of intraspecific struggle and competition and the existence of interspecific struggle and mutual assistance was made, in my opinion, by Kharitonovich in his article, "Interspecific Struggle and Mutual Assistance in the Planting of Steppe Forests," which appeared in No. 6 of the journal *Agrobiologia* for 1948. The article adduced examples showing that if in the steppe rows of oak are sown alternately with rows of ash, the oaks as a rule perish. The ashes choke off the rows of oak while the ashes themselves, owing to the fact that their crowns allow much light to penetrate to the soil, perish from the steppe grasses.

At the same time in those steppe forestries in which the oaks for some reason or other were not choked off by other species, as for instance when planted together with maple, yellow acacia and other underbrush, fine forest massifs were obtained, the dominant species of which was the oak and. under its protection, maple and underbrush.

The long experience, in the past, of steppe-forest cultivation, it seems to me, has finally convinced all foresters that the dryness of the steppe is not an insurmountable obstacle to the creation of good forest massifs. Clear proof of this is furnished by the fine forest massifs, almost 100 years old, which have been grown in the steppes.

In addition to this the past experience of steppe afforestation has convinced all foresters that the so-called Don and "normal" methods of planting forests are absolutely unsuitable as they have failed to produce any favourable results in practice. While criticizing and condemning the nonviability of forests planted by these methods, since such plantations dried up and perished (their span of life was no more than 15-25 years), the foresters did not, however, disclose the fundamental theoretical theses, erroneous from the point of view of biology, upon which the indicated methods of planting forests are based.

These theoretical mistakes were the consequence of recognizing the existence in the plant and animal world of a ruthless intraspecific struggle, nonexistent in nature, and the ignoring of the interspecific struggle and mutual assistance that actually does exist in nature. This theory served as the basis for the recommendation to plant single saplings of various species of woody plants under the Don method and the so-called "normal" method of planting. After numerous failures the foresters discarded the above methods of planting in practice, but in the science of forest cultivation, as has already been stated, nothing was changed.

As a result of this situation some experts in the science of forest management, acting on their book knowledge of the history of steppe-forest cultivation during the past century, quite correctly reject in toto the so-called "normal" method of forest planting. At the same time it seems to me they do the wrong thing when to this day they likewise keep on recommending that trees of various species be row-planted singly and in rotation for the growing of forest belts and forest massifs. These forest experts likewise proceed from erroneous theoretical premises.

These comrades may raise the objection that under the so-called "normal" method of forest planting only tall woody plants were used without any underbrush admixture whereas now the introduction of a certain percentage of underbrush in forest planting is recommended. Yet no forester will deny that underbrush is, by and large, necessary, not to ensure the longevity of the forest, but to cover the soil as quickly as possible, to make it unnecessary as soon as possible to cultivate the soil for the purpose of weed control. The "normal" method of tree planting practised in the past by government forestries was discarded not because these forests required soil cultivation for a lengthy period of time but because they did not live long.

These forests perished not because they did not contain a certain percentage of brushwood but because they consisted of a mixture of singly planted trees which strongly competed with each other. Foresters know that in the majority of our steppe districts planted forests will not live long unless oak is the dominant species. Single oak saplings, however, will always be obstructed, as a rule, by any other variety growing alongside of it.

The dominant species must not be planted singly but in groups or hills so as to prevent other species from oppressing the dominant species—the oak, and in sandy soil the pine—in their early years. When the dominant species, in the case at hand the oak, has grown to a certain height, shade-enduring species such as Norway maples, limes and various bushes, will thrive under its protection. This is the very reason why we propose that instead of rotating the various species of trees singly the dominant species be sown thickly, in groups, or hills.

In order to create better conditions for the young forest tree seedlings in their struggle with the wild steppe grasses it is proposed that suitable annual agricultural crops be planted in the experimental forest belts.

We have taken almost all the individual elements of the proposed system of sowing and tending forests from literature on forestry practice, but both in the selection of these elements and in the grouping of them into a single system we proceeded from the fact that in nature there is no intraspecific struggle; what there is is interspecific struggle and mutual assistance.

The proposed system of experimental forest planting is based on the necessity of providing the best conditions for the creation of permanent forests in steppe areas with a minimum expenditure of energy and material on rearing them. But this system of planting forest belts and tending them in steppe and forest-steppe areas has never yet been given a practical test. It is therefore submitted for experimental testing in scientific and experimental agricultural institutions as well as in state forest farms.

No later than the spring of 1949 all scientific and experimental agricultural institutions as well as forest farms must sow large experimental forest belts by the hill method as explained in the instructions.

In fulfilling this program scientific and experimental institutions must strictly adhere to the principal theoretical propositions upon which the system we recommend is based.

At the same time in selecting tree and shrub varieties as well as annual agricultural crops local soil and climatic conditions are the basic consideration. Thus, if the soil is sandy the pine and not the oak should be the dominant tree. The same applies to tree species that accelerate the development of the oak and to brushwood that shades the soil.

The main point in the proposed system is to sow the dominant species of trees in hills and correctly to distribute these hills over the assigned area so that the sowing or planting of annual agricultural plants can be made the principal method of combating wild grasses.

Special attention must be paid to the selection and growing of the dominant species which is to ensure the longevity of the forest and the good quality of its wood. The secondary but at the same time shade-enduring species of trees and underbrush, necessary for steppe forests, will be distributed so as to come under the protection of the dominant species. *In many steppe and forest-steppe areas the oak should be the dominant species, and in sandy soil the pine.*

That explains why scientific research institutions and forest farms must be particularly careful at the present time properly to store in winter the acorns procured in the autumn so as to prevent them from losing their germinating capacity. Early in spring an additional supply of acorns must be gathered in the forests.

Another very important measure for successful steppe-forest management is the infection of the acorns and sowing areas with suitable fungi—mycorrhiza, for if these are not implanted into the roots of the young oak seedlings the latter will not grow. In the steppe, oak seedlings on whose roots mycorrhiza does not develop will perish during the first or second year. It is therefore necessary during winter and early spring to secure a supply of earth from healthy old oak plantations or nurseries where oak seedlings have been raised. A volume of earth must be prepared at least twice as great as that of the acorns to be sown.

Acorns must be sown in spring as early as possible, to avoid dry soil. We recommend that after the strips intended for the planting of forests are marked off the oaks be sown on patches in the following manner. A small hole is made with a hoe at the centre of the patch where the lines of the marker intersect. For the purpose of causing mycorrhiza infection a bit of earth taken from an oak forest is thrown in; 7-8 acorns are then planted in this earth. Thereupon the hole is closed with damp earth which is pressed down slightly with the foot and covered with a layer of loose earth 1-2 cm. thick. The acorns should be planted about 5-7 cm. deep. Around this hole, at distances of 30 cm. from it, four more holes are dug and supplied with acorns in similar fashion. Consequently, a patch of 1 sq. m. will contain five holes, in each of which 7-8 germinative acorns have been planted.

Each scientific and experimental institution must sow in the manner prescribed by us no less than 70% of the shelter-belt area assigned to its fields. The remaining 30% of forest belt area should, for purposes of comparison, be planted in the manner now accepted in their respective zones.

By the spring of 1949 a total of several thousand hectares of experimental forest belt must be hill-sown in all the scientific and experimental agricultural institutions subordinate to the various ministries, as well as in the various state forest farms. We must expedite the performance of this experiment in order to be able to ascertain as soon as possible the practical utility of the recommendations submitted.

NEW DEVELOPMENTS IN THE SCIENCE OF BIOLOGICAL SPECIES

TO THIS day no clear-cut definition of the term *species* exists in the science of biology. Yet every biologist as he observes living nature—and particularly the practical farmer, the agriculturist dealing with plants, animals or microorganisms—is struck first of all by the fact that all interconnected organic nature consists of separate, qualitatively distinct forms. For instance, in practical agriculture it is self-evident that the horse, the cow, the goat, the sheep, etc., and wheat, rye, oats, barley, carrots, etc., are separate, qualitatively distinct forms of animals and plants, respectively. The same thing is true of the wild animals and plants in free nature that environs us. Everybody can distinguish between the oak, the birch and the pine, for instance, as separate and distinct forms.

It is of such separate forms of plants, animals and also microorganisms, as has already been stated, that interconnected living nature consists. These forms of organisms, which do not interbreed under the ordinary conditions of life that are normal for them or when they interbreed do not produce normally fertile offspring, i.e., forms which are physiologically incompatible, are *species*. In practical farming, and still more so in free nature, there are many cases where the same name is applied to forms, i.e., species of plants and animals, which, although closely related, are known to be separate and distinct and ordinarily not to interbreed. For instance, ordinary soft wheat, durum wheat, one-grained wheat, emmer wheat and others are all called wheat. Besides the dandelion proper, several other separate and distinct forms, i.e., species, which ordinarily do not cross are also called dandelion. Therefore, in order to draw lines of demarcation between the concepts of these forms, i.e., species, a binomial Latin nomenclature was long ago introduced into the systematics of botany and zoology by *Linnaeus* (1707-78). Thus *Triticum vulgare* is common (soft) wheat, *T. durum*—durum wheat, *T. monococcum*—one-grained wheat, etc. The first part of the designation, the noun, for instance "wheat" (*Triticum*), is the generic name common to all closely related species which in practice or science (in systematics) constitute one genus. The second part of the designation—

the adjective, for instance, "common" (*vulgare*), or "durum" (*durum*)—serves to describe the precise form, the species of the plant or animal.

In practice, when only one species of plant or animal is dealt with, species are called only by their generic names, such as wheat, pine, etc., or horse, sheep, goat, etc. If several closely related species are dealt with in practice, either both names are given: common wheat (*Triticum vulgare*) or durum wheat (*T. durum*); or one of the species is called by its generic name; for instance, common (soft) wheat is designated as wheat and the other species are called by different names; thus *T. dicoccum* may be called emmer.

The very structure of living nature, consisting, as it does, of groups of species similar in many respects yet at the same time separate, delimited, distinct, not interbreeding under ordinary conditions of life, suggested to naturalists ages ago that species originate from other species, that closely related species have much in common and that this which they have in common and which indicates that they are connected in origin is what characterizes them as a genus. Hence living nature itself dictated to science the binomial nomenclature of species.

Before the advent of Darwinian biology a metaphysical, antiscientific view of *species* prevailed. *Species* were considered *immutable and by no manner of means interconnected in origin and development.* It was argued that species could not have descended from each other, that a separate act of creation had brought each one into existence independently of all the others.

Lamarck, and more particularly Darwin with his theory of evolution, utterly refuted the false assertion of the metaphysical biologists that species are eternal and fixed and that they originate independently of each other.

Darwin in his doctrine of evolution demonstrated that plant and animal forms, species, originate from each other. He thus supplied the proof that living nature has its history, its past, present and future. This is one of the immortal services performed by Darwin's theory.

But Darwinism is based on a one-sided and flat evolutionism. Darwin's theory of evolution proceeds from a recognition of quantitative changes only: it refuses to take cognizance of the compulsory, law-governed nature of transformations, of transitions from one qualitative state to another. Yet without the conversion of one qualitative state into another, without the genesis of a new qualitative state within the old, there is no development but only increase or decrease of quantity, only what is usually called growth.

Darwinism firmly established in the science of biology the idea that organic forms have their origin in other such forms. However, development in living nature was conceived of by Darwinism as only a continuous, unbroken line of evolution. In biological science—precisely science and not practice—species therefore ceased to be considered as real, separate qualitative states of living nature.

Thus, in his *The Origin of Species* Darwin wrote:

"From these remarks it will be seen that I look at the term species as one arbitrarily given, for the sake of convenience, to a set of individuals closely resembling each other, and that it does not essentially differ from the term variety, which is given to less distinct and more fluctuating forms. The term variety, again, in comparison with mere individual differences, is also applied arbitrarily, for convenience's sake."[1]

K. A. Timiryazev wrote to the same effect: "Variety and species represent merely a difference in time. No line of demarcation is conceivable here."[2]

Thus, according to the theory of Darwinism, there should be no natural border lines, no discontinuity between species in nature.

According to the theory of evolutionism the development of the organic world may be reduced to mere quantitative changes, without anything new being born within the old, without the development of a new quality, a different totality of properties. This theory holds that so great an interval of time is required for one species to arise from another that the entire history of the human race has not been long enough for the emergence of one species from another to be observed.

Yet, organic nature has been in existence for aeons of time. One would therefore suppose that this was ample time for a new species to arise from an old and that as a result of such prolonged changes the appearance, the birth of new species should be observable by now.

But the same theory declares that actually there should be no dividing line between the new, nascest species and the old, procreating species, for which reason it is supposed to be altogether impossible to observe the generation of a new species within an old one.

In spite of the theory of gradualness throughout, which recognizes no break in development, no transition from one quality to another, and which therefore asserts that there can be no boundaries between species, such boundaries do exist in actual fact, and every naturalist has long been fully aware of this. Therefore, Darwinism was forced to invent so-called intraspecific competition, intraspecific struggle, to explain the gap between species. According to this theory all intermediate forms, which, it is maintained, completely filled the gaps between the species and thus constituted an unbroken gradation of forms in organic nature, dropped out in the process of the struggle as being less adaptive.

Thus, Darwin had recourse to the reactionary, pseudoscientific Malthusian doctrine of intraspecific struggle to gloss over the obvious incongruity between the theory of evolutionism and the real development of the plant and animal world. This struggle is supposedly called forth by the fact that always in nature more individuals of a given species are born

---

[1] Ch. Darwin, *The Origin of Species*, London 1901, p. 39.

[2] К. А. Тимирязев, *Сочинения*, т. VII, стр. 97.

than the conditions available for their existence permit. This is the basis on which Darwin built his so-called theory of divergence, i.e., divergence of characters, the appearance of breaks or discontinuities in the continuous range of organic forms, as a result of which easily distinguishable groups—species of plants and animals—are supposed to have arisen. Consequently, boundaries, breaks between closely related species, came about, according to Darwinism, not as a result of qualitative changes or the emergence of qualitatively new groups of organisms—species of plants or animals—but in consequence of a mechanical dropping out, of a mutual extermination of forms which are qualitatively indistinguishable and constitute an unbroken series.

This explains why all adherents of continuous evolution arrive at the conclusion that *species* in theory *are not a result of the process of development of living nature* discovered by science and practice but a convention employed for convenience in classification.

Thus, a palpable contradiction has always existed and still exists between the theory of evolution and reality, i.e., the development of organic nature. Darwinism could therefore only explain somehow or other the development of the organic world. But the explanation given could not serve as an effective theoretical basis for practical transformation, could not supply the theoretical foundation for a planned alteration of living nature in the interests of practical life.

Although unable in his day to overcome Darwinian evolutionism in science, the eminent biologist K. A. Timiryazev, an ardent fighter against idealism and reaction in science, clearly perceived that species are not conventions but real phenomena of nature. He therefore wrote: "These border lines, these sundered links of the organic chain were not introduced by man into nature but forced upon him by nature. This real fact requires a real explanation."[1]

But no such real explanation could be forthcoming from the standpoint of continuous evolution, and Timiryazev himself did not go beyond the erroneous Darwinian statement that this fact was the result of the supposed existence of intraspecific competition.

Only in our time and country, in the land of victorious socialism, where dialectical materialism, developed in the works of Comrade Stalin, is the dominant world outlook, has it become possible to give a real explanation of real biological facts such as *species*. Collective- and state-farm agriculture affords every opportunity for the unlimited development of materialist biological science, of Michurin's teaching—creative Darwinism. I. V. Michurin wrote: "We have as yet no correct exhaustive conception of how nature has created and still incessantly creates innumerable species of plants. At the present time it is of much greater benefit to us to realize that we have entered that stage of our historical development in which we

[1] *Ibid.*, Vol. VI, p. 105.

are able personally to intervene in the actions of nature and, in the first place, can *considerably accelerate and numerically increase the form-building of new species, and*, in the second place, *artificially divert the building of their qualities in a direction more advantageous to man*. We must furthermore appreciate the fact that such work, jointly performed by us and nature, represents *progress of a very high order* and of global significance. This will become evident to all from the results which the development of this undertaking will bring in the future—an undertaking powerfully impulsed by the Revolution that aroused millions of creative minds in the Land of Soviets. For here a considerable portion of the population has been given the opportunity to improve life round about by deliberate action."[1]

Michurin's teaching, creative Darwinism, does not regard development as continuous evolution but as the genesis of a new quality within the old, of a quality that contradicts the old, which undergoes a gradual quantitative accumulation of its peculiar features and in the process of its struggle against the old quality constitutes itself into a new, fundamentally different totality of properties with its own distinct law of existence.

Dialectical materialism, developed and elevated to a new high plane by the works of Comrade Stalin, is the most valuable, most potent theoretical weapon in the hands of Soviet biologists, Michurinists, and this is the weapon they must use in solving the profound problems of biology, including the problem of the origin of one species from another.

In agricultural practice as well as in nature relative but quite definite boundaries between species have always existed. By relative but quite definite specific boundaries we mean that parallel with similarity between species there always exists *specific* distinctness, which divides organic nature into qualitatively distinguishable yet interlocking links, or *species*.

No continuous, unbroken series of forms between species—different qualitatively definite states of living matter—have been found. This is so not because the intermediate forms in a continuous range have died out as a result of mutual competition, but because there is no such continuity in nature, nor can there be. Unbroken continuity does not exist in nature; continuity and discontinuity always form a unity.

A species is a distinct, qualitatively definite state of living matter. Definite intraspecific interrelations between its members are an essential characteristic of each species of plant, animal and microorganism. These intraspecific interrelations differ qualitatively from the interrelations between individuals of different species. Therefore, *the qualitative difference between intraspecific and interspecific interrelations is one of the most important criteria for distinguishing between species and varieties.*

It is wrong to state that a variety is an incipient species and a species a sharply defined variety. For if this erroneous formulation were taken as

---

[1] И. В. Мичурин, *Сочинения*, Сельхозгиз, 1948 г., т. I, стр. 614-15.

the starting point it would follow that there is no qualitative difference, no line, between species and varieties and that the species is not a reality existing in nature but something contrived for convenience of classification, for systematics. Here, and of this mention has been made above, lies one of the basic contradictions between the theory of continuous evolution and the realities of the organic world. Varieties intermediate between species do not exist, not because these varieties dropped out in the process of an intraspecific struggle but because they never did and do not now take form in free nature.

Varieties are forms of existence of a given species and not steps in its transformation into another species. The profusion of varieties is the result of the many-sided ecological adaptivity of the species concerned; it promotes the well-being of the species and tends to preserve it.

The more varieties within a species and the more diversified its intraspecific populations, the more certain the species and all its varieties are to thrive, through the agency of, for instance, cross-pollination.

The interrelations between individuals of the same species are, we have said, of a quality different from that of the interrelations between individuals of different species. The term *species* is therefore fundamentally different in biology from other botanical or zoological terms, such as genus, family, and the like.

It can easily be noticed that the interrelations between individuals of different species belonging to the same botanical or zoological genus not only do not promote the well-being of the species concerned but, on the contrary, are competitive, antagonistic. It is therefore usually difficult to find in nature or practical agriculture instances of prolonged coexistence in populations of individuals belonging to different but closely related species, i.e., of the same botanical genus. Joint existence of plant species may frequently be observed, but these species are distantly related, belong to different botanical genera. Joint existence of species of the same botanical genus is possible, however, only if the members of each species are distributed in beds or hills.

Hence, the concept *genus* in botany and zoology does not imply ordinary ties of kinship such as intraspecific ties but indicates solely that all the species of any genus have a common origin. The term genus serves to specify morphologically similar but qualitatively distinct species.

In spite of their external similarity the members of the different species of a genus do not cross under the living conditions to which they are habituated or when crossed fail to produce normally fertile offspring, i.e., they are physiologically incompatible. Moreover, the interrelations between species of the same genus are competitive, mutually exclusive, as we have already stated.

Species are links in the chain of living nature, stages of qualitative distinctness, steps in the gradual historical development of the organic world.

Botanical and zoological taxonomy includes a number of so-called doubtful species. These are species of which systematists are unable to say whether the diverse plants or animals concerned form one or two species. But such species are doubtful only because these forms are little known or because biologists have found no scientifically objective criterion by which to distinguish species and therefore substitute for such criterion separate characters tentatively accepted for the various species. Proof of this is the fact that in agricultural practice people deal with a variety of animals, plants and microorganisms without a doubt ever arising in the mind of any one as to whether a particular group of plants, animals or microorganisms belongs to one, two or more species. Hence, doubtful species exist only in systematics but not in living nature.

Species in a state of nature are separated by specific qualitative differences, by relative but quite definite lines of distinction. These must be found so that specific forms, groups of plants, animals and microorganisms, may be properly delineated, systematized and classified.

Nor is the thesis correct which maintains that the qualitative specific features of species do not for any length of time remain constant. As a matter of fact species of plants, animals and microorganisms exist in nature as long as the conditions necessary for the preservation of the lives of their respective members exist.

The prime cause of the appearance of species from other species as well as of intraspecific diversity of form is change in the conditions of life of plants and animals, change in the type of metabolism.

The genesis and development of new species is bound up with such alterations in types of metabolism during the process of development of organisms as affect the characteristic features of the species concerned.

This is evidenced by the data obtained during the last few years as a result of research on the problem of speciation in the plant kingdom.

In 1948 V. K. Karapetian observed in his experiments that if 28-chromosome durum wheat (*Triticum durum*) is sown late in the autumn some of the plants are converted rather quickly, in two or three generations, into another species, into 42-chromosome soft wheat *(T vulgare)*.

On the basis of the genetic qualitative heterogeneity of the plant organism's body, a heterogeneity previously established by Michurinist biology, it was decided to search for grains of soft, 42-chromosome wheat in the spikes of experimentally grown durum wheat. As a result, individual grains of soft wheat were quite easily observed in the spikes of durum wheat, i.e., grains of one botanical species were found in the spikes of another species.

When grains of this soft wheat (*Triticum vulgare*) taken from spikes of durum wheat (*T. durum*) were sown, they produced, as a rule, soft-wheat plants. In many districts a careful search will reveal each year the presence of soft-wheat grains in some of the durum-wheat spikes also in ordinary farm fields.

In 1949 a search for rye grains in wheat spikes was made in the fields of the foothill districts where winter-wheat crops are frequently found to be adulterated with rye. Until a few years ago scientists did not know the original cause of such adulteration in these districts.

V. K. Karapetian, M. M. Yakubtsiner, V. N. Gromachevsky and several other research workers as well as agronomists and college students found single grains of rye in durum- and soft-wheat spikes, i.e., in the spikes of two wheat species which grew in the fields of various foothill districts. Over 200 such grains of rye were found in 1949. These grains were sown at the Institute of Genetics of the Academy of Sciences of the U.S.S.R., in an experimental field of the Lenin Academy of Agricultural Sciences of the U.S.S.R. at Gorki Leninskiye, and at the K. A. Timiryazev Agricultural Academy in Moscow.

Unthreshed spikes of durum and soft wheat were likewise sent to the Lenin Academy of Agricultural Sciences of the U.S.S.R. from the districts mentioned. While they were being threshed at different biological research institutions several persons found some more grains of rye.

From these grains of rye, which had developed in spikes of durum and soft wheat, a diversity of plants was grown. These plants, with few exceptions, were nevertheless typical rye. Only in a very few cases were wheat plants obtained from ryelike grains.

In all the above cases where grains of one species of plant were found in spikes of another species neither the plants themselves nor their threshed spikes showed any signs whatever of being intermediate forms. They seemed to be typical, ordinary spikes of durum or soft wheat. But the internal state of these wheat plants was no longer the usual one, was no longer qualitatively homogeneous in respect to species. This is indicated by the fact that these wheat plants produced not only grains of wheat but also some few grains of rye, i.e., grains of another species.

In 1949 the Lenin Academy of Agricultural Sciences of the U.S.S.R. received samples of oats whose panicles contained single grains of wild oats alongside of the grains of cultivated oats, that is to say, the plants of one species, *Avena sativa*, brought forth individual grains of another species, *A. fatua*. Publications abroad as well as in our country have likewise repeatedly referred to cases where wild oats were found in pure-line oats.

It has been observed year after year when cultivating branched wheat (*Triticum turgidum*) on experimental plots of the Lenin Academy of Agricultural Sciences of the U.S.S.R. and in a number of other localities that admixtures of soft and durum wheat, oats, 2- and 4-rowed barley and also spring rye appear in the crops.

All our observations led us to conclude that the original source of these admixtures was the branched wheat (*Triticum turgidum*) itself.

In 1950 it was discovered in several cases that barley plants which were growing as an admixture in branched-wheat crops had developed from grains which in external appearance could not be distinguished from branched-wheat grains.

In practical farming it has long been assumed and repeatedly asserted that one kind of agricultural plant can be converted or transformed into another, as for instance wheat into rye. A great controversy was waged in print on this subject in our country as early as the first half of the previous century. Therefore, the conversion of durum wheat into soft or the conversion of durum and soft wheat into rye would seem by itself to be nothing new. However, all the new facts we have adduced were obtained in a systematic way or as the result of a systematic search.

As regards the past, before our investigation started, the facts were as follows. In fields sown to durum wheat individual plants of soft wheat were discovered. When this wheat was resown the soft-wheat plants multiplied more and more and finally ousted the durum wheat. Similarly, individual rye plants were found amidst winter wheat. When the seeds obtained from crops grown in such fields were resown the rapidly multiplying rye plants pushed out the wheat. But scientists refused as a matter of principle to consider any such discoveries of plants of one species in the stands of other species as a result of the conversion of one species into another. Legitimate doubts were always voiced. It was not established whether or not the prime cause of this adulteration was ordinary mechanical admixture so frequently met with. There was no assurance that the original seeds really did not contain an admixture of a few seeds of another species, or that seeds of another species had not been carried to the sown field in question by water, wind, birds or some other agency; nor could one be sure that seeds of the admixed breed had not been in the soil of that field for a long period of time, etc.

This explains why it was impossible to prove by facts relating to the past that *the emergence of one plant species from another species* might also be an original source of the various crop admixtures and adulterations, besides their frequent introduction into crops by mechanical means.

All the enumerated objections to the idea of one species giving rise to another fall to the ground in the cases referred to by us. Individual grains of rye discovered in spikes of wheat which had grown for several generations under definite conditions could not possibly have been introduced into these spikes from without by either birds or man or in any other way.

These grains of rye were generated by wheat plants and developed in spikes of wheat.

The supposition that these seeds might be of hybrid origin also goes by the board. It is a known fact that wheat can be crossed with rye, though seldom. However, in these cases the product obtained is an obvious rye-wheat hybrid which can readily be distinguished from wheat and rye by its external appearance. Besides, rye-wheat hybrids, as a rule, are self-sterile;

they yield no seeds unless they are pollinated with the pollen of one of their parents, preferably the wheat. In the case at hand the grains of rye from the wheat spikes produced ordinary rye plants with normal fertility. The said plants manifested no hybrid properties whatever.

The same applies to the other facts we have mentioned.

The above examples of the generation of particular plant species by others are especially valuable because analogous cases may be observed any year in suitable fields. Similar results may likewise be obtained by cultivating plants specially sown under experimental conditions for this purpose.

The factual material so far obtained on the problem of species formation concerns the plant world only. We do not yet have the data essential to demonstrate how species are formed in the animal world. But we may rest assured that before long the development of the theory of Michurinist biology will make it possible to accumulate data also for zoological objects analogous to the data taken from the world of plants.

The material available on the problem of speciation in the plant world affords grounds for belief that many, if not all, existing species of plants can arise de novo at the present time, and under suitable conditions repeatedly do arise from other species. Moreover, one plant species may give rise to several species closely related to it. For example, a single species, durum wheat (*Triticum durum*) can produce both soft wheat (*T. vulgare*) and rye (*Secale cereale*).

A change in the environmental conditions essential to the specific nature of the particular organisms sooner or later changes this specificity perforce—certain species originate other species. Under the influence of the changed conditions, which have become deleterious to the natures (heredities) of the organisms of the plant species growing here, rudiments of bodies of other species more fit for the changed environmental conditions arise and take shape in the bodies of the organisms constituting these species. Such qualitative heterogeneity in the body of a plant organism which is characteristic of various other species may in some cases be detected even by the naked eye.

The appearance under the influence of suitable environmental conditions of specific qualitative heterogeneity in the bodies of plants explains the often repeated creation of some species by others that have long been in existence. When plants of a particular species somehow or other come under the influence of conditions relatively unfavourable for the normal development of the peculiar features of their species, enforced alteration takes place, and rudiments of another species with peculiar features more in accordance with the new environmental conditions, appear in the plant organisms of that particular species. As they are more responsive to the particular conditions, the isolated specimens of the other species generated within the old species rapidly multiply and are capable under these conditions of extruding the species which gave them birth. If this goes on in free nature the emergent species will rapidly multiply and completely oust the species that gave rise to it from the habitat.

Things are otherwise in practical agriculture where the plants cultivated are shielded and protected from weed species by agrotechnical methods.

Scientists have long known that many weed species grow only in cultivated fields and that in free nature they not only do not but cannot exist. Thus, if a field overgrown with numerous species of weeds is abandoned, remains uncultivated and unsown, it will soon enough, in about 20 or 30 years, be completely rid of its many weed varieties. Such a field will no longer grow species of weeds but other plant species which are the peculiar product of ordinary unbroken, untilled plots in the particular locality.

Weed species are generated partly by species existing in free nature and partly by cultivated plant species. For instance, cultivated oats may give rise to wild oats, one of the worst of weeds.

Not a single plant species at home on virgin soil will, when that soil is broken, find the conditions requisite for its normal development. Therefore, the species that grew on the virgin soil change sooner or later with greater or less rapidity but with absolute certainty into other species suited to the conditions created by the tilling of the soil. The same takes place with cultivated plants when they encounter unfavourable climatic or agrotechnical conditions. They are also certain to change sooner or later into other species better adapted to these conditions.

Some weed species have long been introduced into cultivation. Rye, for instance, begotten under certain conditions by wheat, is under these conditions a pernicious weed which drives the wheat from the field. In such districts special measures are therefore taken—crop weeding, sorting wheat seeds from rye seeds—to protect wheat at all times from extrusion by rye. In other districts, on the contrary, rye has long been a cultivated plant. The same can be said of soft wheat. It is frequently produced by durum wheat and in that event adulterates it. Durum wheat is therefore protected against such adulteration by weeding the seed crops. Soft wheat, on the other hand, is a crop that man has cultivated for ages.

Many other species of cultivated plants are the products of other cultivated plant species. This will explain why no wild, ancestoral forms have been found so far for many species of cultivated plants.

Bad agrotechnique, which does not create in the fields the good conditions that cultivated plants require, leads to a deterioration of the nature of these plants with respect to yield and quality of crop. Simultaneously, bad agrotechnique promotes the multiplication of various species of weeds the seeds and other rudiments of which are to be found in the soil or are introduced into it by badly sorted sowing material. Finally, bad agrotechnique may also create the conditions for the generation de novo, by cultivated plants, of isolated rudiments of a number of weeds.

To ascertain the original sources of the emergence of particular species of weeds and discover the environmental conditions which determine such emergence constitutes a task of paramount importance to agronomic biology. Research work conducted to this end will not only facilitate the control of weeds now existing in the fields but also enable us to preclude the possibility of weed species being brought into existence by other such species or by cultivated plants.

The creation of new conditions for organisms or the withdrawal of these organisms from the action of certain existing environmental conditions makes it possible to produce new plant species useful to practical agriculturists and also to preclude the possibility of generating weed species harmful to agricultural practice.

This is one, but not the only one by far, of the practically important tasks involved in the theoretical elaboration of the problem of speciation.

VITALITY OF PLANT AND ANIMAL ORGANISMS[1]

GERMS, embryos and organisms in general possess viability. Different plant and animal organisms possess viability in different degrees. Some possess more, others less viability. Young organisms possess greater viability, greater pontentiality of life, than old ones. Viability in process of realization is life, the process of life. The intensity of this process constitutes the vitality of the organism. Hence, vitality is the measure, the degree, of viability.

Let us illustrate this proposition by a generally known example. The crop qualities of seeds, cereal seeds for instance, are characterized not only by the percentage of seeds that sprout but unfailingly also by the energy with which they sprout. In the case in question the germination percentage indicates how many living, viable seeds there are in the particular lot of seeds. On the other hand, the degree of viability, i.e., the vitality in practical seed control is expressed in germinating power and is determined by the relative quantity of seeds that sprout within the period of time peculiar to the given plant species.

Idealist, vitalistic biology has been endeavouring to explain vitality, and life itself, by the presence of a "vital force" independent of matter, i.e., of the organism's body—a mythical force which its followers of that trend have themselves conjured up.

The "vital force," the idealists hold, is the origin of the body's life. It somehow penetrates the nonliving body from without, in consequence of which the body becomes alive. In other words, the mythical "vital force" engenders the material processes of life. That is what the vitalists claim.

This idealist explanation of life is absolutely wrong. It erects an impassable barrier between living and nonliving nature; it blocks the road to efficacious knowledge of the objective laws of living nature which are of importance to science and practice. Michurinian biology holds that not the idealists with their fictitious conceptions, but the materialists with their opposite assertions that correspond to reality are right. It is not an imaginary

---

1 This article was written for the *Bolshaya Sovetskaya Entsiklopedia* (second edition).

"vital force" that gives birth to material life processes. Quite the contrary. A definite state of matter, a definite state of the body determines, necessarily gives rise in various cases and in various degrees, to a pronounced possibility, capacity, of the body to live. From the unity of such bodies with the environmental conditions indispensable to their life, their life processes take their origin, bodies become alive.

The immense factual material that has accumulated in the practical breeding of related and unrelated plants and animals was left without a correct theoretical explanation by the Weismannist biologists, nor could it have been otherwise. No clear reason was advanced by them why closely related breeding is attended very often by a sharp decrease in the vitality of the respective organisms and a drop in the fertility of the plants and animals. It was likewise unclear why in unrelated crosses within a variety or breed, and more particularly between varieties or breeds, the vitality and fertility of plants and animals increase.

As practical stockbreeders have pointed out, on stock farms producing for the market the offspring of closely related parents prove, as a rule, to have low productivity because of their lowered vitality. Therefore, closely related animals and plants should not be mated or bred on commodity stock farms. But practice also shows that in a number of cases such propagation is not only possible to a certain extent but even necessary when producing new breeds in pedigree stock farms of research institutions and in pedigree state and stud farms.

The Morganist-Weismannists were wrong in attributing the sharp drop in vitality and fertility when inbreeding cross-pollinating plants and crossing consanguineous animals to defects in the heredity of the parental forms. They believed that the heredity of the parental forms of plants and animals contains so-called "lethal genes." According to the Weismannist teaching there are two genes (one in each paired chromosome) for each property and character of a plant or animal. If one of such a pair of genes is not lethal, the parental organism will be viable and the effect of the lethal gene will not be manifested. But in propagating offspring of closely related parents, the homologous (paired) chromosomes which carry the two lethal genes unite in the fertilized egg cell derived from the viable father and mother. As a result the progeny obtained is unviable.

With this false conception of the heredity of organisms as their point of departure and ignorant as they therefore were of the difference between vitality and heredity, the Weismannist-Morganists proposed the fundamentally defective method of testing and culling sires in practical stockbreeding according to whether they had "lethal genes" or not. They suggested that such tests be made by mating sires with their daughters and that if the progeny of a sire contained stillborn or frail, undergrown organisms, he should be culled even if he is the most valuable improver of an unrelated herd, on the ground that he is a carrier of concealed "lethal genes."

That these recommendations were wrong appears from the very fact that such a test would make it necessary to cull all sires in the animal world and all cross-pollinating plants. After all, it has long been known to practice and science that in the propagation of closely related organisms, especially if it is continued for a long time, all animals without exception, as well as cross-pollinating plants, are bound to turn into degenerate forms possessing little vitality and fertility.

For this highly important law of living nature discovered by Darwin the Weismannist-Morganists substituted the false theory of lethal genes and suggested, as above indicated, that the carriers of the lethal genes in the animal and plant world should be sifted out. They wanted to improve the nature of organisms and yet they ignored the laws that govern the nature of these organisms.

The decrease in vitality and fecundity of animals and plants when closely related pairs are crossed and the corresponding increase when breeds and varieties are crossed bear no relation to heredity. *Though they are closely interrelated properties of one and the same living body, vitality and heredity, the breed peculiarities of organisms, are separate and distinct.*

The truth of this can readily be seen, for instance, from the fact that any species of animal or plant (particularly cross-pollinators), no matter what genus, family or class it may belong to, i.e., no matter what heredity it may possess, in closely related breeding becomes, as has already been stated, unfertile and loses vitality, degenerates. On the other hand, the offspring of unrelated intravarietal and intrabreed crosses are fruitful, full of vitality, vigorous.

Consequently, *any* breed, regardless of its hereditary peculiarities and distinctions, if inbred, may likewise turn into an unfertile organism of low vitality, may degenerate.

Generally known biological facts support the proposition that vitality and the hereditary features of the breed of a living body are two different properties and not one and the same property. Let us mention some of these facts.

Female and male sex cells (gametes) of plants and animals usually are in full possession of the heredity inherent in the plant variety or animal breed the product of whose vital functions they are. At the same time female sex cells of those plant and animal species which do not reproduce without fertilization do not by themselves possess, though they are living, sufficient vitality for embryos and subsequently adult organisms to develop from them without fertilization.

That vitality and heredity are two different properties, different aspects of a single living body, can readily be observed from the inbreeding and outbreeding of bisexual cross-pollinators; rye, for instance.

If the egg cells of a rye spike are fertilized with its own pollen, seeds are obtained only in very rare instances; when pollinated with the pollen of another spike of the same plant, fertilization results rarely but more frequently than in the first instance.

If, however, a young tillering rye plant is divided at the tillering node into 10-15 parts, if these parts are rooted separately and plants grown from them under relatively different conditions and if at the time of flowering they are united into one group, their cross-pollination will result in the formation of seeds. The spikes may contain slightly fewer seeds than is normal, but their number will be incomparably greater than in the first two cases.

Normal fertility may be observed in rye plants when the egg cells in a given spike are fertilized with pollen obtained from other plants which have been grown from other seeds of the same variety, i.e., seeds of relatively identical heredity.

Finally, if the spikes of a given variety are pollinated with pollen taken from spikes of other varieties the fertility will, as a rule, be greater than usual, above normal.

The above concerns solely the question of fertility.

What kind of vitality, then, is possessed by the embryos of rye seeds obtained by the method indicated?

Here their vitality corresponds largely to the degree of fertility. The lower had been the fertility, i.e., the lower had been the percentage of fertilized egg cells, the less the vitality of the embryos in the seeds obtained.

The aforesaid is in complete accord with the conclusions Darwin drew on the basis of his many years of research in cross-pollinating plants.

Vitality is at its lowest in plants produced by seeds obtained from the self-pollination of a single spike; next, from the self-pollination of a single plant. Vitality is at its highest in plants produced by seeds obtained from intervarietal crossing.

When a spike of rye is pollinated with pollen from the same spike, i.e., is self-pollinated, the number of seeds obtained is usually very small and the plants grown from these seeds are very feeble, undergrown, sickly, and perish easily. If a young tillering rye plant is divided into parts and plants are grown from them which at the time of flowering are united into one group, their cross-pollination (fertilization) results, as has already been stated, in almost normal seed formation. The plants obtained on sowing these seeds are neither puny nor feeble, but normal. Yet *the paternal and maternal plants which produced these seeds sprang from one embryo, one seed.* These plants were thus most closely related to each other; their heredity was identical.

The illustrations given show that vitality and heredity are different properties and it is incorrect to identify them in science.

Heredity (breed) is the property of a living body, or organism, to develop in a relatively definite direction, to have a definite type of metabolism, which requires definite conditions of life, definite environmental conditions.

But for a body to possess heredity and for heredity to manifest itself the body must be living. It must develop, must undergo conversion during its life process.

The characteristic feature of a viable body, the feature that distinguishes it from a nonviable body, i.e., from bodies of nonliving nature, is the inherent necessity of existing in inseparable unity with definite environmental conditions, conditions of life. The more necessary it is for a living body to be able to form a unity with its conditions of life and the more necessary it is for it to be able to assimilate definite environmental conditions, the greater is the viability of that particular body, the more intensive is its life process, the stronger is its vitality.

What calls forth, what brings into existence, the viability of a body, its property to form a unity with the conditions of life, to assimilate nonliving substances, i.e., nutriment and convert it into part of its own living self?

Materialist dialectics supplies the answer to this question.

"Contrary to metaphysics, dialectics holds that internal contradictions are inherent in all things and phenomena of nature," Comrade Stalin points out in his work, *Dialectical and Historical Materialism.*

And right there Comrade Stalin adds:

" 'In its proper meaning,' Lenin says, 'dialectics is the study of the contradiction *within the very essence of things*.' (Lenin, *Philosophical Notebooks*, p. 263.)

"And further:

" 'Development is the "struggle" of opposites.' (Lenin, Vol. XIII, p. 301.)"

An analysis of the factual material made from thi- angle will produce convincing proof that the viability of a body depends upon internal contradictions inherent in it. And the contradictory character of a viable body is due to its heterogeneity. The more heterogeneous, to a certain extent, an integral living body is, the greater is its contradictoriness and therefore the greater is its vitality.

If the essence of an organism's *vitality* is understood in this way the biological role of the sexual process, the process of fertilization, becomes clear. The union into a single cell of sex cells (female and male) which differ to a certain extent and the union of the nuclei of sex cells into a single nucleus engender the heterogeneity of the living body, its contradictory character, on the basis of which there arises self-motion, self-development, the life process—assimilation and dissimilation, i.e., metabolism. Hence fertilization creates vitality, the impulse of life.

Sex cells usually do not develop without fertilization, do not produce embryos, organisms, because their bodies are insufficiently heterogeneous. It has already been pointed out that in rye, for instance, the heterogeneity of the body of the egg cell is insufficient to develop a normal seed not only when there is no fertilization but even when pollen from the same plant is taken for pollination. But if a grain of rye is brought forth by a wheat plant, a phenomenon observed in the transformation of one species into another, there are grounds for assuming that such a grain could have developed also from an unfertilized egg cell.

This supposition is confirmed by the fact that rye plants grown from rye grains produced by wheat plants yield seeds quite readily when self-pollinated and even when castrated and the castrated flowers are isolated, i.e., produce seeds without any fertilization, though in small quantities.

V. K. Karapetian, a research worker of the Institute of Genetics, Academy of Sciences of the U.S.S.R., castrated 12 spikes of rye grown from grains which he had found in wheat spikes, and without fertilization obtained 14 grains of rye, which when sown produced plants of normal vitality.

In the second generation the rye obtained from grains produced by wheat seeds without fertilization of the egg cell are of still rarer occurrence. The subsequent generations of this rye are, however, like ordinary rye, obviously altogether incapable of producing seeds without fertilization.

The formation of rye seeds without fertilization can only be explained by the fact that when rye grains emerge in wheat organisms, i.e., when a rye egg cell develops in a wheat spike, remnants of the wheat body are still preserved in the body of the rye egg cell, which gives rise to heterogeneity of the integral body of the egg cell sufficient for the development of a seed embryo and subsequently, when that is sown, of a plant.

The possibility of preserving remnants, granules, of the wheat body in the rye body brought forth by a wheat plant is confirmed by the fact that when these rye grains are sown wheat and not rye plants may be obtained, though such cases are rare.

A correct understanding of vitality is of importance both to theoretical biology and for the practical breeding of plants and animals.

Proper intervarietal and intravarietal crossing in cereal seed growing may result in the obtaining of seeds that produce plants more vital and more resistant to climatic inclemencies than the maternal variety.

Intra- and intervarietal crossing followed by selection of typical plants of the maternal variety during the first two generations will increase the vitality of cereal varieties assigned to the various districts without doing violence to or changing their heredity.

The crossing of suitable breeds in nonpedigree stockbreeding will likewise yield favourable results.

Once it is known that viability and its degree, i.e., vitality, are created by fertilization, as the result of uniting two slightly different sex cells (even of relatively like heredity), the method of inbreeding may be applied with considerably greater success in producing new varieties of cross-pollinating plants and new breeds of animals. This method may be used not only to preserve but also enhance desirable hereditary properties and characters of the original parental forms. To prevent a weakening of vitality in closely related breeding, related organisms intended for crossing must be reared under differing conditions.

Thus, a scientific analysis of the phenomenon of greater or less vitality of plants and animals observed in practical farming shows the following:

First, that vitality and heredity peculiarities are different properties of the living body.

Second, that vitality of organisms is usually created by the sexual process, the process of fertilization. The degree of viability, i.e., the vitality of plants and animals within a species, depends upon the extent to which the sexual elements that were united on fertilization differ. A living body has vitality as long as it remains heterogeneous. As the heterogeneity of the living body of a given plant or animal gradually wanes the assimilation-dissimilation process peters out, the vitality of the body normally approaches extinction, the body grows old.

Vitality may be increased in other ways than sexually, as for instance through the assimilation by plants or animals of environmental conditions which are new to them. Such assimilation also creates heterogeneity; hence the contradictory nature of the living bodies of plant and animal organisms.

Third, the conditions of life, the environmental conditions assimilated by the organisms of the ancestors, and especially the parental organisms that directly produced the given sex cells, are likewise a prime cause of difference in the sex cells that in the process of fertilization of the egg cell create the vitality of the embryo and subsequently of the organism.

The environmental conditions, as is well known to Michurinist biologists, are also largely responsible for changes in the old heredity of plants and animals and for its conversion into a new heredity, for alterations of old breeds and their transformation into new breeds.

THE CONVERSION OF NONWINTERING SPRING VARIETIES INTO WINTER-HARDY WINTER VARIETIES

OUR SOVIET biological science quite some time ago disclosed the hereditary distinctions between the winter and spring forms of plants. These distinctions manifest themselves in the *different demands* which winter and spring forms make upon the external environmental conditions in order to be able to pass through one of the vitally important processes of their individual development, namely, the process of vernalization.

Winter plants sown in the autumn usually complete their passage through the vernalization phase when steady winter frosts set in. During the same period of time winter plants acquire hardiness, i.e., develop resistance to winter inclemencies.

Passage through the vernalization phase and the development of resistance of winter cereal plants to winter inclemencies (hardiness) are different processes. But in the field both processes are interconnected and take place under the external environmental conditions of the autumn period.

Experiments conducted by Soviet investigators have shown that winter cereals whose plants undergo the vernalization phase before they are sown, in the seeds which have just begun to germinate, cannot during the autumn period acquire resistance to winter frosts as well as do plants of the same varieties for the sowing of which ordinary seeds, not vernalized before sowing, are used. It is also well known that in winter, after lengthy thaws, winter plants begin to sprout slightly and lose their hardiness. Since such plants, as a rule, are completely vernalized by that time they cannot, after the thaws, reacquire resistance to severe frosts. All this goes to show that in winter cereals the hardening process is closely connected with the passage through the phase of vernalization.

The inseparability of the hereditary properties of winter habit and resistance to the climatic inclemencies of winter depends in the field upon the fact that the formation of these properties of cereals takes place under the same conditions. Degree of winter habit and degree of resistance to winter inclemencies are hereditary properties of identical protoplasms and of identical phasic states of it.

Michurinist genetics has demonstrated that the environmental conditions which a plant form requires for its normal individual development were at one time the original cause that gave rise in the plant form to the requirement of these conditions. In other words, it has already been proved that plant forms require those environmental conditions from which or under the influence of which they have been and are being created.

It is now an established fact that if with plants the process called vernalization formed itself under spring conditions hereditary spring forms are obtained. But if the vernalization process developed under autumn conditions hereditary winter forms are obtained. The comparatively large experimental material received of late shows that in the given case the difference in the photo factors under spring and under autumn conditions is of paramount importance for the creation of spring or winter forms. We assume that light here plays the part of a substance assimilated in photosynthesis by the green leaves of the young cereal plants. The spring or autumn light is converted into an inseparable part of the living body as the result of its assimilation by the plants. When spring light is assimilated the living body of the cereals obtained exhibits spring-habit properties and consequently is incapable of hardening to winter inclemencies. If autumn light is assimilated living bodies of cereals possessing the properties of winter habit are obtained, bodies capable of hardening in the autumn to winter inclemencies; yet not of any district but, as a rule, only of the district out of the autumn conditions of which it was created.

It is the assimilation of autumn and not the influence of winter conditions that constitutes the principal factor in the building up in plants of the property of winter habit and the property of autumn-hardening to the climatic inclemencies of winter. This is borne out by the following factual material.

For many years N. A. Byelozerova, a research worker of the Siberian Institute of Grain Husbandry awarded the Order of the Red Banner of Labour, conducted experiments in late-autumn sowing of spring wheat whose seeds only reach the stage of sprouting before cold winter weather sets in. They showed that crops sown late in the autumn do not yield winter plants. Only 5 to 10% of the seeds were of the winter form and these were found most often in crops sown a second time late in the autumn. But seeds from third- and fourth-generation crops sown late in the autumn usually did not produce winter plants.

At the same time we knew that in the experiments with autumn-sown spring plants conducted by N. A. Byelozerova in Omsk, in A. T. Trukhinova's experiments first made in the Chelyabinsk State Plant-Breeding Station and subsequently at the Institute of Genetics of the Academy of Sciences of the U.S.S.R. in Moscow, in the experiments performed by V. F. Khitrinsky at the All-Union Institute of Selection and Genetics in Odessa, as well as by a number of other scientific workers, hereditarily stable winter varieties were obtained from spring varieties of wheat. These facts proved beyond all doubt that spring varieties can be converted into winter varieties by repeated autumn sowings.

But it has been stated above that spring forms are not always converted into winter forms when sown repeatedly late in the autumn and that instances are not rare when winter forms obtained from second-generation seeds peter out and disappear in the third generation after one more late-autumn sowing. Such a crop yields only spring forms.

Why is it that in some cases when spring wheat varieties are sown repeatedly hereditarily stable winter forms are obtained while in other cases, namely, in repeated late-autumn sowings, winter forms are obtained, but only in small percentages and only in the second sowing, while in the late-autumn sowings following it the winter forms found in the crop not only do not increase but decrease, peter out and vanish?

An analysis of the factual material from the point of view of Michurinist genetics warranted our conclusion that success in the conversion of spring forms into winter forms depends entirely on the *time when* the spring forms are sown in the *autumn* for the second time. The first autumn sowing of the spring form is necessary to eliminate the old spring-habit heredity.

The experimental material shows that the autumn sowing of the spring variety had best take place the first time as late as possible so that when winter sets in the seeds may get no further than full sprouting. The seeds obtained from such a crop will by no means be of the winter form as yet but in respect to their vernalization phase they will no longer be the spring forms they were. If such seeds are sown in the spring they can easily be made to produce spring plants with a new, a spring vernalization phase; and if sown in the autumn they can be made to produce winter forms with a winter vernalization phase. The second autumn sowing of spring varieties that are being changed into winter varieties should therefore under no circumstances take place late. Such sowings should be so timed that the plants, acting through their green leaves, will be able in photosynthesis to assimilate, convert the autumn conditions, the main factor of which is autumn light in the case in question, into a living body of the winter type.

It would not be very inaccurate to say that in spring plants obtained from the above-mentioned repeated autumn sowing it is possible to observe directly how the body of a winter plant originates and grows.

The sprouts of spring varieties of almost all the soft wheats in our districts have pubescent leaves while the winter varieties, as a rule, have glabrous leaves. One can therefore tell whether wheats are of the winter or spring variety by the presence or absence of pubescent leaves in the wheat sprouts. When spring wheat is autumn-sown for a second time all the sprouts without exception have such pubescent leaves as are characteristic of the given spring variety. This is convincing proof that the sprouts have not yet turned into winter forms. But when these young plants begin to assimilate autumn conditions the growing lower parts of the leaves (in wheat the leaves grow from the base) are at times glabrous or, as is more frequently the case, pubescent sections of the leaves alternate with glabrous sections. This would indicate that with the plants in question glabrous sections of the body possessing properties of winter habit arise from the assimilated autumn food.

The fact that after repeated late-autumn sowings spring forms are not converted into winter forms goes to prove that light as a substance is what counts most in building up a winter plant in the autumn sowing. In late-autumn sowing either the sprouts do not have sufficient time to appear and consequently there is no photosynthesis or even if green sprout leaves do appear almost no photosynthesis takes place because of the low temperatures. In both the first and second cases of late-autumn sowing photosynthesis can begin only under spring illumination conditions, for which reason a living body possessing the properties of spring habit builds itself up in these plants.

This explains why not only is there no conversion of spring plants into winter plants when the former are repeatedly sown late in the autumn, but even why the small number of winter plants which in several cases do make their appearance, disappear and are lost in the offspring as a result of repeated autumn sowing.

Michurinist biologists have established long ago that in order to change the nature (heredity) of plants their metabolism must be changed. But plant and animal forms are conservative, are physiologically coordinated. They elect from the external environment only those conditions that are suitable for their heredity, their nature, and actively do not include, do not assimilate, conditions that are not specific for them. The explanation for this is to be found in the fact that every living body requires precisely those conditions and substances for its life and growth of which it (a living thing) was composed until it became alive. At the same time it actively counteracts the inclusion of other conditions whose assimilation would lead to a change in its nature, its heredity.

Consequently, science has been faced with the task of finding increasingly efficient methods of breaking down the conservatism of the heredity of plant and animal forms. Plant organisms whose conservative heredity has been broken down are able to assimilate environmental conditions in whose conversion into living matter we are interested. Thus, organisms are created whose heredity is of the type desired by us.

The method of eliminating spring and creating winter heredity has been so well elaborated and concretized by now that it has served as the basis for an agrotechnical scheme which makes it possible to turn into winter forms in any district any variety of spring wheat, barley or other plants of species which can have both the spring and winter form.

The method of converting spring varieties into winter ones is as follows: the spring-wheat variety must be sown in the autumn, about 20-25 days after the optimal sowing date for winter plants in the particular district. The sowing must be so timed that with a relatively high temperature to start with the seeds will just have sprouted when cold weather sets in. Sown in this way the process of vernalization begins in these plants under the relatively high temperature conditions under which spring wheat normally passes through it. But when winter conditions set in and temperatures fall spring varieties will not have the high-temperature conditions they need for the vernalization process. But since the process has already begun and is only a little short of completion it will reach its end during the long period before spring even under low temperatures. In such event the vernalization process terminates in an unusual way for spring varieties.

This method, by which a spring variety is compelled to complete its vernalization process under low-temperature conditions that are unsuitable for it, is in the case in question precisely the means of eliminating or, to be more exact, of breaking down the old, i.e., conservative heredity of the vernalization phase.

Numerous experiments have clearly demonstrated that late-autumn-sown spring varieties produce seeds in which the conservative old heredity of the process of vernalization has been removed. Plants grown from such seeds quite readily accept both spring and autumn conditions for the building up of the new vernalization process, of the new heredity. Spring-sown plants obtained from such seeds possess a new, a spring heredity. If autumn-sown the plants obtained are of winter heredity. Before the second wintering they must be sown in the field without fail on the date set for the given district.

This date is fixed empirically. For this purpose it is necessary to sow the said seeds several times in the autumn at intervals of 5-7 days. The first sowing must take place about 10 days after the optimal date for the sowing of winter plants in the given district; the second sowing, 5-7 days after the first, and the third 5-7 days after the second. One of these sowing dates will strike the autumn conditions during whose assimilation by the green leaves in the process of photosynthesis a body possessing the properties of winter habit will build itself up. The bodies of these plants will be heterogeneous with respect to all their parts when winter sets in. Their bodies will consist of winter-habit sections and spring-habit sections. Such plants winter much better than pure spring plants but worse than winter plants. Their seeds will produce a large percentage of winter plants capable of wintering well in the district by whose autumn conditions the spring plants were converted into winter ones.

Let me remind the reader that the winter-hardiness of varieties obtained in this way will largely correspond to the conditions of the winter inclemencies prevailing in the district whose autumn conditions created the varieties in question or, more exactly, created their property of winter habit and their property of winter-hardiness.

The method indicated now permits every agronomist and collective farmer to convert within two years any spring variety into a winter variety which will winter well in the particular district.

We attach great importance to experiments in the conversion of spring varieties into winter varieties:

First, because they are of substantial interest to theory since they graphically show that changes in plant forms correspond to the influence exerted by the environmental conditions. In the given case the spring forms turn into winter forms due to the action of the autumn illumination conditions.

Second, these experiments are important because our collective and state farms, as well as plant-breeding institutions, are being provided with a method for creating winter forms of wheat, barley and a number of other cultivated plants which winter well in the district whose autumn conditions create these forms.

In the preceding articles in which I dealt with the conversion of spring grains into winter ones under the influence of suitable environmental conditions, I frequently voiced the erroneous view, which I had formed on the basis of Darwin's one-sided theory of evolution, that with cereal plants the properties of winter or spring habit, as the case may be, gradually accumulate or diminish. I assumed that under the influence of the autumn-winter conditions spring-wheat plants become winter plants gradually, from generation to generation, and that the degree of winter habit of such plants must rise from generation to generation. I believed that in the beginning the plants obtained were of slight winter habit and that in the next generation the winter habit became more intense and subsequently still more so until their heredity became stable, conservative.

The comprehensive factual material that has been assembled on changing the heredity of spring plants into that of winter plants has proved that this theoretical postulate was incorrect. It appeared that in changing spring plants into winter plants the action of autumn field conditions alone is required and not of both autumn and winter field conditions; and that winter forms are not obtained by an increase in the degree of winter habit from generation to generation. The experiments have not disclosed any plants possessed of low winter habit, which intensifies in succeeding generations, i.e., plants whose winter habit increases. In all experiments known to us the plants obtained in a particular plot were either all spring ones that had not changed into winter ones, or were only winter ones, or both kinds; but the degree of winter habit of the winter plants did not increase in the succeeding generations.

The winter habit of plants is obtained at once, without subsequent increase or decrease in the degree of winter habit formed. Moreover, the degree of winter habit corresponds to the influence of those autumn conditions which called forth the particular conversion of spring plants into winter plants. But while the degree of winter habit of the changed plants establishes itself or, more correctly, manifests itself at once, this does not mean that the process of conversion of spring plants into winter plants takes place instantaneously.

It has already been pointed out that spring-wheat seeds obtained from a crop once sown in late autumn yield spring and not winter plants. We are still unable to distinguish such spring plants from ordinary spring plants by their morphological characters. But one need only sow, under Moscow Region conditions during the first few days of September, spring-variety seeds obtained from a crop sown late in autumn and seeds of the same variety obtained from an ordinary spring-sown crop and gather a harvest from the two sowings to find a great difference between the two crops that will be reaped. The seeds obtained from crops sown twice in autumn will give a considerable number of winter plants, while those obtained from crops sown once in autumn will produce only spring plants. This means that while the progeny of a crop sown once before winter sets in does not differ in any respect externally from that of a spring-sown crop, still changes have already taken place in the seed embryos without which upon a second autumn sowing the property of winter habit cannot come into existence.

When spring plants that have been sown once late in the autumn are changed to winter plants, granules of living body, no longer of the spring habit but of the new, winter habit, arise, emerge in the bodies of these plants under the influence of the autumn conditions. The heterogeneous bodies of these changing plants clearly confirm the above. This explains why seeds produced by these plants are, as a rule, of different kinds: some are of spring habit, others of winter habit.

In our previous works we paid attention to the differences in the vernalization phases of various winter varieties, but only with respect to length of time required to pass through this phase at 0°-2° C. Today it is no longer sufficient to differentiate vernalization phases of different varieties only in this respect. Such a differentiation is too one-sided.

The vernalization phase of each winter-wheat variety has its own specific features, its own qualitative protoplasmic state characteristic of the given variety. That is why the experiments showed that the degree of winter habit of the plants in question cannot be enhanced or diminished by mere increase or decrease of the vernalization phase. It has been found that the degree of winter habit can be diminished or augmented only by abolishing the old and creating a new vernalization phase. That is precisely what the law of change of the heredity of the organism's properties and characters amounts to, namely:

First break down the old and then build up the new heredity.

Michurinist biologists are aware of the fact that plant characters and properties are changed only under and in accordance with the influence of external environmental conditions. Taking this correct postulate as their starting point many scientific workers believe that if a variety of winter wheat, for instance, does not overwinter or overwinters with difficulty in any particular district where winter conditions are severe, one need only help the plants of such varieties to withstand the winter by artificially warming them during several generations for the specific variety to increase its degree of winter habit, adapt itself to the particular conditions and survive the winter in fine shape. But as a matter of fact it has long been known that regardless of the number of generations an insufficiently winter-hardy variety has been growing in a particular district, its resistance to winter conditions will not, as a rule, increase.

At first blush these facts seem to contradict the basic thesis of Michurinist biology but in reality this is not so. If a spring variety which possesses no winter-hardiness at all is sown in the same district in which insufficiently winter-hardy winter varieties will neither overwinter nor enhance their winter-hardiness during a number of generations, it will in two years be converted, as has been said above, into a winter variety that overwinters well in the district in question. This can be readily understood: to compel a plant to change its heredity, in the particular case its vernalization-phase heredity, it must be given an opportunity to start its passage through the vernalization processes under the environmental conditions peculiar to its nature. Thereafter these conditions must be excluded and those conditions substituted the requirement for (adaptation to) which it is desired to create in the plant.

If winter varieties which for the particular district possess little winter-hardiness are sown a second time under the same conditions, their winter-hardiness does not increase because the environmental conditions of autumn are conducive to their normal passage through the vernalization phase and to its completion. Therefore, no matter how many generations of them are sown under these conditions their vernalization phase does not change and their winter habit is not intensified.

The degree of winter habit of the particular winter variety may be raised by first eliminating its existing winter habit (heredity). This can be accomplished by sowing vernalized seeds early in the spring and then sowing the seeds obtained from this crop in the autumn in order to build up the new winter habit that will correspond to the conditions of the district.

Practically it is simpler to create winter-hardy winter varieties by converting spring plants into winter plants.

It is also important to note that in the experiments conducted by Academician A. A. Avakian, and research workers A. T. Trukhinova, B. D. Feinbron and others a sowing of the offspring of separate spikes taken from spring wheat that is being converted into winter wheat has shown that many of them behave like ordinary sexual hybrids of spring and winter wheats. The progeny of a number of the spikes segregated with respect to winter and spring habit. Some showed typical winter habit; the rest were purely of spring habit. Generally speaking, we have been unable to distinguish this material from ordinary first-generation sexual hybrids of spring and winter forms. Hence, on changing the nature of plants by allowing the environmental conditions, in the given case autumn conditions, to influence the vernalization phase of spring wheats, the same heterogeneity of the plant body results as in sexual hybridization. In both cases the indicated heterogeneity explains the diversity, i.e., the segregation of these plants' progeny.

Progress in Michurinist theory implies the continual discovery of new laws that may serve as the basis for creating methods for changing hereditary properties of plant organisms by excluding the habitual conditions from the process of assimilation and including conditions whose assimilation will create the characters and properties we want.

The sum and substance of our practical task is to find ways and means of compelling plant organisms to assimilate the requisite conditions and substances and to assimilate them when they are in such a state that as a result of the assimilation the desired plant characters and properties are obtained. That is the object of raising to a still higher plane the scientific mastery of directed change of the natures of plant organisms.

Expressed in general terms, the multifarious experimental material collected on the conversion of spring wheat into winter wheat clearly testifies that the building up of a vernalization-phase heredity, its specific features and the qualitative state that will be inherent in the variety concerned is the result of corresponding photosynthesis. If after the elimination of the old vernalization-process heredity a new process forms itself out of spring conditions a hereditary spring-vernalization phase is obtained, and if it is formed out of autumn conditions a hereditary winter-vernalization phase is obtained. As autumn conditions differ in different districts a hereditarily different process of vernalization arises in the different winter varieties, each having its own specific features, its own properties and a vernalization phase created by the different autumn conditions.

Let us bear in mind that such hereditary properties as the process of vernalization and the process of hardening plants to winter inclemencies are properties of one and the same protoplasm, of one and the same state called vernalization phase. Consequently, by mastering the method of eliminating the old vernalization-process heredity and of creating a new heredity of this process we learn to master the method of converting non-winter-hardy varieties into winter-hardy varieties.

The winter climatic inclemencies vary for winter cereals with each district. That explains why varieties that can stand certain winter inclemencies of one district may prove incapable of standing other winter inclemencies in other districts. I. V. Michurin therefore often advised creating particular varieties for each district by bringing the influence of the conditions prevailing in the respective districts to bear upon them.

In the case under examination—the creation of winter-hardy cereals for a particular district—the heredity of the vernalization process, of the vernalization phase, of these varieties must be created by the autumn conditions of that particular district.

There is every reason to assume that the application of the method of converting spring varieties into winter ones set forth in the present article will make it possible in two years to create—for instance, in the northern and northwestern districts of our country with their deep snows, where the winter-grain sector to this day consists almost entirely of rye—well wintering wheat varieties which, unfortunately, are not yet to be seen in these districts. My assertion that this can be done is borne out by the fact that in these districts many winter forms of local cereals produced here overwinter well, without fear of deep snows or excessive moisture.

The above method may also be used in these districts to create well wintering varieties of winter barley, winter-hardy clover, winter vetch and other plant species.

BIBLIOGRAPHICAL INDEX

SUBJECT INDEX

BIBLIOGRAPHICAL INDEX

COLLECTIONS OF WORKS

I *Articles on Selection and Genetics.* Voronezh Regional Publishing House, 1939, 229 pp., portrait.

II *A Biology of Plant Development.* A collection of articles on questions of genetics, plant breeding and seed growing. Kiev-Kharkov. State Publishing House of Collective- and State-Farm Literature of the Ukrainian S.S.R., 1940, 332 pp., drawings and portrait.

III *The Food Resources of the Soviet State Must Be Increased.* Moscow-Omsk. Publishing House of the People's Commissariat of Agriculture of the U.S.S.R., [1942], 80 pp.

IV *Agrobiology.* Essays on Problems of Genetics, Plant Breeding and Seed Growing. Moscow, Selkhozgiz [State Publishing House of Agricultural Literature], 1943, 352 pp., drawings, tables.

4th enlarged edition, Moscow, Selkhozgiz, 1948, 683 pp., drawings, tables, and portrait.

5th stereotyped edition. Moscow, Selkhozgiz, 1949.

V *Agrobiological Science in Wartime Agriculture.* Frunze, Kirghiz State Publishing House, 1943, 94 pp. (Academy of Sciences of the U.S.S.R., Institute of Selection and Genetics.)

VI *Works During the Great Patriotic War.* Articles and speeches. Moscow, Selkhozgiz, 1943, 248 pp.

VII *Cultivation of Winter Plants in the Steppes of Siberia.* Moscow, Selkhozgiz, 1945, 83 pp., tables.

2nd enlarged edition, Moscow, Selkhozgiz, 1949, 102 pp., tables.

SEPARATE EDITIONS AND PUBLICATIONS[1]

1923

1. "Sugar-Beet Grafting." [Written in collaboration with A. S. Okonenko. Report to the Third Congress of Plant Breeders.] *Bulletin of the Variety Breeding and Seed Board of the State Sugar Trust*, No. 4, 1923, pp. 77-80.

2. Technique and Methods of Tomato Breeding at the Byelaya-Tserkov Plant-Breeding Station, The." Same report. *Ibid.*, pp. 73-76.

1928

3. "The Influence of the Thermal Factor on the Length of the Developmental Phases of Plants." An experiment with cereals and cotton. Baku, 1928, 168 pp., tables. (*Publications of the Orjonikidze Central Experiment Plant-Breeding Station*, Ganja, Azerbaijan, Vol. 3.) 2nd edition, Moscow, Selkhozgiz, 1949, 210 pp., tables.

---

[1] The general year indicated refers to the original publication. Important new editions and reprints are specified separately.—*Ed.*

1929

4. "Concerning the Essence of Winter Habit." [Written in Collaboration with D. A. Dolgushin.] In the book *Works of the All-Union Congress on Genetics, Plant Breeding, Seed Growing and Pedigree Stockbreeding Held in Leningrad on January 10-16, 1929,* Vol. III. A study of cultivated plants. Leningrad 1929, pp. 189-99.

5. "Gist of the Hypothesis of the 'Winter Habit' of Plants, The" *Selskokhozyaistvennaya gazeta* [*Agricultural Newspaper*], December 7, 1929.

6. "Influence of the Thermal Factor on the Developmental Phases of Plants, and a Program of Work with Sugar Beet on this Problem, The." Summary of report made by T. D. Lysenko. In the book *Materials Pertaining to the All-Union Conference on Problems of Agronomic Research in the Sugar Industry, Convened by the Central Research Institute of the Sugar Industry and the Variety Breeding and Seed Board of the State Sugar Trust in Kiev, December 12-19, 1928.* Moscow 1929, pp. 34-36. (Publications of the Central Research Institute of the Sugar Industry, Vol. 2.)

1930

7. "Vernalization of Economic Crop Plants." Newspaper *Sotsialisticheskoye zemledeliye* [*Socialist Agriculture*], July 2, 1930.

1931

8. "New Methods of Changing the Length of the Vegetative Period of Plants." [Written in collaboration with D. A. Dolgushin.] *Sotsialisticheskoye zemledeliye,* March 1, 1931.

9. "Problem of Regulating the Vegetative Period of Agricultural Plants, The." Journal *Semenovodstvo* [*Seed Growing*], No. 13-14, 1931, pp. 22-23; 29-34.

10. "Theses for a Report on 'The Vernalization of Agricultural Plants and the Fight Against Drought.'" Moscow, Selkolkhozgiz [Agricultural Collective-Farm State Publishing House], 1931, 7 pp. (Organizational Bureau on the Convocation of the All-Union Conference on Drought Under the People's Commissariat of Agriculture of the U.S.S.R.)

11. "Vernalization and the Fight Against Drought." [From his address at the All-Union Conference to Combat Drought.] Newspaper *Izvestia,* October 29, 1931.

12. "Warmth-Loving Plants no Less Than Wheat Can Be Vernalized." *Sotsialisticheskoye zemledeliye,* November 1, 1931.

13. "What Are the Perspectives of Vernalization?" [Abridged Report to the All-Union Conference to Combat Drought.] Newspaper *Pravda,* October 30, 1931.

1932

14. "Accelerated Development of Potatoes Under Field Conditions in Socialist Economy." (A posture of the question.) [Written in collaboration with D. A. Dolgushin.] *Byulleten yarovizatsii* [*Vernalization Bulletin*], No. 2-3, 1932, pp. 35-45.

15. "Brief Instructions on the Vernalization of Cotton." Odessa 1932, 8 pp. (People's Commissariat of Agriculture of the U.S.S.R., Lenin Academy of Agricultural Sciences of the U.S.S.R., Ukrainian Institute of Selection and Genetics.)

2nd edition, Odessa 1935, 16 pp. (in Ukrainian).

16. "Brief Instructions on the Vernalization of Various Wheat Varieties." Odessa 1932, 15 pp. (People's Commissariat of Agriculture of the Ukrainian S.S.R., Lenin Academy of Agricultural Sciences of the U.S.S.R., Ukrainian Institute of Selection and Genetics.)

17. "Chief Results of the Vernalization of Agricultural Plants, The." *Byulleten yarovizatsii*, No. 4, 1932, pp. 1-57, drawings, tables.

18. "Cotton Ripens Even When the Summer Is Cold." [From a table giving the latest data on the vernalization of agricultural plants.] *Sotsialisticheskoye zemledeliye*, September 26, 1932.

19. "Instruction on the Vernalization of Hard Spring Wheat." *Byulleten yarovizatsii*, No. 1, 1932, pp. 62-70.

20. "Is the Demand for Photo Periods Inherent in the Nature of Agricultural Plants?" *Byulleten yarovizatsii*, No. 2-3, 1932, pp. 16-34.

21. "Let Us Accelerate the Ripening of Cotton by Means of Vernalization!" [Vernalization technique.] *Sotsialisticheskoye zemledeliye*. April 9, 1932.

22. "Physiology of Plants at a New Stage, The." [On Vernalization.] *Sotsialisticheskoye zemledeliye*, November 12, 1932.

23. "Preliminary Communication on the Vernalization of Wheat Sown in the State and Collective Farms in 1932." *Byulleten yarovizatsii*, No. 2-3, 1932, pp. 3-15.

24. "Problem of Regulating the Length of the Vegetative Period of Agricultural Plants, The." *Byulleten yarovizatsii*, No. 1, 1932, pp. 5-13.

25. "Problem of Vernalizing Maize, Millet, Sudan (Sorghum vulgare sundanense), Sorghum and Soya, The." *Byulleten yarovizatsii*, No. 2-3, 1932, pp. 46-64.

26. "Results of the 1930 Experiments in Vernalized Sowing in the Collective and State Farms of the Ukrainian S.S.R." *Byulleten yarovizatsii*, No. 1, 1932, pp. 57-61.

27. "Results of the 1931 Experiments in Vernalized Sowing in the Collective and State Farms of the Ukrainian S.S.R." In his book *Vernalization of Agricultural Plants*. Kharkov 1932, pp. 20-25.

28. "Vernalization—a Powerful Weapon in the Battle for Big Crops." [Wheat Vernalization Technique.] *Sotsialisticheskoye zemledeliye*, February 1, 1932.

29. "Vernalization of Agricultural Plants." [Abridged report read in 1931 at the All-Ukrainian Conference on Plant Breeding.] *Byulleten yarovizatsii*, No. 1, 1932, pp. 14-29. Same (re-edited) and Concluding Remarks. Kharkov 1932, pp. 1-20. (In Ukrainian.)

30. "Vernalization of Agricultural Plants and the Fight Against Drought." In the book *The Fight Against Drought*. Publications of the All-Ukrainian Conference to Combat Drought. Kharkov 1932, pp. 31-37.

31. "What Is Vernalization?" In the book *To the Young Communist League Members on Vernalization*. Edited stenographic report of a conference on vernalization, held under the auspices of the Agricultural Department of the C.C. of the Ukrainian Y.C.L. and the editors of the Journal *Molodoy Bolshevik* [*Young Bolshevik*], Lysenko and representatives of the Ukrainian Institute of Selection and Genetics attending. Kharkov-Odessa 1932, pp. 5-34.

32. "Wheat Vernalization in Collective-Farm Fields." [Preliminary Results.] *Sotsialisticheskoye zemledeliye*, September 5, 1932.

1933

33. "Vernalization and Budding of Potatoes, The." [Written in collaboration with A. A. Baskova.] Kharkov, Gosselkhozizdat [State Agricultural Publishing House], 1933, 15 pp. (People's Commissariat of Agriculture of the Ukrainian S.S.R.—All-Ukrainian Academy of Agricultural Sciences.)

34. "Vernalization Has the Floor." Newspaper *Komsomolskaya pravda*, February 26, 1933.

35. "Vernalization of Agricultural Plants, The." *Pravda*, March 9, 1933.

36. "Vernalization of Agricultural Plants and How It Is Done, The." [With instructions on the vernalization of various wheat varieties.] Moscow, Selkhozgiz, 1933, 32 pp., drawings, tables. (People's Commissariat of Agriculture of the U.S.S.R., Lenin Academy

of Agricultural Sciences of the U.S.S.R., Ukrainian Institute of Selection and Genetics.) 5th revised and enlarged edition entitled *Vernalization of Agricultural Plants*, Moscow, Selkhozgiz, 1937, 63 pp., drawings, tables.

37. "Vernalization of Potatoes, The." Odessa 1933, 16 pp., drawings. (People's Commissariat of Agriculture of the U.S.S.R., Lenin Academy of Agricultural Sciences of the U.S.S.R., Ukrainian Institute of Selection and Genetics.

2nd edition, Moscow, Selkhozgiz, 1936, 40 pp., drawings.

Same, Kiev 1947, 32 pp.

38. "Vernalization of Spring and Winter Vetch, The." Odessa 1933, 4 pp. (People's Commissariat of Agriculture of the U.S.S.R., Lenin Academy of Agricultural Sciences of the U.S.S.R., Ukrainian Institute of Selection and Genetics.)

1934

39. "B. Y. Krivoshlyk's Theoretical Fantasies." [A review of Krivoshlyk's *The Vernalization of Agricultural Plants and How It Is Done.*] *Sotsialisticheskoye zemledeliye*, April 17, 1934.

40. "Do Not Distort the Theory of Vernalization!" *Sotsialisticheskoye zemledeliye*, December 26, 1934.

41. "Let Us Master the Principles of Vernalization!" [Written in collaboration with A. M. Favorov.] *Sotsialisticheskoye zemledeliye*, February 14, 1934.

42. "Physiology of Plant Development in Plant Breeding, The." [Paper read at a session of the Scientific and Technical Council of the All-Union Seed-Growing Association, held in Moscow on January 16, 1934.] *Semenovodstvo*, No. 2, 1934, pp. 20-31.

43. "Place Vernalization at the Service of Socialist Agriculture!" [Written in collaboration with D. A. Dolgushin.] In the book *Agricultural Science in the U.S.S.R.*, Moscow-Leningrad 1934, pp. 96-114.

44. "Sowing Vernalized Millet in Stubble." Newspaper *Chornomorska komuna* [*Black Sea Commune*], Odessa, June 15, 1934.

45. "Vernalization." A chapter in the book *Agricultural Primer*. 2nd edition. Prepared by a brigade composed of Y. F. Gomon (brigade leader), Academician T. D. Lysenko and others. Kharkov 1934, pp. 294-323.

46. "Vernalization and Fruit Growing." [From a speech delivered at a conference of experimenting fruit growers held in Michurinsk.] Journal *Plodoovoshchnoye khozyaistvo* [*Fruit and Vegetable Growing*], No. 11, 1934, pp. 50-51.

47. "Vernalization of Stubble-Grown Millet Is Part of the Fight for a Bigger Crop." *Sotsialisticheskoye zemledeliye*, June 26, 1934.

1935

48. "Combine Vernalization with High Agricultural Technique!" *Sotsialisticheskoye zemledeliye*, March 21, 1935.

49. "Creator of the Theory of Distant Hybridization, The." [On I. V. Michurin. Written in collaboration with I. I. Prezent.] *Sotsialisticheskoye zemledeliye*, June 8, 1935.

50. (Excerpt from his speech at the First Session of the Lenin Academy of Agricultural Sciences of the U.S.S.R.) *Byulleten VASKhNIL* [*Bulletin of the Lenin Academy of Agricultural Sciences of the U.S.S.R.*], No. 7, 1935, pp. 1-3.

51. "Experimental and Practical Planting of Seed Potatoes in the Summer of 1935, The." Journal *Sad ta gorod* [*Orchard and Vegetable Garden*], Kiev, No. 3-4, 1935, pp. 4-6; Journal *Khata-laboratoria* [*Collective-Farm Laboratory*], Kiev, No. 4, 1935, pp. 5-6.

52. "Germination of Young Potatoes, The." [For Mass Experimentation.] Odessa, Institute of Selection and Genetics, 1935, 19 pp.

53. "How Seed Potatoes Should Be Planted in Summer for Experimental and Practical Farming." Odessa, published by the newspaper *Chornomorska komuna*, 1935, 8 pp. (Institute of Selection and Genetics.)

54. "Immediate Tasks of Vernalization, The." [Address at the session of the Lenin Academy of Agricultural Sciences of the U.S.S.R., held on October 28 to November 3.] *Sotsialisticheskoye zemledeliye*, October 29, 1935.

55. "In Memory of the Great Revolutionist in Science." [Concerning I. V. Michurin. Written in collaboration with I. I. Prezent.] *Chornomorska komuna*, Odessa, June 14, 1935.

56. "Let Us Give the Collective Farms Good and Regenerated Seed Varieties!" [A Talk.] *Chornomorska komuna*, Odessa, July 21, 1935.

57. "Let Us Grow Our Own Steppe Potatoes!" Newspaper *Komunist* [*Communist*], Kiev, March 21, 1935.

58. "Most Honourable Assignment, A." Tasks of Agricultural Science. Newspaper *Visti* [*News*], Kiev, October 8, 1935.

59. "New Achievements in Agricultural Science." [Address at a conference of agronomists and zootechnicians of the Kiev area held on November 28, 1935.] *Komunist*, Kiev, December 1, 1935.

60. "New Varieties of Early-Ripening Wheat." [Report to the C.C., C.P.S.U.(B.), the People's Commissariat of Agriculture of the U.S.S.R. and the Lenin Academy of Agricultural Sciences of the U.S.S.R., on the fulfilment of the obligation undertaken to create a new variety of spring wheat for the Odessa area.] *Sotsialisticheskoye zemledeliye*, August 3, 1935; Journal *Yarovizatsia* [*Vernalization*], No. 1, 1935, pp. 3-4.

61. On the article "Some Critical Remarks." Journal *Selektsia i semenovodstvo* [*Plant Breeding and Seed Growing*], No. 2, 1935, pp. 21-30.

62. "Plant Breeding and the Theory of Phasic Development of Plants." [Written in collaboration with I. I. Prezent.] Moscow, Selkhozgiz, 1935, 64 pp., drawings.

63. "Planting Potatoes in the Summer." [Address at a regional conference on vernalization in practice.] *Chornomorska komuna*, Odessa, December 30, 1935.

64. "Potatoes in the South." [The degeneration of potatoes and measures against it.] *Izvestia*, September 15, 1935.

65. "Potatoes in the South and the Theory of Phases." *Sotsialisticheskoye zemledeliye*, October 9, 1935.

66. "Potato Vernalization in the South." *Sotsialisticheskoye zemledeliye*, March 26, 1935.

67. "Preface." To the book *A World-Wide Collection of Wheat with Vernalization in the Background*, by D. A. Dolgushin, Moscow 1935, pp. 1-3.

68. "Regeneration of Seeds, The." [Cross-pollination.] *Izvestia*, July 15, 1935.

69. "Regeneration of the Soil, The." [Achievements of agricultural science in plant breeding.] *Pravda*, November 7, 1935.

70. "Regeneration of the Variety. Pure lines, self-pollination and methods of seed growing." *Sotsialisticheskoye zemledeliye*, June 30, 1935.

Same (slightly abridged). Journal *Kolkhoznoye opytnichestvo* [*Collective-Farm Experimentation*], No. 8, 1935, pp. 13-16.

71. "Reorganization of Seed Growing, The." [A Paper Read at the Visiting Session of the Lenin Academy of Agricultural Sciences of the U.S.S.R., held in Odessa, June 27, 1935.] *Yarovizatsia*, No. 1, 1935, pp. 25-64.

72. "Research Work Must Be Planned." *Chornomorska komuna*, Odessa, June 4, 1935.

73. "Some Vernalization Results." *Pravda*, October 27, 1935.

74. "Stakhanov Movement and the Tasks of Soviet Agrobiology, The." [Written in collaboration with I. I. Prezent.] *Yarovizatsia*, No. 3, 1935, pp. 1-12.

75. *Theoretical Principles of Vernalization, The*. Moscow-Leningrad, Selkhozgiz, 1935, 152 pp., drawings, table and portrait.

2nd revised and enlarged edition. Moscow, Selkhozgiz, 1936, 192 pp., drawings, and table.

76. "Theory of Plant Development and the Fight Against Potato Degeneration in the South, The." [Stenographic report of his address at a meeting of the Fruit and Vegetable Sector of the Lenin Academy of Agricultural Sciences of the U.S.S.R., held in Moscow on September 13, 1935.] *Yarovizatsia*, No. 2, 1935, pp. 3-22.

77. "Utilizing Our Knowledge of the Phasic Development of Plants in Selection and Genetics." [For the visiting session of the Cereal, Grain and Leguminous, Oil Bearing and Fodder Crops Section of the Lenin Academy of Agricultural Sciences of the U.S.S.R., held on June 26-29, 1935 in Odessa.] *Byulleten VASKhNIL*, No. 6, 1935, pp. 24-25.

78. "Vernalization." In *Agricultural Encyclopedia*, Vol. IV, Moscow 1935, pp. 1045-47.

79. "Vernalization." [Paper read at a regional conference of collective-farm laboratory managers.] *Chornomorska komuna*, Odessa, March 10, 1935.

80. "Vernalization—a Powerful Means of Increasing Yield." [Abridged stenographic report of his speech at the Second All-Union Congress of Collective-Farm Shock Workers.] *Pravda*, February 15, 1935; *Izvestia*, February 15, 1935; *Sotsialisticheskoye zemledeliye*, February 15 and 21, 1935 (varying texts).

81. "Vernalization on Collective and State Farms." *Komunist*, Kiev, March 9, 1935.

82. "Vernalization of Potatoes, The." [Works of 1934-35. A talk.] *Pravda*, June 25, 1935.

83. "Vernalization Results Must Be Fully Taken into Account, The." *Sotsialisticheskoye zemledeliye*, April 28, 1935.

84. "We Must Master the Technique of Vernalization," *Sotsialisticheskoye zemledeliye*, March 27, 1935.

1936

85. "About 'Ologies' and Real Science." [Written in collaboration with I. I. Prezent. On A. S. Serebrovsky's article "The Hybridization of Animals as a Science."] *Pravda*, June 26, 1936.

86. "Address." [Held at a conference of activist agronomists of Kiev Region. On the life and development of agricultural plants.] In the book *A Conference of Activist Agronomists of Kiev Region*, Kiev 1936, pp. 60-71.

87. "Are They Confused or Spreading Confusion?" [Reply to the article "A Few Words on the Work of the Odessa Institute of Selection and Genetics," by Academician P. N. Konstantinov, Academician P. I. Lisitsyn and Doncho-Kostov.] *Yarovizatsia*, No. 5, 1936, pp. 30-44.

88. "Autobiography." In the book *Works of Academician T. D. Lysenko*. A bibliographical index, Kiev 1936, pp. 7-12.

89. "Autumn Sowing of Unharvested Cotton Fields." Journal *Bavovnitstvo i novi tekh. kulturi* [*Cotton Growing and New Industrial Crops*], Kiev 1936, No. 2, p. 15.

90. *Brief Instructions on Vernalization of Sugar Beets.* [For collective-farm laboratories.] Odessa, Institute of Selection and Genetics, 1936, 12 pp.

91. *Collaboration Between Research Institutes and Collective-Farm Laboratories.* An Address at the First Ukrainian Conference of Collective-Farm Laboratory Managers, held on March 2, 1936. Kiev-Kharkov. State Publishing House of Collective- and State-Farm Literature of the Ukrainian S.S.R., 1936, 32 pp.

92. "Collect a Ton of Cotton per Hectare Before the Frosts Set In." [The Importance of Cotton Topping Under Ukrainian Conditions.] *Khata-laboratoria*, Kiev 1936, No. 6, pp. 42-44.

93. "Cotton Crop Topping" [in the South of the Ukrainian S.S.R.]. *Pravda*, October 6, 1936; *Izvestia*, October 12, 1936; the journal *Borba za khlopok* [*The Struggle for Cotton*]. Tashkent 1936, No. 7-8, pp. 61-65.

94. "Cotton Plant Topping Experiments Should Be Developed on a Mass Scale." [Speech delivered at a meet of cotton field team leaders of Odessa Region.] *Komunist*, Kiev, March 24, 1936.

95. "Cultivation of Seed Potatoes Under the Conditions of the South of the U.S.S.R., The." In the book *Potato Cultivation in the South and Southeast of the U.S.S.R. Reports and Decisions of the First Plenary Meeting of the Fruit and Vegetable Section*. Moscow 1936, pp. 8-20. (*Publications of Lenin Academy of Agricultural Sciences of the U.S.S.R.*, Vol. 12.)

96. "Five Pivotal Questions. The unity of science and practice and the work of collective-farm laboratories." [An outline of the work of the Institute of Selection and Genetics in Odessa. An address at a session of the Lenin Academy of Agricultural Sciences of the U.S.S.R.] *Sotsialisticheskoye zemledeliye*, March 6, 1936.

97. "Greater Theoretical Clarity" [in plant breeding. Speech held at a session of the Lenin Academy of Agricultural Sciences of the U.S.S.R.] *Sotsialisticheskoye zemledeliye*, March 3, 1936.

98. "Intravarietal Crossing" (of wheat). *Sovkhoznaya gazeta* [*State-Farm Newspaper*], October 26, 1936.

99. "Intravarietal Crossing of Self-Pollinating Plants, The." [Edited stenographic report of an address delivered at the Omsk session of the Grain Section of the Lenin Academy of Agricultural Sciences of the U.S.S.R., held on August 29, 1936.] *Selektsia i semenovodstvo*, No. 11, 1936, pp. 13-27; Journal *Sotsialisticheskaya Rekonstruktsia selskogo khozyaistva* [*Socialist Reconstruction of Agriculture*], No. 10, 1936, pp. 70-86.

100. "It Is in Man's Power. Let Us Wrest From Nature the Key to the Variability of Plant Forms." *Sotsialisticheskoye zemledeliye*, November 7, 1936.

101. "Keeping Abreast of the Times." [The new tasks confronting agricultural science in connection with the development of socialist economy.] *Khata-laboratoria*, No. 2, 1936, pp. 7-11.

102. "Make Up For Lost Time. On Regenerated Varieties." [Intravarietal crossing of wheat.] *Sotsialisticheskoye zemledeliye*, June 4, 1936.

Same. Revised and enlarged edition. *Yarovizatsia*, No. 2-3, 1936, pp. 9-18.

103. "Most Important Results of Wheat Vernalization, The." [Report of the Second Session of the Lenin Academy of Agricultural Sciences of the U.S.S.R., held on October 29, 1935.] In the book *Wheat Cultivation*. Materials of the Second Session of the Lenin Academy of Agricultural Sciences of the U.S.S.R., Moscow 1936, pp. 48-54. (*Publications of Lenin Academy of Agricultural Sciences of the U.S.S.R.*, Vol. 8.)

104. "Multiply the Ranks of Michurinists." [On the anniversary of the death of Michurin. Written in collaboration with I. I. Present.] *Komsomolskaya pravda*, June 6, 1936.

105. "Necessary Warning, A." Anent an article by agronomist I. Balaya. [*Victories of Vernalization.*] *Sovkhoznaya gazeta*, March 4, 1936.

106. "Physiology of Plant Development and the Problem of Winter-Hardiness of Winter Grain, The." [An address delivered in Dniepropetrovsk at an All-Union conference on winter-hardiness.] In the 1935 issue of *Agriculture in the U.S.S.R.*, an annual publication. Moscow 1936, pp. 12-17.

107. "Reorganization of Potato Seed Growing in the Southern Ukraine, The." *Journal of the Botanical Institute of the Academy of Sciences of the Ukrainian S.S.R.*, Kiev 1936, No. 8, pp. 3-13.

108. "Role of Agricultural Science in Solving the Problems of Yield, The." [In connection with the task assigned by Comrade Stalin of obtaining an annual grain crop of 7-8 thousand million poods in the very near future.]

Journal *Front nauki i tekhniki* [*The Science and Technique Front*], No 2, 1936, pp. 60-61.

109. "Science and Crops." Speech delivered at a meeting of leaders of regional and city Party and Soviet organizations and factory Stakhanovites with decorated agricultural personnel from the Odessa area. *Chornomorska komuna*, Odessa, January 11, 1936.

110. "Science in the Struggle for a Stalin Crop." *Sovkhoznaya gazeta*, August 2, 1936.

111. "Some Results of Vernalization." In the book *Vernalization—a Powerful Means of Raising Yields*. Materials of a conference on crop yields held under the auspices of the newspaper *Kommuna* [*Commune*], Voronezh 1936, pp. 8-11.

112. [Speech at the Third Session of the Lenin Academy of Agricultural Sciences of the U.S.S.R. held on February 25, 1936, dealing with problems of cotton-plant breeding and seed growing.] In the book *The Fight for Better Cotton Crops*. (Session material.) Moscow 1936, pp. 89-95. (*Publications of Lenin Academy of Agricultural Sciences of the U.S.S.R.*, Vol. 26, section 2.)

113. Speech [held at a conference of leaders in the fight for big cereal yields, and of tractor drivers and thresher operators with Party and Government leaders]. *Komunist*, Kiev, January 1, 1936. (In Ukrainian); *Pravda*, January 2, 1936; *Izvestia*, January 3, 1936; *Sovkhoznaya gazeta*, January 3, 1936; *Sotsialisticheskoye zemledeliye*, January 3, 1936; *Selektsia i semenovodstvo*, No. 2, 1936, pp. 10-17.

114. "Stakhanov Movement and Science, The." *Krestyanskaya gazeta* [*The Peasants' Newspaper*], September 1, 1936.

115. "Summer Planting of Potatoes, The." [*Instructions*.] Tiraspol, State Publishing House of Moldavia, 1936, 44 pp., drawings, tables. (Lenin Academy of Agricultural Sciences of the U.S.S.R., Institute of Selection and Genetics.)

2nd [5th] revised edition [written in collaboration with A. M. Favorov], Moscow, Selkhozgiz, 1939, 32 pp., drawings.

116. "Topping of Cotton Plants, The." [Advice to team leaders.] Odessa, Institute of Selection and Genetics, 1936, 20 pp., drawings.

117. "Topping of Cotton Plants, The." *Chornomorska komuna*, Odessa, June 3, 1936. (In Ukrainian); Journal *Sovetsky khlopok* [Soviet Cotton], No. 7, 1936, pp. 49-53.

118. "Vernalization of Cereals and Potatoes in Moscow Region." Newspaper *Rabochaya Moskva* [Workers' Moscow], April 20, 1936.

119. *Vernalization of Spring Vetch, The*. Leningrad 1936, 6 pp. (Lenin Academy of Agricultural Sciences of the U.S.S.R., Ukrainian Institute of Selection and Genetics.)

120. "Vernalization of Sugar Beet, The." Newspaper *Za pishchevuyu industriyu* [*For a Modern Food Industry*], April 9, 1936.

121. "Vernalization of Sugar-Beet Seeds, The." *Komunist*, Kiev, March 27, 1936.

122. "We Must Learn the Michurin Method of Creating Varieties." *Sotsialisticheskoye zemledeliye*, January 24, 1936 [abridged].

Same, in full, entitled "A Brilliant Geneticist and Plant Breeder," in I. V. Michurin's book *Results of Sixty Years' Work*, 4th edition. Moscow 1936, pp. VII-XVI. 5th edition, Moscow 1940, pp. IX-XVI.

123. "What Are We Working On?" [A work of the Institute of Selection and Genetics] *Izvestia*, March 24, 1936.

124. "What We Have Accomplished." [Report on the Fulfilment of the Targets Pledged by the Odessa Institute of Selection and Genetics, delivered at a conference at the Lenin Academy of Agricultural Sciences of the U.S.S.R. convened October 10, 1936 to discuss the results of the socialist emulation between research institutes. Abridged edition.] *Izvestia*, October 12, 1936.

Same in full, entitled "First Results." *Yarovizatsia*, No. 5, 1936, pp. 3-14.

125. "When to Commence Steeping Seed?" (Vernalization of cereals.) *Sotsialisticheskoye zemledeliye*, March 5, 1936.

126. "Collective-Farm Laboratories and Agronomic Science." *Yarovizatsia*, No. 5, 1937, pp. 12-32.

Same, Moscow, Selkhozgiz, 1949, 32 pp.

127. "Enlarge the Area Under Vernalized Potatoes!" [Suggestion to experimenters and agronomists of Moscow Region on the technique of vernalizing potatoes. From the stenographic report of a conference held on November 1, 1935 at the editorial offices of the *Moskovskaya kolkhoznaya gazeta* [*Moscow Collective-Farm Newspaper*]. *Moskovsky agronomichesky jurnal* [*Moscow Agronomic Journal*], No. 1, 1937, pp. 22-23.

128. "Essence of the Influence of the Stock on the Scion, The." [Paper read at the Fourth Plenum of the Fruit and Vegetable Section of the Lenin Academy of Agricultural Sciences of the U.S.S.R., held June 25 to July 1, 1936 at Michurinsk.) Moscow-Leningrad, Lenin Academy of Agricultural Sciences of the U.S.S.R. Publishing House, 1937, 12 pp.

129. "Fundamentals of Intravarietal Crossing, The" [of self-pollinators. The cross-pollination of wheat.] *Sotsialisticheskoye zemledeliye*, May 21, 1937.

130. "Improve the Work of the Collective-Farm Laboratories!" Journal *Zernove gospodarstvo* [*Seed Management*], No. 12, 1937, pp. 19-27; *Khata-laboratoria*, No. 12, 1937, pp. 28-36.

131. "Let Us Master the Michurinian Methods." *Visti*, Kiev, June 10, 1937.

132. "Multiplication of Regenerated Seeds, The" [of winter wheat.] *Sotsialisticheskoye zemledeliye*, August 20, 1937.

133. "Potatoes in the Southern Districts of the U.S.S.R." [A discussion of the problems of the Third Five-Year Plan.] *Pravda*, June 27, 1937.

134. "Practical Confirmation of Darwin's Theory." [Results of Intravarietal Crossing. Written in collaboration with I. I. Prezent.] *Komunist*, Kiev, April 18, 1937, *Rabochaya Moskva*, April 20, 1937. (Re-edited.)

135. "Remaking the Nature of Plants." [From an address at a conference of Ukrainian collective farm laboratory managers held in Odessa.] *Sotsialisticheskoye zemledeliye*, November 12, 1937.

136. "Remaking the Nature of Plants." Moscow, Selkhozgiz, 1937, 48 pp., drawings and portrait. (*New Developments in Agriculture*, Vol. 1.)

137. "Road that Brought Me to Science, The." *Pravda*, October 1, 1937, *Selektsia i semenovodstvo*, No. 11, 1937, pp. 16-18.

138. "Spinelessness of the Academy on Questions of Genetics and Selection, The. [*From a speech at a meeting of Academy activists.*] *Byulleten VASKhNIL*, No. 4, 1937, pp. 11-14.

139. "Strength of Soviet Science, The." *Komunist*, Kiev, November 7, 1937; *Sotsialisticheskaya rekonstruktsiya selskogo khozyaistva*, No. 11-12, 1937, pp. 69-85.

140. "Summer-Planted Potatoes." *Visti*, Kiev, July 12, 1937.

141. "Ten Million Hectares Under Vernalized Crops." [Grain Vernalization Methods.] *Sovkhoznaya gazeta*, March 6, 1937.

142. Topping of Cotton Plants, The. [Written in collaboration with A. A. Avakian.] Moscow, Selkhozgiz, 1937, 22 pp., drawings. (*The New in Agriculture*, Vol. 12.)

2nd ed., Moscow, Selkhozgiz, 1949, 24 pp., drawings.

143. "Transformation of the Nature of Plants, The." [Abridged stenographic report of an address at the Fourth Session of the Lenin Academy of Agricultural Sciences of the U.S.S.R., December 23, 1936.] *Izvestia*, December 24, 1936; *Sovkhoznaya gazeta*, December 24, 1936; Newspaper *Sovetskoye zemledeliye* [*Soviet Agriculture*], December 24, 1936.

Same, in full entitled "Two Trends in Genetics." *Yarovizatsia*, No. 1. 1937, pp. 29-75.

144. "Transformer of Plant Nature, A." [I. V. Michurin.] *Sotsialisticheskoye zemledeliye*, June 6, 1937.

145. "Vernalization of Millet, The." Odessa, Institute of Selection and Genetics, 1937, 8 pp.

146. "We Must Proceed in Earnest to Intravarietal Crossing." A letter to all agronomists of district land departments, machine and tractor stations and state farms; to all collective-farm laboratory managers. *Chornomorska komuna*, Odessa, March 27, 1937.

147. "What 'Conclusions' Disturb Academician Konstantinov?" [In connection with articles written by P. N. Konstantinov against vernalization.] *Sotsialisticheskoye zemledeliye*, April 4, 1937; *Yarovizatsia*, No. 2, 1937, pp. 19-31.

148. "What Shall We Display at the All-Union Agricultural Exhibition?" [The work of the Institute of Selection and Genetics on the intravarietal crossing of plants.] *Pravda*, April 6, 1937.

1938

149. "Achievement of an Advanced Science, The." [Experience of the Institute of Scientific Methods of Sowing. The Ukraine.] *Sotsialisticheskoye zemledeliye*, August 28, 1938; *Zernove gospodarstvo*, Kiev, No. 8, 1938, pp. 21-25.

150. "Agrotechnique of Summer-Planting Potatoes, The." *Komunist*, Kiev, March 29, 1938; *Sad ta gorod*, Kiev, No. 5, 1938, pp. 5-9.

151. "Cleaning of Barley Seeds of Pieces of Smut Before Vernalization, The." *Bolshevistskoye znamya*, Odessa, March 9, 1938; *Chornomorska komuna*, Odessa, March 9, 1938 (in Ukrainian); *Sotsialisticheskoye zemledeliye*, March 22, 1938.

152. "Directed Change in the Nature of Plants by Growing Them Under Appropriate Conditions." *Visti AN URSR* [*Herald of the Ukrainian Academy of Sciences*], Kiev, 1938, No. 5, pp. 24-35.

153. "Fighter for Advanced Agronomic Science, A." [On the 75th anniversary of the birth of Academician V. R. Williams.] *Sotsialisticheskoye zemledeliye*, October 10, 1938.

154. *Guide to Intravarietal Crossing of Winter and Spring Wheats*. [Written in collaboration with D. A. Dolgushin.] Moscow, Selkhozgiz, 1938, 15 pp., drawings.

155. "Intravarietal Crossing and Mendel's So-Called 'Law' of Segregation." [Revised stenographic report of a lecture delivered on April 15, 1938, at a seminar on seed growing at the All-Union Institute of Selection and Genetics.] *Yarovizatsia*, No. 1-2, 1938, pp. 114-26; *Selektsia i semenovodstvo*, No. 8-9, 1938, pp. 1-6.

156. "Intravarietal Crossing of Winter and Spring Wheat." [Written in collaboration with D. A. Dolgushin.] Newspaper *Bolshevistskoye znamya* [*Bolshevik Banner*], Odessa, March 24, 1938; *Selektsia i semenovodstvo*, No. 8-9, 1938, pp. 7-8.

157. "Seed Growing Must be Based On Michurin's Theory." *Yarovizatsia*, No. 4-5, 1938, pp. 24-39.

Same, Moscow, Selkhozgiz, 1949, 24 pp.

158. "Mentor—a Powerful Means of Plant Breeding, The." [Preface to the collection of works of I. V. Michurin. Revised and enlarged preface to I. V. Michurin's book, *Results of Sixty Years' Work*, see No. 122.] *Yarovizatsia*, No. 3, 1938, pp. 35-44.

159. "Scheme for the Production of Elite Seeds by State Plant-Breeding Stations, A." [Proposals of a commission presided over by Academician T. D. Lysenko.] *Yarovizatsia*, No. 1-2, 1938, pp. 143-46.

160. "Stakhanovite Methods of Work." [Science in agriculture.] *Sotsialisticheskoye zemledeliye*, August 30, 1938.

161. "Summer Planted Potatoes in the South of the Ukraine." *Pravda*, July 4, 1938.

162. "Summer Planting of Potatoes, The." [Experiments of the All-Union Institute of Selection and Genetics.] *Bolshevistskoye znamya*, Odessa, April 1 and 2, 1938; *Sotsialisticheskoye zemledeliye*, April 6, 1938.

163. "Treading New Paths." [The tasks of reorganizing the work of the Lenin Academy of Agricultural Sciences of the U.S.S.R. From a speech delivered at a session of the Presidium of the Academy.] *Pravda*, April 9, 1938; *Sotsialisticheskaya rekonstruktsiya selskogo khozyaistva*, No. 5, 1938, pp. 38-42.

164. "Valuable Contribution to Agronomic Science, A." [A method of artificially pollinating cross-pollinating plants, proposed by agronomist A. S. Musiiko.] *Sotsialisticheskoye zemledeliye*, October 20, 1938.

165. "What Will the All-Union Institute of Selection and Genetics Work on in 1939?" (Explanatory Notes for a thematic plan.) *Yarovizatsia*, No. 6, 1938, pp. 21-26.

166. "Works of I. V. Michurin Form the Basis of Soviet Genetics, The." *Pravda*, June 6, 1938.

1939

167. "Achievements of Science Should Be Made Available to Collective-Farm Practice, The." *Sotsialisticheskoye zemledeliye*, January 1, 1939.

168. Agronomic Science and the Struggle Against Drought." [Revised stenographic report of a speech at a Saratov regional conference on drought control, held December 6, 1938.] *Sotsialisticheskoye zemledeliye*, December 18, 1938, portrait; Journal *Sotsialisticheskoye zernovoye khozyaistvo* [*Socialist Grain Management*], Saratov, No. 6, 1939, pp. 20-27.

169. "Azerbaijan Will Be Able in the Very Near Future to Secure a Plentiful Supply of Home-Grown Potatoes." [A talk in the C.C., of the C.P.(B.) of Azerbaijan on the re-

sults of the summer planting of potatoes.] Newspaper *Bakinsky rabochy* [*The Baku Worker*], December 7, 1939.

170. "Creative Role of Selection, The." [From a paper read at a session of the Lenin Academy of Agricultural Sciences of the U.S.S.R. (December 1-3, 1939) dedicated to the 80th anniversary of the publication of Charles Darwin's *The Origin of Species*] *Sotsialisticheskoye zemledeliye*, December 16, 1939.

171. "Creator of Soviet Agrobiology, The." (On the occasion of the fourth anniversary of the death of I. V. Michurin.) *Sotsialisticheskoye zemledeliye*, June 6, 1939 [abridged]; *Yarovizatsia*, No. 3, 1939, pp. 15-26. [In full.]

172. "Directed Change of the Hereditary Nature of Plants." *Vestnik AN S.S.S.R.* [*Herald of the Academy of Sciences of the U.S.S.R.*]. No. 11-12, 1939, pp. 120-26; Journal *Sovetskaya agronomia* [*Soviet Agronomy*], No. 12, 1939, pp. 33-37.

173. "Great Heritage of Darwin and Michurin in Action, The." Newspaper *Bilshovitska pravda* [*Bolshevik Truth*], Vinnitsa, June 7, 1939.

174. "High-Yield Millet Plots." [In the collective farms of Saratov Region.] *Sotsialisticheskoye zemledeliye*, July 9, 1939.

175. "How to Keep Potatoes in Earth-Covered Trenches." *Sotsialisticheskoye zemledeliye*, May 5, 1939.

176. "How to Release in the Fields the Telenomus to Devour the Eggs of the Tortoise Beetle." Newspaper *Sotsialisticheskaya Moldavia* [*Socialist Moldavia*], Tiraspol, May 12, 1939 (in Ukrainian); *Bolshevistskoye znamya*, Odessa, May 14, 1939.

177. "Let Us Grow and Harvest a Big Millet Crop!" Newspaper *Kommunist* [*Communist*], Saratov, July 8, 1939.

178. "Michurin's Theory in Action at the Review of the Outstanding Victories of Socialist Agriculture." [On the All-Union Agricultural Exhibition.] Newspaper *Kolgospnik Ukraini* [*Ukrainian Collective Farmer*], Kiev, July 30, 1939.

179. "On the work of Agronomist A. S. Musiiko." [Artificial pollination of cultivated cross-pollinating plants.] *Yarovizatsia*, No. 1, 1939, pp. 74-75.

180. (Paper read at a conference on genetics and plant breeding called by the editors of the journal *Pod znamenem Marxisma* [*Under the Banner of Marxism*], October 7-14, 1939). *Pod znamenem Marxisma*, No. 11, 1939, pp. 146-68.

181. Preface to I. D. Kolesnik's book *We Must Obtain Millet Crops of No Less than 15 Centners a Hectare*. [Agronomic instructions for the collective and state farms of the southeastern section of the U.S.S.R.] Moscow 1939, pp. 3-6.

182. "Science that Breaks Down Obsolete Traditions, A." [On the occasion of the opening of the All-Union Agricultural Exhibition.] *Sotsialisticheskoye zemledeliye*, August 1, 1939.

183. "Should Seeds Be Vernalized When the Sowing Is Exceptionally Early (February)?" *Sotsialisticheskoye zemledeliye*, February 23, 1939.

184. "Use the Telenomus to Fight the Tortoise Beetle." Newspaper *Bolshevik Poltavshchiny* [*The Poltava Bolshevik*], May 6, 1939 (in Ukrainian); *Bolshevistskoye znamya*, Odessa, May 11, 1939.

185. "Vernalization of Millet, The." [Agrotechnical advice] Newspaper *Priuralskaya pravda* [*Ural Truth*], Uralsk, April 10, 1939.

186. *Vernalization of Spring Cereals, The*. Izhevsk, Udmurtia State Publishing House, 1939, 24 pp.

187. "Ways of Rearing Winter-Hardy Varieties of Winter Plants in the East." [From a speech delivered at a conference of plant-breeding station personnel of the eastern districts of the U.S.S.R. which took place at the Lenin Academy of Agricultural Sciences of the U.S.S.R. on March 7, 1939.] *Yarovizatsia*, No. 2, 1939, pp. 55-60.

188. "We Must Constantly Improve Plant Nature." [A speech held at a meeting of practical-farming personnel participating in the All-Union Agricultural Exhibition with scientists of our country.] *Sotsialisticheskoye zemledeliye*, August 12, 1939.

189. "With reference to Academician N. I. Vavilov's Article 'How to Arrange a Course in Genetics, Plant Breeding and Seed Growing.'" *Sotsialisticheskoye zemledeliye*, February 1, 1939; *Yarovizatsia*, No. 1, 1939, pp. 136-40.

1940

190. "Anent Academician A. N. Kolmogorov's Article." ["A New Confirmation of Mendel's Laws."] *Doklady A.N. S.S.S.R.* [Papers read at the Academy of Sciences of the U.S.S.R.] Vol. 28, No. 9, 1940, pp. 834-35.

191. "Combating the Tortoise Beetle Pest." [Abridged stenographic report of a speech held at a conference called to fight the tortoise beetle, September 24, 1940.] *Sotsialisticheskoye zemledeliye,* October 23, 1940. Journal *Za ustoichivy urojai na jugo-vostoke* [*For a Stable Crop in the Southeast*], Saratov, No. 11-12, 1940, pp. 1-6.

192. "Consolidate the Successes Achieved in Raising Millet Yields!" [From a speech held at a conference on millet at the Lenin Academy of Agricultural Sciences of the U.S.S.R. on January 13, 1940.] *Yarovizatsia,* No. 1, 1940, pp. 28-31.

193. "Fine Theoretician of Darwinism, A." (On the 20th anniversary of the death of K. A. Timiryazev.) Journal *Partiynoe stroitelstvo* [*Party Development*], No. 9, 1940, pp. 16-23.

194. "Following in the Tracks of the Great Russian Darwinist." [I. V. Michurin]. *Sotsialisticheskoye zemledeliye,* June 6, 1940.

195. "Genuine Darwinist, A." [K. A. Timiryazev.] *Sotsialisticheskoye zemledeliye,* April 28, 1940.

196. "Great Biologist and Transformer of Nature, A," [on the 85th anniversary of the birth of I. V. Michurin.] *Pravda,* October 27, 1940.

197. "How to Get Plenty of Potatoes." [The task of introducing summer planting into the collective farms of the region.] *Moskovsky bolshevik* [*Moscow Bolshevik*], *March 12,* 1940.

198. "How to Increase Potato Crops in Moscow Region." [Abridged stenographic report of an address delivered at a conference held under the auspices of the editors of the *Moskovsky bolshevik* on March 27, 1940. *Moskovsky bolshevik* (special issue), April 17, 1940.

199. "Let Us Increase the Millet, Grass and Potato Crops!" [From a speech at a conference of Peasant Youth Department Managers of Regional and Territorial Committees and the C.C. of the Y.C.L. of the Union Republics held at the Y.C.L. C.C.] *Komsomolskaya pravda,* June 1, 1940.

200. "Main Thing in Vernalization, The." Newspaper *Leninskoye znamya* [*Leninist Banner*], Tosno, April 11, 1940.

201. "Making Practical Use of the Theory of the Phasic Development of Plants." In the book *The Experience of Advanced Workers in Agriculture.* A symposium of lectures and papers read by scientists and advanced collective- and state-farm workers at the All-Union Agricultural Exhibition of 1939, Vol. 1, Moscow 1940, pp. 12-17.

202. "Michurin's Theory at the All-Union Agricultural Exhibition." Journal *Vestnik selskokhozyaistvennoy nauki. Plodovoyagodnye kultury.* [*Herald of Agricultural Science. Fruits and Small Fruits*], Vol. 1, 1940, pp. 3-11.

203. "New Achievements in Controlling the Nature of Plants." [Stenographic report of an address delivered at the All-Union Conference of heads of Marxism-Leninism departments on July 6, 1940. Newspaper *Sovetskaya Ukraina* [*Soviet Ukraine*], Kiev, October 16, 1940; *Pod znamenem Marxisma,* No. 10, 1940, pp. 110-26.

Same, Moscow, Selkhozgiz, 1949, 30 pp.

204. "New Method of Sowing Kok-Saghyz, A." *Sotsialisticheskoye zemledeliye,* April 11, 1940.

205. "Powerful Means of Improving Planting Material, A." [Summer planting of potatoes.] *Sotsialisticheskoye zemledeliye,* April 15, 1940.

206. "Preparing the Field for the Summer Planting of Potatoes." Newspaper *Bolshevistskaya smena* [*The Bolshevik New Generation*], Rostov-on-Don, May 26, 1940.

207. "Soviet Agrobiology at the Exhibition." Newspaper *Vsesoyuznaya selskokhozyaistvennaya vystavka* [*The All-Union Agricultural Exhibition*], November 19, 1940.

208. "Speech..." [delivered at a republican conference of advanced workers in Ukrainian agriculture]. *Sovetskaya Ukraina,* Kiev, February 21, 1940.

209. "Strive for a Big Potato Crop." [Speech delivered at a Rostov Region conference on vegetable and potato growing.] Newspaper *Molot* (*Hammer*), Rostov-on-Don, February 28, 1940.

210. "Stubble-Sowing of Alfalfa in the South" [of the Ukraine]. Newspaper *Zhivotnovodstvo* [*Animal Husbandry*], February 21, 1940; *Vestnik A.N. S.S.S.R.*, No. 3, 1940, pp. 42-44.

211. "Summer Planted Potatoes Should Be Picked on Time." Newspaper *Moskovsky bolshevik*, September 20, 1940.

212. "Triumph of Michurin's Ideas in Agrobiology, The." *Vsesoyuznaya selskokhozyaistvennaya vystavka*, June 6, 1940.

213. "Vernalization" [of cereals]. *Sotsialisticheskoye zemledeliye*, March 24, 1940.

214. "Ways of Controlling Plant Organisms." [A paper read at the first annual Timiryazev Reading, April 28, 1940.] *Yarovisatsia*, No. 3, pp. 27-44; *Vestnik A.N. S.S.S.R.*, No. 6, 1940, pp. 13-28.

Same, Moscow, Selkhozgiz, 1949, 32 pp.

215. "What Is Michurinian Genetics?" [Revised stenographic report of a public lecture delivered on October 15, 1940, at the Leningrad State University.] Newspaper *Leningradsky universitet* [*Leningrad University*], November 23, 1940. *Yarovizatsia*, No. 6, 1940, pp. 3-19.

1941

216. Collective- and State-Farm Practice Is the Basis of Advance Agronomic Science." [On the Opening of the All-Union Agricultural Exhibition.] *Pravda*, May 25, 1941.

217. "Engels and Certain Problems of Darwinism." [Abridged stenographic report of an address read on November 28, 1940, at a meeting of the History and Philosophy Department of the Academy of Sciences of the U.S.S.R., dedicated to the 120th anniversary of the birth of Frederick Engels.] *Sotsialisticheskoye zemledeliye*, January 15, 1941; *Vestnik A.N. S.S.S.R.*, No. 1, 1941, pp. 1-11; *Doklady VASKhNIL* [Reports of the Lenin Academy of Agricultural Sciences of the U.S.S.R.], Vol. 3, 1941, pp. 3-10.

218. "Extermination of the Tortoise Beetle, The" *Chornomorska komuna*, Odessa, June 8, 1941.

219. "Fight Against the Weevil Pest, The." [Stenographic report of a speech held at a Republican conference called in Kiev on January 25, 1941, to raise the yield of sugar beets.] *Sovetskaya Ukraina*, Kiev, February 12, 1941; *Sotsialisticheskoye zemledeliye*, February 14, 1941; *Doklady VASKhNIL*, Vol. 5, 1941, pp. 3-10.

220. "Fight for High-Quality Seeds, The." *Sotsialisticheskoye zemledeliye*, November 29, 1941.

221. "Good Kok-Saghyz Sprouts Ensure a Big Crop." *Izvestia*, March 14, 1941; Pravda, March 14, 1941; *Sotsialisticheskoye zemledeliye*, March 14, 1941; *Yarovizatsia*, No. 3, 1941, pp. 3-11.

222. "Harvesting in the East." *Sotsialisticheskoye zemledeliye*, August 24, 1941.

223. "High-Grade Seeds Are an Earnest that a Stalin Crop Will Be Reaped." Newspaper *Omskaya pravda* [Omsk Truth], December 11, 1941.

224. "How a Winter-Hardy Wheat Variety Is Being Created for Siberia." *Pravda*, October 3, 1941.

225. "More Seed Potatoes Must Be Prepared." *Omskaya pravda*, October 30, 1941; *Pravda*, November 16, 1941.

226. "No Time Must Be Lost in the Fight Against the Harmful Tortoise Beetle." [Written in collaboration with G. Pavlov.] *Sotsialisticheskoye zemledeliye*, March 30, 1941.

227. "Once More About Good Kok-Saghyz Sprouts." *Pravda*, April 4, 1941; *Sotsialisticheskoye zemledeliye*, April 4, 1941; *Doklady VASKhNIL*, Vol. 9, 1941, pp. 3-6.

228. "Preparing Newly-Harvested Potato Tubers for Summer Sowing." *Sotsialisticheskoye zemledeliye*, June 6, 1941.

229. "Self-Sacrificing Work Will Ensure Big Crops." *Pravda*, June 29, 1941.

230. "Some Urgent Problems of Agricultural Production." *Pravda*, August 7, 1941.

231. "Shortcomings in the Struggle Against the Weevil Pest [in the sugar-beet districts of the Ukraine] Must Be Quickly Eliminated, The." *Pravda*, May 19, 1941.

232. "Special Features of the 1941 Grain Harvest" [in Siberia and North Kazakhstan]. *Omskaya pravda*, August 21, 1941.

233. "Technique of Seed Vernalization, The," [of spikes. Text of a poster for the People's Commissariat of Agriculture, U.S.S.R.], *Sotsialisticheskoye zemledeliye*, May 9, 1941.

234. "These Are the Crucial Days for the Annihilation of the Weevil Pest." [Tasks assigned to the collective farmers of the sugar-beet growing districts of Kiev Region.] *Pravda*, April 28, 1941.

235. "Use the Tops of Food Potato Tubers for Planting." Omsk State Publishing House, 1941, 16 pp.

236. "Using Root Cuttings to Plant Kok-Saghyz." *Molot*, Rostov-on-Don, April 19, 1941.

237. "Vernalization Under the Conditions of a Late Spring." *Pravda*, May 8, 1941; *Sotsialisticheskoye zemledeliye*, May 9, 1941.

1942

238. "Agrotechnical Significance of Autumn Ploughing, The." (Speech at a meet of young ploughmen of Yaroslavl Region.) *Komsomolskaya pravda*, October 1, 1942.

239. "Autumn Ploughing Means Fighting for a Good Crop in 1943" *Pravda*, October 11, 1942; *Sotsialisticheskoye zemledeliye*, October 13, 1942.

240. "Cultivation of Potatoes and Sugar Beets in Uzbekistan, The." [Corrected stenographic report of a lecture delivered at a meeting of agronomic workers in Tashkent on December 19, 1941.] Journal *Kolkhoznoye proizvodstvo* [*Collective-Farm Production*], No. 1, 1942, pp. 11-17; Journal *Sotsialisticheskoye selskoye khozyaistvo* [*Socialist Agriculture*], No. 1, 1942, pp. 30-37.

241. "Elaboration of the Problem of Sowing Winter Wheat in Siberia Must Be Completed in Time, The." *Doklady VASKhNIL*, Vol. 7-8, 1942, pp. 7-12.

242. "Extend the Area Sown to Potatoes and Increase the Crop." *Pravda*, October 7, 1942; *Sotsialisticheskoye zemledeliye*, October 8, 1942; *Sovkhoznaya gazeta*, October 8, 1942; *Doklady VASKhNIL*, Vols. 9-10,1942, pp. 3-6; *Kolkhoznoye proizvodstvo*, No. 12, 1942, pp. 15-16.

243. "Fighting the Weevil Pest." *Sotsialisticheskoye zemledeliye*, February 1, 1942.

244. "Grain Crop Must Be Harvested in Time and Without Loss, The." *Pravda*, August 4, 1942; *Sotsialisticheskoye zemledeliye*, August 6, 1942; *Sovkhoznaya gazeta*, August 6, 1942; Doklady VASKhNIL, Vols. 7-8, 1942, pp. 3-6.

245. "Let Us Prepare More Potato Tuber Tops." [Methods of preparation and storing.] Newspaper *Trud* [*Labour*], November 26, 1942.

246. "More Food-Potato Tuber Tops Must Be Prepared for Planting!" Newspaper *Uralsky rabochy* [*Ural Worker*], Sverdlovsk, November 20, 1942.

Same, Moscow Publishing House of the People's Commissariat of Agriculture of the U.S.S.R., 1942, 12 pp.

247. "More Potatoes and Vegetables." Journal *V pomoshch fabrichno-zavodskim i mestnym komitetam* [*Assistance to Factory and Local Trade Union Committees*], No. 6, 1942, pp. 5-8.

248. "Potato Tuber Tops." *Moskovsky bolshevik*, May 7, 1942.

249. "Seeds of Good Quality Guarantee a Bumper Crop." *Pravda*, March 20, 1942.

250. "Some Fundamental Tasks of Agricultural Science." [Revised verbatim report to the general meeting of members and corresponding members of the Academy of Sciences of the U.S.S.R., held in Sverdlovsk on May 6, 1942, *Vestnik A.N. S.S.S.R.*, No. 5-6, 1942, pp. 49-59; *Doklady VASKhNIL*, Vols. 5-6, 1942, pp. 3-12.

251. "Some Special Features of the Grain Harvests in the Eastern Districts of the U.S.S.R." *Kolkhoznoye proizvodstvo*, No. 5-6, 1942, pp. 21-22.

252. Speech on Closing the Session [of the Lenin Academy of Agricultural Sciences of the U.S.S.R.] on December 17, 1942; *Sotsialisticheskoye zemledeliye*, December 19, 1942.

253. "Storing of Cut-Off Potato Tuber Tops in Spring and Their Preparation for Planting, The." *Omskaya pravda*, April 23, 1942; *Pravda*, April 24, 1942; *Sotsialisticheskoye zemledeliye*, May 5, 1942; *Doklady VASKhNIL*, Vols. 1-2, 1942, pp. 3-7.

254. "Summer Planting of Sugar Beets and the Planting of Freshly Picked Potato Tubers, The." Newspaper *Pravda Vostoka* [*Truth of the East*], Tashkent, July 17, 1942.

255. "Summer-Sown Potato Area Must Be Increased, The." Newspaper *Sovetskaya Kirghizia* [*Soviet Kirghizia*], Frunze, July 24, 1942.

256. "Vegetable Gardens Are a Matter of Great Importance." [From a speech held at a meeting of the Presidium of the All-Union Central Council of Trade Unions.] *Trud*, March 25, 1942.

257. "Ways of Raising the Sprouting Capacity of Grain Seeds." *Omskaya pravda*, April 17; *Doklady VASKhNIL*, Vols. 3-4, 1942, pp. 3-6.

258. "We Want Bumper Grain Crops in Siberia." *Kolkhoznoye proizvodstvo*, No. 2, 1942, pp. 21-23. Continuation: *Pravda*, March 2, 1942; *Sotsialisticheskoye zemledeliye*, March 3, 1942; *Kolkhoznoye proizvodstvo*, No. 3-4, 1942, pp. 21-24.

259. "Work of the Laboratory I Am in Charge of, The." *Sotsialisticheskoye zemledeliye*, June 6, 9 and 10, 1942.

1943

260. "Cultivation of Millet, The." *Kolkhoznoye proizvodstvo*, No. 2-3, 1943, p. 15.

261. "Don't Allow Potato Tops to Rot; Restore Them!" *Moskovsky bolshevik*, May 12, 1943; *Pravda*, May 12, 1943; *Sotsialisticheskoye zemledeliye*, May 13, 1943.

262. "Heredity and Its Variability." *Sotsialisticheskoye selskoye khozyaistvo*, No. 1-2, 1943, pp. 47-69; No. 3-4, pp. 36-51.

Same, Omsk, People's Commissariat of Agriculture of the U.S.S.R. Publishing House, 1943, 75 pp.

263. "Hill-Sowing Kok-Saghyz by Hand." *Pravda*, April 17, 1943; *Sotsialisticheskoye zemledeliye*, April 17, 1943; *Doklady VASKhNIL*, Vol. 2, 1943, pp. 18-20.

264. "How to Prevent the Winter Plants in Siberia from Perishing in the Winter." Newspaper *Sovetskaya Sibir* [*Soviet Siberia*], Novosibirsk, August 20, 1943; *Sotsialisticheskoye selskoye khozyaistvo*, No. 10-11, 1943, pp. 44-49.

265. "Increase the Field Germinating Capacity of Cereal Seeds!" *Pravda*, April 29, 1943; *Sotsialisticheskoye zemledeliye*, April 29, 1943; *Doklady VASKhNIL*, Vol 2, 1943, pp. 3-6.

266. "Lose no Time in Getting the Spring Under Way in Siberia." *Pravda*, April 22, 1943, *Sovkhoznaya gazeta*, April 24, 1943; *Sotsialisticheskoye zemledeliye*, April 22, 1943.

267. "More Attention to Millet Cultivation." *Pravda*, May 17, 1943; *Sotsialisticheskoye zemledeliye*, May 17, 1943.

268. "Organism and Environment." [Stenographic report of a lecture delivered in the Polytechnical Museum on January 11, 1941.] In his collection of works *Agrobiology*, Moscow 1943, pp. 298-312.

269. "Plant Vegetables on Every Free Patch of Land." [Paper read at a conference devoted to the problem of individual and collective vegetable gardens, held on April 5 in Omsk.] *Omskaya pravda*, April 7, 1943.

270. "Prepare More Food Potato Tuber Tops for Planting!" In the book *Advice to the Rural Youth on Spring Sowing*, Moscow 1943, pp. 19-22.

271. "Preparing and Storing Potato Tuber Tops." *Moskovsky bolshevik*, March 18, 1943.

272. "Preparing Potato Tops." Newspaper *Altaiskaya pravda* [*Altai Truth*], Barnaul, March 27, 1943; *Sovkhoznoye proizvodstvo* [*State-Farm Production*], No. 3-4, 1943, pp. 25-7.

273. "Problem of Preparing Potato Tuber Tops, The." *Doklady VASKhNIL*, Vol. 4, 1943, pp. 15-16.

274. "Problem of Sowing Rye in Stubble, The." *Omskaya pravda*, September 4, 1943.

275. "Science Should Be Placed at the Service of Spring Sowing." [Tasks assigned to research institutions and agronomist experimenters.] *Pravda*, March 25, 1943.

276. "Spring Agrotechnique in Arid and Semiarid Districts." *Kolkhoznoye proizvodstvo*, No. 4, 1943, pp. 10-12.

277. "Start Collecting Potato Tops!" *Trud*, November 24, 1943.

278. "K. A. Timiryazev and the Tasks of Soviet Agrobiology." *Sotsialisticheskoye zemledeliye*, June 3, 1943 [abridged].

Same [in full]. *Doklady VASKhNIL*, Vol. 4, 1943, pp. 3-14; *Pod znamenem Marxisma*, No. 6, 1943, pp. 53-63.

279. "Kliment Arkadievich Timiryazev." [On the 100th anniversary of his birth. Written in collaboration with I. I. Prezent.] Journal *Krasnoarmeyets* [*Red Armyman*], No. 11, 1943, pp. 21-22.

280. "We Must Raise the Yield of the Vegetable Gardens!" *V pomoshch fabrichno-zavodskim i mestnym komitetam*, No. 3-4, 1943, pp. 6-9.

281. "Urgent Problems of Soviet Agricultural Science." [On rendering assistance in the spring sowing. Corrected stenographic report of an address at the session of the Lenin Academy of Agricultural Sciences of the U.S.S.R., December 13, 1942.] *Pravda*, January 29-30, 1943; *Sotsialisticheskoye zemledeliye*, January 30 and February 2, 1943.

Same, Moscow, Selkhozgiz, 1943, 36 pp.

282. "Yield of Individual and Collective Vegetable Gardens Must Be Increased, The." [Speech at an enlarged sitting of the Presidium of the A.U.C.C.T.U. held on March 11, 1943.] *Trud*, March 18, 1943.

Same, Moscow, Transport and Railway Publishing House, 1943, 14 pp.

283. "Cultivation of Winter Plants in the Siberian Steppes, The." *Sovkhoznoye proizvodstvo*, No. 10-11, 1944, pp. 16-27; Journal *Planovoye khozyaistvo* [*Planned Economy*], No. 2, 1944, pp. 51-68.

284. "Gist of Our Proposal to Sow Winter Grains in Stubble in the Siberian Steppes, The." *Kolkhoznoye proizvodstvo*, No. 4, 1944, pp. 16-23.

285. "Growing Maize in Workers' Vegetable Gardens Near Moscow." In the book *Railway Workers' Vegetable Gardens*. [Compiled by N. Sorokin and V. Chaban; edited by D. E. Kamyshchenko.] Moscow 1944, pp. 14-16.

286. "Maize in the Vegetable Gardens of the Vicinity of Moscow." [Advice.] *Trud*, May 12, 1944.

287. "Preparing Potato Tuber Tops for Planting." *Trud*, April 11, 1944.

288. "Raising the Germinative Capacity of Cereal Seeds." *Sotsialisticheskoye zemledeliye*, April 1, 1944.

289. "Soil Moisture Preservation Before Sowing." [Abridged stenographic report of a speech held at a conference of *Sotsialisticheskoye zemledeliye* correspondents.] *Sotsialisticheskoye zemledeliye*, March 23, 1944.

290. "Some Agrotechnical Questions Relating to Spring Sowing in 1944." *Pravda*, March 30, 1944; *Sovkhoznaya gazeta*, April 1, 1944.

Same, Moscow, Selkhozgiz, 1944, 28 pp.

291. "Speech ... [held at the 12th Plenum of the A.U.C.C.T.U. on measures for the further spread of the individual and collective vegetable garden movement.] *Trud*, March 23, 1944.

292. "We Must Grow More Potatoes and Vegetables in the Vegetable Gardens of the Workers." *Pravda*, April 13, 1944.

293. "We Must Sow at Such Times and in Such Manner as to Gather in Big Crops." [Replies to an open letter of collective-farm chairmen in Ryazan Region.] *Sotsialisticheskoye zemledeliye*, March 14, 1944; *Doklady VASKhNIL*, Vols. 2-3, 1944, pp. 3-8.

1945

294. "Alliance Between Science and Practice, The." [Raising yields in collective and individual vegetable gardens.] *Trud*, March 3, 1945.

295. "Collective and State Farms of Siberia Can Already Grow Winter Wheat, The." In his book *Cultivating Winter Grains in Siberian Steppes*, Moscow, 1945, pp. 78-83.

296. "Good Means of Fighting the Beet Weevil, A." [The use of chickens to exterminate pests.] *Pravda*, March 12, 1945.

297. "Let Us Develop Michurin's Theory." [On the 10th anniversary of the death of I. V. Michurin.] *Pravda*, June 7, 1945; *Sotsialisticheskoye zemledeliye*, June 7, 1945; *Doklady VASKhNIL*, Vol. 6, 1945, pp. 3-5.

298. "Short Summary of Our Work in 1941-43." [A study of the laws of plant heredity and its variability.] In his book *Papers on Research Work in 1941-43*, Moscow-Leningrad 1945, pp. 230-31. (Academy of Sciences of the U.S.S.R. Biological Sciences Department.)

299. "Raw Material Base of Natural Rubber Should Be Expanded, The." *Pravda*, September 1, 1945.

1946

300. "Advice to Vegetable Garden Holders." *Trud*, March 12, 1946.

301. "Agricultural Science Is Fighting for the Fulfilment of the Stalin Program." [From a paper read at an open Party meeting of the Lenin Academy of Agricultural Sciences of the U.S.S.R.] *Izvestia*, March 6, 1946; *Sovkhoznaya gazeta*, March 7, 1946.

302. "Agronomic Advice to the Collective Farms of the South." *Pravda*, June 8, 1946.

303. "Big Vegetable Garden Crops Are Needed." Speech [delivered at a meeting of the Moscow railway net workers]. Newspaper *Vechernyaya Moskva* [*Evening Moscow*], February 27, 1946.

304. "Genetics." [Article written for the 3rd edition of the *Agricultural Encyclopedia*], Journal *Agrobiologia* [*Agrobiology*], No. 5-6, 1945, pp. 32-47.

305. "My Advice to Vegetable Garden Holders." Newspaper *Gudok* [*Whistle*], March 1, 1946.

306. "Natural Selection and Intravarietal Competition. [Lecture read on November 5, 1945, at the qualification improvement courses for state plant-breeding station workers.] *Sotsialisticheskoye zemledeliye*, January 5, 6, 8, 10 and 12, 1946; *Agrobiologia*, No. 2, 1946, pp. 3-27.

307. "Soviet Darwinism." [Stenographic report of an address read March 27, 1941, at a scientific theory conference of the K. A. Timiryazev Agricultural Academy in Moscow on Marxism and Darwinism.] *Agrobiologia*, No. 1, 1946, pp. 7-18.

308. "Spatial Isolation of Varietal Crops." [From a speech delivered at a conference of technical school personnel under the People's Commissariat of State Farms of the R.S.F.S.R.] *Sovkhoznaya gazeta*, January 17, 1946.

309. "Stick to Your Last!" [Reply to Prof. P. M. Zhukovsky's article "Darwinism in a Crooked Mirror."] *Pravda*, June 28, 1946.

310. "Millet Yields Must Be Greatly Increased!" *Sotsialisticheskoye zemledeliye*, April 15, 1947.

311. "Millet Yields Must Be Increased Sharply by Careful Tillage." *Kolgospnik Ukraini*, Kiev, May 29, 1947.

312. "Old and New Genetics." *Agrobiologia*, No. 4, 1947, pp. 3-5.

313. "Preface." To G. A. Babajanyan's book *The Elective Capacity of the Fertilization of Agricultural Plants*, Yerevan 1947, pp. VII-VIII.

314. "Reserves for Increasing the Supply of Seed Potatoes. More attention Must Be Paid to Preparing Potato Tops." *Sotsialisticheskoye zemledeliye*, April 15, 1947.

315. "Tasks of the Lenin Academy of Agricultural Sciences of the U.S.S.R., The." [Revised stenographic report of an address delivered at an open Party meeting of the Academy on the tasks in the light of the decision of the Plenum of the C.C., C.P.S.U.(B.)

entitled "On the Measures to Be Taken to Improve Agriculture During the Postwar Period."] *Sotsialisticheskoye zemledeliye*, March 21, 1947, *Agrobiologia*, No. 2. 1947, pp. 3-18; *Selektsia i Semenovodstvo*, No. 3, 1947.

316. "Using Food Tuber Tops to Plant Potatoes." [A talk.] *Trud*, February 13, 1947; *Sotsialisticheskoye zemledeliye*, February 14, 1947.

317. "We Must Prepare More Potato Tops," Moscow 1947, 16 pp.

318. "Why Bourgeois Science Is Up in Arms Against the Works of Soviet Scientists." [Reply to questions of *Literaturnaya gazeta (Literary Gazette)* correspondents.] *Literaturnaya gazeta*, October 18, 1947.

1948

319. "Agrobiological Theory Should Be Closely Linked Up with Collective- and State-Farm Practice." In the book *Collective-Farmer's Calendar for 1949*, Moscow, Selkhozgiz, 1948, p. 19.

320. "Experimental Hill Sowing of Forest Belts." [A paper read at a conference of scientific workers of the Lenin Academy of Agricultural Sciences of the U.S.S.R., held on November 23, 1948.] In his collection of works *Agrobiology*, 4th edition, Moscow 1948, pp. 655-73.

321. "Situation in Biological Science, The." [Address and Concluding Remarks at the August Session of the Lenin Academy of Agricultural Sciences of the U.S.S.R.] *Pravda*, August 4, 5 and 10, 1948; *Sotsialisticheskoye zemledeliye*, August 4, 5 and 10, 1948.

Same, Moscow, Selkhozgiz, 1948, 64 pp.

Same. In the book *The Situation in Biological Science*. A verbatim report of the session of the Lenin Academy of Agricultural Sciences of the U.S.S.R., July 31, to August 7, Moscow, Selkhozgiz, 1948, pp. 4-41 and 512-23.

322. "Sow Siberia's Fields with Winter Wheat!" *Izvestia*, February 18, 1948; *Agrobiologia*, No. 2, 1948, pp. 3-9.

1949

323. "Advice from Academician T. D. Lysenko to the collective farmers of Kuibyshev Region" [at a conference in the Lenin Academy of Agricultural Sciences of the U.S.S.R., on January 20, 1949, on questions relating to the development of the region's agriculture.] Newspaper *Volzhskaya kommuna* [*Volga Commune*], Kuibyshev, February 1, 1949.

Same, in the book bearing the same title. Kuibyshev 1949, pp. 15-46.

324. Advice from Academician T. D. Lysenko to the collective farmers of Chkalov Region [at a conference held on March 5, 1949. Problems of drought control, etc.] In the book *Chkalov Collective Farmers at the Lenin Academy of Agricultural Sciences of the U.S.S.R.*, Chkalov 1949, pp. 17-34.

325. "Collection and Winter Storage of Acorns, The." [Advice.] *Sotsialisticheskoye zemledeliye*, October 8, 1949; *Sovetskaya agronomia*, No. 11, 1949, p. 87.

326. "Hill Sowing of Forests, The." [Talk with a correspondent of the journal.] Journal *Ogonek* [*Scintilla*], No. 10, 1949, pp. 6-7.

327. "Hill Sowing of Forest Shelter Belts, The." *Agrobiologia*, No. 5, 1949, pp. 9-27; *Doklady VASKhNIL*, Vol. 10, 1949, pp. 3-21.

328. "Instructions for the 1950 Hill Sowing of Forest Shelter Belts." *Sovkhoznaya gazeta*, October 22, 1949, with diagrams; *Sotsialisticheskoye zemledeliye*, October 22, 1949, with diagrams.

Same, Moscow, Ministry of Agriculture of the U.S.S.R. Publishing House, 1949, 16 pp., with diagrams.

Same; Moscow, Ministry of Forestry of the U.S.S.R. Publishing House, 1951, 20 pp., with diagrams.

Same (for 1951, 2nd corrected and enlarged edition). Journal *Les i step* [*Forest and Steppe*], 1951, No. 6, pp. 7-18.

329. "Results of the Work of the Lenin Academy of Agricultural Sciences of the U.S.S.R. and the Tasks of Agricultural Science." [Paper read on October 28, 1949 at the Jubilee Session in Celebration of the Twentieth Anniversary of the Academy.] *Sovkhoznaya gazeta,* November 3, 1949; *Sotsialisticheskoye zemledeliye,* November 3, 1949.

Same, Moscow, Selkhozgiz, 1949, 32 pp.

330. "Some Problems of Field Grass Sowing." [Stenographic report of a speech delivered at a conference under the auspices of the Ministry of Agriculture of the U.S.S.R. on mastering lea crop rotation, grass sowing and seed growing of perennial grasses. February 1949.] *Pravda,* February 15, 1949; *Sotsialisticheskoye zemledeliye,* February 15, 1949; *Sovetskaya agronomia,* No. 4, 1949, pp. 9-18.

331. "Specific Features of the Agrotechnique of Field Work in Spring in the Southeast." [Reply to grain growers of Kuibyshev, Chkalov and other southeastern regions.] *Sotsialisticheskoye zemledeliye,* April 3, 1949; *Sovkhoznaya gazeta,* April 5, 1949.

332. "J. V. Stalin and Michurinian Agrobiology." *Izvestia,* December 15, 1949; *Sotsialisticheskoye zemledeliye,* December 16, 1949; Journal *Nauka i zhizn* [*Science and Life*], No. 12, 1949, pp. 40-44. Also in the book *To Joseph Vissarionovich Stalin From the Academy of Sciences of the U.S.S.R.,* Moscow 1949, pp. 89-126

333. "Summer Planting of Potatoes and the Sowing of Alfalfa, The." *Bolshevistskoye znamya,* Odessa, June 26, 1949; Newspaper *Pravda Ukrainy* [*Truth of the Ukraine*], Kiev, June 25, 1949.

334. "Theoretical Substantiation of the Hill Sowing of Forest Shelter Belts." *Les i step,* No. 4, 1949, pp. 22-29.

335. "Three-Year Plan for the Development of Public Collective- and State-Farm Meat and Dairy Stockbreeding and the Tasks of Agricultural Science, The." [Paper read at a session of the Lenin Academy of Agricultural Sciences of the U.S.S.R. on May 5, 1949.] *Sovkhoznaya gazeta,* May 21, 1949; *Sotsialisticheskoye zemledeliye,* May 21, 1949.

Same, Moscow, Selkhozgiz, 1949, 32 pp.

336. "Twenty Years' Existence of the Lenin Academy of Agricultural Sciences of the U.S.S.R." [A talk.] *Pravda,* October 28, 1949.

337. "We Need Stable Big Crops in the Volga Area." [Abridged stenographic report of a speech delivered at a conference of collective farmers and agronomists of Kuibyshev Region at the Lenin Academy of Agricultural Sciences of the U.S.S.R. in January, 1949.] *Izvestia,* January 30, 1949; *Sovetskaya agronomia,* No. 4, 1949, pp. 3-8.

1950

338. Advice of Academician T. D. Lysenko to the stockbreeders of Kuibyshev Region. [From a paper read at a conference of agricultural personnel of Kuibyshev Region.] *Sotsialisticheskoye zemledeliye,* February 14, 1950; *Kolkhoznoye proizvodstvo,* No. 4, 1950, pp. 20-35.

339. "Agronomic Teaching of V. R. Williams, The." *Pravda,* July 15, 1950.

Same, Moscow, Selkhozgiz, 1950, 32 pp.

340. "Hill Sowing of Forest Shelter Belts Must Be Carried Out With Success! The." *Pravda,* March 25, 1950.

341. "New Developments in the Science of Biological Species." *Pravda,* November 3, 1950; *Selektsia i semenovodstvo,* No. 11, 1950, pp. 6-12.

342. "Partial Change in the Instructions on the Hill Sowing of Forest Shelter Belts for 1950." *Sotsialisticheskoye zemledeliye,* August 19, 1950; *Sovkhoznaya gazeta,* August 19, 1950.

343. "Toward New Successes in the Realization of the Stalin Plan for the Transformation of Nature." [Speech held at a meeting with Odessa Region collective farmers in the Lenin Academy of Agricultural Sciences of the U.S.S.R. Abridged stenographic report.] *Pravda Ukrainy,* March 26 and 28, 1950.

Same, Moscow, Selkhozgiz, 1950, 24 pp.

1951

344. "Advice to Collective Farmers in the Vicinity of Moscow." [Speech delivered at a conference of Moscow Region collective-farm chairmen. Abridged stenographic report.] Newspaper *Moskovskaya pravda* [*Moscow Truth*], December 16, 1951.

345. "Duty of Soviet Scientists, The." *Pravda*, October 5, 1951.

346. "For the Good of the Country, for the Sake of Peace." [On the Third Anniversary of the Adoption of the Stalin Plan for the Transformation of Nature.] Newspaper *Lesnaya promyshlennost* [*Timber Industry*], October 21, 1951.

347. "Onward to Further Successes!" [Tasks of Soviet agrobiologists.] *Izvestia*, January 1, 1951.

348. "Potatoes Should Not Be Planted in Summer Before the Dates Fixed." *Sotsialisticheskoye zemledeliye*, June 17, 1951.

349. "Talk with collective farmers and agricultural specialists of the Byelorussian S.S.R., A." *Agrobiologia*, No. 6, 1951, pp. 141-49; *Doklady VASKhNIL*, No. 12, 1951, pp. 3-11; *Sovetskaya agronomia*, No. 12, 1951, pp. 3-11.

350. "Work of O. B. Lepeshinskaya, Member of the Academy of Medical Sciences of the U.S.S.R., The." [Paper read at a conference at the Biological Sciences Department of the Academy of Sciences of the U.S.S.R., on the problem of living substance and cellular development, May 22-24, 1950.] *Literaturnaya gazeta*, September 13, 1951; *Agrobiologia*, No. 6, 1951, pp. 3-5.

SUBJECT INDEX